SEPARATIONS FOR BIOTECHNOLOGY

2

Papers presented at the Second International Symposium on 'Separations for Biotechnology' held at the University of Reading, UK, 10–13 September 1990

Organiser and Sponsors

Biotechnology Group and Solvent Extraction Group of the SCI

in association with

BioIndustry Association
European Federation of Biotechnology Working Party on Downstream Processing

Organising Committee

Prof. D. L. Pyle (*Chairman*)	University of Reading
Dr J. A. Asenjo	University of Reading
Dr J. R. Birch	Celltech Limited
Dr H. Chase	University of Cambridge
Dr M. J. Hudson	University of Reading
Dr A. Lyddiatt	University of Birmingham
Dr D. Moss	ICI plc
Dr A. Rosevear	Harwell Laboratory
Dr A. Thomson	Kodak Limited
Mr M. Verrall	SmithKline Beecham Pharmaceuticals
Mr S. P. Vranch	Celltech Limited

SEPARATIONS
FOR BIOTECHNOLOGY

2

Edited by

D. L. PYLE

Biotechnology Group, University of Reading UK

Published for SCI
by
ELSEVIER APPLIED SCIENCE
LONDON and NEW YORK

SCI

ELSEVIER SCIENCE PUBLISHERS LTD
Crown House, Linton Road, Barking, Essex IG11 8JU, England

Sole Distributor in the USA and Canada
ELSEVIER SCIENCE PUBLISHING CO., INC.
655 Avenue of the Americas, New York, NY 10010, USA

WITH 97 TABLES AND 254 ILLUSTRATIONS

© 1990 SCI
© 1990 JOHN BROWN ENGINEERS & CONSTRUCTORS LTD—pp. 29–37
© 1990 UNITED KINGDOM ATOMIC ENERGY AUTHORITY—pp. 83–92, 217–226

British Library Cataloguing in Publication Data

Separations for biotechnology 2.
1. Biotechnology. Separation. Techniques
I. Pyle, D.L.
660.6

Library of Congress CIP data applied for

ISBN-13: 978-94-010-6839-0   e-ISBN-13: 978-94-009-0783-6
DOI: 10.1007/978-94-009-0783-6

Softcover reprint of the hardcover 1st edition 1990

Preface

The challenge of bioseparations is to isolate and purify identified products from the dilute product broth produced from cell culture. Innovation in bioseparations technology is increasingly driven by the requirements imposed by the growing importance of production on a process scale of injectable-grade products, and economic pressures to improve the efficiency of downstream processing. As in other areas of technical change, science does not necessarily precede new technology: progress results from a complex and messy mixture of advances in understanding, ingenious ideas, novel techniques and chance discoveries. What is certain is that close interaction between academics and practitioners, biological scientists and process engineers is needed to solve the problems of bioseparations. The Second International Conference on Separations for Biotechnology at Reading, UK, in September 1990 set out to provide a critical multidisciplinary forum for the discussion of bioseparations. This volume contains the papers presented at the meeting.

The meeting was organised around six themes with oral and poster presentations on the science and practice of bioseparations technology, and the same structure has been kept for this book. We have also included the texts of the keynote review paper by Professor Alan Michaels and the introductory review papers specially commissioned for the conference. Within each part of this book the review paper is followed by the contributed papers grouped alphabetically by their first author. All the original papers published here were accepted for publication after scientific refereeing.

This book would not have been possible without the help, unsung and invariably unpaid, of many colleagues. I am especially grateful to the SCI, the Organising Committee—the session chairmen in particular—and to the many anonymous reviewers. I owe a special debt to Tom Arnot and Gary Lye for their sterling work on the index and, finally, to Jean Davis for her outstanding and patient secretarial support in creating order out of paperwork whose complexity rivalled the biological soup which provides the basis for this volume.

D. L. PYLE

Contents

Part 2: Concentration Processes

Part 3: High Resolution Separations

Part 4: Product Finishing

Plenary Paper

FRONTIERS OF BIOSEPARATIONS TECHNOLOGY:
UNSOLVED PROBLEMS AND NOVEL PROCESS CONCEPTS

Alan S. Michaels
President, Alan Sherman Michaels, Sc. D., Inc.
Chestnut Hill, Massachusetts
and
Distinguished University Professor, Emeritus
North Carolina State University
Raleigh, North Carolina

INTRODUCTION

The unique properties, and formidable processing problems associated with the isolation and purification of bioactive proteins and other biologically derived substances have become a major challenge to the chemical and biochemical engineer. The extraordinarily high standards of purity demanded of such products to satisfy regulatory requirements of safety, coupled with their dependence upon precisely controlled tertiary and quarternary structure for proper biological activity, require the development and use of separation/purification techniques uncommon to traditional bioprocessing practices, and of novel processing concepts which call upon the collective expertise of life scientist and engineer.

The basic separation strategies employed for bioproduct recovery are well-recognized, and generally observed in all bioprocessing operations today. These comprise (1) separation of the product(s) of interest in relatively impure form and low concentration from the biomass (either intact cells or cell debris) which produced them; (2) concentration of the product-bearing extract; (3) separation of the product in relatively high purity and still higher concentration from the majority of the adventitious impurities; and (4) isolation of the product from all but miniscule amounts of inactive or potentially toxic biologically-derived impurities. However, of the alternative technologies available for pursuing these strategies today, few are ideal for these purposes. Their limitations include poor product yield, inadequate separation selectivity, loss in product bioactivity due to denaturation or other structural change, inadequate process biosafety, or excessive processing cost. Efforts to circumvent these limitations have involved both attempts to modify existing separations processes, and development of radically new processes better suited to meeting these stringent requirements. Some recent examples of these efforts as summarized below.

RECOVERY OF INSOLUBLES

In the area of solid/liquid separations for recovering extracellular bioproducts from whole cell suspensions, or intracellular products from dispersions of lysed cell fragments, UF/MF membrane separation processes continue to receive much attention. The recent development of surface-modified microfiltration/ultrafiltration membranes which resist fouling by biopolymers and colloids has been noteworthy, as has the development of novel fluid-management strategies (e.g., pulsatile flow, rotating-cylinder devices) which markedly improve the flux-performance, separation efficiency, and service lifetime of UF/MF separation systems. Of particular promise has been the development of ceramic and ceramic/metallic composite MF/UF membranes, which display unique fouling resistance, facile cleanability and sterilizability, and extraordinary durability.

Where genetically engineered proteins form inclusion bodies within recombinant organisms, and conventional solid/liquid separations are ineffective, it has been found that, by carefully controlled cell lysis, appropriate control of suspension pH and ionic strength, and precisely monitored centrifugation, efficient recovery of the inclusion bodies with relatively little contamination by cell fragments and other contaminants can be accomplished.

A novel process which involves treatment of a whole fermentation broth by fluidization with a selective particulate adsorbent permits efficient simultaneous recovery and purification of an extracellular metabolic product without need for any dewatering step. This process lends itself particularly well to cell-recycle, and semicontinuous whole broth extraction.

A unique single-step combination of membrane filtration and adsorption has been demonstrated: This makes use of a hollow fiber membrane ultrafiltration module, the shell space of which is filled with a particulate selective adsorbent; a cell-suspension containing a low concentration of desired protein is recirculated through the fiber lumens, with a small permeate stream being removed from the shell space. The cell-free, protein-bearing permeate is stripped of protein by the adsorbent; when the adsorbent is saturated, the shell-space of the module is flushed free of permeate, and eluted with saline or other displacing mixture to release and permit recovery of the purified protein.

EXTRACTION OF PROTEINS

Two-phase aqueous separation techniques for recovery of proteins from whole broths and lysed cell slurries continue to receive intensive study. A broadening selection of water soluble polymers and hydrocolloids, which yield improved selectivity and phase partitioning, and are economically practicable, should make this technology increasingly useful for large-scale protein recovery. The use of surfactants which micellize in non-aqueous media, and which in the micellar state are capable of sorbing specific proteins from aqueous solution, are the basis for the novel process of "inverse micelle" liquid/liquid separation for protein recovery and purification. When combined with the relatively new process of "membrane solvent extraction", which utilizes microporous membranes to greatly facilitate liquid/liquid extraction of "dirty" or emulsion-forming feedstreams, these techniques may constitute an important new dimension in bioproduct recovery.

ADSORPTIVE SEPARATIONS

A simple procedure for removal and isolation of a single protein species from a complex biopolymer mixture remains a major goal of modern separation technology. While preparative elution chromatography continues to be the preferred procedure from the standpoint of ultimate product purity, the complexity, tediousness, and high cost of the process are powerful deterrents to its use for industrial scale applications. Moreover, in many cases where column chromatography does permit efficient product recovery, the conditions of loading or elution often result in complete or partial denaturation of the protein of interest; this is tantamount to winning the battle and losing the war. The problems of restoring a denatured protein to its active, natural conformation (about which more below) are sufficiently apprehending to encourage the process engineer to seek other less drastic alternatives.

Recently, there has been rapidly increasing interest in the use of functionalized, selective solid-phase adsorbents for treatment of aqueous multicomponent protein solutions for the sorptive removal of one (or several members of a common class of) protein. In this process (sometimes termed "flash chromatography" or "bind/release"), the solution is contacted with the adsorbent until the solid is virtually saturated with the adsorbate, the solid phase flushed free of unadsorbed solutes, and then treated with a desorbing solution which displaces the adsorbed species in concentrated form, and regenerates the adsorbent. The solid phase may be functionalized in a number of ways: Attachment of ionogenic functions renders it an ion exchanger, which can differentiate between proteins of differing isoelectric points; attachment of "generic" affinity ligands such as triazine dyestuffs can render it selectively sorptive for certain classes of enzyme; attachment of cell surface-immunoproteins such as Protein A can render it selectively sorptive for classes of immunoglobulin such as IgG; or attachment of specific monoclonal antibodies can render it uniquely sorptive for one antigenic protein.

Several physical forms of such adsorbent are now available and under evaluation. These include particulate solids whose surfaces are activated; microporous particles whose internal pore surfaces have been functionalized; microporous membranes or plastic sheets, whose internal surfaces have been activated; and microporous hollow fiber membranes whose internal pore surfaces have been so treated. The objective is to produce a high-sorptive- capacity solid in the most convenient physical form for rapid permeation by the feed solution containing the product of interest, and for equally rapid flushing and elution, with minimal fluid requirement. While the bind/release process may often not be as selective as elution chromatography, it has the benefit of very rapid recovery (particularly important for labile products), and very high production capacity per unit of adsorbent.

Ion exchange bind/release adsorbents are among the most versatile media for protein separation, since virtually any protein (or protein class) displaying a unique isoelectric point vis-a-vis others in the mixture will usually be amenable to selective recovery by appropriate adjustment of solution pH and ionic strength; moreover, conditions for binding and desorption are often mild enough to avoid denaturation complications. Such adsorbents are usually quite stable, and also relatively inexpensive to produce. The fact, however, that the degree of purity achievable by bind/release ion exchange sorption is in many cases limited is a constraint of significance. Terminal purification of the product, by either affinity sorption or elution chromatography, may be the ultimate processing step.

Specific affinity sorbents are surely the most selective candidates for this technique, although they are usually costly, and frequently have short service-lifetime. Moreover, where proteins or other bioactive substances are employed as affinity ligands, there are serious regulatory concerns about the possibility of detachment of such ligands (or their fragments) from the solid phase, and their contamination of the product. The possibility that these contaminants might be immunogenic or carcinogenic or otherwise toxic, must be considered and properly addressed.

MEMBRANES

Prevailing wisdom teaches that membrane ultrafiltration is relatively ineffectual for separation of biopolymers of differing molecular size, both because of inherent lack of sieving discrimination by such membranes, and polarization by larger molecules which impede passage of smaller ones. Recent research, however, suggests that this may not be the general rule -- that membranes displaying low protein sorptivity, if operated (paradoxically) under polarizing conditions, can efficiently separate mixtures of proteins whose molecular weights differ by less than a factor of two. Also, improved membrane fabrication techniques have made it possible to produce membranes with much narrower pore-size distribution in the nanometer and submicron size range than has been heretofore possible; such membranes may well be capable of further improving molecular-size-based protein separations.

PRECIPITATION PROCESSES

Protein separation by fractional precipitation, which is probably one of the oldest bioprocessing practices, has recently received a new lease on life by virtue of an increased understanding of the mechanisms of molecular aggregation, organization, and crystallization of proteins from solution. It now appears that, by sophisticated control of such variables as temperature, agitation, pH, ionic strength, ion valency, and solvent/nonsolvent ratio, and the dynamics of time-variation of these parameters, it may be possible selectively to precipitate single protein species as small, dense, quasicrystalline particles virtually devoid of entrapped contaminants. Separation of these particles from the suspending medium by membrane filtration or centrifugation allows recovery of single protein species in extraordinarily high purity and concentration. Adaptation of these findings to large-scale bioseparations remains to be accomplished, but if it can be done successfully, it may revolutionize the practice of protein purification.

PROTEIN RENATURATION

One of the knottiest problems confronting the protein chemist and bioprocessing engineer today is the failure of recombinant DNA technology to yield bioactive proteins which possess the proper conformation for their intended biological activity, and/or the propensity of natural bioactive proteins to undergo conformational changes (i.e., denaturation) with attendant loss in bioactivity during the biorecovery process. This has led to intensive efforts to attempt to treat denatured or malconformed proteins to encourage their reconformation into native, bioactive structures.

Central to achievement to success in this task is a thorough understanding of the intermolecular forces which participate in establishing protein molecule conformation, and the environmental variables which dictate whether and when these forces come into play. This understanding is the primary domain of the protein chemist and polymer physicist. There is growing evidence to support the belief that, if conditions are established first to fully denature a protein to eliminate intramolecular covalent cross- linkages, hydrophobic, ionic, and hydrogen-bonding interactions, and then time-controlled adjustments in solution composition are initiated to allow slow re-establishment of these interactions, the natural (bioactive) conformation of the molecule will be the preferential consequence. This approach to renaturation, having been demonstrated successfully in the laboratory, remains to be proved out in bioproduct manufacture. It will probably prove to be useful for the production of recombinant oligopeptides, and relatively low molecular weight proteins, but will be much more difficult to utilize for large protein molecules, or glycosylated peptides and proteins.

QUALITY CONTROL

Establishment of biological safety of the final product resulting from the bioseparations process is, of course, the sine qua non of success in bioproduct manufacture. This means the absence from the product of pathogens (virus), pyrogens (principally microorganism-derived polysaccharides or glyco- proteins), immunogens (mainly foreign proteins and lipoproteins), and nucleic acids. Successful removal of trace amounts of these contaminants (even the detection of which is often difficult and tedious) is a major unsolved problem today. Recent advances in ultrafiltration membrane fabrication offer promise of yielding a "positive" virus filter capable of effecting a seven-log reduction in virus particles; if this can be confirmed, this may prove to be the long-awaited solution to the virus-removal problem for mammalian-cell-derived proteins and peptides. For removal of pyrogens and nucleic acids, anion-exchange adsorbents may provide the answer. Specific affinity adsorbents based on monoclonal or polyclonal antibodies as affinity ligands are the most promising prospects for removal of immunogenic proteins.

In conclusion, it can be said with some confidence that many of the remaining obstacles to effective large scale bioproduct recovery are likely to be surmounted by recent progress in bioseparations technology involving constructive collaboration between the life science and bioengineering disciplines. Many difficult problems remain, but their probability of resolution is high. Much is to be gained from continuing and improved communication between these disciplines, both in industry and academia.

Part 1

Cell Disruption and Removal of Insolubles

CELL DISRUPTION AND REMOVAL OF INSOLUBLES

J.A. Asenjo

Biochemical Engineering Laboratory
University of Reading
P.O. Box 226, Reading RG6 2AP, England

ABSTRACT

Disruption and separation of insolubles are unit operations used in the initial stages of product recovery/isolation. This paper reviews the main methods available both for cell disruption and removal of insolubles. Main insolubles considered are whole cells, cell debris and particles such as inclusion bodies. Separation and renaturation of inclusion bodies and other recombinant particles pose important challenges to modern process biotechnology. This paper critically assesses some of the main shortcomings of existing methodology used in the large scale today and points out recent novel developments that show potential for scale-up. Emphasis is given to methods that show improved selectivity and to novel process concepts in terms of their suitability for bio-processing.

INTRODUCTION

Cell disruption is a necessary operation for the release of intracellular proteins that cannot be secreted by the cells whereas removal of insolubles is important for both intracellular and extracellular products. These operations are both used in the initial stages of a protein separation and purification process which is known as the isolation or recovery subprocess. Two typical recovery subprocesses are shown in Figs. 1 and 2. The recovery subprocess comprises taking the broth out of the biochemical reactor system (e.g. a fermenter, a hollow fibre reactor) and processing it until a cell free solution is obtained and where the total protein concentration including the product is around 60 to 70 grams per liter [1, 2].

Cell separation or harvesting is always required. The variety of equipment found in industrial practice is not very large and the selection depends on the microbial source, equipment availability, equipment efficiency and economics. If the product of interest is secreted (extracellular) then the liquid part is kept (Fig. 1.); if the product is intracellular the solid fraction of the harvesting is kept (Fig. 2.). When a mammalian cell culture is used, the product will usually be extracellular. Typical harvesting operations for separation of insolubles (whole cells) are centrifugation (mainly for yeast but also for mammalian cells and bacteria), rotary vacuum filtration (mainly fungii) and microporous filtration (bacteria, yeast, mammalian cells and also fungii).

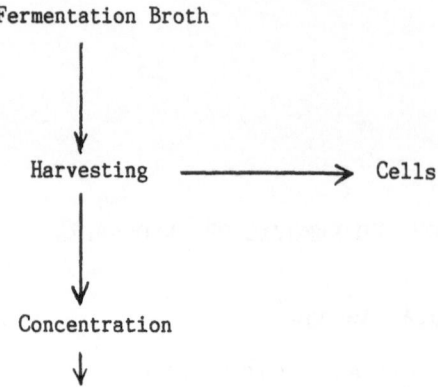

Figure 1. Recovery subprocess of extracellular product (yeast,
mammalian, bacterial) [1].

Cell disruption is required when the product is intracellular. The
equipment is selected mainly on the basis of the microbial source, as each
type of microbe presents particular resistance to disruption. The choice of
disruption technique determines the size of the resulting debris which in
turn has an influence on operations used for the separation of cell debris.
Typical cell disruption operations used are pressure homogenization (most
bacteria including E. coli and also yeast) and bead milling (gram positive
bacteria and specific yeast applications, eg. hepatitis B vaccine).
Mechanical disruption releases nucleic acids which need to be precipitated.
There is one standard method to achieve nucleic acid precipitation which is
the use of polyethyleneimine

Separation of cell debris from the proteins in solution has to be
undertaken once the cells are disrupted. As a result of this step, the
product will be in a solution with other proteins but without solids. A
novel method for the efficient separation of cell debris is the use of
aqueous two-phase systems. The insolubles will partition to the lower phase
and the target protein to the upper phase. The small size of the insolubles
present in cell debris makes them difficult to separate either by
centrifugation of filtration.

If the intracellular product is manufactured in E. coli, high
expression of heterologous proteins will usually accumulate in the form of
insoluble inclusion bodies. This makes necessary the processing of the
inclusion bodies into the native protein by denaturing and refolding. If
the intracellular product is manufactured in yeast in many instances the
protein is present in homogeneous particulate form, typically 30 - 60 nm
particles such as VLP's or 'virus like particles'. Although the processing
of intracellular particulate recombinant proteins is an important aspect of
downstream processing there are not many satisfactory methods for large
scale separation, denaturation and refolding of the particulate proteins.
Recent developments in the use of reverse micelles for protein refolding
[3] and of two-phase aqueous systems for separation of VLP's from yeast
homogenates [4] appear particularly attractive.

CELL DISRUPTION

Cell disruption is a unit operation necessary for the isolation of
intracellular proteins that the cell will not secrete. Recent developments
in recombinant DNA technology have resulted in the cloning of many

heterologous proteins in bacteria and yeast. For many of them secretion mechanisms have been found, however, for a substantial number this has not been possible. Hence, efficient and, hopefully, selective disruption techniques are an important step in producing many intracellular proteins.

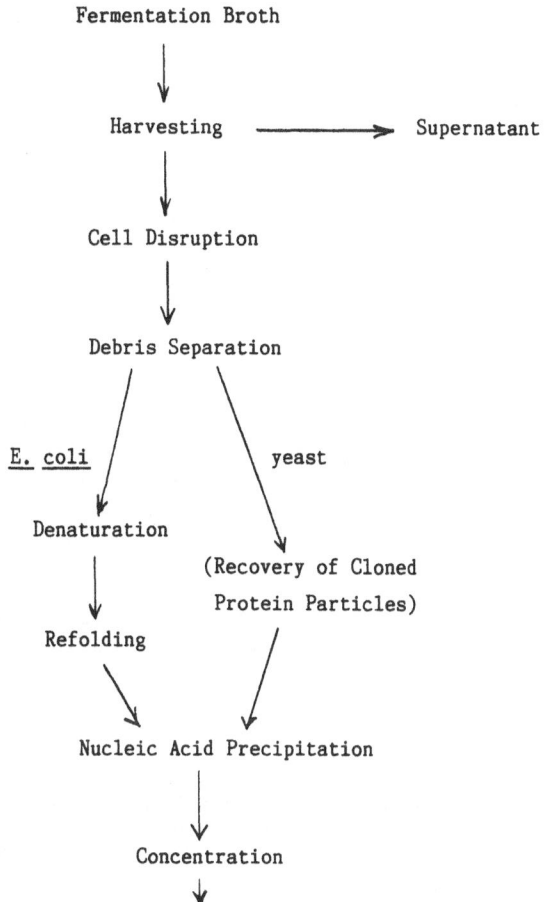

Figure 2. Recovery subprocess of intracellular product
(E. coli, yeast) [1].

A wide range of cell disruption techniques are used in the laboratory; they include chemical methods like alkali tratment, enzymatic treatment and the use of detergents and solvents, and mechanical methods like sonication, use of a pressure cell, homogenization and the use of a bead beater. Few of these can be used on a large scale. The techniques most suitable for large scale application today are the homogenizer (Gaulin homogenizer) which uses a very large pressure drop (800 bar or more) over a very narrow gap to disrupt the cells and the high speed bead mill.

Homogenization
In a homogenizer (Gaulin or pressure homogenizer) a cell suspension of up to 15 - 20% dry weight of microbial cells is forced under high pressure (up to 1200 bar in the latest models (18,000 psig)) through a very narrow annulus or valve over which the pressure drops to atmospheric. The almost instantaneous large pressure drop over an extremely short distance produces

the breakage of the cells. The concentration of cells does not affect the efficiency of disruption or protein release. The rate of protein release is a first order rate process which is a function of the number of passes in the homogenizer (residence time in the case of a bead mill). In homogenization, the equation for the rate of protein release has been discussed in detail [5, 6]. The first order rate constant is a function of temperature, the number of passages through the valve after time t and the upstream pressure applied. The feedstream is usually precooled, and cooled again immediately after passing through the homogenizer. Multiple passes through the machine are usually necessary to obtain the desired level of protein release. Cooling the stream betweem them is also essential. Refrigerated cooling down to 4 – 5 °C is usually necessary. The heat generated can raise the temperature of the stream by 2.2– 2.4°C per 100 bar (1.5 °C per 1,000 psi) of operating pressure. This will result in a temperature increase of ca. 20°C after a single passage operating at 800 bar. The energy input is ca. 0.35 kw per 100 bar operating pressure [5].

The cell debris produced by mechanical breakage often consists of small fragments making the solution difficult to clarify [6]. Complete product release often requires more than one pass through the disruption device, increasing the problem by further reducing the size of the fragments. These fragments are difficult to remove by continuous centrifugation because the throughput is inversely related to the square of particle diameter. Filtration is also complicated by the gelatinous nature of the homogenate [7]. In addition, many proteins will be denatured by the heat generated unless the device is efficiently cooled and, although shear by itself does not seem to be detrimental to proteins, shear at gas–liquid interphases, which will occur, can be very harmful [8]. The largest available homogenizers can handle a maximum throughput of 6,000 L/h. At present homogenization is a widely used large scale disruption technique. However, it is not always the most efficient one. It is not suitable for some highly filamentous microorganisms (which will block the discharge valve) nor for m.o. particularly resistant to disruption. In these cases bead milling is more efficient.

Bead Milling

In a bead mill a cell suspension is agitated at a high rate in the presence of small glass beads (0.3 – 0.4 mm in diameter). Cell disruption and protein release are described by a similar equation as that used for homogenization [5]. The rate constant k which represents the efficiency is a function of agitation speed, concentration of cells, concentration of glass beads, bead diameter, bead density and temperature [5].

Different agitator designs exist; they will have an effect on the efficiency of disruption and amount of backmixing within the chamber. An important difference with homogenization is that the concentration of cells will affect the efficiency of cell disruption. This effect is bigger at low agitator speeds and smaller at high agitator speeds. The optimimum concentration will vary from case to case but is usually between 30 – 60% cells wet weight. The largest available bead mills can handle throughputs of up to 2000 L/h. Similarly to the homogenizer, a large amount of heat has to be removed, however no recycle stream is usually present so this is carried out by means of a cooling jacket. A very substantial cooling facility is necessary.

Chemical and Enzymatic Permeabilization and Lysis

Although protein release from microorganisms on the large scale is usually accomplished by mechanical disruption, such methods have several drawbacks. Since cells have to be broken completely to obtain a high yield, all intracellular materials are released. Hence the product of interest has to be separated from a complex mixture of contaminants (proteins, nucleic acids and cell wall fragments). In addition nucleic acids will considerably increase the viscosity.

Chemical and enzymatic cell permeabilization and lysis have not been widely used for large scale intracellular product release and disruption of microbial cells. However, both operations are presently beeing investigated as powerful techniques for this purpose [7, 9, 10, 11] particularly since they offer the very attractive potential of differential protein release from different cell compartments (Andrews, et al, paper in this book) as well as selective membrane permeabilization [7] and the release of recombinant proteins [12].

Lysozyme is a commercial lytic enzyme that has been used for a number of years in industry for different large scale purposes including enzyme extraction. However this enzyme is only active on bacterial cells. Other bacteriolytic and yeast lytic enzymes (e.g. Zymolyase and enzymes from Oerskovia and Cytophaga) are only available as laboratory reagents, so it is not possible to make an accurate cost comparison with other breakage methods on an industrial scale. However, design calculations based on enzyme production data in a 1000 liter fermenter and more recent work on regulation of enzyme synthesis demonstrate the economic viability of this cell rupture technique [10].

Several chemical methods have been employed to extract intracellular components from microorganisms by permeabilizing outer wall barriers. An excellent review of this topic, which includes permeabilization of bacteria, yeast and mammalian cells, is that by Naglak et al [7]. Most work has been carried out with gram negative bacteria (mainly E. coli) using chelating agents (e.g. EDTA), solvents (e.g.5% toluene), anionic and non ionic detergents (such as sodium dodecyl sulphate, SDS, and Triton X- 100) and chaotropic agents (guanidine and urea). Mammalian cells can be permeabilized using the steroid glycoside digitonin (digitin) and the nonionic detergent saponin. One of the very few large scale chemical extraction of protein though involves gram positive cells. 0.5% triton was used to release about 50% of the cholesterol oxidase from Nocardia cells [13].

Challenges

As has been shown in Fig. 2 when the intracellular product is manufactured in E. coli, heterologous proteins will usually accumulate as large insoluble particles called inclusion bodies. If the intracellular product is manufactured in yeast in a number of cases the protein is present in particulate form, typically 30 - 60 nm particles such as virus like particles [4, Riveros-Moreno paper in this book].

Inclusion bodies have to be solubilized, in many cases chemically or enzymatically modified, and correctly refolded otherwise the process will produce large quantities of inactive product. This is usually the case when bacteria are used for the manufacture of human proteins. A very recent study of a process with E. coli [14] has shown that denaturation and solubilization of inclusion bodies with, for example, guanidine HCl are the steps that account for most of the downstream raw materials cost (77 % for guanidine HCl and carboxypeptidase only). This clearly shows that there are no satisfactory methods for large scale denaturation and refolding of particulate proteins at present.

Finally, some of the advantages of chemical and enzymatic permeabilization and lysis methods discussed in the previous section should be investigated in greater detail particularly since mechanical disruption techniques have several drawbacks related to obtaining high product yield (micronized wall materials, nucleic acids, high viscosity, complex mixture of contaminants, partially damaged product) which are difficult to overcome. Release of recombinant intracellular proteins by chemical permeabilization and enzymatic lysis techniques has been successfully achieved [12] and the release of recombinant particles from yeast is presently under investigation [4]. For this however, greater availability of specialized reagents (e.g. wall lytic enzymes) will be necessary as presently these are almost only available as laboratory reagents.

REMOVAL OF INSOLUBLES

The main operations used for the separation of microbial and mammalian cells from their supernatants are centrifugation, conventional filtration (also called "dead end filtration" given the operation procedure usually used) and membrane filtration, mainly microporous but also ultrafiltration (also called "cross flow filtration" which is the most common procedure used). Comparison of centrifugation and filtration methods must take into account capital, operating and maintenance costs, as well as recovery yield and processing time. The overall yield of active product is often a function of the processing time and other factors such as the processing temperature.

One of the main factors influencing centrifuge economics is the size of the particles to be separated. Filtration operations will also be affected by particle diameter. However, filtration economics are not such a strong function of particle size as they are for centrifugation. As particle size increases, separation costs via centrifugation are substantially decreased, while filtration costs are less dependant on size (Fig. 3). The crossover point is in the 1- 2 μm range. Microfiltration will produce a sterile stream which centrifugation cannot produce and is important for containment but cell fragments and debris will not necessarily be retained.

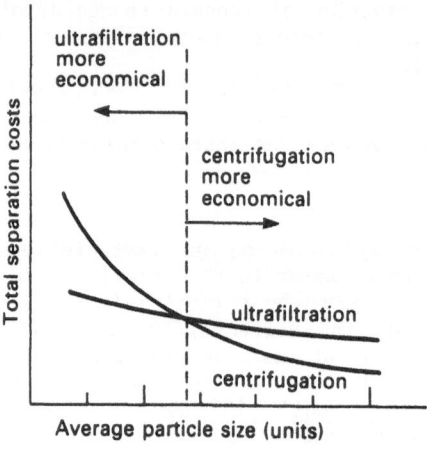

Figure 3. Centrifugation costs depend more on particle size than do membrane filtration costs [2].

Centrifugation
One of the major factors controlling centrifuge capacity for biological separations is the size of the particles involved. Recovery of yeast has been carried out for many years by centrifugation, however yeast cells are ten times larger than bacteria. This means that the theoretical centrifuge capacity is reduced by 100 fold when separating bacteria.

A number of different centrifuges are available and they all follow the same physical principles for cell separation. Equations for the rate of settling of particles are described in the literature [5] and they are important to know the variables that affect centrifuge design. These are density difference between the cells and the liquid, diameter of the cells and viscosity of the liquid.

The major centrifuge manufacturers have developed new machines specially for bacterial cell applications mainly by increasing the rotational speed (typically resulting in 14,000 - 15,000 g). The most common centrifuge for the large scale separation of bacteria as well as cell debris is the disc stack centrifuge. A properly operated large scale disk centrifuge should separate 99% of the solids from the liquid stream and produce an 80 - 90% wet solids concentrate. Particles down to 0.5 μm can be separated with low flow rates of 300 - 500 L/h, which is sufficient to remove cell debris from suspension. To separate whole cells from suspension with this machine flow rates of 3,000 - 5,000 L/h can be used. Some of the important limitations for centrifuge operation revolve around the generation of heat, aerosols and noise as well as the lack of containment and the difficulty of sterile operation. This is particularly true for the very high speed machines. Short residence times are necessary as well as cooled units. Aerosols are generated either by air entrainment through the centrifuge seals or by the use of compressed air for actuating the solids discharge mechanism. These aerosols must be contained at all times and particularly so in the presence of toxic compounds and recombinant cells. Containment implies protection of the environment from biological materials in aerosol form generated within the machine. This has particular implications for the safety of the operators working the process. It also implies that the product stream within the centrifuge is protected from the outside environment and the microbiological contamination that this implies. Contained centrifuge systems are available today. They are hermetically sealed and provided with sterile filters [2].

Filtration
The separation of solids from liquids by filtration ranks amongst the most familiar of procedures in the chemical and biological sciences. The consideration of filtration procedures progresses
from depth filters to microporous filters to ultrafilters.

Depth filtration
Although centrifugation is the most widely used cell separation method it is very costly. Other techniques that have been used for cell separation fall into the area of filtration and amongst these the rotary vacuum filter has been a popular one. Various types of rotary vacuum filters have been used. They are mainly used to remove bacterial and filamentous microorganisms from fermentation broths in large scale enzyme and antibiotic plants. Typical feed rates of fermenter broths are 100 - 200 litres/m^{2}h. Filter sizes are up to 100 m^2 filter area for the largest production units. The utility of depth filters in large scale processing of protein solutions lies in their ability to collect relatively large amounts of suspended solids like cell debris, insoluble denatured protein and precipitates prior to further filtration or chromatographic processing.

Membrane filtration
Membrane filtration systems are increasingly being used to separate whole cells and cell debris from cell suspensions, fermentation broths and cell homogenates. Removal of bacterial cells and cell debris from fermentation broths at pilot plant scale has been the major application of microfiltration processes in biotechnology. They are particularly suited

for containment and sterile separations, better than disc stack centrifuges which may also be more expensive. Membrane filtration also appears to offer lower operating costs than rotary vacuum filtration mainly since the need for expensive filter aids is eliminated [15]. A large variety of materials have been used for the manufacture of filtration membranes. These include for microfiltration porous thermoplastics (eg. PTFE, nylon), ceramic (inorganic oxides eg. Al oxide) and sintered metal [16, 17] and for ultrafiltration polysulphones and polyacrylamides. Configurations of membrane units can be tubular, spiral sheet (spiral wound), flat sheet and hollow fibre, all have been used in large scale operations. Typical flux rates for cross-flow operation are between 15 to 50 L/m^{2}h [16]. Operation of the system in a cross flow mode [17] seems to have a strong effect in increasing filtration rates by lowering concentration polarization on the membrane surface. Still the main dissadvantage of membrane filters seems to be the inability to handle a high concentration of cells, which dramatically lowers the filtration rate. It has thus been suggested to use a "cascade" of three cross flow units in series, the first one producing a 10% solids suspension, the second one a 20% solids suspension and the third one a 30% one [18]. Higher concentrations are virtually impossible to handle practically.

Microfiltration (MF) will retain very fine particulate material but not proteins in solution, whereas ultrafiltration will be able to retain proteins in solution. Microporous membranes used in microfiltration can retain particles as small as 0.01 μm. They are usually absolute filters which means they will retain virtually all particles above a particular size. Typical filter sizes used in biotechnology to retain bacteria are 0.22 or 0.45 μm. These filters will produce sterile streams.

Debris Separation

Separation of cell debris is one of the important challenges of modern biotechnology. Centrifugation has been used for the separation of cell debris after disruption but is not a particularly adequate operation since it cannot handle satisfactorily particles with a diameter smaller than 0.5 μm. This results in extremely low flow rates with machines that are already operating at very high centrifugal forces. Alternatives for the separation of broken cell debris from cell homogenates are ultrafiltration and microfiltration membranes and the use of cell debris remover, CDR, which is a selective adsorbant. CDR however, is usually used after most cell debris (ca. 98%) has been removed (e.g. by centrifugation) in order to transform the cell lysate into a "clear", column ready, solution. It will bind lipids, nucleic acids, polyphenols, colloids, and other fine material (Whatman). CDR is normally not recycled. If ultrafiltration is used a large m.w. cutoff has to be chosen (300,000 - 500,000 or larger) to insure that all the protein of interest is in the filtrate and only large material and debris is retained. This operation will usually produce a clear supernatant suitable for use in further concentration and purification operations, such as chromatographic steps. Microporous membranes that can go down in size below 0.1 μm are becoming increasingly popular for debris separation, particularly when operated in the cross flow mode. The only disadvantage of membrane filtration appears to be fouling which does not allow the processing of high cell concentrations and thus makes it necessary to wash the debris if one requires a high yield or recovery of product.

A technique that is particularly attractive for the removal of cell debris is aqueous two-phase separation. Aqueous two-phase partitioning or two-phase aqueous liquid-liquid extraction is an operation that is gaining interest not only for primary protein separation but can also simultaneously separate the broken cells and cell debris which will normally partition to the lower phase whereas the proteins of interest will partition to the upper phase. This can be achieved with an extremely high efficiency.

Nucleic Acids Removal

When intracellular proteins are released by cell breakage, nucleic acids are also released adding not only a contaminant which is probably difficult to separate but which also increases the viscosity of the suspension quite dramatically. Techniques used to remove nucleic acids include the use of nuclease enzymes and precipitation. It is also possible to use a cellulose based adsorber called cell debris remover, CDR. CDR will remove cell debris as well as nucleic acids, lipids, polyphenols and other small suspended substances. Precipitation of nucleic acids is usually carried out using a longchain cationic polymer with a molecular weight of about 24,000 called polyethileneimine. This precipitation can be used to remove both nucleic acids and cell debris by centrifugation. Typical large scale conditions are to use a 1% of a 1% solution of polyethileneimine, mix and nucleic acids should precipitate almost instantaneously. Polyethileneimine is usually not recovered.

Challenges

Many of the new and important developments in membrane technology for the separation of insolubles have already been discussed. This is starting to have wide ranging implications in the developments of present and future biotechnology equipment and process design. The separartion of cell debris and impurities generated during cell breakage still represents an important challenge.

Separation of inclusion bodies from debris can be achieved on a large scale by the use of centrifuges even if the material is small (ca. 1.0 μm) mainly due to the relatively large density of inclusion bodies (e.g. 1.3 g/ml)[14]. However, flow rates have to be reduced several fold compared to the separation of whole E. coli cells where flow rates are already low. This results in large capital requirements.

Aqueous two-phase systems are a very attractive alternative for the separation of cell debris from target product proteins [2,4,19]. The separation of recombinant particles from yeast has been demonstrated using this technique [4, Riveros-Moreno, paper in this book]. In the presence of debris and recombinant particles two stages were more appropriate, the first to separate the cell debris and the second for separation of contaminant proteins [4].

REFERENCES

1. Asenjo, J.A. Selection of Operations in Separation Processes in 'Separation Processes in Biotechnology'. Ed.: J.A. Asenjo, Marcel Dekker, N.Y., 1990, p. 3–16.

2. Asenjo, J.A. and Patrick, I. Large Scale Protein Purification in 'Protein Purification Applications: a Practical Approach', Eds.: E.L.V. Harris and S. Angal, IRL press, U.K., 1990, p. 1 – 29.

3. Hagen, A.J., Hatton, T.A. and Wang, D.I.C. Biotechnol. Bioeng., 1990, 35, 955 – 965.

4. Huang, R.-B. Ph.D. thesis, University of Reading, 1990.

5. Kula, M.R. Recovery Operations, in 'Biotechnology', Ed.: H.J. Rehm and G. Reed, vol.2: Fundamentals of Biochemical

Engineering, Verlag Chemie Mannheim, 1985.

6. Schutte, H. and Kula, M.R., Pilot and Process Scale Techniques for Cell Disruption. 32 International IUPAC Congress, Stockholm, 2-7 August, 1989.

7. Naglak, T.J., Hettwer, D.J. and Wang, H.Y., Chemical Permeabilization of Cells for Intracellular Product Release in 'Separation Processes in Biotechnology'. Ed.: J.A. Asenjo, Marcel Dekker, N.Y., 1990, p. 177 - 205.

8. Fish, N.M. and Lilly, M.D., Bio/Technology, 1984, 2, 623 - 627.

9. Hettwer, D.J. and Wang, H.Y. in 'Separation, Recovery and Purification in Biotechnology'. Ed.: J.A. Asenjo and J. Hong. ACS books, 1986, 314, p. 2 - 8.

10. Andrews, B.A. and Asenjo, J.A., Trends Biotech., 1987, 5, 273 -277.

11. Hunter, J.B. and Asenjo, J.A. in 'Separation, Recovery and Purification in Biotechnology'. Ed.: J.A. Asenjo and J. Hong, ACS books, 1986, 314, p. 9 - 31.

12. Asenjo, J.A., Andrews, B.A. and Pitts, J.M., Ann.N.Y.Acad. Sci., 1988, 542, 140 - 152.

13. Buckland, B.C., Richmond, W., Dunnill, P. and Lilly, M.D. in Industrial Aspects of Biochemistry. Ed: B. Spencer, Elsevier, Amsterdam, 1974, 30, 65 - 79.

14. Datar, R. and Rosen, C.-G., Downstream Process Economics in 'Separation Processes in Biotechnology'. Ed.: J.A. Asenjo, Marcel Dekker, N.Y., 1990, p. 741 - 793.

15. Brocklebank, M.P. in Food Biotechnology, vol.1, ed.: R.D. King and P.S.J. Cheetham, Elsevier, Holland, 1987, 1, p. 139 - 192.

16. Bjurstrom, E., Chem. Eng., 1985, 95,(4), 126 - 158.

17. Brown, D.E. and Kavanagh, P.R., Process Biochem., 1987, 22, 96 - 101.

18. Hedman, P., Intern. Biotechnol. Laborat., May/ June, 1984.

19. Andrews, B.A. and Asenjo, J.A., Aqueous Two-Phase Partitioning in 'Protein Purification Methods: a Practical Approach'. Eds. E.L.V. Harris and S. Angal, IRL Press, Oxford, 1989, p. 161 - 174

DIFFERENTIAL PRODUCT RELEASE FROM YEAST CELLS
BY SELECTIVE ENZYMATIC LYSIS

B.A.ANDREWS, R.-B. HUANG and J.A.ASENJO
Biochemical Engineering Laboratory
University of Reading, P.O.Box 226
Reading, RG6 2AP, England

ABSTRACT

Differential product release from yeast cells has been achieved. This is a novel technique which has wide ranging potential as it allows the release of proteins from different cell locations with a diminished level of contaminants. This facilitates downstream processing. It has been possible to release proteins from the wall, the cytoplasm and the mitochondria in three different stages, each with a high yield for the specific protein. The overall extraction of each component was usually larger than with mechanical breakage and the total level of protein contaminants was much lower.

INTRODUCTION

Yeast has become increasingly important as a recombinant DNA protein host and bioproduct source in biotechnology. As with other microorganisms the rigid cell wall is the main barrier to effective recovery of the target product. Although protein release from microorganisms on the large scale is usually accomplished by mechanical disruption, such methods have several drawbacks. As cells have to be broken completely to obtain a high yield, all intracellular materials are released. Hence the product of interest has to be separated from a complex mixture of contaminants (proteins, nucleic acids and cell wall fragments). In addition nucleic acids will considerably increase the viscosity and mechanical disruption will damage a proportion of the protein product.

Enzymatic and chemical cell permeabilization and lysis have not been widely used for large scale intracellular product release and disruption of microbial cells. However, both operations are presently being investigated as powerful techniques for this purpose [1,2] particularly as they offer the very attractive potential of differential protein release from individual cell compartments as well as selective membrane permeabilization and the release of recombinant proteins with virtually no product damage. Lytic enzymes have been successfully used for the release of a number of intracellular proteins including recombinant human serum albumin and hepatitis B surface antigen and other intracellular recombinant proteins [3].

A novel process has been developed for the separation of bioproducts from yeast cells. The method uses a combination of physical, chemical and biological agents such as lytic enzymes, osmotic supports, spheroplast stabilizers and membrane permeabilization. Using this technique products (proteins, enzymes) can be released from different locations in the cell at distinct process stages.

The use of specific lytic enzymes for Differential Product Release (DPR) from microbial cells is thus a very powerful and novel technique for selective cell lysis. Our present work which includes the differential release of wall proteins, cytoplasmic proteins and organelle proteins is presented and discussed in this paper.

MATERIALS AND METHODS

Reagents: invertase, alcohol dehydrogenase (ADH), fumarase, bovine serum albumin (BSA), NAD^+, O-dianisidine, glucose oxidase, peroxidase, L-malic acid and DEAE-Dextran were obtained from Sigma Chemical Co.. All other chemicals were analytical grade. Cakes of compressed Bakers yeast were obtained from the United Yeast Co. (Reading, UK). The yeast was frozen in aliquots until needed then thawed at room temperature and washed three times with distilled water prior to use.

Lytic Enzyme: from a continuous culture of Oerskovia xanthineolytica as described previously [4]. The enzyme system was concentrated using a Sartocon Mini Crossflow System with a 5000 MW cutoff module (Sartorius). The retentate was further concentrated by precipitation with 70% w/v ammonium sulphate. After centrifugation the pellet was dialysed against 0.05 M phosphate buffer, pH 7 before use. This preparation contained 1856 U/1 of glucanase activity and 414 U/ml of protease activity [4].

The methods involved in Differential Product Release (DPR) have been described in detail elsewhere [5]. Briefly, formation of spheroplasts and invertase release was achieved by incubation of yeast cell suspension with lytic enzyme in osmotic support (sorbitol) with biomembrane stabilizers at 30 C for 60 min. After centrifugation the pellet contained the spheroplasts and the supernatant the wall-associated proteins. Spheroplasts were disrupted by incubation with DEAE-Dextran and glucose for 15 min at 30 C. After centrifugation the pellet contained the cellular organelles and the supernatant the cytosol proteins, Cellular organelles were disrupted by suspension in 0.25% Triton X-100 for 60 min in an ice bath. After centrifugation the supernatant contained the mitochondrial proteins.

Mechanical disruption was done using a bead mill (Biospec., USA) with 0.5 mm diameter beads as described previously [5].

Enzyme assays: invertase activity was assayed using the method of Goldstein and Lampen [6] with glucose as standard. ADH was assayed by the method of Racker [7] and fumarase by the method of Stitt [8]. Protein was assayed using the dye binding method [9] with BSA as standard.

RESULTS AND DISCUSSION

The release of enzymes from different locations within cells of Baker's yeast has been used to demonstrate Differential Product Release (DPR) as shown in figure 1. The release of invertase, alcohol dehydrogenase (ADH) and fumarase as well as total protein were monitored in the presence of different combinations of osmotic supports, membrane stabilizers and lytic enzyme. Invertase is a wall-associated enzyme and can be extracted by wall disruption leaving intact spheroplasts (fig. 1A). The wall-associated proteins are released by enzymatic lysis and the lytic enzymes together with the wall proteins are removed at this stage. After separation of the wall-associated products and wall debris the spheroplasts are lysed to release cytosol products of which ADH is an example (fig. 1B). Cell organelles remain intact at this stage and are disrupted after removal of cytosol products to yield enzymes such as fumarase from mitochondria (fig. 1C).

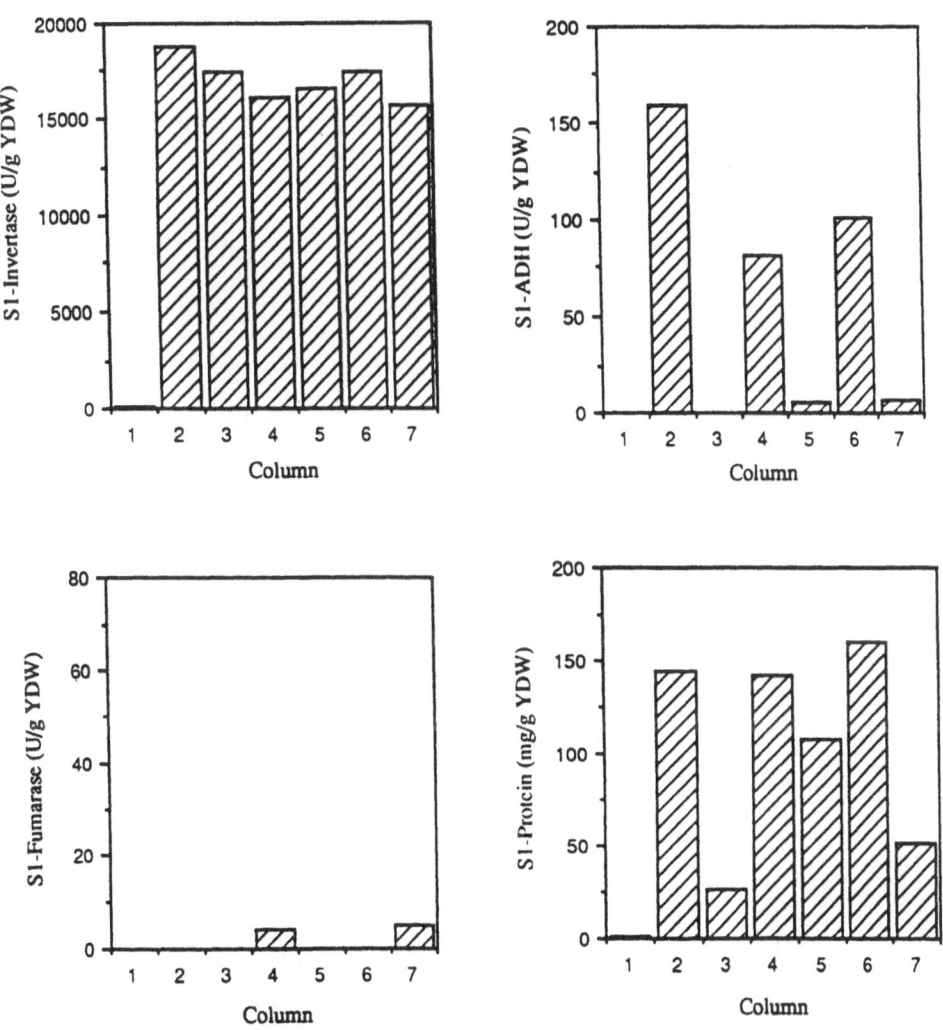

FIGURE 1A. Cell Wall Lysis using Lytic Enzymes (for legend see fig. 1C)

The process of DPR can be used with yeast concentrations up to 150 g dry weight/1. In figure 1 each of the 7 columns shows the results of treatment with different combinations of osmotic supports and membrane stabilizers. Column 1 is an autolysis control without lytic enzyme. From fig. 1A it is evident that the components used in columns 2, 4 and 6 do not result in intact spheroplasts as both invertase and ADH are released, therefore it is necessary to use membrane stabilizers to prevent the plasma membrane from breaking. High concentrations of total protein were released in these incubations (ca. 150 mg/g yeast dry weight). In columns 3 and 7 the release of total protein is small and there is no release of ADH, the Zn^{2+} ions protect the plasma membranes and the spheroplasts remain intact. In column 5 release of total protein is high, Mg^{2+}_2 does not seem to function as an osmotic support in the presence of Zn^{2+} ions. Almost no fumarase was released at this stage.

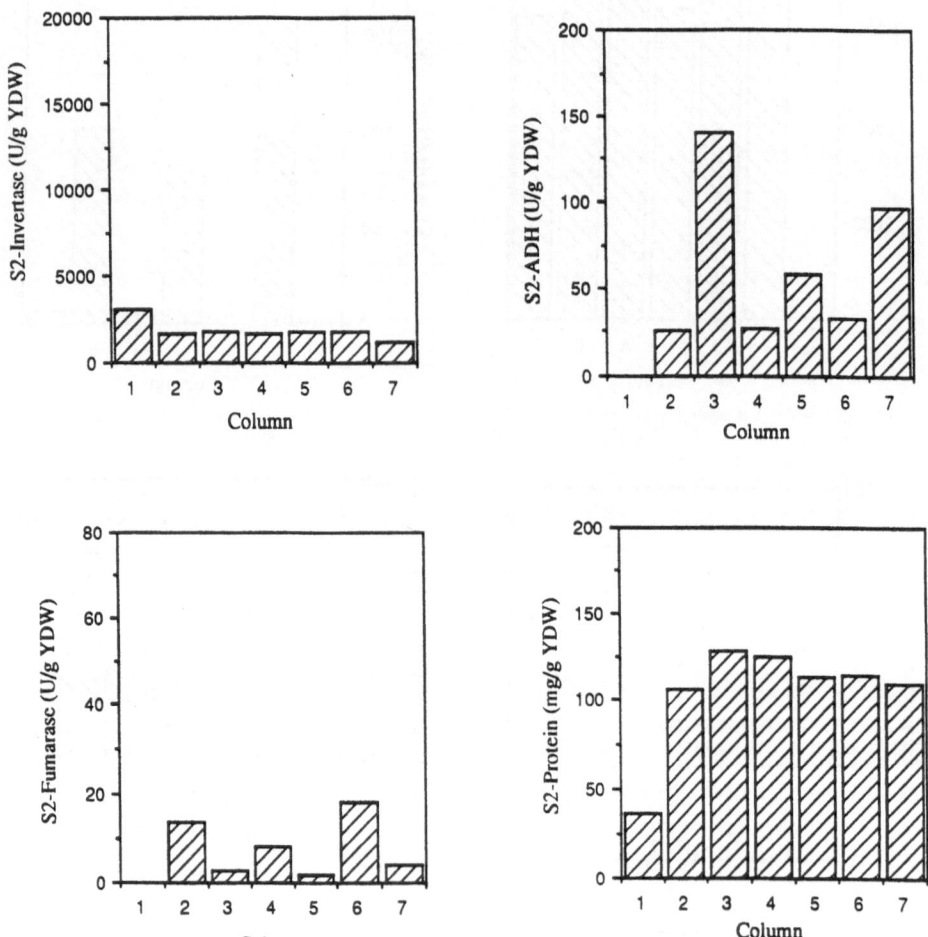

FIGURE 1B. Spheroplast Disruption with DEAE–Dextran and Glucose
(for legend see fig. 1C)

Figure 1B shows the incubation of the spheroplasts with DEAE–Dextran and glucose. Columns 3 and 7 show highest release of ADH, little fumarase was released indicating that the mitochondria remained intact at this stage. Almost 50 % of the total protein was released at this stage.

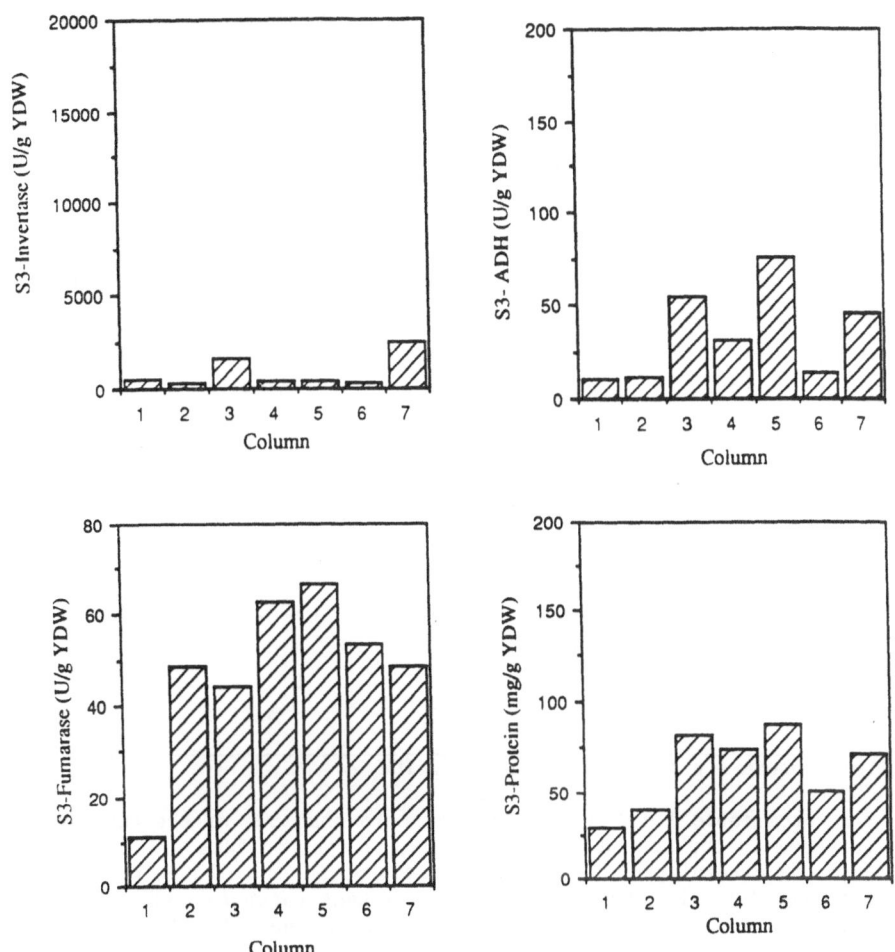

FIGURE 1C. Disrupion of Mitochondria with Triton X–100
          Column 1. 0.6M KCl (autolysis control)
                  2. 0.6M KCl + Lytic Enzyme
                  3. 0.6M KCl + Lytic Enzyme + 20mM $ZnSO_4$
                  4. 0.6M $MgSO_4$ + Lytic Enzyme
                  5. 0.6M $MgSO_4$ + Lytic Enzyme + 20mM $ZnSO_4$
                  6. 0.6M sorbitol + Lytic Enzyme
                  7. 0.6M sorbitol + Lytic Enzyme + 20mM $ZnSO_4$
      Yeast Dry Weight (YDW) – 67 g/l
      Units – Invertase, fumarase and ADH activity – Units/g YDW
      Protein – mg/g YDW

In the next step, fig. 1C, the cellular organelles were disrupted with detergent. The release of protein was low indicating that most of the protein has been extracted in former stages. In this step between 80 and 97 % of the total fumarase activity was released.

Table 1 shows in detail an example of Differential Product Release from yeast cells. The results are from column 3 in figure 1 using 0.6M KCl as osmotic support, lytic enzyme and Zn^{2+} as membrane stabilizer. The release of enzymes and protein is shown for each of the three steps and the total of all three. Fumarase release was higher using $MgSO_4$ as the osmotic support instead of KCl (ca. 70 U/gYDW).

TABLE 1
Differential Product Release from Yeast Cells

	DPR 1 Step A	DPR 2 Step B	DPR 3 Step C	TOTAL
Protein	0.026	0.128	0.082	0.236
Invertase	17,500	1,730	1,600	20,830
ADH	0	141	55	196
Fumarase	0	3	44	47

Units: Protein – g/g YDW, Enzymes – U/g YDW

From Table 1 it is evident that more than 50% of the total protein is released in the second step (B) when the spheroplasts are disrupted to release the cytosol proteins. Very little protein is released when the cell wall is lysed if the spheroplasts remain intact (step A). 30% of the total protein is released in the third step (C) when the subcellular organelles are ruptured. 83% of the total invertase is releaed in the first step (A) and 94% of the total fumarase in step C. No fumarase or ADH activity was detectable after the first stage and very little fumarase after the second stage. More than 60% of the total ADH was released in the second stage.

Table 2 shows the results of Differential Product Release (from table 1) in comparison with mechanical cell disruption in a bead mill and enzymatic lysis using lytic enzymes.

During mechanical disruption of yeast cells substantially more protein is released than during enzymatic lysis or differential product release, this is expected due to the harsh process conditions in a bead mill. The amount of protein released during mechanical disruption is almost twice that released during differential product release and three times that released during enzymatic lysis.

Wall-associated enzyme (invertase) release is very similar during enzymatic lysis and mechanical disruption; 25% more invertase is released with differential product release. The level of ADH released is highest with differential product release and lowest with mechanical disruption.

TABLE 2

Protein and Enzyme Release from Yeast Cells by Mechanical, Enzymatic and Differential Product Release

	Mechanical	Enzymatic (60 minutes)	DPR
Protein	0.43 (20 min)	0.14	0.236
Invertase	14,800 (9 min)	14,500	20,830
ADH	141 (4 min)	169	196
Fumarase	73	7	47 (70)*

Units: Protein - g/g YDW, Enzymes - U/g YDW
* amount of fumarase released in DPR using $MgSO_4$ as the osmotic support (see Figure 1)

Fumarase release after enzymatic lysis is very low compared with the other disruption methods, this indicates that the mitochondria are not lysed with this lytic enzyme system. Nearly 7 times more fumarase is released by differential product release using KCl as the osmotic support and 10 times more by mechanical disruption or by differential product release using $MgSO_4$ as the osmotic support.

Release of Recombinant Proteins

Enzymatic lysis and differential product release have been used for the recovery of recombinant proteins from yeast cells including human serum albumin [3], hepatitis B surface antigen, active human immunodeficiency virus reverse transcriptase [10] and presently their use in the release of virus like particles is beeing investigated.

When using lytic enzymes for differential product release it is important to determine whether the lytic enzyme system has any components that may have activity against the product (usually proteases) and if so how to prevent or inhibit this action. An important recent development is the purification of a pure glucanase fraction with no protease activity but with lytic activity towards whole yeast cells [11]. In the release of recombinant human serum albumin (HSA) from yeast cells using a lytic enzyme system from Oerskovia xanthineolytica [3], a comparison of the release of HSA from cells disrupted using a bead mill and by lytic enzyme has been made. In addition the effect of the protease component on the product (HSA) has been studied. No evidence of degradation of the HSA was found [3].

CONCLUSIONS

Differential Product Release from yeast cells has been achieved. It has been possible to release proteins from the wall, the cytoplasm and the mitochondria in three different stages, each with a high yield for the specific protein. The overall extraction of each enzyme was usually larger than with mechanical or enzymatic breakage. The total level of protein contaminants was much lower.

This novel technique has wide ranging potential for the release of recombinant products from yeast.

REFERENCES

1. Andrews, B.A. and Asenjo, J.A., Trends in Biotechnology, 1987, **5**, 273 - 277

2. Naglak, T.J., Hettwer, D.J. and Wang, H., Chemical Pearmeabilization of Cells for Intracellular Product Release. In Separation Processes for Biotechnology, ed. J.A. Asenjo. Marcel Dekker, 1990, pp. 177 - 205

3. Asenjo, J.A., Andrews, B.A. and Pitts, J.M., Ann. N.Y. Acad. Sci., 1988, **542**, 140 - 152

4. Andrews, B.A. and Asenjo, J.A., Biotechnol. Bioeng., 1987, **30**, 628 - 637

5. Huang, R.-B., Ph.D. Thesis, University of Reading, 1990

6. Goldstein, A. and Lampen, J.O., Methods. Enzymol., 1975, **42**, 504 - 511

7. Racker, E., J. Biol. Chem., 1950, **184**, 313

8. Stitt, M., in Methods Enzymol. Anal., ed. H.U. Bergemeyer, Verlag Chemie, 1984, **IV**, 359 - 362

9. Sedmak, J.J. and Grossberg, S.E., Anal. Biochem., 1977, **79**, 544

10. Barr, P., Power, M., Lee-Ng, C., Gibson, H. and Luciw, P., Bio/ Technology, **5**, 486-489.

11. Ventom, A.M. and Asenjo, J.A. Characterization of Yeast Lytic Enzymes from Oerskovia xanthineolytica LL-G109, Enz. Microb. Technol., 1990 (in press)

CONCEPTUAL DESIGN OF A NOVEL FLOCCULATION DEVICE.

M.H.J. Ashley.
John Brown Engineers & Constructors Limited,
1 Buckingham Street,
Portsmouth. PO1 1HN.

Abstract.

A design concept for a flocculating device has been derived by considering the effects of hydraulic conditions between corrugated baffles which are held in a spiral arrangement. Performance predictions suggest that this device may be more efficient than conventional stirred tank flocculators. High intensity phase separation applications may also be enhanced utilising this design principle.

Introduction.

Flocculation generally involves the mixing of a process stream containing finely divided suspended particles and a flocculant which brings about a change to the particles such that they become attracted to each other. This mixing is preferably carried out at high shear so as to form many active sites on the particles for subsequent agglomeration and ageing. Floc growth is achieved by reducing shear to allow agglomeration but maintaining mixing to bring fresh sites in contact. Ideally, this should subject the flocs to cycling but gradually attenuated shear so as to break delicate flocs but retain strong ones thereby bringing stronger flocs together to form even larger ones.

Conventionally flocculation is achieved by flowing a process stream through a series of increasingly larger stirred tanks each with less intense agitation than the previous one. The ideal would be to achieve controlled fluid shear and mixing which is cycling and attenuated in a small and simple continuously operating device in order to form flocs of optimum properties for subsequent processing. It is proposed that such conditions may be set up by combining the effects of gradually decreasing velocity (both radial and superficial) on a process stream flowing outwards between spiral baffles and the effects of alternating acceleration and deceleration as it flows between baffles which are corrugated in the same direction as the flow but are set at an angle to each other so that the area open to flow increases and therefore superficial velocity decreases. This may be simply achieved by using an assembly of spiral corrugated baffles and this paper attempts to describe the design principles which have been adopted to develop the concept at least to a stage to allow predictions of shear conditions to be made.

Description of the Design.

The design features of a continuous corrugated spiral baffle flocculator are described as follows and are illustrated in Figure 1. The process stream and flocculant are introduced as a mixture or may be introduced separately and mixed in a central chamber and by means of spiral baffles the flow of the mixture is directed outwards towards a volute casing or tangential outflow such that progressively less acceleration and shear are exerted on the mixture. The flocculant may be introduced into the process stream in a variety of ways but the preferred method is via an injection tube into the centre of the entry port. The central mixing chamber may be empty, contain static in-line mixing elements or preferably a turbine agitator. The spiral baffles are corrugated along their length.

Figure 1: Diagram of the flocculator design.

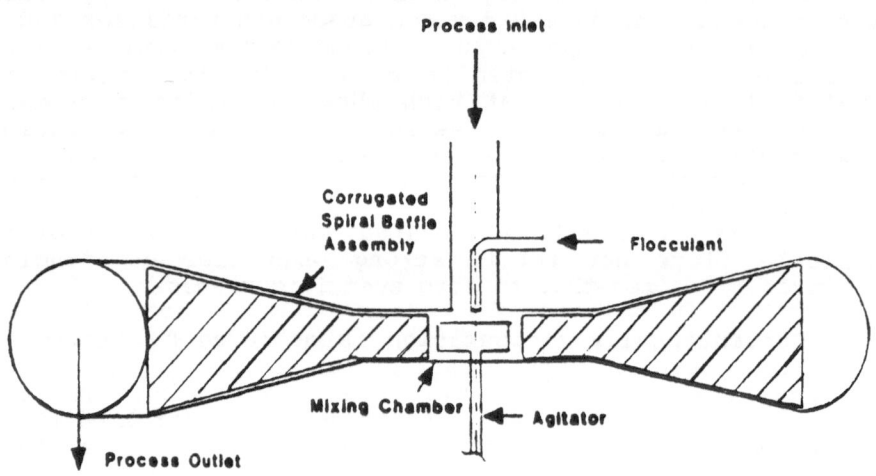

The size of the port for the inflow of the process stream would be an economic choice and therefore this dimension can be used as a means of specifying the nominal size of the unit. All other design dimensions can then be scaled to this size. The design of the mixing chamber may be adjusted to effect optimum mixing but typically has the same diameter as the entry port and is half this in depth. The diameter and speed of the agitator and its shape may be adjusted to be optimal for mixing but it is typically a turbine with straight blades 0.75 times the diameter of the entry port with a speed giving a shear rate of about 1,000 s^{-1}.

The number and shape of the spiral baffles can be adjusted to be optimal for controlling shear rate as the process mixture flows between them and should be regularly spaced. Each one typically traverses 180 degrees from its inner diameter which defines the periphery of the mixing chamber to its other diameter. The change in diameter with radial progression should be non-linear so that the area for flow is progressively increased. The spiral baffles terminate such that they discharge the process stream to a volute casing or collection chamber with an outlet port which is tangential to the axis of the spiral. The outlet diameter of the spiral is typically about 20 times the diameter of the inlet port.

The height of the chamber containing the spiral baffles may be constant from its inner to outer diameter or it may increase. Typically it is constant for half of the traverse of the spiral and then increases from the height of the mixing chamber to at least twice the diameter of the inlet port which may also be the diameter of the exit port emanating from the volute casing. This change in height may be linear with increasing spiral diameter or it may be non-linear.

The corrugations of the spiral baffles may have a variety of combinations of amplitude and frequency both of which may vary along the traverse from inner to outer diameter. The preferred amplitude progressively varies along the spiral from half the diameter of the inlet port at the inner part of the spiral to half this value at the end of the traverse. The preferred wavelength is twice the diameter of the inlet port and is constant along the traverse of the spiral baffles.

Process materials and flocculating conditions in terms of chemical and physical properties dictate appropriate materials of construction as well as the range and variation of shear rates and residence times necessary for the formation and conditioning of flocs which have suitable properties for subsequent processing. Consequently it is not possible to generally define a simple design specification and a unique correlation by which all dimensions are fixed relative to each other. However, the following section provides an attempt to theoretically establish the design principle. Then, as an example, results are predicted for a particular combination of critical dimensions.

Design Principles.

The formation of flocs depends on the rate of collision of smaller particles which may be expressed as:

Collision rate = f (number of flocs x volume of flocs
x velocity gradient)

This disruption of flocs depends on the rate of shear. It is believed that flocs are formed and destroyed at very high rates. The small difference between these rates provides the overall effect of growth and as a consequence their reduction in numbers. This may be expressed in the following form:-

$$\frac{dN}{dt} = -xN \ GD^3 \qquad\qquad (A)$$

where

N	=	particle number concentration
x	=	collision effectiveness constant
G	=	mean velocity gradient (shear)
and D	=	mean particle diameter

For the conservation of the total mass of floc particles:-

$$\frac{ND^3 \ pi}{6} = \text{constant and substituting in equation (A)}$$

$$\frac{d(D^{-3})}{dt} = \text{constant . G}$$
$$\text{or D} = \text{constant} \int f(G) dt \qquad\qquad (B)$$

Current design practice for flocculators is based on the assumption that floc size is inversely proportional to shear providing a minimum "ageing time" is incorporated. Conventional stirred tank flocculators are designed to achieve growth of flocs from 1 micron to 10 microns in a volume equivalent to about 20 minutes residence time with shear values that range from 1,000 s^{-1} for flocculant mixing, 200 s^{-1} in a second tank and 20 s^{-1} in the final tank. However, it is not possible to simply reduce shear in one stage because this would result in very open and weak flocs.

It is also generally assumed that floc strength is proportional to the product of shear and time. Such assumptions do not have a theoretical basis and it is the contention of this paper that it would be preferable to subject flocs to cycling shear whilst progressively reducing mixing intensity in a large number of stages. Moreover, it is suggested that the cycling of shear rate at each stage avoids the need for substantial residence volumes.

In a continuous processing device such as a spiral baffle system (see Figure 2) the progress of an element of flocculating material can be approximated to Ø, the angle it has traversed around the spiral from its entry point. Again for the conservation of mass this element must accelerate and decelerate depending on the area of flow (height of baffles times separation between baffles) and so providing the geometry of the spiral system can be expressed as a function of Ø as follows:-

Velocity   = Flow/Area and
Area   = hp
where baffle height h   = f(Ø) and
baffle separation p   = f(Ø) and so
shear G   = Flow/Area x p

$G = \text{Flow} \cdot f(\emptyset)$ and for a given flowrate:

$D = \text{constant} \cdot \int_{\emptyset 1}^{\emptyset 2} f(\emptyset)\, d\emptyset$

where Ø1 and Ø2 are the positions along the traverse of adjacent spirals.

A spiral may be generally expressed in polar co-ordinates as:-

$r = ce^{k\emptyset}$

However, the superimposition onto a spiral of corrugations that may also vary in amplitude make the calculation of baffle separation difficult in this form. An alternative expression for a spiral is:-

$r' = a\emptyset^2 + b\emptyset + c$

and the transverse length is approximately r"Ø where r" is the average radius of the transverse.

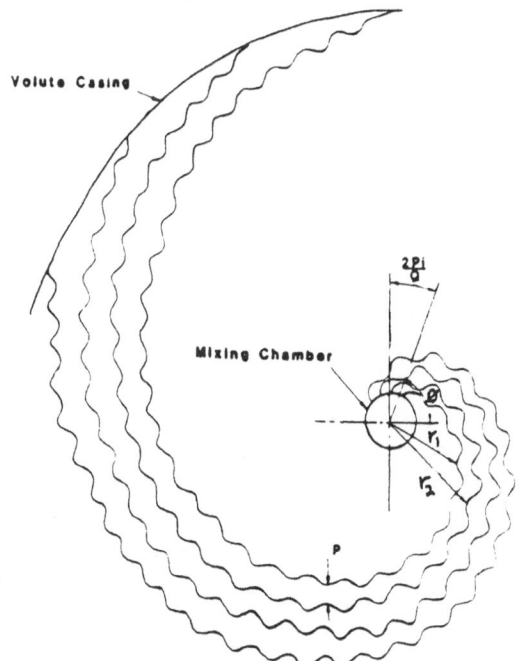

Figure 2: Corrugated
          spiral
          configuration.

A superimposed sine wave of the form y = msinx can be used to calculate the locus of corrugations so that r = r' + y where r defines the actual radius for any value of Ø and m is the factor which defines the amplitude where x is the transverse along the spiral which is r"Ø. To account for variations in amplitude m can also be expressed as m = f(Ø).

Separation between baffles (t) can now be calculated since:-

$$r_1 = f(Ø)$$

$$r_2 = f(Ø + 2pi/Q)^{-1}$$

for Q regularly radially spaced baffles and to a first approximation:

$$p = r_2 - r_1$$

Consequently p = f(Ø) and similarly baffle height (h) may be expressed as h = f(Ø). In other words for a given flowrate it is possible to calculate and plot shear (G) versus the angle of progression round a spiral (Ø).

Calculations of this type have been performed over a large range of corrugated spiral baffle configurations and it has been found that a device designed according to these principles does indeed impose cycling and attenuated shear. In order to optimise the design it is advisable to mount the mathematical expressions for each of the critical dimensions on a micro-computer spreadsheet and to modify the many constants until a satisfactory shear plot is obtained. There follows, as an illustration of the design method, definitions of the critical dimensions for a system based on an inlet flowrate of 36 m^3/h and having a nominal inlet port size of 75mm.

Volumetric flowrate		m^3/s	0.01
Inlet port diameter		mm	75
Mixing chamber diameter		mm	100
Mixing chamber height		mm	50
Agitator blade diameter		mm	80
Agitator speed		rpm	3000
Number of baffles	Q		8
Traverse angle	Ø	rads	2 pi
Spiral radius	r	mm	50 to 1000 (1)
Spiral outer height		mm	2000
Spiral height	h	mm	50 to 300 (2)
Corrugation amplitude	a	mm	50 to 25 (3)
Corrugation wavelength		mm	150
Outlet port diameter		mm	300

(1) Spiral radius r = 50 + 325(Ø/pi) + 75(Ø/pi)^2

(2) Spiral height d = 50 for 0 < Ø < pi
            and d = 50 + 250(Ø-pi)/pi for pi < Ø < 2pi

(3) Corrugation amplitude a = 50 + 12.5(Ø/pi)

The values of shear over a 2 pi baffle traverse angle have been calculated and a plot is included in Figure 3. This shows that in a system volume of 0.75 m^3 (with residence time of 75 s) shear may be reduced from approximately 1000 s^{-1} to 1 s^{-1} in about 30 cycles each of diminishing attenuation.

According to conventional flocculator design practice sustained changes in shear between 1000 s^{-1} and 1 s^{-1} result in growth of stable or "aged" aqueous based inorganic flocs from 1 micron to 10 microns. This paper proposes that cycling of shear takes the place of "ageing time" and so a similar increase in floc size can be predicted. Numerical analysis of many similar sets of results to derive a plot of shear G with spiral angle Ø (Figure 3) indicates that the predicted operating performance of the system critically depends on the spiral radius equation (1). This governs baffle separation and hence maximum cycle shear values. Variations in corrugation amplitude (3) have a profound impact at the inner diameter (start) of the spiral but have less effect in other parts of the spiral. In practice the spiral shape achieved will depend on the relative stiffness between the short inner baffle sections and the taller sections outside and this is determined by the spiral height equation (2). Careful optimisation of the constants in all three equations is required. Fine control of process performance may be possible by slightly rotating the inner chamber to wind or unwind the spiral. This alters relative phase changes as well as the distance between adjacent baffles consequently changing the ratios between maximum and minimum shear.

The foregoing description illustrates how it is possible to develop a model of a spiral flocculator system based on relatively simple shapes. Process application of these principles may demonstrate how the design principle can be further refined and no doubt will lead to pragmatic design development based on process performance in much the same way that conventional stirred tank systems have evolved.

Figure 3: Shear versus Traverse Angle for 36 m^3/h of water through an 8 baffle corrugated spiral of 2.0 m diameter and 0.75 m^3 volume.

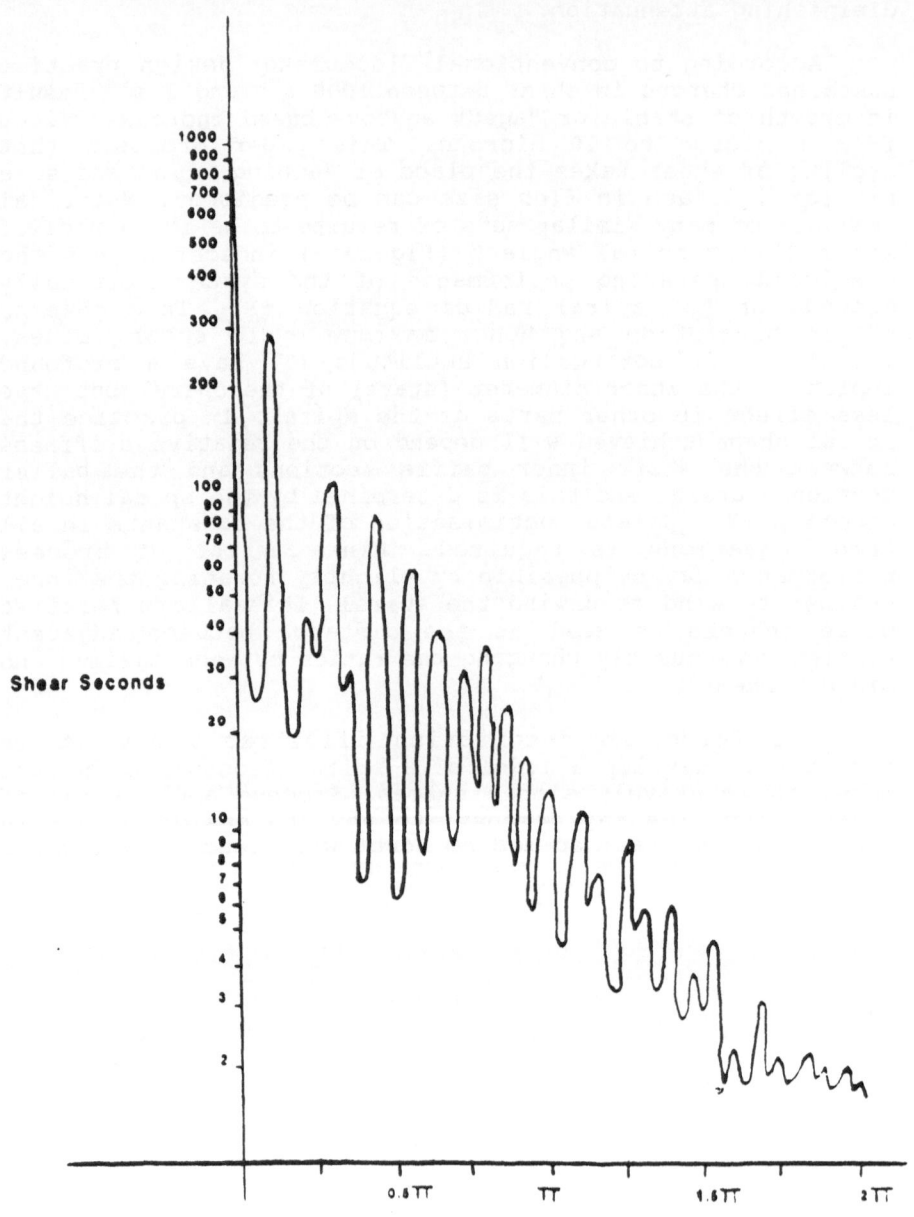

Shear Seconds

Traverse Angle of Spiral

Conclusions.

An attempt has been made to theoretically develop the design of a flocculating device which is both simple in concept and provides more intensive and so potentially more effective process performance than conventional stirred tank flocculators. It is premature to make process performance claims for this design until substantial practical work has been carried out. Nevertheless predictions of the shear conditions in such a device indicate that residence time may be reduced from 20 minutes to about 1 minute with the process flow being subjected to cycling shear which is progressively attenuated over three orders of magnitude. This should produce stable flocs of ten times the original particle size. It is anticipated that the design principle of flow through a corrugated spiral baffle system may be exploited over a range of applications for flocculation and for phase separation duties.

THE EFFECT OF CULTURE HISTORY ON THE DISRUPTION OF
ALCALIGENES EUTROPHUS BY HIGH PRESSURE HOMOGENISATION.

SUSAN T. HARRISON, JOHN S. DENNIS[‡] and HOWARD A. CHASE.
Department of Chemical Engineering, University of Cambridge,
Pembroke St., Cambridge, CB2 3RA.
[‡]Consultant Chemical Engineer, 20 Church Lane, Dullingham,
Newmarket, Suffolk, CB8 9DX

ABSTRACT

To date, rigorous study of the use of high pressure homogenisation in cell rupture over a wide range of operating conditions has been confined to the yeasts. This investigation concentrates on the rupture of the Gram-negative bacterium *Alcaligenes eutrophus* by an industrial homogeniser. The parameters affecting cell disruption such as operating pressure, temperature and biomass concentration as well as the effect of cell characteristics such as growth rate, size, shape and wall strength were studied. Bacterial cultures which are growing rapidly in the logarithmic phase exhibit less resistance to rupture than nutrient limited cultures in the stationary phase. This is accompanied by a decreased dependence of the degree of cell rupture on operating pressure with increased culture age. The extent of accumulation of the storage product poly-β-hydroxybutyrate during stationary phase has a less significant effect on the resistance to cell rupture than the growth phase of the micro-organism. This is particularly interesting owing to the concomitant change in the shape and size of the bacterium on the accumulation of storage product. A variety of chemical and physical treatments of the stationary phase micro-organisms prior to rupture have been considered in an attempt to decrease cell wall strength. Cell rupture efficiency is increased by any of the following pretreatments: incubation at 45°C, alkaline pH shock, addition of the detergent sodium dodecyl sulphate or addition of sodium chloride.

INTRODUCTION

Products of recombinant technology, the overproduction of proteins as inclusion bodies and the potential use of the intracellular storage product poly-β-hydroxybutyrate (PHB) as a thermoplastic have led to increased interest in efficient and cost effective cell rupture to enable the recovery of intracellular microbial products. The use of the high pressure homogeniser for the disruption of yeasts is well documented in the literature [1,2]. While disruption of bacterial cells by homogenisation is reported [3,4,5], a thorough investigation has not been documented.

This paper sets out to report a rigorous study of the disruption of Gram-negative bacteria by homogenisation. *Alcaligenes eutrophus* known for its ability to accumulate high levels of PHB is used as the model organism. It is widely suggested that mechanical cell disruption is influenced by cell wall type and strength, cell size, shape and growth rate. An introductory study of the combined influence of size, shape and growth phase on disruption is therefore presented. This is facilitated by the ability to effect alterations in the size and shape of *A.eutrophus* by changes in the PHB content. Finally, the idea of broth pretreatment with the intention of weakening the cell wall is considered. This provides the practical opportunity to achieve greatly improved rupture with reduced energy requirements as well as giving an insight into the importance of cell wall strength in rupture.

MATERIALS AND METHODS.

The Micro-organism: A selected strain of *Alcaligenes eutrophus* H16 was obtained from ICI Biological Products Business, Billingham, U.K. The organism was grown on minimal medium using glucose as the carbohydrate source at pH 6.8 and 30°C. PHB accumulation occurred under phosphate limitation and a fed-batch configuration. Unless otherwise stated, stationary phase cultures containing approximately 70% PHB (by mass on a dry basis) were employed in the disruption studies.

Disruption Equipment: The APV-Gaulin laboratory homogeniser 15M 8BA used in this study employs a high pressure positive displacement pump to pass the cell suspension through an adjustable, restricted orifice discharge valve. This model may be operated continuously at 0 to 55 MPa or intermittently to 69 MPa. A flowrate of 1.5×10^{-5} m^3s^{-1} is maintained. The ceramic cell disruption (CD) valve with lengthened knife edge has been used predominantly in this study. Use of the Stellite cell rupture (CR) valve showed similar cell rupture efficiency.

Analysis of Cell Disruption: Release of soluble protein is the most commonly reported method for monitoring cell rupture. Soluble protein release was determined by means of the Lowry assay [6] modified for use in combination with a Technicon autoanalyser. Extreme rupture conditions may lead to variations in protein solubility. However, subjection of protein-containing supernatants to homogenisation and subsequent protein determination has shown that while power inputs were maintained below 1 kW and the suspension temperature does not exceed 35°C, a maximum error of 3% results. While the initiation of protein release requires only the permeabilisation of the cell, DNA release necessitates more extensive disruption. DNA release has been monitored by the diphenylamine assay of Burton [7]. Cell disruption was directly observed by phase contrast microscopy at 1000 x magnification. Quantitative data was obtained microscopically by the use of a Thoma fixed volume counting chamber (Weber, England).

PHB content of the cells was determined as hydroxybutyric acid by HPLC following hydrolysis with perchloric acid. A column of cross-linked sulfonated polystyrene-divinyl-benzene copolymer resin in the protonated form is used in conjunction with a dilute sulphuric acid mobile phase. Eluted species are detected by UV absorbance at 210 nm. A size and shape distribution of the cells was obtained by analysis of scanning electron micrographs by means of an Optomax Image Analyser. Viscosity measurements of the disrupted broth were conducted by means of the Haake VT181 concentric cylinder viscometer. Shear rates in the range 13.25 to 424 s^{-1} were used. Measurements were taken at 25°C.

RESULTS AND DISCUSSION

Effect of Energy Input and Broth Parameters on Cell Rupture.

The energy input required for cell rupture by homogenisation is affected by operating pressure, number of passes through the system and input broth temperature. Variation of cell rupture on a single pass is shown as a function of operating pressure in Figure 1. A threshold value of the pressure difference across the valve must be exceeded for cell rupture to occur. This varies with the assay procedure used as each technique detects a different degree of disruption of the cell. Cell permeabilisation resulting in the onset of soluble protein release is observed at pressures exceeding 10 MPa. Release of DNA requiring extensive cell disruption is observed at pressures exceeding 25 MPa. Cell rupture as observed by phase contrast microscopy occurs at pressures of 20 MPa . In general, cell rupture is a sigmoidal function of operating pressure.

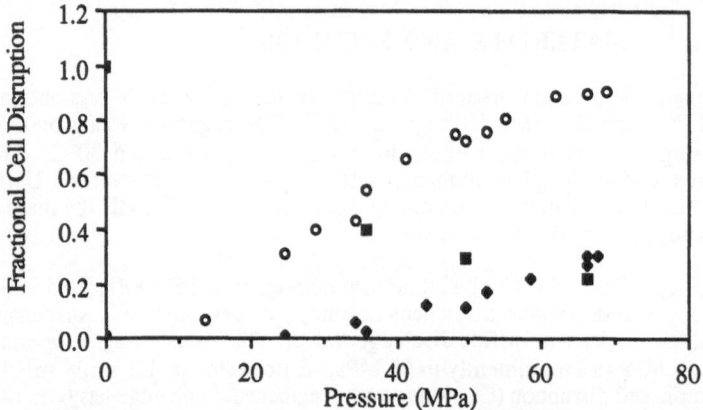

Figure 1. Cell disruption on a single pass as a function of operating pressure.
   o   Soluble Protein Release   •   DNA Release
   ■   Intact cells visible as a fraction of cells visible at 1000X magnification.

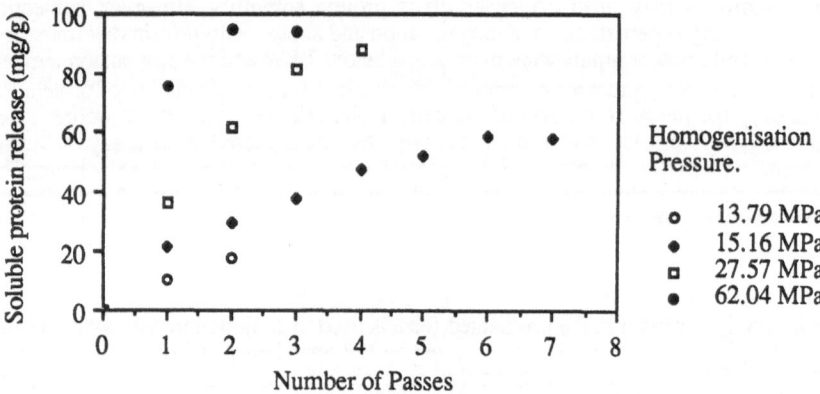

Figure 2. Soluble protein release on homogenisation is shown as a function of the number of passes at a constant operating pressure.

    Complete cell rupture is not achieved on a single pass in the pressure range 0 to 62 MPa (Figure 2). A maximum, in terms of soluble protein release, is obtained on two passes at 62 MPa . However, three such passes are required for maximum DNA release (result not shown). Soluble protein release is first order with respect to number of passes (Figure 3). While the use of lower operating pressure necessitates more passes to achieve the same degree of rupture, an operating pressure in excess of 15 MPa is required for maximum release (Figure 2). The efficiency of the rupture process in terms of soluble protein release is maximised by use of the highest operating pressure attainable by this apparatus and the minimum number of passes required to fulfil a particular energy input (Figure 4).

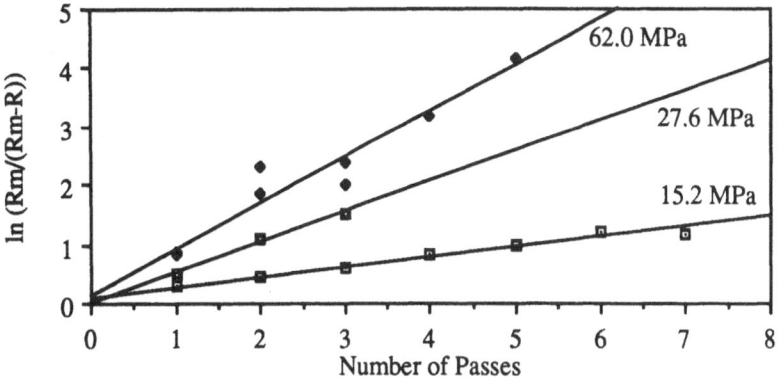

Figure 3. Illustration of First Order Relationship between Protein Release and Number of Passes through the Homogeniser at Constant Pressure.

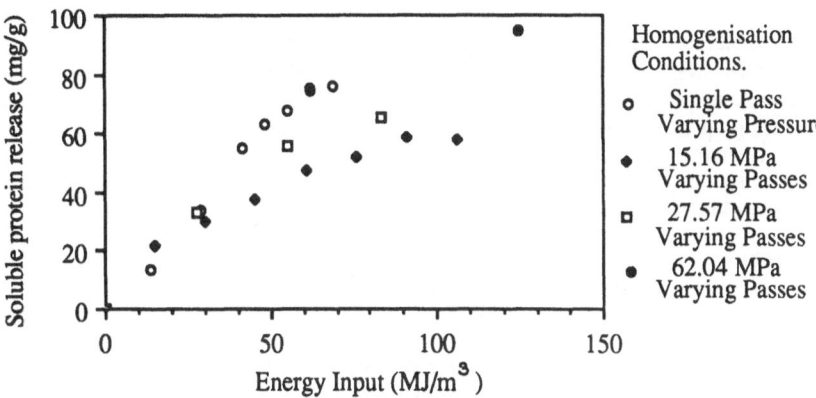

Figure 4. Energy input for homogenisation is a function of operating pressure and number of passes. Variation of these parameters independently as detailed indicates that rupture efficiency is improved by the use of higher operating pressure and fewer passes.

Most hypotheses put forward for the mechanism of cell rupture by homogenisation express the logarithm of the extent of cell rupture as a function of pressure raised to an exponent. This, together with the first order relationship between protein release and number of passes forms the basis of the correlation of Hetherington et al [1]:

$$\ln \frac{R_{max}}{R_{max} - R} = KNP^a \qquad (1)$$

where R_{max} is the maximum protein available for release (mg/g), R is the soluble protein release (mg/g), K is the rate constant, N is the number of passes, P is the operating pressure and a is the pressure exponent. This correlation has been shown to provide a reasonable fit to the data obtained using A.eutrophus (Figure 5).

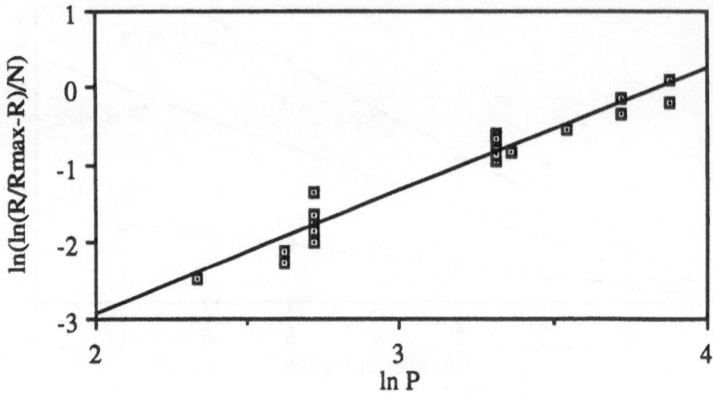

Figure 5. Illustration of the suitability of the correlation of Hetherington *et al* [1] to the disruption of *A.eutrophus* by high pressure homogenisation in the range 0 to 50 MPa.

As shown by Hetherington *et al* [1], cell rupture by homogenisation is improved by increased broth input temperature. For *A.eutrophus*, a 1.5 fold increase in soluble protein release is obtained by increasing the broth input temperature from 13.5°C to 26°C. This is represented by an increase in the rate constant from 1.336×10^{-4} to 2.356×10^{-4} MPa^{-2} for a pressure exponent of 2. During homogenisation, temperature increases as a linear function of operating pressure by 0.22°C/MPa and 0.18°C/MPa for the CD and CR valves respectively owing to energy dissipation across the valve. The extent of this temperature increase is independent of the input temperature, the nature of the suspension and its concentration in the range investigated.

Various biomass concentrations were obtained by concentration of the fermentation broth in a laboratory centrifuge. Biomass concentrations in the range 95 to 260 gm dry biomass/l show no effect on the efficiency of rupture (Table 1). A large effect is, however, seen on the apparent viscosity of the ruptured broth. This ruptured broth is non-Newtonian, characterised by shear-thinning behaviour. Reduction of viscosity of the ruptured broth is required if the use of high biomass concentrations is to be practical.

TABLE 1
The Effect of Biomass Concentration on Cell Rupture and Broth Viscosity.

Biomass Concentration (g dry mass/l)	Soluble Protein Release at 52.4 MPa (mg/g)	Apparent Viscosity (mPa.s) of Ruptured Broth at Shear Rate of	
		53 s^{-1}	424 s^{-1}
96.5	49.6	3.8	3.5
152.2	49.5	192.0	19.1
257.0	52.2	292.0	33.9
307.0	43.9	385.0	62.0

Effect of Cell Parameters on Disruption.

It is well known that cell size and shape will influence the ease of cell disruption in mechanical systems [8]. *A.eutrophus* grown under conditions for the accumulation of PHB offers a good opportunity to study the influence of these parameters while using the same micro-organism. Growth conditions [4] and growth rates [5] will, however, also influence the ease of rupture. A preliminary study is reported here in which four broths of increasing culture time and PHB content (defined in Table 2) are disrupted by high pressure homogenisation.

TABLE 2
Fermentation Broths used to study the Effect of Cell Shape, Size and Growth Phase.

Broth No.	Growth Phase	Nutrient Supply	Time of Glucose Feed (hours)	PHB Content (%)
1	exponential phase	balanced	0.0	< 2
2	late exponential phase	onset limit.	2.5	~ 6
3	early stationary phase	limited	8.0	~ 20
4	late stationary phase	limited	65.0	~ 70

TABLE 3
Changes in Size and Shape of *A.eutrophus* on PHB Accumulation.
Units of length are quoted in μm. ρ is the standard deviation about the mean.
ρ/M provides a measure of the spread of the distribution.

Feed Time (hours)	PHB Content (%)	Sample Number	Longest Dimension Mean ρ ρ/M	Spherical Diameter Mean ρ ρ/M	Area Mean ρ ρ/M	Form Factor Mean ρ ρ/M
0.0	4	257	0.9 0.3 0.3	0.7 0.1 0.2	0.4 0.1 0.4	0.81 0.12 0.15
3.5	24	324	1.1 0.4 0.3	0.8 0.2 0.3	0.5 0.2 0.4	0.82 0.13 0.16
9.0	31	283	1.2 0.4 0.4	0.8 0.2 0.3	0.6 0.3 0.4	0.78 0.14 0.17
23.7	37	203	1.1 0.5 0.5	0.7 0.2 0.3	0.5 0.3 0.6	0.78 0.16 0.20
50.0	55	127	1.1 0.4 0.4	0.8 0.2 0.3	0.6 0.2 0.4	0.82 0.19 0.23
71.7	73	176	1.3 0.4 0.3	1.0 0.2 0.2	0.8 0.3 0.4	0.89 0.12 0.13

During exponential growth *A.eutrophus* is a small, short rod. PHB accumulation necessitates an increase in cell size. The results of a study of cell shape and size through the fermentation are correlated in terms of the PHB content of the cells and fermentation time in Table 3. Trends in these data are confirmed by scanning electron micrographs in Figure 6. While a small change is seen in the longest dimension, cell size, shown by both spherical diameter and area, increases significantly with the PHB content. This is shown by a 75% rejection of the hypothesis that the mean values of the sample parameter are equivalent to the mean value across all samples using a statistical t-test with a confidence limit of 95%. The most marked changes occur on the transition from exponential growth to

44

nutrient limited growth and over the PHB accumulation range of 50 to 75%. The form factor, F defined as:

$$F = \frac{4\,\pi\ area}{perimeter^2} \qquad (2)$$

quantifies the sphericity of the cells. This decreases in early stationary phase owing to the observed lengthening of the rods. At high PHB content, the cell becomes increasingly spherical to accommodate the maximum amount of PHB within the constriction of the cell envelope.

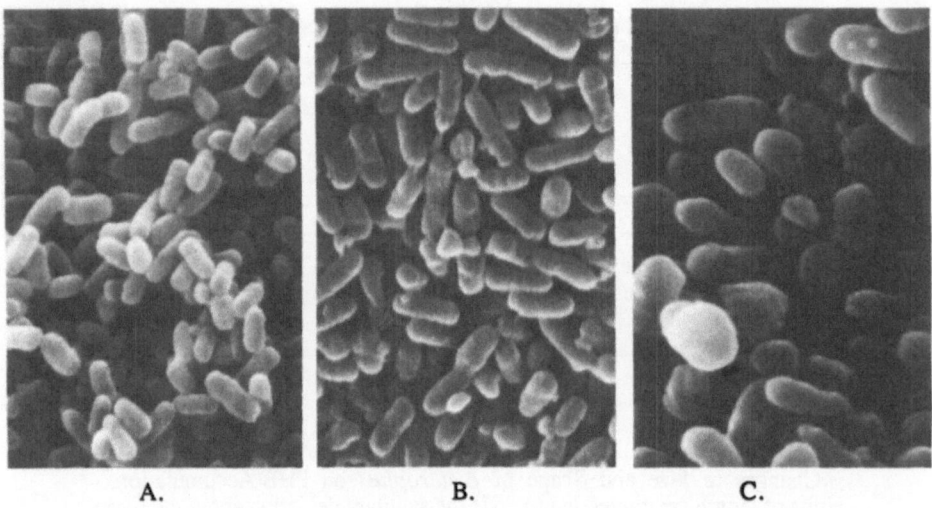

A.   B.   C.

Figure 6: Scanning electron micrographs of *A.eutrophus* through the course of fermentation:
A: 17.2 hours, 4% PHB; B: 26.3 h, 31% PHB; C: 89 h, 73% PHB.
Length bar represents 1 μm: ⊢——⊣ .

The data obtained on the homogenisation of the youngest and the oldest of these broths is presented in Figure 7. Determination of the rate constant K and pressure exponent a is detailed in Table 4. Data obtained at an energy input below 100 MJ.m^{-3} results from a single pass through the homogeniser. Data corresponding to an energy input exceeding 100 MJ.m^{-3} results from multiple passes at constant pressure. Figure 7 shows that although 2 to 3 passes at pressures exceeding 60 MPa are required for maximum disruption of stationary phase cells (measured by protein release), actively growing *A.eutrophus* cells can be disrupted on a single pass. The disruption of late stationary phase cells is, however, initiated at lower pressures than are required for the disruption of log phase cells. It is postulated that this increased rupture at low pressures is due to increased cell size. Table 4 shows a marked effect of cell history on the pressure exponent,a. Exponentially growing cells show a significantly greater susceptibility to changes in pressure (a = 3.1) than nutrient limited cells (a < 1.7). Since large changes are not seen on further increases in cell size with PHB accumulation (broths 3 and 4), it appears that the pressure exponent is primarily affected by growth phase.

Differing R_{max} values are obtained for exponential phase and stationary phase broths owing to the PHB content per unit biomass. Correction for the PHB present results in a constant value of R_{max} of 330 mg protein/ g non-PHB cell mass on a dry basis for *A.eutrophus*.

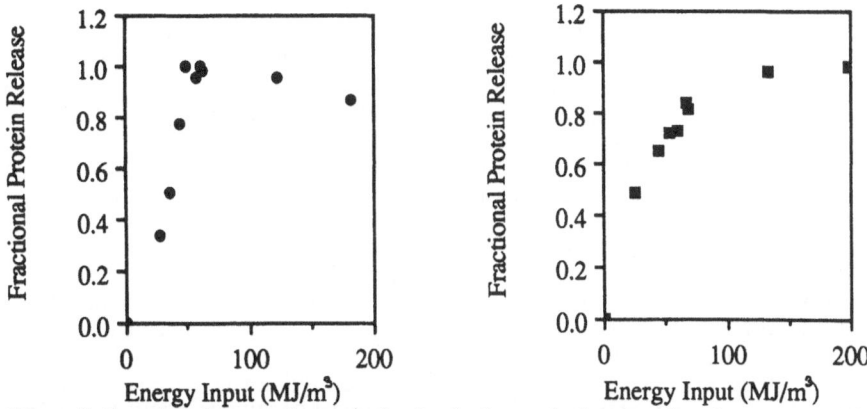

Figure 7. Protein release on homogenisation is shown for broths taken from exponential phase growth (left) and late stationary phase growth following the accumulation of PHB (right).Where the energy input exceeds 100 MJ.m^{-3}, multiple passes at 60.7 to 62.0 MPa were employed. Data at energy inputs below 100 MJ.m^{-3} originate solely from single pass studies using pressures in the range 0 to 62.0 MPa.

TABLE 4

Effect of growth conditions on the exponent of pressure and the rate constant describing cell rupture by high pressure homogenisation.

Broth	Pressure Exponent a	Rate Constant K (MPa^{-a})	Correlation Coefficient r^2
1	3.08	1.35 E-5	0.964
2	2.80	1.83 E-5	0.921
3	1.69	1.18 E-3	0.965
4	1.59	2.16 E-3	0.951

Effects of Pretreatment of the Bacterial Cell on the Ease of Cell Rupture.

The strength of the cell wall is expected to have a significant influence on the efficiency of cell rupture by mechanical means. This strength is reported to be diminished by subjection to operating conditions insufficient to cause cell rupture [9]. A series of pretreatments has therefore been considered in an attempt to alter cell wall strength prior to homogenisation. The aim is two-fold: firstly, it is desirable to disrupt a greater proportion of cells at a specific pressure or to achieve total disruption on a single pass; secondly, an understanding of the extent of the role of cell wall strength in mechanical cell disruption is sought. Treatments investigated include elevated temperature, alkaline conditions, the additions of the detergent sodium dodecyl sulphate (SDS) and of sodium chloride. In all cases, homogenisation was carried out at neutral pH and ambient temperature.

Thermal broth pretreatment at 45°C for 1 hour resulted in protein release equal to 17% R_{max}. Following subsequent homogenisation at 62 MPa, protein release exceeded that obtained for the untreated broth by 1.8 fold. Use of higher temperatures could not be satisfactorily studied by monitoring the release of protein or DNA owing to decreases in the solubility of these components.

Studies in this laboratory (unpublished) have shown that alkaline conditions of pH 10.0 to 11.5 cause a marked and rapid release of soluble protein from *A.eutrophus*. At 8°C and pH 10, 60 to 80% of the reaction occurs within 30 seconds. The protein release obtained is equivalent to approximately 30% cell disruption. This is insufficient for effective downstream processing. Subsequent homogenisation of neutralised cultures at 43.4 MPa and 64.8 MPa shows an increasing trend in both soluble protein release and DNA release with increased pH in the range pH 7.0 to 11.0. A maximum increase of 25% in protein release following homogenisation is achieved with alkaline pH shock compared to untreated cells.

The use of sodium dodecyl sulphate (SDS) to rupture cells is a common technique in molecular biology [10]. Disruption is achieved through the dissociation of the outer membrane, protein denaturation and the removal of lipopolysaccharide. Conditions of 1% (m/v) SDS, 70°C and 20 minutes have been identified as optimal in reducing cell integrity (unpublished results). While the presence of detergent and heat alter the solubility of the protein and DNA, it was clearly seen that maximum disruption was achieved by a single pass at 34.5 MPa. Use of SDS at 0.1% (m/v) under similar conditions required a single pass at 62 MPa for maximum cell breakage. Use of 0.1% SDS at low temperature (8°C) had no marked effect on the ease of homogenisation.

The addition of sodium chloride (8 g/l) to the fermentation broth followed by incubation at 60°C for one hour resulted in a 20% release of the available soluble protein. DNA release was negligible. Maximum cell rupture in terms of protein release is achieved on a single pass at 64.8 MPa. In terms of DNA release, approximately 66% rupture is obtained. This is compared with 67% protein release and 28% DNA release achieved for similarly treated broths in the absence of NaCl. Measurement of cell rupture by soluble protein release following elevated temperature is susceptible to inaccuracies owing to protein precipitation. Cell disruption achieved may indeed be higher than indicated. Comparison with similarly treated broths to which no NaCl is added confirms the benefit of NaCl addition. The use of NaCl at ambient or reduced temperatures has not been considered. The effect of NaCl addition is independent of the biomass concentration. This effect may be directly mediated through changes in ionic strength altering the components of the cell envelope. It has been suggested that NaCl may alter the size of the micro-organism and thereby aid rupture [9].

CONCLUSIONS

A study of both the operating parameters of the homogeniser and the broth conditions during homogenisation indicates that the relationships reported for the disruption of yeasts [1] are also appropriate for the rupture of *A.eutrophus*. The process of cell rupture can be satisfactorily represented by the equation (1). Small increases in the temperature of the broth at the inlet to the homogeniser cause significant increases in the disruption efficiency. This is quantified by an increase in the rate constant, K at constant values of the pressure exponent, a. Biomass concentration is irrelevant in the range 95 to 270 g dry mass/l. Inter-batch variations (~10%) in the degree of cell rupture achieved are attributed to wear of the homogeniser valve and fermentor broth history.

An introductory study on the effect of cell growth phase, size and shape shows that rapidly growing cells are ruptured more easily than stationary phase cells. Maximum rupture can be obtained with a single pass for exponentially growing cells whereas 2 to 3 passes are required for the equivalent rupture of stationary phase cells. The ease of rupture of exponential cells is borne out by an increase in the pressure exponent, a. Substantial increases in cell size at high PHB contents (approximately 70%) do not cause a change in the pressure exponent with respect to nutrient limited cultures of low PHB content. However, this increase in cell size appears to be accompanied by the initiation of disruption at lower pressures than are required for the rupture of exponentially-growing cells.

Pretreatment of the cells prior to homogenisation has shown that substantial cell weakening can be achieved such that the ensuing cell rupture by high pressure homogenisation results on a single pass, possibly at reduced operating pressure. This is advantageous both in terms of a reduction in energy requirements of the process and the minimisation of the micronisation of the cell debris which may hinder further separations. The pretreatments described here have been chosen with consideration of the conditions withstood by the PHB storage granules. A further study will be required to identify similar treatments which would not denature the more labile components such as proteins.

Acknowledgements. The authors wish to acknowledge Marlborough Biopolymers Ltd for valuable discussion and financial assistance. STH acknowledges the support of a Trinity College, Cambridge External Research Studentship.

SYMBOLS.

ρ	standard deviation about mean
a	pressure exponent
F	form factor
K	rate constant (MPa^{-a})
M	mean value
N	number of passes through homogeniser
P	operating pressure (MPa)
R_{max}	maximum soluble protein available for release (mg/g)
R	soluble protein release (mg/g)

REFERENCES:

1. Hetherington, P.J., Follows, M., Dunnill, P. and Lilly, M.D., Release of protein from Baker's yeast (*Saccharomyces cerevisiae*) by disruption in an industrial homogeniser. Trans. Instn. Chem. Engrs, 1971, **49**, 142-148.

2. Engler, C.R. and Robinson, C.W., Disruption of *Candida utilis* cells in high pressure flow devices. Biotechnol.Bioeng., 1981, **23**, 765-780.

3. Keleman, M.V. and Sharpe, J.E.E., Controlled cell disruption: a comparison of the forces required to disrupt different micro-organisms. J. Cell Sci., 1979, **35**, 431-441.

4. Gray, P.P.P., Dunnill, P. and Lilly, M.D., The continuous-flow isolation of enzymes. In Fermentation Technology Today, ed. G.Terui, Society for Fermentation Technology, Japan,1972, pp. 347-351.

5. Higgins, J.J., Lewis,D.J., Daly, W.F., Mosqueira, F.G., Dunnill, P. and Lilly, M.D., Investigation of the unit operations involved in the continuous flow isolation of β-galactosidase from *Escherichia coli*.. Biotechnol.Bioeng., 1978, **20**, 159-182.

6. Lowry, O.H., Rosebrough, N.J., Farr, A.L. and Randall, R.J, Protein measurement with the Folin phenol reagent. J. Biol. Chem., 1951, **193**, 265-275.

7. Burton, K., Determination of DNA concentration with diphenylamine. Methods in Enzymology, 1968, **12B**, 163-166.

8. Wase, D.A.J. and Patel, Y.R., Effect of cell volume on disintegration by ultrasonics. J. Chem. Tech. Biotechnol., 1985, **35B**, 165-173.

9. Doulah, M.S. and Hammond, T.H., A hydrodynamic mechanism for disintegration of *S.cerevisiae* in an industrial homogeniser. Biotechnol. Bioeng., 1975, **17**, 845-858.

10. Hancock, I.C. and Poxton,I.R., Bacterial Cell Surface Techniques, John Wiley and Sons, 1988, pp. 59-62, 112-114.

SEPARATION BY BIOPOLYMER
SEPARATION OF SUSPENDED SOLID BY MICROBIAL FLOCCULANT

Ryuichiro KURANE

Fermentation Research Institute, Agency of Industrial Science
& Technology.

1-1-3,higashi, Tsukuba City, Ibaragi 305, JAPAN.

ABSTRACT

Rhodococcus erythropolis produces a kind of the microbial
flocculant. This bioflocculant could efficiently flocculate
all suspended solids in aqueous solution tested. Among those
tested were microorganisms such as E. coli and alchhole yeast
activated sludge, algae, kaolin clay, muddy water, river
bottom sediment, ash and chacoal powder. The bioflocculant
from Rhodococcus erythropolis have a wide flocculating activity
against both organic and inorganic materials.

Culture conditions for producing the bioflocculant were
also tested. More than 90 % of the flocculating activity
was in the culture broth and less than 10 % in the cells.

INTRODUCTION

Many kinds of flocculating agents have been used in
a wide range of industrial fields such as waste water treat-

ment, dreding, industrial process, and so on. Inorganic flocculant such as polyaluminium chloride and organic synthetic high-polymer such as polyacrylamide derivatives are frequently used as economical and powerful flocculating agents. However, it has been also reported that the monomer of acrylamide is a strong carcinogen and a strong nervous toxin.

Therfore, biodegradable, safer bioflocculant, which cause no problems in human body and environmental pollution, are attracting wide interest and are urgently called for.

RESULTS

1 Screening for flocculant producing microorganisms

Cultures stocked in my Labo. were tesyed for the ability to flocculate kaolin clay. Flocculation was found in Rhodococcus erythropolis, Nocardia restricta, Nocardia calcarea, Nocardia rhodnii, Corynebacterium sp., Alcaligenes cupidus and Alcaligenes latus. The microorganism which have the strongest flocculating activity among them tested was Rhodococcus erythropolis S-1.

2 Characteristics of the bioflocculant of Rhodococcus erythropolis

The rates of flocculation of various suspended solids in aqueous solution with the cultures of R. erythropolis S-1 are summarized in Table I. The bioflocculant from R. erythropolis S-1 could flocculate all matrials tested, including microorganisms such as E. coli and alcohol yeast (Saccharomyces sp.), activated sludge, Microcystis aeruginosa (a kind of green algae) which cause serious problems especially in lakes and marshes during the summer time, river bottom sediment that cause also serious environmental problems, muddy water, river dreding muddy water, ash from a steam-power station, charcoal powder and kaolin clay.

TABLE I   FLOCCULATION OF VARIOUS SUSPENDED SOLIDS IN AQUEOUS SOLUTION
WITH CULTURE OF *R. erythropolis* S-1

Suspended solids	Flocculating activity			
	Precipitate volume (%)		Clarification of supernatant (OD_{660})	
	1 min	3 min	1 min	3 min
Microorganisms				
E. coli				
E. coli	0	0	1.00	1.00
E. coli + Ca^{2+}	0	0	1.00	1.00
E. coli + Ca^{2+} + cultures	5	16	0.45	0.18
Alcohol yeast (*Saccharomyces* sp.)				
Yeast	0	0	1.00	1.00
Yeast + Ca^{2+}	0	0	1.00	1.00
Yeast + Ca^{2+} + cultures	14	9	0.15	0.05
Microcystis aeruginosa (AOKO)[a]				
AOKO	0	0	3.8	3.8
AOKO + Ca^{2+}	0	0	3.8	3.8
AOKO + Ca^{2+} + cultures	75	38	0.42	0.17
Soil solid liquid (Muddy water)[b]				
Muddy water	0	0	>2.00	>2.00
Muddy water + Ca^{2+}	0	0	>2.00	>2.00
Muddy water + Ca^{2+} + cultures	15	20	0.85	0.10
River dredging muddy water (RDMW)[c]				
RDMW	0	0	1.80	1.80
RDMW + Ca^{2+}	0	0	1.80	1.80
RDMW + Ca^{2+} + cultures	20	18	0.10	0.06
River bottom sediment (HEDORO)[d]				
HEDORO	0	0	>2.00	>2.00
HEDORO + Ca^{2+}	0	0	>2.00	>2.00
HEDORO + Ca^{2+} + cultures	55	19	0.35	0.25
Coal ash from a steam-power station[e]				
Coal ash	0	0	1.15	1.15
Coal ash + Ca^{2+}	0	0	1.15	1.15
Coal ash + Ca^{2+} + cultures	16	12	0.01	0.01
Activated charcoal powder[f]				
Charcoal	0	0	7.4	7.4
Charcoal + Ca^{2+}	0	0	7.4	7.4
Charcoal + Ca^{2+} + cultures	4	5	0.82	0.35

[a] *Microcystis aeruginosa* from Kasumigaura Lake during summer time was used.

[b] Soil solid liquid was prepared as followings; 500 g of soil was added and stirred in 1 liter of distilled water, and then upper phase after standing for 3 min was used as soil solid liquid.

[c] River dredging muddy water was estimated to contain 1.4 w/v% of mud.

[d] River bottom sediment "HEDORO" from the Yodo river was used.

[e] Concentration of coal ash in distilled water was 1,000 ppm.

[f] Activated charcoal powder, which was finely ground in a mortar with a pestle, was suspended in distilled water and used.

This results indicated that the bioflocculant from R. erythropolis had a wide flocculating activity against both organic
and inorganic materials.

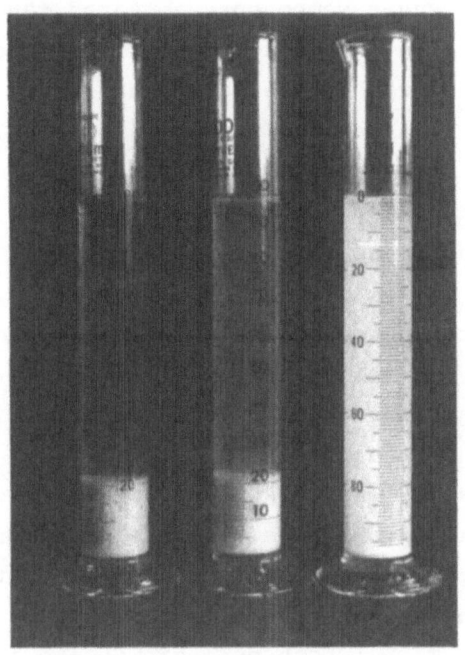

A   B   C

Figure 1. Photograph after 3 min of Standing of Kaolin Clay
          Flocculated with Rhodococcus erythropolis Flocculant

          A : with R. erythropolis S-1 Flocculant
          B : with R. erythropolis 260-2 Flocculant
          C : control
          Concentration of kaolin clay was 5,000 ppm.

3 Culture conditions for production of microbial flocculant by Rhococcus erythropolis

(1) Effects of carbon sources

Glucose and fructose were effective for the flocculant
formation among the various water-soluble carbon sources
tested. These two sugar components appeared favorable for
cell growth as well as flocculant formation. The maximum
flocculant formation was in the early stationary phase.
On the other hand, mannose, galactose, arabinose, xylose,
lactose, maltose, cellobiose, and sucrose were not favorable
for either flocculant formation and cell growth. Among
water-insoluble carbon sources tested, more cell growth
was observed with olieve oil than fructose. R. erythropolis
S-1 also assimilated n-hexadecane, super heavy normal
paraffin and heavy normal paraffin well. However, these
water-insoluble carbon sources reduced flocculation activity.

(2) Effect of initial pH

The growth rate and flocculant production were affected
by the culture initial pH. The alkaline pH, around pH
8 - 9 , greatly stimulated the flocculant production.
Under more alkaline initial pH up to 11.0, this strain
S-1 was able to grow, but both cell growth and flocculant
formation were decreased compared with initial pH 8.5.

(3) Large scale cultivation

Large scale cultivation was done using 30 and 1,000 liter
tank at an agitation of 100 - 200 rpm and an aeration rate
of one-thenth volume per medium volume per minute.
A representative course of flocculant formation and a cell
growth curve when fructose surup (fructose and glucose
mixture)was used as the carbon source is presented in
Fig. 2.

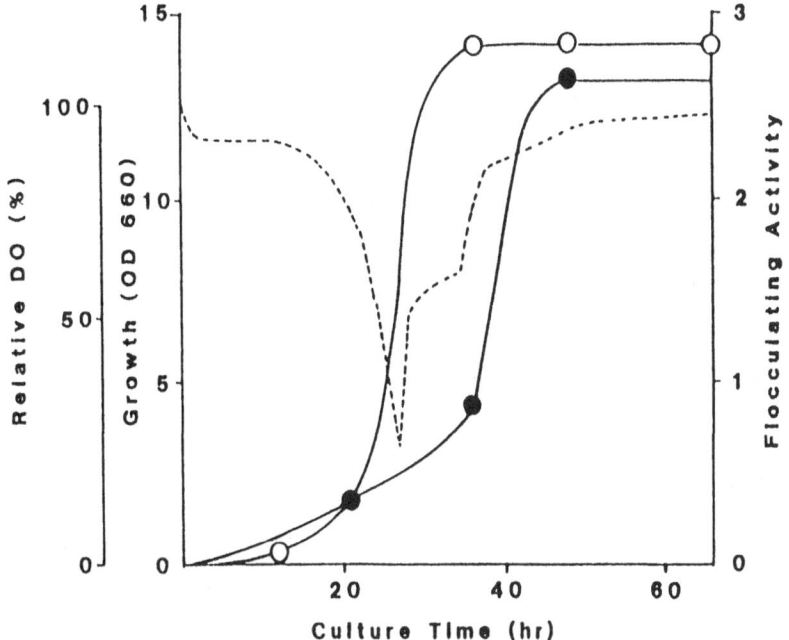

FIG. 2   Course of Flocculant Production by *R. eryth-ropolis* S-1.

O—O, growth (OD$_{660}$);   ●—●, flocculating activity (1/OD$_{550}$); ---, DO (dissolved oxygen) (%).

4 Distribution of the flocculating activity

The distribution of the flocculating activity inthe cultures was investigated. Both the culture broth and cells apeared to have the flocculating activity. Even when ten-fold concentrated cells was used, the flocculating activity was lower than that of culture broth. It has concluded that more than 90 % of the flocculating activity was in the culture broth and less than 10 % in the cells.

DISCUSSION

The wide-spread distribution of <u>Rhodococcus</u> <u>erythropolis</u>
as a major gram positive bacterium in japanease soils[1]
and greater numbers of R. erythropolis which cells counted
about 2 - 3 % normally against total cells in the normal
activated sludge[2] mean that humans have been contract with
this microorganism for a long period in the open environments.
It might be this bioflocculant from <u>Rhodococcus</u> <u>erythropolis</u>
has an advantage in its application to open systems such
as wastewater treatment system and even in the downstream
processing in the field of biotechnology.

The finding that <u>R</u>. <u>erythropolis</u> S-1 was able to grow
well at alkaline initial pH up to 11 suggested that this
microorganism might be rather an alkali-resistant bacterium
than an alkaliphilic bacterium.

REFERENCES

1. Katoh, K. and Suzuki, T. , Microflora in japanease
    soil, Bulletin National Inst. Agric. Science, (B),
    30, 73, (1979) (in Japanease)
2. Kurane, R., Suzuki, T. and Takahara, Y., Microbial
    population and identification of phthalate ester-utilizing
    microorganisms in activated sludge inoculated with
    microorganisms. <u>Agric</u>. <u>Biol</u>. <u>Chem</u>., 43, 907 (1979)
3. Kurane, R., Takeda, K. and Suzuki, T., Screening
    for and characteristics of micrbial flocculant.
    <u>Agric</u>. <u>Biol</u>. <u>Chem</u>., 50, 2301 (1986)
4. Kurane, R., Toeda, K., Takeda, K. and Suzuki, T.,
    Culture conditions for production of microbial flocculant
    by <u>Rhodococcus</u> <u>erythropolis</u>.   <u>Agric</u>. <u>Biol</u>. <u>Chem</u>.
    50, 2309 (1986)

PROTEIN RELEASE FROM THE YEAST *PICHIA PASTORIS* BY CHEMICAL PERMEABILIZATION: COMPARISON TO MECHANICAL DISRUPTION AND ENZYMATIC LYSIS

THOMAS J. NAGLAK AND HENRY Y. WANG
Department of Chemical Engineering
The University of Michigan
Ann Arbor, MI 48109

ABSTRACT

Conventional large scale recovery methods for intracellular proteins involve mechanical disruption of the cells, a nonselective process which yields a solution contaminated by nucleic acids and cell wall fragments. We have investigated chemical permeabilization of microorganisms as an alternative to mechanical disruption. For the yeast *Pichia pastoris,* we have compared chemical permeabilization to two more traditional means of protein release, mechanical disruption and enzymatic lysis. The chemical treatment agents in our permeabilization scheme, guanidine hydrochloride and Triton X-100, combine synergistically to release more protein from *P. pastoris* than either acting individually. We have demonstrated overall protein recovery by chemical permeabilization of this host to be on the same order attainable by mechanical disruption or enzymatic lysis, but each process has its own advantages and disadvantages.

INTRODUCTION

Cell disruption for intracellular product release is an important part of many protein production processes involving microorganisms. If a protein, native or recombinant, must be recovered from inside the producing cells, the selection of a cell disruption method will have a profound effect on the remaining downstream processing steps. Traditional industrial means of attaining product release based on mechanical disruption result in lysis of the cells and complete release of all intracellular material. The result is a complex mixture which includes cell fragments, nucleic acids, and contaminating proteins, as well as the product of interest.

One alternative to mechanical disruption which has shown promise as a means of attaining selective protein release from microorganisms is chemical permeabilization. In this approach, chemicals which are known to solubilize membrane components are added to a cell suspension to alter cell wall permeability and allow intracellular proteins to diffuse to the extracellular medium.

The purpose of this paper is to describe initial feasibility studies in applying the concept of chemical permeabilization to the eucaryotic host *Pichia pastoris*. This model system was selected for several reasons. The industrial microbiology of yeasts is well understood, and yeast cells are capable of carrying out post-translational modifications of proteins. Furthermore, although some recombinant proteins expressed in yeast have been successfully engineered for secretion to the extracellular medium, many have not, and it is not yet known whether all genetically engineered proteins will be able to be secreted from yeast [1]. The particular genus and species *P. pastoris* was selected because its alcohol oxidase gene regulation system AOX1 has been used for high level expression of foreign proteins in this strain, such as hepatitis B surface antigen [2].

EXPERIMENTAL PROCEDURES

Culture and Conditions

Pichia pastoris (NRRL Y-11430, Northern Regional Research Center, Peoria, IL) was grown on modified yeast medium (MYM) at 25°C with constant shaking. MYM contains, per liter distilled water, 3 g yeast extract, 3 g malt extract, 5 g peptone, and 20 g cerelose. Cell growth was monitored by turbidity, measured using a Klett-Summerson Photoelectric Colorimeter 800-3 equipped with a red filter. Fresh MYM was inoculated with 10% (vol/vol) of an overnight culture and allowed to grow to late exponential phase. The cells were recovered by centrifuging and washed once in cold treatment buffer (0.05M phosphate, pH 6).

Chemical Permeabilization

Cells resuspended in treatment buffer were added to flasks containing similarly buffered solutions of guanidine hydrochloride (practical grade) and/or Triton X-100 to yield final solutions with the indicated concentrations of treatment chemicals and approximately 2-6 g/l total cellular protein. The flasks were left gently shaking at 4°C, and samples were withdrawn at the indicated intervals for assay.

Mechanical Disruption

Cells resuspended in treatment buffer were mixed with 0.2 mm glass beads (approximately 50% by volume) and added to the chamber of a Bead-Beater (Biospec Products, Bartlesville, OK). The suspension was beaten in 30 second intervals, with each interval followed by 30 seconds of cooling the chamber on ice. Samples were withdrawn at the indicated intervals for assay. Reported times refer to accumulated time of beating intervals.

Enzymatic Lysis

Cells were washed with lyticase buffer (0.1M phosphate, pH 7.5). The cells were centrifuged and resuspended in buffer. To 5 ml cell suspension was added 3 ml β-mercaptoethanol (50 mM in buffer), 3 ml lyticase (3 mg/ml in buffer), and 4 ml buffer. The cells were incubated at 30°C with rotation, and samples were withdrawn at the indicated intervals for assay. Lyticase (Sigma Chemical Company, St. Louis, MO) is an enzyme preparation isolated from *Arthrobacter luteus* which hydrolyzes glucose polymers which have β-1,3 linkages, such as yeast cell wall glucan.

Assays

Protein was determined by the method of Bradford [3] using bovine serum albumin as the standard. Total cellular protein was determined by lysing the cells in 1M NaOH at 100°C for 20 minutes [4]. Reported protein values represent the average of two determinations. Cell counts were determined by light microscopy using a Petroff-Hausser bacteria chamber (Hausser Scientific, Blue Bell, PA). Reported cell count data are the average of at least three determinations.

RESULTS

Table 1 shows protein release from *P. pastoris* as a function of concentration of the two treatment chemicals guanidine and Triton X-100. As with *E. coli* [5], these two chemicals exert a synergistic effect whereby their combination releases a greater percentage of intracellular protein than individual doses at the same concentrations. Figure 1 shows this synergistic effect for the combination of 2M guanidine and 0.5% Triton X-100. As Figure 1 shows, most of the protein release is accomplished in the first few hours. The overall protein release attained by 24 hour treatment with the combination of 2M guanidine and 0.5% Triton X-100 is 46% ± 6% (standard deviation based on 6 different batches of cells).

By comparison, two other traditional methods of protein release from yeast, mechanical disruption and enzymatic lysis, release the same order of magnitude of overall protein as the better chemical permeabilization conditions, although the time scale of these processes is on the order of minutes to hours. Figure 2 shows protein release from *P. pastoris* by mechanical disruption in a bead mill. The maximum release of 53% is attained within 4 minutes. Figure 3 shows protein release by enzymatic lysis using lyticase, where the maximum release is 69% and the time scale is less than 2 hours. Higher protein recovery by enzymatic lysis compared to mechanical disruption could be due to release of cell wall proteins by the degradative enzymes.

TABLE 1

Overall percent protein release from *P. pastoris* as a function of treatment chemical concentration for a 48-52 hour exposure. All values represent the average of final protein release values from at least two different batches of cells.

Triton X-100 (%)	Guanidine Hydrochloride (M)			
	0	1	2	3
0.0	0.5	3.3	18.	15.
0.5	3.8	29.	47.	55.
1.0	4.3	27.	55.	46.
1.5	3.2	25.	46.	42.
2.0	4.2	27.	54.	43.

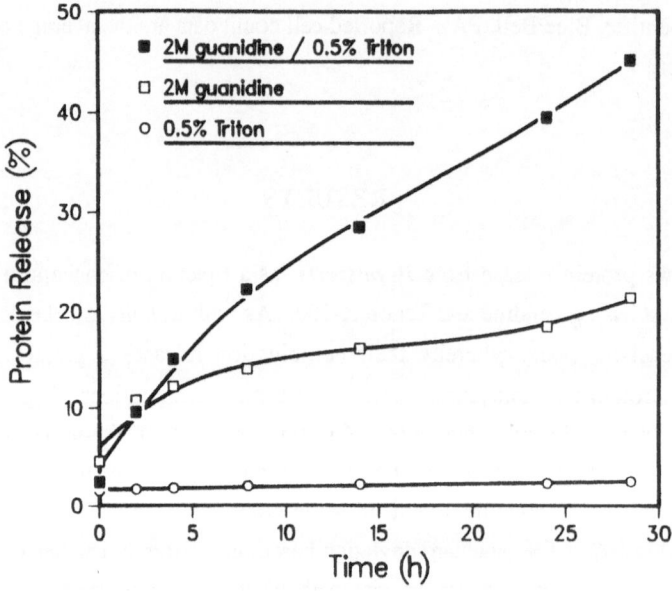

Figure 1. Overall protein release from *P. pastoris* as a function of time for exposure to 2M guanidine, 0.5% Triton X-100, or both. The control treatment (cells in buffer) results in less than 1% protein release. Data represent replicate assays using a single batch of cells.

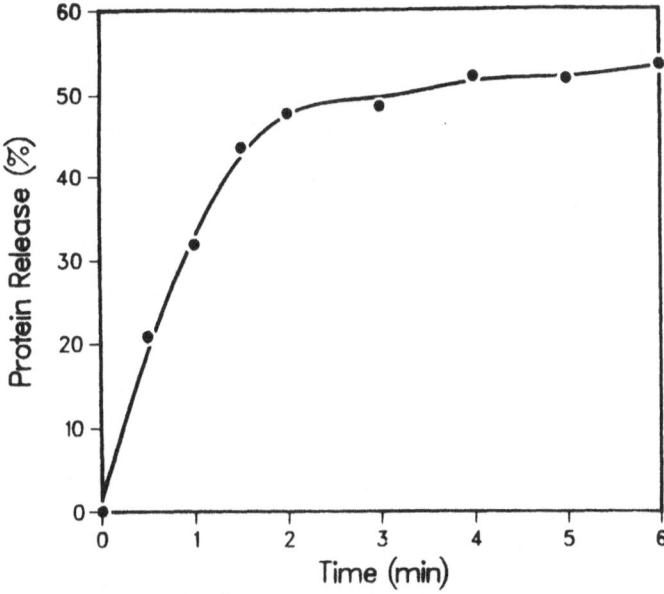

Figure 2. Overall protein release from *P. pastoris* as a function of time for mechanical disruption in a bead mill.

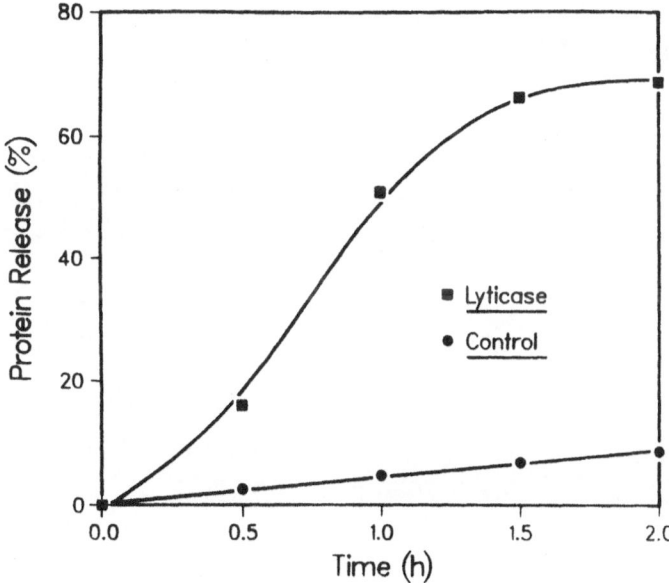

Figure 3. Overall protein release from *P. pastoris* as a function of time for enzymatic lysis with lyticase in the presence of β-mercaptoethanol. The control incubation consists of cells in buffer plus β-mercaptoethanol.

The most important distinction between chemical permeabilization and these other two methods of intracellular product release is that the latter result in lysis of the cells. Figure 4 consists of light micrographs of control cells as well as cells after each of the three protein release treatments. It shows that chemically permeabilized cells retain their morphology. In contrast, mechanical disruption and enzymatic lysis create small cell fragments. Figure 5 shows quantitatively that chemical permeabilization does not result in cell lysis. While all three of the indicated treatments result in overall protein release between 45 and 70%, only chemical permeabilization does not significantly decrease the number of cells present.

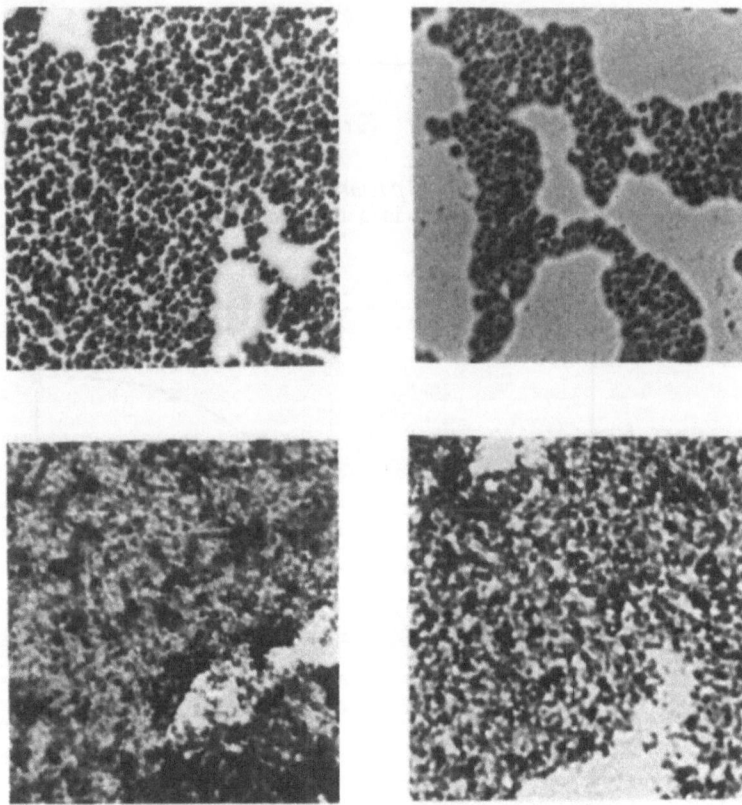

Figure 4. Light micrographs of *P. pastoris* cells before and after various cell disruption processes. Upper left: control cells. Upper right: chemically permeabilized cells. Lower left: mechanically disrupted cells. Lower right: enzymatically lysed cells. Conditions for the three protein release treatments correspond to Figure 1 (2M guanidine and 0.5% Triton X-100, 24 hr treatment), Figure 2, and Figure 3, respectively.

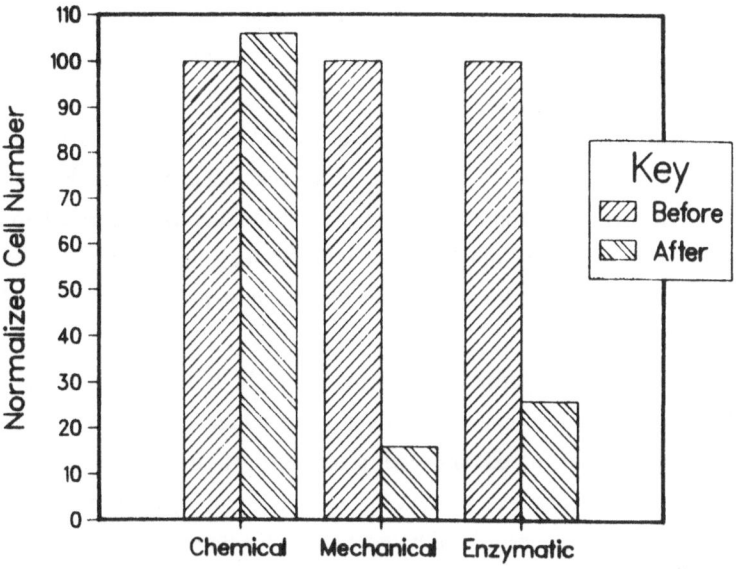

Figure 5. Cell number before and after protein release from *P. pastoris* by chemical permeabilization, mechanical disruption, and enzymatic lysis. Conditions for the three protein release treatments correspond to Figure 1 (2M guanidine and 0.5% Triton X-100, 24 hr treatment), Figure 2, and Figure 3, respectively. Cell number is normalized to 'Before' value of each treatment separately. Cell number before and after chemical permeabilization is the same within 95% confidence limits.

DISCUSSION

Engineered secretion of foreign proteins from microorganisms may not be possible to attain in all cases. Some products may not be efficiently exported by any signal sequence scheme [6, 7]. Even in some instances where secretion is possible, it may be disadvantageous. Some proteins may not be stable in a fermentor due to shear and foaming [8]. Furthermore, concentrating the product by centrifuging the cells prior to disruption may be easier than concentrating it from dilute solution in the culture supernatant [1, 9]. The selection of a cell disruption method for product release has a great impact on the remainder of the downstream processing steps.

Because the cells are not fragmented by chemical permeabilization under the conditions described here, they are easily removed from suspension by centrifugation. In contrast, the small particles of cell wall debris, combined with the gelatinous nature of the suspension created by lysis, make mechanical disruption homogenates difficult to clarify, either by centrifugation or filtration [10]. Figure 4 provides a graphic illustration of the difference between cell envelopes after chemical permeabilization and cell debris after mechanical disruption or enzymatic lysis.

Because chemical permeabilization can be carried out in a simple stirred vessel, capital and operating costs are likely to be lower than for mechanical disruption. Enzymatic lysis, on the other hand, is often inappropriate for uses other than laboratory scale due to its high cost [9], but there are specific examples of economically feasible large scale operations [11]. Economic comparisons of the three methods will depend on the relative yields of the specific protein involved.

Of course, chemical permeabilization is not without its disadvantages. Mechanical disruption and enzymatic lysis extract slightly more protein in a significantly shorter time than chemical permeabilization, based on overall protein assays. It should be pointed out that all three methods do not necessarily extract the same proteins in the same relative amounts. Individual proteins may be released at different rates by the three methods.

For this host system, a relatively high concentration of guanidine is required to bring about overall protein release on the order of 50%. Guanidine concentrations near 2M can denature some proteins, although such denaturation is not necessarily irreversible. Many proteins retain all or part of their biological activity after exposure to guanidine concentrations as high as 6M, as in the solubilization of inclusion bodies [12]. For other host systems, the possibility of loss of recovery due to denaturation may not be a problem at all. For *E. coli*, guanidine concentrations as low as 0.1M act synergistically with 0.5% Triton X-100 to release about 50% of intracellular protein [5]. A recombinant β-lactamase has been recovered with full activity from *E. coli* by chemical permeabilization with 0.2M guanidine [13].

For some products, genetically engineered secretion may be more appropriate than any of these three cell disruption methods. Proper post-translational modifications such as glycosylation depend upon a protein being directed into the secretion pathway through the endoplasmic reticulum and Golgi body [14]. Not all proteins that enter the secretion pathway, however, make it all the way to the extracellular medium. Some heterologous proteins with secretion signals end up in various organelles rather than crossing the plasma membrane [15] or may stay in the periplasmic space [6]. In these cases, some form of cell disruption will still be necessary.

CONCLUSIONS

We have compared three means for attaining intracellular product release from the yeast *P. pastoris*. Chemical permeabilization using guanidine hydrochloride and Triton X-100 is a simple procedure which results in overall protein release on the same order of magnitude as mechanical disruption or enzymatic lysis. The cells maintain their morphology after the chemical permeabilization process, in contrast to the latter two methods, which result in cell lysis. The advantages of chemical permeabilization are based on easy removal of the cell envelopes by centrifugation. Disadvantages include long extraction times and the possibility of denaturation of the product. The optimum method of product release for a given protein depends on the interaction of these factors, which vary for different host cell systems, as well as on their impact on downstream processing steps.

ACKNOWLEDGEMENT

The support of the National Science Foundation, Grant No. ECE-8603976, is gratefully acknowledged.

REFERENCES

1. Wiseman, A., King, D.J. and Winkler, M.A., The isolation and purification of protein and peptide products. In Yeast Biotechnology, ed. D.R. Berry, I. Russell and G.G. Stewart, Allen & Unwin, London, 1987, pp. 433-470.

2. Cregg, J.M., Tschopp, J.F., Stillman, C., Siegel, R., Akong, M., Craig, W.S., Buckholz, R.G., Madden, K.R., Kellaris, P.A., Davis, G.R., Smiley, B.L., Cruze, J., Torregrossa, R., Veliçelebi, G. and Thill, G.P., High-level expression and efficient assembly of hepatitis B surface antigen in the methylotrophic yeast, *Pichia pastoris*. Bio/Technology, 1987, **5**, 479-485.

3. Bradford, M., A rapid and sensitive method for the quantitation of microgram quantities of protein utilizing the principle of protein dye-binding. Anal. Biochem., 1976, **72**, 248-254.

4. Stewart, P.R., Analytical methods for yeasts. In Methods in Cell Biology, Volume XII, Yeast Cells, ed. D.M. Prescott, Academic Press, New York, 1975, pp. 111-147.

5. Hettwer, D. and Wang, H., Protein release from *Escherichia coli* cells permeabilized with guanidine-HCl and Triton X100. Biotech. Bioeng., 1989, **33**, 886-895.

6. Boyd, A., Protein secretion in yeast. In Proc. of BioFair Tokyo '86, 1986, pp. 267-273.

7. Kingsman, S.M., Kingsman, A.J., and Mellor, J., The production of mammalian proteins in *Saccharomyces cerevisiae*. Trends in Biotech., 1987, **5**, 53-57.

8. Fish, N.M, and Lilly, M.D., The interactions between fermentation and protein recovery. Bio/Technology, 1984, **2**, 623-627.

9. Darbyshire, J., Large scale enzyme extraction and recovery. In Topics in Enzyme and Fermentation Biotechnology 5, ed. A. Wiseman, Ellis Horwood Limited, Chichester, 1981, pp. 147-186.

10. Bui, P. T., Recovery and purification of biologically active polypeptides from rDNA microorganisms. Bio/Technology, 1983, 1, 488-490.

11. Asenjo, J.A., Process for the production of yeast-lytic enzymes and the disruption of whole yeast cells. In Advances in Biotechnololgy, Volume III, Fermentation Products, ed. C. Vezina and K. Singh, Pergamon Press, Toronto, 1981, pp.295-300.

12. Naglak, T.J., Hettwer, D.J., and Wang, H.Y., Chemical permeabilization of cells for intracellular product release. In Downstream Processing in Biotechnology, ed. J.A. Asenjo, Marcel Dekker, New York, 1990 (in press).

13. Naglak, T.J., and Wang, H.Y., Recovery of a foreign protein from the periplasm of Escherichia coli by chemical permeabilization. Enzyme Microb. Technol., 1990 (in press).

14. Goodey, A.R., Doel, S., Piggott, J.R., Watson, M.E.E., and Carter, B.L.A., Expression and secretion of foreign polypeptides in yeast. In Yeast Biotechnology, ed. D.R. Berry, I. Russell, and G.G. Stewart, Allen & Unwin, London, 1987, pp. 401-429.

15. Hitzeman, R.A., Leung, D.W., L.J. Perry, Kohr, W.J., Hagie, F.E., Chen, C.Y., Lugovoy, J.M., Singh, A., Levine, L.H., Wetzel, R., and Goeddel, D.V., Expression, processing, and secretion of heterologous gene products by yeast. In Yeast Molecular Biology-Recombinant DNA: Recent Advances, ed. M.S. Esposito, Noyes Publications, Park Ridge, 1984, pp. 173-190.

THE CONCENTRATION OF YEAST SUSPENSIONS BY CROSSFLOW FILTRATION

M. PRITCHARD, J. A. SCOTT and J. A. HOWELL
School of Chemical Engineering, University of Bath,
Avon BA2 7AY, UK

ABSTRACT

Exploiting the full potential of crossflow filtration to clarify or concentrate fermentation broths requires a greater understanding of the influence of filter geometry and suspension rheology upon flux rates. Suspensions of bakers yeast have been concentrated up to 20% dry weight with both tubular and flat sheet membrane modules. Although significantly different flux/concentration profiles were observed with the different modules, they can be related in terms of the shear conditions at the membrane surface.

INTRODUCTION

The use of membranes for sterilisation and clarification of fluids with a low cell concentration using dead–end microfiltration is an established technology in the pharmaceutical industry. However a continual decline in flux is experienced as retained matter is deposited on the membrane making the technique unsuitable for concentration applications. Crossflow microfiltration limits the build–up of the deposited cake through the shearing action of the feedstream on the membrane yet typical product fluxes of only 50 l/m^2h to 150 l/m^2h are observed [1], [2]. Threshold flux values of between 100 l/m^2h and 400 l/m^2h have been quoted as necessary before crossflow filtration will compete with such established operations as centrifugation and precoat filtration [1], [3]. A fuller understanding of the flux limiting phenomena is therefore necessary in order to lead to improvements in the design and/or operating procedures for crossflow filtration.

It is thought that the majority of the resistance to flux is offered by cells and extracellular matter which deposit on the membrane [4] either forming a stagnant cake or blocking the pores. Deposited cells may be removed from the cake layer if the shear stress exerted by the fluid overcomes the adhesive forces between the cells. Above the cake layer will be a concentrated boundary layer where the cells are kept in suspension. Various mass transfer mechanisms have been proposed to explain this transfer of cells away from the membrane which balances the convective flow of cells towards it: shear–induced diffusion [4] (a phenomenon akin to Brownian motion down a 'concentration' gradient due the imbalance of shear–induced collisions between areas of high and low volume fraction), lateral migration [5] (a lift velocity experienced by particles near solid surfaces due to their rotation in a shear field) and eddy diffusion (under turbulent conditions) are the most prevalent.

Flux/Concentration Profiles

Flux through ultrafiltration membranes is often found to be proportional to the logarithm of the bulk concentration for macromolecular solutions. This is generally not so with cell suspensions which instead display a sigmoidal profile comprising an initial decline phase, a concentration–independent plateau phase then a further decline phase at high concentrations. Some or all of these three phases have been reported [1, 2, 6, 7]. More recently a five–phase, double sigmoidal model has been proposed [8] in which there is an intermediate, concentration–dependent flux decline but at cell concentrations some three orders of magnitude less than the lowest concentration used in this study.

Initial decline phase: During the first 15 to 30 minutes of operation, the flux declines markedly from the solvent flux level due to blocking of the membrane pores and build–up of a steady–state cake thickness [9].

Plateau phase: A broad concentration band is encountered where the flux is relatively independent of the bulk concentration. In this phase there must either be no net deposition of cells or else those which are deposited do not influence the cake resistance. Some time–dependent decline may occur due to the consolidation of cake particles and the accumulation of fines.

High concentration decline phase: At high concentrations a decline in flux is reported to occur [1, 8], falling to zero flux when the feed concentration reaches the maximum cell packing density. This has been attributed to mass transfer limitations and compaction of compressible cake layers. An increase in the viscosity of the feed will not, *per se*, cause a decrease in flux as it is the solvent viscosity and not that of the suspension which affects the pressure drop through the membrane and the cake layer.

The main aim of this study was to investigate the role of the shear conditions at the membrane in the transition from the plateau phase to the high concentration phase.

MATERIALS AND METHODS

Yeast Suspension

Fresh bakers yeast (*Saccharomyces cerevisiae*) was obtained from a local supplier as compressed cake of approximately 300 g/kg dry weight (DW). The yeast was diluted down for use with a solution of 10 mM phosphate buffer (pH 7) and 1 g/l bacteriological peptone. A 50 g/kg DW suspension was used as the starting point for the concentration process, having a cell count of 2×10^9 ml^{-1} and a viability of 98% as determined microscopically by the uptake of methylene blue dye.

Filtration System

Two different filter module geometries were employed in this study: wide–bore tubular and thin–channel flat sheet. Ultrafiltration membranes were used in both modules as they have been previously shown to allow flux rates equal or better than that of microporous membranes, [6, 10, 11]. The tubular module was a single tube (1 m long x 12.5 mm ID; filter area of 0.039 m^2) from Paterson Candy International (PCI) which had been modified by the insertion of divisions on the permeate side so as to allow the permeate to be collected from separate axial segments. The membrane used was a PCI FP100 made of PVDF which had a nominal molecular weight cut–off (MWCO) of 100,000 and an average clean membrane resistance of 1.3×10^{12} m^{-1}. The flat sheet system was a Pellicon module from Millipore (UK) Ltd, equipped with open feed channels which had been milled-out to a depth of 1mm to permit operation at higher viscosities, (38 channels @ 145 mm long; filter area of 0.033 m^2). A polysulphone membrane (type PTHK) was used, having a 100,000 MWCO and an average clean membrane resistance of 0.8x10^{12} m^{-1}.

Both modules were connected in turn to a filtration rig of 2.5 l working volume in which the circulation was provided by a rotary lobe pump (AP125, SSP Ltd). The crossflow rate was monitored using a magnetic flowmeter (MAGFLO 1000, Danfoss), inlet pressure by a diaphragm pressure gauge (0-10 b, Millipore) and module pressure drop by a mercury manometer. Flux measurement for both modules was by collection in measuring cylinders over a timed interval. Temperature control was by a coil situated in the feed tank.

Procedure

At start-up, the desired crossflow velocity and average transmembrane pressure drop (200 kPa) were obtained within 20 seconds. The system was then run for 30 minutes at an initial concentration of 50 g/kg DW by recycling the permeate to allow a pseudo-steady state to develop. Concentration then commenced using a simplified closed-loop batch method: a constant working volume was maintained by slowly adding increments of concentrated yeast slurry as the permeate was removed. Adequate mixing and fine temperature control were ensured even at high viscosities through the combination of high feed-tank aspect ratio, low tank residence time and careful design of the feed return lines. A temperature of 30°C $^+/_-$ 0.5°C was used for all runs. Effective cleaning of the membrane was achieved with a 0.25 % solution of the alkali detergent Ultrasil 11 (Henkel Chemicals) at 50°C for one hour.

Rheology

The suspension rheology was studied using a capillary viscometer constructed in this laboratory, full details of which are given in [12]. In brief, samples were driven through various stainless steel tubes (0.55 m long x 1,2,4 mm ID) from a feed reservoir which could be pressurised to 200 kPa. The rate of flow and the pressure drop through the tubes allowed the calculation of the shear rate and shear stress respectively.

RESULTS AND DISCUSSION

Rheology

Full details of the calculation of the suspension viscosities from the capillary viscometer data are given elsewhere [12]. The shear stress /shear rate plots for the three different viscometer tube diameters did not coincide (Figure 1), indicating the occurrence of apparent slip at the tube walls, *i.e.* the movement of cells away from the walls creating a zone of lower viscosity thus giving a higher wall shear rate for a given shear stress. This is a common phenomenon with suspensions flowing in non-porous ducts which has been observed with blood [13] and mycelial broth [14]. It has not yet been established if slip occurs to the same extent in porous ducts such as a membrane channel where the lateral migration effect moving the cells away from the membrane is counteracted by the convection of cells towards it. A slip correction procedure was applied [15] to establish the 'true' suspension rheology which made the data for the three viscometer tubes more coincident (Figure 2). The influence of the slip correction upon the measured viscosity can be seen in Figure 3 for the 1mm tube which, being the narrowest, gave the largest discrepancy. Up to 150 g/kg DW the suspensions were Newtonian becoming mildly dilatant above this (power-law index of 1.14 at 212 g/kg).

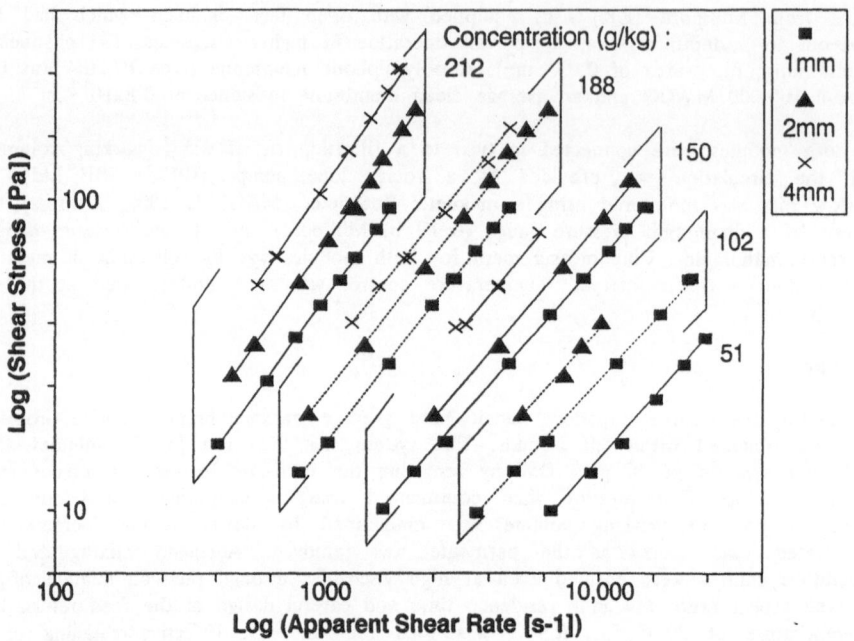

Figure 1. Bakers yeast rheogram showing the influence of tube diameter.
(no correction for slip.)

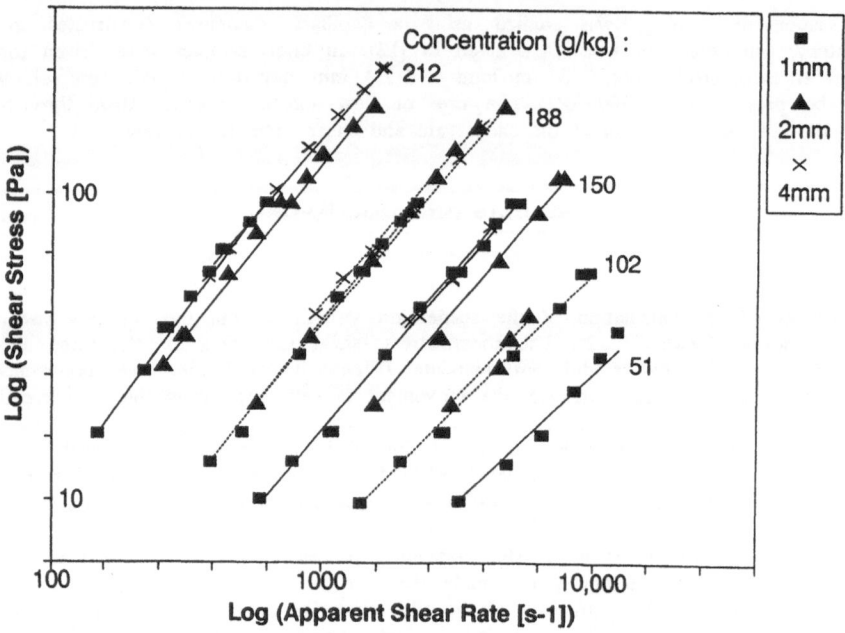

Figure 2. Bakers yeast rheogram incorporating the slip correction.

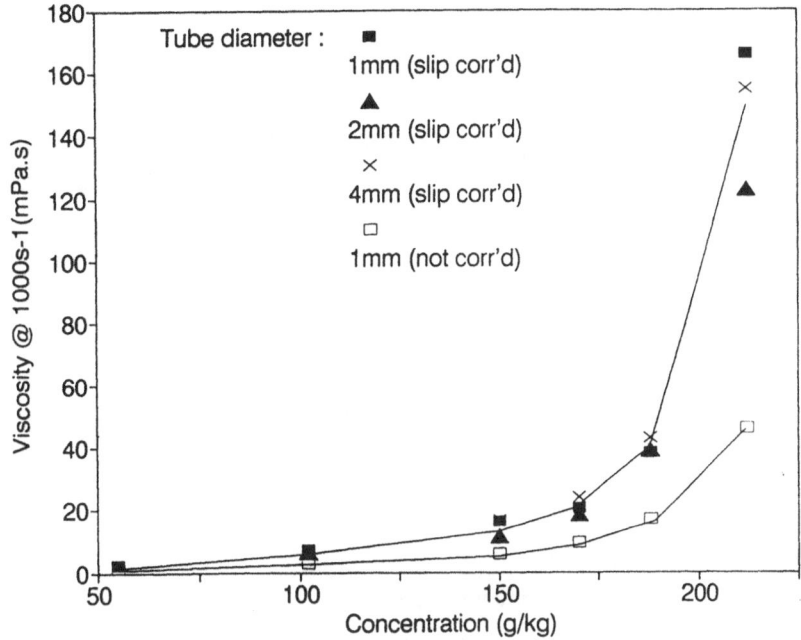

Figure 3. Comparison of bakers yeast viscosities
with and without the slip correction.

Concentration by the Tubular Module

Figure 4 shows the initial time–dependent flux decline phase (A–B) at constant concentration followed by the concentration stage from point B onwards. The fall in Reynolds number due to increasing viscosity during the concentration stage is also shown (the Reynolds numbers were calculated using slip–corrected viscosities as these better reflect the true behaviour of the bulk suspension).

There was a rapid decline from the clean membrane flux (700 l/m^2h at 200 kPa) during the first two minutes of operation (earliest measurement). The initiation of the concentration phase by the addition of concentrated yeast slurry halted the decline in flux. With the 2 m/s run there was even a slight flux increase at this point. In contrast, runs allowed to proceed at a constant concentration past the first 30 minutes (data not shown) continued to show a flux decline. The beneficial effect of an increased shear stress at the membrane arising out of the increase in viscosity as the concentration stage proceeded appears to offset the negative effects of the decrease in turbulence, as shown by the falling Reynolds number, or the increased convection of cells towards the membrane.

The flux plateau extended up to point C when a decline began, the concentration at which this occurred varying with the crossflow velocity. The Reynolds number at Point C was about 2300 at each crossflow velocity (Figure 4) which lies within the normal range for the transition from a turbulent to a laminar flow regime [16]. This phenomenon was reproducable although duplicates have not been shown in Figure 4 for the sake of clarity. The flux profile along the length of the membrane (Figure 5) changed markedly from that before (line C) to that after (line D) the flow transition. Under turbulent conditions eddy diffusion provides a high shear rate at the membrane with the boundary layer rapidly

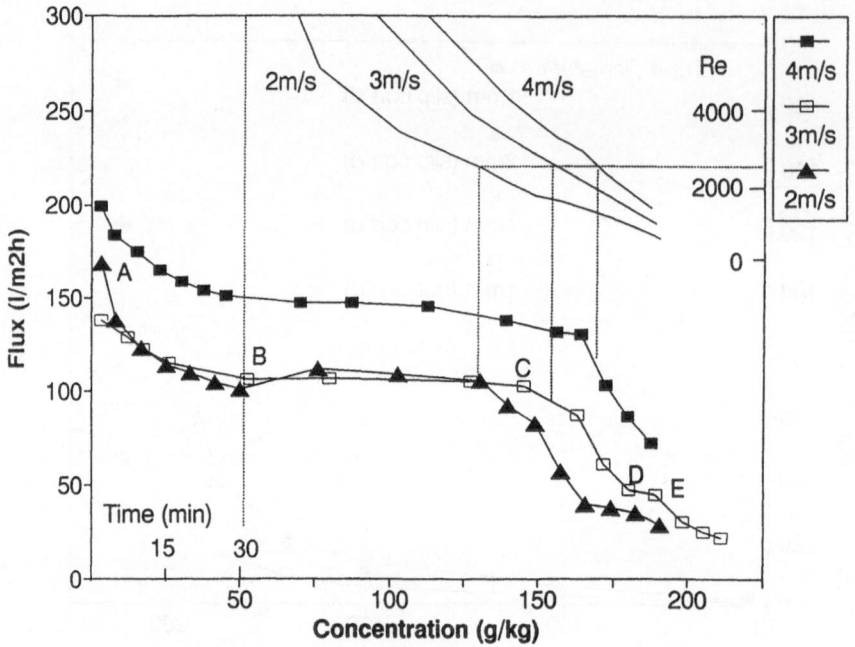

Figure 4. Flux and Reynolds number for the tubular module as a function of time and yeast concentration at various crossflow velocities.

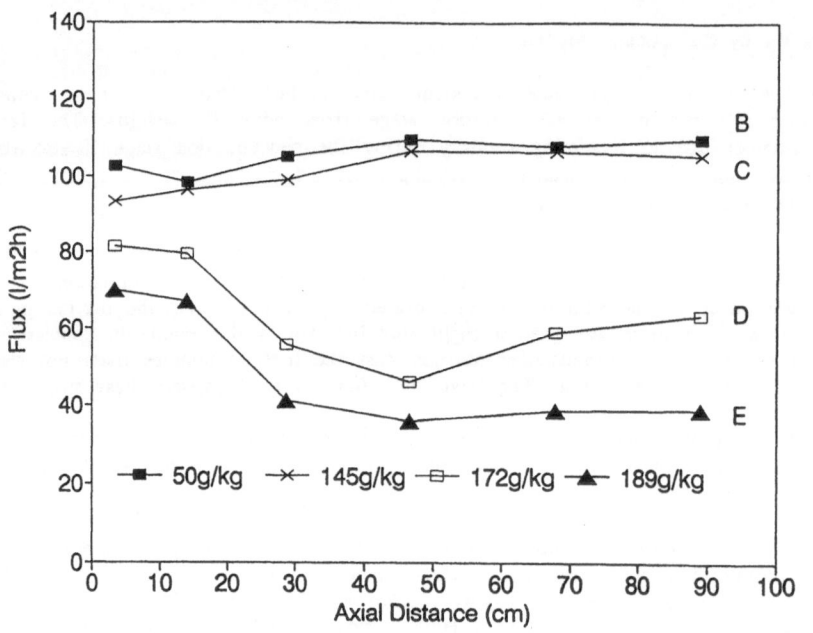

Figure 5. Flux as a function of axial distance along the tubular module at various yeast concentrations. (Letters refer to points on fig. 4)

attaining an equilibrium thickness thus leading to an axially independent flux [17]. With the transition to laminar flow the loss of eddy diffusion allows the boundary layer to thicken along the membrane and thus the flux declines with length. The increase in flux near the filter outlet for line D, reproduced for all other crossflow velocities, is possibly due to turbulence stemming from exit effects or less cake compression due to a lower transmembrane pressure.

The flux decline in the transition region was found not to continue indefinitely. A second laminar plateau can be seen for 2 and 3 m/s from point D onwards, where the high shear stress at the membrane counteracted the high cell convection rate. When the pressure drop in the system became too high to maintain the crossflow velocity the runs were stopped (point E) except the 3 m/s run. For this run the velocity was allowed to fall and the flux can be seen to drop off with the poorer shear conditions.

Concentration by the Flat-Sheet Module

The flux/concentration profiles for the flat sheet module are given in Figure 6. As with the tubular module, there was a rapid initial decline in flux which levelled-off after 15 to 20 minutes and was arrested as the concentration stage began at point B. As the thinness of the channels ensured laminar flow conditions from the outset there was no flow regime transition as the viscosity rose. At point C, the flux began to rise again. This phenomenon has not been reported before although oscillations in flux, which were attributed to the shearing-off of the cake layer, have been observed over long-term runs [9]. It can be seen from Figure 6 that the flux increase occured in each case at the same value of wall shear stress. This indicates that there existed a threshold value of shear stress above which the upper layers of the cell cake were progressively sheared away. The observed threshold of 50 Pa falls within the range of 20-60 Pa reported for the removal of bacterial cells from glass surfaces [18]. It should be noted that the shear stresses shown in Figure 6 were also calculated using slip-corrected viscosities. Had the viscosities not been corrected for slip, then the experimental threshold shear stress, although lower, would still fall within the range quoted above.

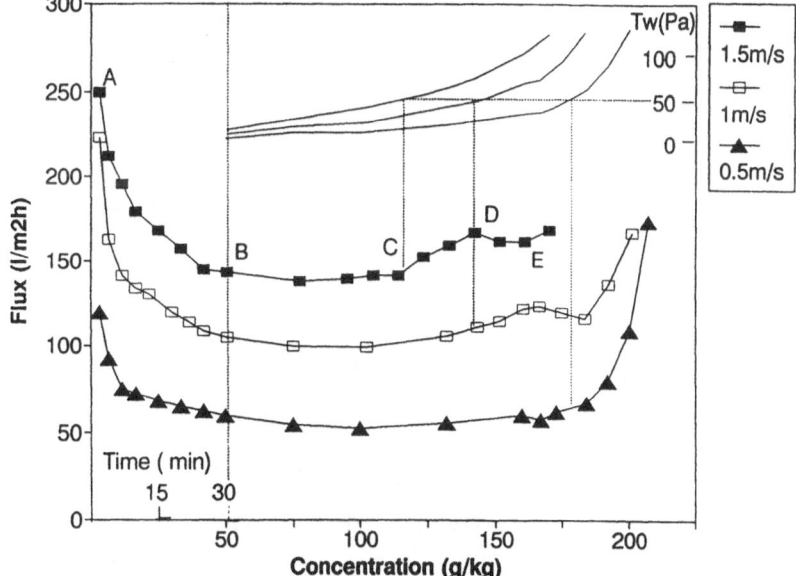

Figure 6. Flux and wall shear stress (Tw) for the flat sheet
module as a function of time and yeast concentration at various crossflow velocities.

At point D there was a temporary drop in flux for the 1 and 1.5 m/s runs. This may be a consequence of the way in which the transmembrane pressure was controlled: as the module pressure drop rose the inlet pressure was maintained at the expense of the outlet pressure hence the average transmembrane pressure fell. If the primary resistance to flux resides with irreversibly deposited cells there should be no reduction in this resistance when the transmembrane pressure driving force is reduced. The downstream sections of the filter module would thus suffer a flux decline as a result although this cannot be substantiated with this flat sheet module as the flux was not collected from separate axial segments. However once the back–pressure valve at the module outlet was fully open, the inlet pressure was allowed to rise with a concomitant increase in flux again.

CONCLUSIONS

The permeate flux during the concentration of bakers yeast suspensions was found to be controlled by both the shear stress and the shear rate at the membrane surface. The increase in shear stress which accompanied the rise in suspension viscosity was sufficient to halt the initial time–dependent flux decline. For a flat sheet module, a marked increase in flux occurred at high concentrations when the shear stress exceeded a threshold value whereas in a tubular module, the loss of turbulent shear which occurs with the transition from turbulent to laminar conditions led to a step decrease in flux. The maximum concentration which could be obtained at a given crossflow velocity was dictated by the module pressure drop limit. The previously reported tendency for the flux to decline to zero in the high concentration phase was not observed before this pressure drop limit was reached.

ACKNOWLEDGEMENTS

The contribution of equipment and materials by the following companies is gratefully acknowledged : PCI; Millipore and Henkel Chemicals. Mark Pritchard is supported by a British Commonwealth Scholarship.

REFERENCES

1.   Kroner, K.H., Schutte, H., Hustedt, H. and Kula, M.R., Pro. Biochem., April 1984, 67–74.

2.   Scott, J.A., Pro. Biochem., October 1988, 146–148.

3.   Gutman, R.G., Membrane Filtration The Technology of Pressure–Driven Crossflow Processes, Adam Hilger, Bristol, 1987.

4.   Romero, C.A. and Davis, R.H., J. Memb. Sci., 1988, 39, 157–185.

5.   Green, G. and Belfort G., Desalination, 1980, 35, 129–147.

6.   Henry, J.D. and Allred, R.A., Dev. Ind. Microbiol., 1972, 13, 177–190.

7.   Le, M.S., J. Chem. Tech. Biotechnol., 1987, 37, 59–66.

8.   Nagata, N., Herouvis, K.J., Dziewulski, D.M. and Belfort, G., Biotech./Bioeng., 1989,

34, 447–466.

9. Riesmeier, B., Kroner, K.H. and Kula, M.R., J. Memb. Sci., 1987, **34**, 245–266.

10. Gatenholm, P., Fell, C.J. and Fane, A.G., Desalination, 1988, **70**, 363–378.

11. Bell, D.J. and Davies, R.J., Biotech./Bioeng., 1987, **19**, 1176–1178.

12. Pritchard, M. and Howell, J.A., (In Preparation).

13. Nubar, Y., Biophysical J., 1971, **11**, 252–264

14. Allen, D.G. and Robinson, C.W., in Biochemical Engineering 5, Ann. N.Y. Acc. Sci., Vol. 506, eds. M.L. Shuler and W.A. Weigand, N.Y. Acc. Sci., New York, 1987, pp. 589–599.

15. Skelland, A.H.P., Non–Newtonian Flow and Heat Transfer, John Wiley and Sons, New York, 1967, pp. 28–36.

16. Coulson, J.M. and Richardson, J.F., Chemical Engineering, Vol.1, Pergammon, Oxford, 1985, 3rd ed., pp.39–40.

17. Blatt, J.F., Dravid, A., Michaels, A.S. and Nelson, L., in Membrane Science and Technolgy, ed. J. E. Flinn, Plenum, New York, 1970, pp. 47–97.

18. Powell, M.S. and Slater, N.K.H., Biotech./Bioeng., 1982, **24**, 2527–2537

HETEROLOGOUS FLOCCULATION OF ESCHERICHIA COLI AND YEAST

*R G PROTHEROE, R H CUMMING AND A J MATCHETT
North East Biotechnology Centre, Department of Chemical Engineering
Teesside Polytechnic, Middlesbrough, Cleveland, TS1 3BA

*Current address: Department of Chemical Engineering
Loughborough University of Technology, Loughborough, Leicestershire

ABSTRACT

Mixed cell suspensions of Escherichia coli and Saccharomyces sp. spontaneously flocculated to an extent dependent upon environmental ionic strength. When separately suspended, the bacteria or yeast maintained their mono-dispersity over the range of ionic strengths investigated. Maximal expression of flocculation was observed at an ionic strength of 0.015 sodium chloride and above. At ionic strengths of zero and 0.0015, mixed cell populations did not flocculate. Eventual floc size distributions were a function of the initial concentrations of each species. Flocculation persisted irrespective of the presence of nutrient broth growth medium.

INTRODUCTION

Following industrial fermentation, flocculating agents are commonly used to aid the separation of cells from fermenter broth, a process refered to as de-watering [1]. Flocculants increase the mean particle size of the suspended population, and consequently improve the efficiency of separation processes such as settling, centrifugation and filtration. Such flocculating agents are usually polyelectrolytes and/or amphiphilic polymers, which rely upon charge and hydrophobic interactions to bridge cells to each other [2]. Hetrologous flocculation, on the other hand, occurs between particles of complementary physical properties and has been used for the flocculation of microbes by the addition of bentonite to fermentation broths. In this case divalent cations were found to be essential for the process. They probably exerted their effect through a bridging mechanism. Different sized latex spheres have been hetro flocculated and the mechanism extensively studied.

This study reports heteroflocculation between Escherichia coli and baker's yeast, and the environmental factors which influence the process.

MATERIALS AND METHODS

Organisms and growth conditions

Organisms were Escherichia coli and a Saccharomyces sp. (baker's yeast). E.coli was aerobically grown in nutrient broth (Oxoid) shake flasks at 37°C for 16 hours. Cultures were then harvested by centrifugation (23,600 g for 10 minutes). Pellets were washed with 0.15 M sodium chloride (British Drug Houses, AnalaR) before a second centrifugation. Bacterial pellets were then resuspended with test solution before flocculation. Baker's yeast was reconstituted with 0.15 M sodium chloride and harvested by centrifugation (14,000 g for 5 minutes). Pellets were washed with 0.15 M sodium chloride before a second centrifugation, and subsequent pellet resuspension in test solution was 7.5.

Measurement of population cell size distributions

The cell size distributions of populations of E.coli and Saccharomyces sp., were determined using a Coulter Counter model ZM, with 30 and 50 µm tubes respectively.

Flocculation assay

Flocculation was effected by mixing suspensions of E.coli and yeast, each suspended by solutions of identical composition. Suspension were mixed in 20 ml disposable plastic universal bottles (asay volume 4ml). Bottles were secured on their sides on a shaking platform (Grant Instruments), and shaken at 200 strokes per minute at room temperature (22 ± 2°C). With the exception of experiments relating shaking time with floc formation, universals were shaken for 30 minutes.

Following shaking, population size distributions were determined immediately. A representative 200µl sample was diluted approximately ^{1}/$_{100}$, by test solution. The population size distribution of this suspension was assessed using a 2200 series Malvern Particle Sizer. This sizing technique was used in preference to the Coulter Counter, because it did not cause floc disruption.

For experiments relating floc formation to initial E.coli concentration, in the presence and absence of nutrient broth, the remaining floc assay suspension was centrifuged for 30 seconds (7,000 g), to sediment flocs. A supernatant sample was then removed, and a Coulter Counter was used to determine the final concentration of free E.coli not associated with flocs. This centrifugal force did not sediment control suspensions of E.coli. Population size distribution analysis revealed that supernatnat cell suspensions were not contaminated by free yeast cells or flocs.

RESULTS AND DISCUSSIONS

Time of floc formation

Heterologous flocs formed spontaneously and rapidly following mixing (figure 1). Stable floc population size distributions were established following 10 minutes of mixing. Prior to this time it is assumed that insufficient collisions had occurred to establish stable equilibrium. Floc population size distributions remained constant for at least 30 minutes, before some floc distribution became evident at 65 minutes.

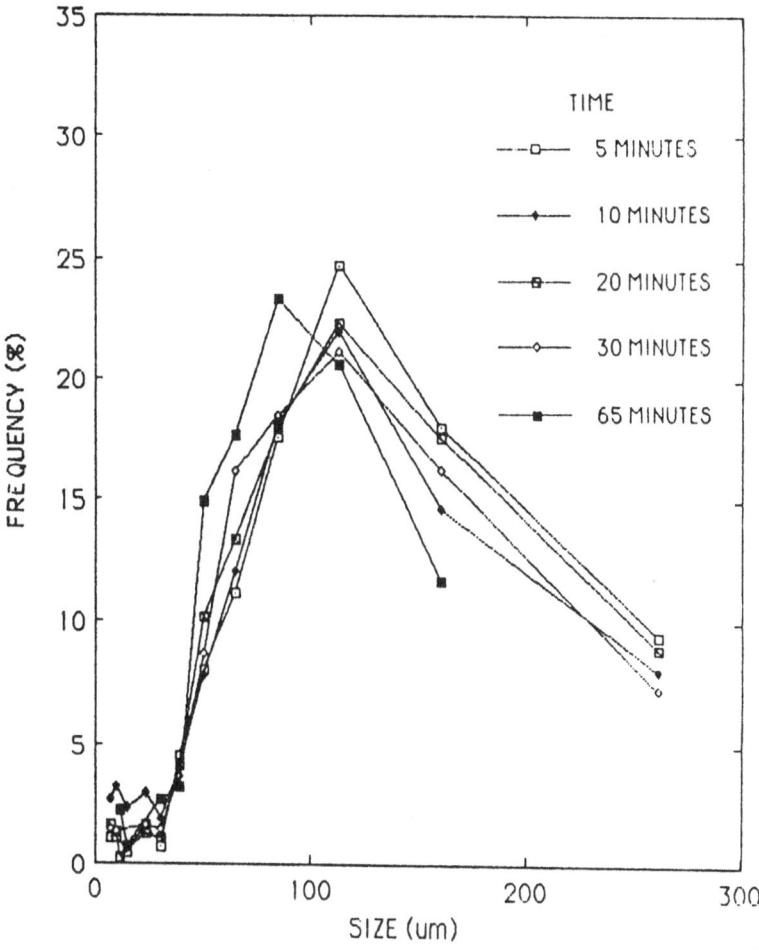

Figure 1. Time course of flocculation 10.21 x 10^9 cells mL^{-1} E coli and 1.34 x 10^7 cells mL^{-1} of yeast, 0.15 M saline.

Ionic strength and flocculation

The extent of flocculation was dependent upon ionic strength (figure 2). When mixed populations were suspended by distilled water or 0.0015 M NaCl, each species remained mono-disperse. With increase in ionic strength, and constant initial concentration of bacteria and yeasts, population floc distributions changed as larger flocs formed. Floc size distributions became insensitive to ionic strength at electrolyte concentrations of 0.015 m and above. The tendency for size distributions to remain constant at the critical concentration of 0.015 m, suggests maximal exertion of this influencial parameter at this concentration.

Ionic strength effects of this type are usually recognised to be a consequence of charge interactions between surfaces [5]. E.coli and yeast cells would be negatively charged under the environmental conditions of this study [6]. Increasing ionic strength suppresses the extent of electric double layer width around charged surfaces. At short distances of separation, electric double layers reduce the magnitude of electrostatic repulsion generated between surfaces of like sign. As repulsion is reduced, a larger proportion of the mixed cell suspension collide, thereby assisting floc formation. At a critical ionic strength, electric double layers become incapable of further suppression, and collision rate proceeds at a maximum. For mixtures of cells of E.coli and yeast, this value is 0.015 M NaCl.

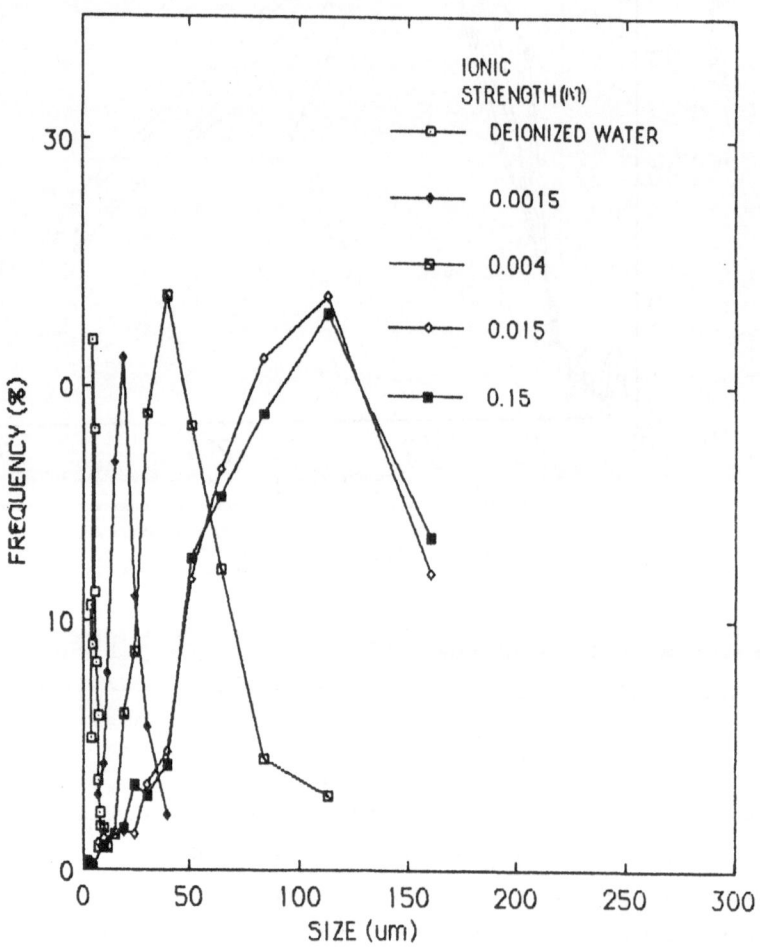

Figure 2. Effect of ionic strength of floc size, cell numbers as in figure 1.

Floc size as a function of initial cell concentrations.

At constant initial yeast concentrations of 1.34 x 10^7 cells mL^{-1}, eventual floc size distribution was dependent upon the initial concentration of <u>E.coli</u> (in the range 0.26 - 16.32 x 10^9 cells mL^{-1}). As the concentration of <u>E.coli</u> increased the range of floc size distribution widens (figure 3). Floc size distributions did, however, become insensitive to <u>E.coli</u> concentration, at bacterial densities of approximately 2.56 x 10^9 cells mL^{-1} and above.

Presumably a saturation mechanism is occurring, wherby the extent of flocculation is governed by the number of contacts between <u>E.coli</u> and yeast. When all the yeast surface is 'filled' by <u>E.coli</u> no further

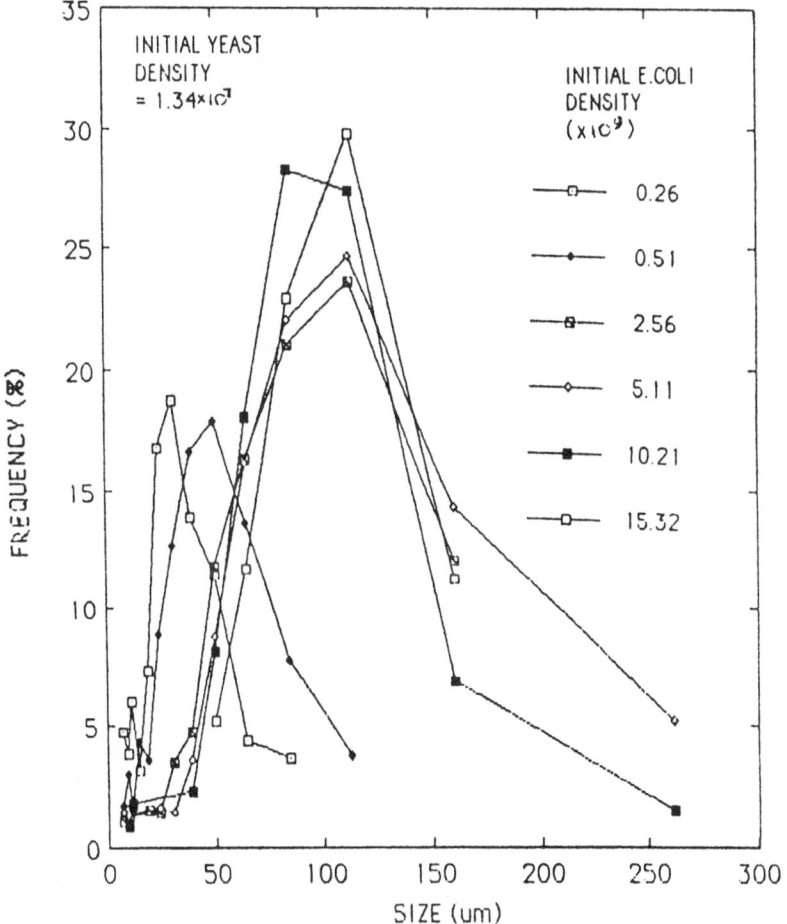

Figure 3. Effect of bacterial cell numbers on floc size.

flocculation occurs. The maximum floc size appears to stabilise at 120 μm when sufficient E.coli cells have been added (figure 3). A high ratio of bacterial cells to yeast cells would lead to strong flocs which in turn would withstand the mild shear conditions employed and lead to the formation of larger flocs.

Floc formation was insensitive to the presence of nutrient broth growth medium, when used at a concentration of microbial cultivation. The inability of medium to inhibit the flocculating process, indicates that medium components did not significantly affect influencial physico-chemical parameters of the environment, and have no affinity for the cell surface binding sites which may be operative in flocculation.

An interesting observation of floc populations was the overwhelming variation in the proportions of bacteria and yeast incorporated into flocs. Irrespective of E.coli concentration, all yeast cells became floc components, with no free yeasts remaining. Contrastingly, significant proportions of the bacterial population remained in free suspension.

Theoretical considerations.

When bacteria and yeasts are present in mixed suspension, other interactions operate between dissimilar cells in addition to those of pure suspension. The inability of ionic strength to flocculate pure suspensions of either species, is a clear pointer to the operation of destabiling force which may be physico-chemical and/or biological in nature.

The cell size difference of bacteria and yeasts is one additional physical parameter, capable of influencing the mode of physico-chemical interaction. Interaction energies [3,4] of pure and mixed suspensions are represented by figure 4. It is apparent that at short distances of separation, potential energy barriers exist between the converging surfaces of similar cells. This repulsive force maintains population stability. However, in mixed suspension under identical environment conditions, no barrier to the collision of dissimilar cells is generated. This is a consequence of the greater penetrating ability of charged surfaces with narrow radii (E.coli), which encounter broader electric layers (yeast) [8].

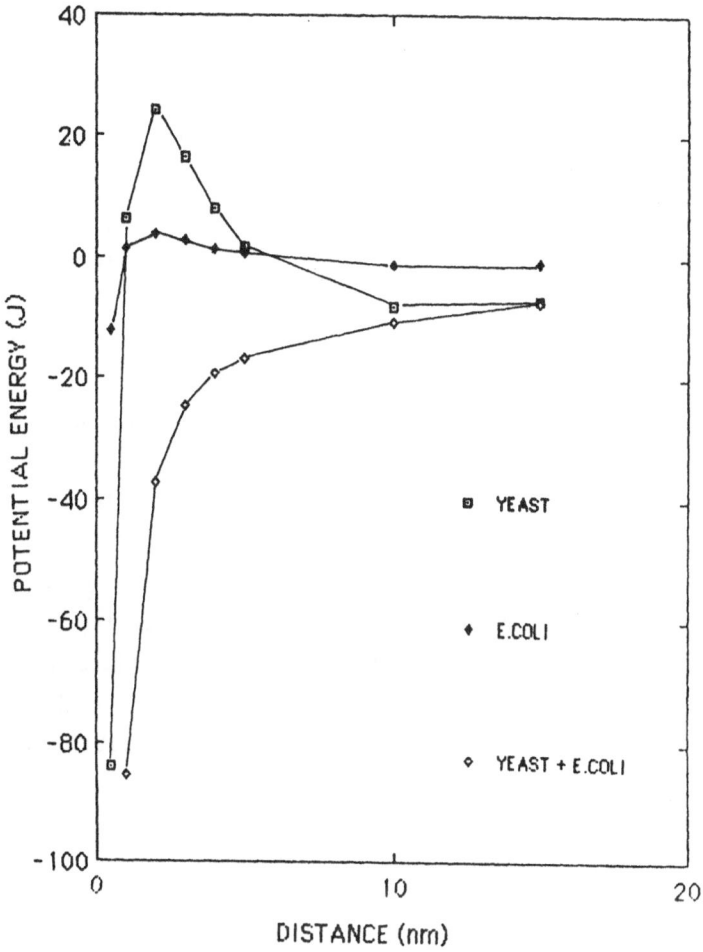

Figure 4.   Potential energy diagram for yeast and E.coli.   Ionic strength
0.015.

CONCLUSION

Mixed cell suspension of E.coli and baker's yeast readily flocculated
under favourable environmental conditions of ionic strength.   Associated
increase in the mean particel size of the suspended population, would
greatly   increase   the   settling   rate   and   thus   markedly   improve   the
efficiency of separation processes.   The simplicity of operation and the
food compatible acceptability of the process, are both advantages which
contribute to the usefulness of heteroflocculation as a dewatering aid.

REFERENCES

1   Wang, D.I.C., Separation for Biotechology. In Separations for Biotechnology, ed. M. S. Verall and M. J. Hudson, Ellis Horwood, 1987, pp.30-48.

2   Vincent, B., The Effect of Adsorbed polymers on dispersion stability. Advances in Colloid and Interface Science, 1974, **4**, 193-277.

3   Verway, E.J.W., and Overbeek, J.Th.G, Theory of stability of lyophobic colloids. Elsevier, Amsterdam, 1948.

4   Shaw, D.J., in Introduction to Colloid and Surface Chemistry. Butterworths, London, 1980.

5   Visser, J., Adhesion of colloidal particles. Surface and Colloid Science, 1976, **8**, pp 3-84.

6   James., The electrical properties and Topochemistry of Bacterial Cells. Adv Colloid Interface Science, 1982, **15**, pp 171-221.

7   Vrij, A., Polymers at interfaces and the interactions in colloidal dispersions. Pure and applied chemistry, 1976, **48**, 471-483.

8   Vincent, B., Young, C.A., and Tadros, Th,F.T., Equilibrium aspects of heterologous flocculation in mixed sterically stabilised dispersions. Faraday Discussions of the Chemical Society,

LATERAL MIGRATION: A laminar fluid flow mechanism suited to Biotechnology separations

G.J. PURDOM and C.A. LAMBE
Chemical Engineering Division, Harwell Laboratory,
AEA Technology, Oxon, OX11 ORA.

ABSTRACT

This paper describes how a fluid mechanism provides a new method for the separation of drops and particles from the type of dilute suspension common in biochemical processes.

Lateral migration is a low Reynolds number fluid phenomenon observed in laminar duct flow. The mechanism causes the dispersed phase of drops or particles to migrate across flow streamlines to a characteristic equilibrium position. This paper considers only the behaviour of single particles. The equilibrium streamline is dependent on the settling velocity of the dispersed phase, the bulk velocity and the ratio of particle diameter to channel height. Preliminary results, using a scaled-up channel, are presented.

The technique is particularly attractive for the continuous separation of dilute suspensions of droplets and particles $\geq 50 \, \mu m$ with very low settling terminal velocities and which may be shear sensitive. This type of separation is not effectively fulfilled by continuous centrifuges. The proposed system is both easily and cheaply constructed and since it has no moving mechanical parts is ideally suited to operation behind containment or in aseptic conditions as required by the biotechnology industry.

INTRODUCTION

When bubbles, drop and particles are entrained in laminar flow between two parallel closely-space plates, ie. within 2-d Poiseuille flow, they migrate across streamlines to a stable position dependent on the direction of buoyancy forces (as shown schematically in figure 1).

This motion is known as lateral migration. Lateral migration may find application, when coupled to a flow splitter, for:

- dewatering of cell and floc suspensions.
- separating two-phase aqueous-aqueous solutions.
- separation of organic-aqueous suspensions.

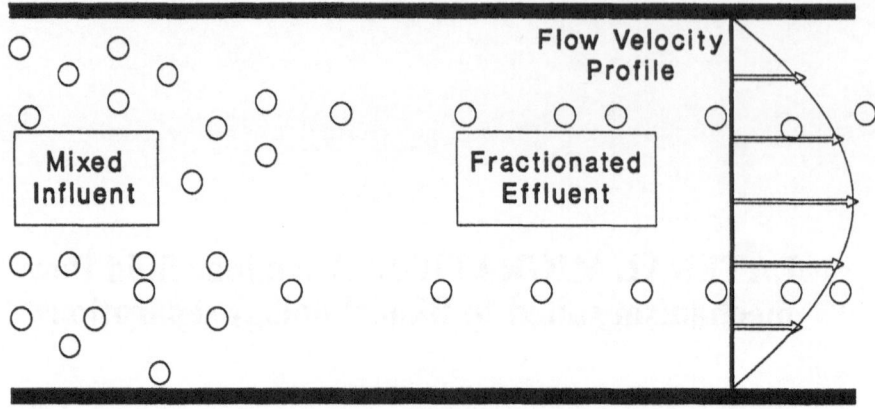

Figure 1. Lateral migration in 2-d Poiseuille flow

This paper first considers published experimental evidence and the limitations of existing theory, primarily for single particles. It then describes how application data is being generated to determine the value of lateral migration as a new Biotechnology separation tool.

EXPERIMENTAL EVIDENCE

Lateral migration was first observed when red blood cells were seen to align at the centre-line of blood capillaries. This was termed the 'Sigma phenomenon' and was explained by the particle moving to a position of minimum entropy production rate where viscous dissipation is reduced. In 1962 Segré and Silberberg (20) observed that neutrally buoyant spheres in Poiseuille flow within tubes, migrated to 0.6 of the tube radius from the centre-line. They labelled the phenomenon the 'tubular-pinch effect'. Segré and Silberbergs' observations were followed by a rush of papers generally confirming the validity of this empirical evidence and attempting to explain its origin. The same mechanism, now generically termed 'lateral migration', was found to also cause migration across stream-lines in both Couette (7,14) and in 2-d Poiseuille flow (ie between two parallel infinite planes). Table 1 lists the experimental conditions where lateral migration has been observed either in tubular or 2-d Poiseuille flow. This area of study has been the subject of a number of comprehensive reviews (8,13).

The following important parameters used in the table and in this report are defined as:

$$\text{Particle Reynolds number, } Re_p = \frac{\rho U_{max} a^2}{d\mu} \tag{1}$$

$$\text{Tube Reynolds number, } Re_T = \frac{\rho U_{max} d}{\mu} \text{ or } = \frac{2\rho U_{max} R}{\mu} \tag{2}$$

$$\text{Terminal velocity, } V_\infty = \frac{2}{9}\frac{a^2 \rho g}{\mu} \tag{3}$$

where $\rho, \mu, U_{max}, d, 2R, a$ are the density, viscosity, centre-line undisturbed channel velocity, channel height, tube diameter and particle radius.

Table 1. Experimental Results For Single Particles.

Observer	Particle Reynolds number Re_p	Tube Reynolds number Re_T	Tube diameter $2R$ (mm)	Particle radius a (mm)
TUBE POISEUILLE				
Segré/Silberberg (18)	<0.01	2-30	11.2	0.16-0.78
Goldsmith/Mason (6)	$2.2\text{-}5.3\times10^{-6}$	$0.80\text{-}1.6\times10^{-3}$	8.00	0.49-0.54
Oliver (16)	1.0-13.0	100-500	9.4	1.15-1.70
Karnis et al (12)	$.09\text{-}1.9\times10^{-2}$	0.01	4.00	0.5
Small/Eichhorn (21)	80-247	860-2400	10.7	0.78-1.61
Denson (4)	6.0-120	208-890	25.8	1.54-2.45
Jeffery/Pearson (11)	0.010-0.11	11.2-76.8	32.5	0.75-1.44
	0.37-1.3	22.7-116	32.5	0.75-1.44
	0.28-1.6	22-180	32.5	0.75-1.44
COUETTE FLOW				
Halow (7)	.0085-.307	4-322	5.9-18.2	0.30-0.85
2-D POISEUILLE				
Repetti (19)	0.23-1.89	9.4-238	25.4	0.79-6.35
Yanizeski (24)	>3.5	350-800	1.37	0.14-0.58
Tachibana (22)	.006-0.78	15-70	30	0.62-3.17
Vasseur (23)	~ 0.093	~ 554	30.5	0.31-0.56
Hiller (9)	<0.001	$1\times10^{-4}\text{-}1.0$	0.125	4×10^{-3}
THIS WORK	0.001-5	1-500	15-30	0.5-3.2

Experiments have also been carried out using non-spherical and non-rigid particles. Goldsmith and Mason (6) observed lateral migration of rods, oblate and prolate spheroids, but doubted its existence for spheres. Goldman *et al* (5) investigated the motion of a droplet in Poiseuille flow within tubes. They noted that droplets undergo slow radial migration dependent on the distortion parameter. Karnis and Mason (12) and Hiller and Kowalewski (9) have confirmed this finding both in tubes and between two plates. McTigue *et al* (15) have considered the rheological effect of lateral migration on dilute suspensions and Rakow and Chappell (18) measured lateral migration of microalgae at a concentration of 0.365mg solid/g solution.

THEORY

The majority of theoretical research on particles in low Reynolds number flow use the approximation that inertial terms are negligible compared to viscous terms. This allows the non-linear Navier-Stokes equation to be greatly simplified to the linear Stokes equation. However Bretherton (1) has shown that if the inertia terms of the equations of motion are neglected, no lateral migration can occur for any axisymmetric body in any arbitrary uni-directional flow. This implies that the mechanism is inertia induced and consequently an answer is dependent on resolving the full steady-state Navier Stokes equations.

Cox and Brenner (2) used a matched asymptotic expansion to tackle the problem of lateral migration of neutrally buoyant and non-neutrally buoyant rigid spheres in Poiseuille flow. They expanded the fluid velocity velocity, U, and the pressure, P, in power series in Re_p of the form:

$$U=U_0+Re_pU_1+......$$ (4)

$$P=P_0+Re_pP_1+......$$ (5)

Assuming that

$$\frac{a}{d}\ll1 \quad \text{and} \quad Re_p=\frac{\rho U_{max}a^2}{d\mu}\ll\left\{\frac{a}{d}\right\}^2\ll1$$ (6)

However Cox and Brenner were unable to evaluate the complex coefficients contained in their investigation. Ho and Leal (10) were able to use this general approach, but using an analysis based on the method of reflections, and managed to derive the coefficients for both Couette and 2-d Poiseuille systems when using single rigid neutrally buoyant spheres.

Cox and Hsu (3) used the method developed by Cox and Brenner for the lateral migration of particles near a plane. Vasseur and Cox (22) extended the analysis to buoyant and neutrally buoyant, freely rotating and non-rotating spheres between two walls.

The above collective work predicts that, firstly, single neutral density particles in 2-d Poiseuille flow will migrate to an equilibrium position 0.62 of the distance from the centre-line for a particle which is free to rotate. If the particle is not free to rotate then the corresponding equilibrium position is 0.48 of the distance from the centre-line. Secondly, denser particles in upflow or buoyant particles in downflow are predicted to migrate to the centre-line. Similarly heavier particles in downflow or less dense particles in upflow will migrate to the walls. Thirdly, and most interestingly, particles of similar density to the fluid and suspended in 2-d Poiseuille flow are predicted to experience an equilibrium position which may be at any position between the centre-line and the wall. As a consequence in certain circumstances, particles of differing density ought to migrate to different streamlines. The investigations of Repetti and Leonard (17) and others have not been sufficient to unequivocally verify or refute this third point.

In the applications mentioned at the beginning of this report the settling velocity V_∞ is small because either the particles are of a similar density to the fluid and/or because they have small diameters. Therefore it is likely they will fall into the third case defined above of 'nearly' neutrally buoyant spheres.

The experimental evidence given in table 1 indicates that lateral migration occurs at

$$\frac{a}{d} > 0.4 \text{ and} \tag{7}$$

$$\frac{\rho U_{max} a^2}{d\mu} > 1 \tag{8}$$

But existing theory is dependent on the conditions:

$$\frac{a}{d} \ll 1 \quad \text{and} \quad Re_p = \frac{\rho U_{max} a^2}{d\mu} \ll 1 \tag{9}$$

Consequently the most obvious way to further develop theoretical understanding is through numerical analysis since this approach is not necessarilly restricted by these limitations (17). However this is outside the scope of this paper.

METHOD AND RESULTS

Although the limitations of published theory have been highlighted the theory does provide a useful datum against which experiments can be compared. Using a large aspect ratio rectangular channel, the extent of lateral migration in 2-dimensional Poiseuille flow was measured using various combinations of channel heights, particle diameters, fluid viscosities, relative densities and velocities. The apparatus was designed to be dimensionally scaled-up relative to the intended applications to assist observation but to dimensionlessly mimic potential applications.

Each experiment involved the release of individual plastic spheres from an injection tube at a predetermined position between two plates into a fully developed flow profile. Figure 2 shows the channel configuration.

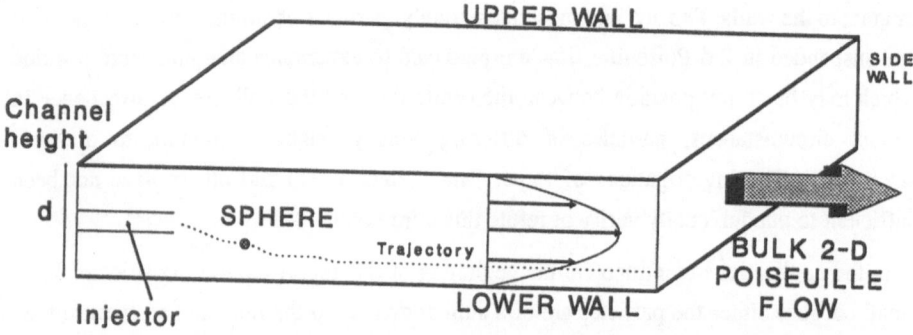

Figure 2. Channel configuration

A period of time was allowed for the sphere to attain its rotational and axial velocity in the entraining fluid, then positional measurements of the particle relative to the channel were started. The trajectory of the particle within the channel was measured using a CCD video camera viewing a shadow-graph image of the particle through clear plastic side walls. The camera and parallel light source were mounted on either side of the channel on a trolley which was propelled by a micro-stepper motor parallel to the flow direction. The trolley was suspended in linear bearings which only allow movement parallel to direction of flow. The image was analysed in real-time to determine the position of the particle relative to the moving camera. The image consists of a silhouette of the particle on a white background which, because it is highly contrasted, requires no image processing and as a result 'grabbed-frames' can be processed upto eight times a second. This information was both stored for post-processing and used on-line to control the speed and position of the camera. Only the central 1.8m of the 3.0m channel length was used for tracking purposes in order to take account of entry and end-length effects. Two plastics with different densities, polystyrene and cellulose acetate, were used for the spheres and glycerol/water mixtures were made to match the density of each particle type. (A third plastic, nylon, was also going to be used but X-ray inspection revealed voids within the spheres which made them unsuitable).

Figure 3. Lateral migration experimental rig

The magnitude and direction of the buoyancy force is significant when interpreting the results. This force was determined by measuring the settling velocity of the particle in a large isothermal still tank of the relevant fluid. The flow rig can be mounted horizontally (as illustrated in Figure 3), on its side or vertically so that the buoyancy force acts either perpendicularly to the upper and lower walls or perpendicularly to the plane of interest or alternatively parallel to the direction of flow.

Using this equipment lateral migration has been observed to occur to a varied extent dependent on the conditions used. Figure 4 illustrates two cases where lateral migration is and is not significant within the trajectory length. Each graph represents the trajectories of a number of 4.0mm diameter particles released under identical conditions but at different points between the upper and lower plates, ie. the channel height. The continuous lines are the theoretical predictions of Ho and Leal (10) for a particle released from very close to either the centre-line or the wall. These trajectories are for effectively neutrally buoyant particles where the buoyancy force is acting out of the plane of interest. These results show that both the axial distance to migration equilibrium is underestimated and that the dimensionless channel height equilibrium position is different from Ho and Leal (10), where from Ho and Leal (10):

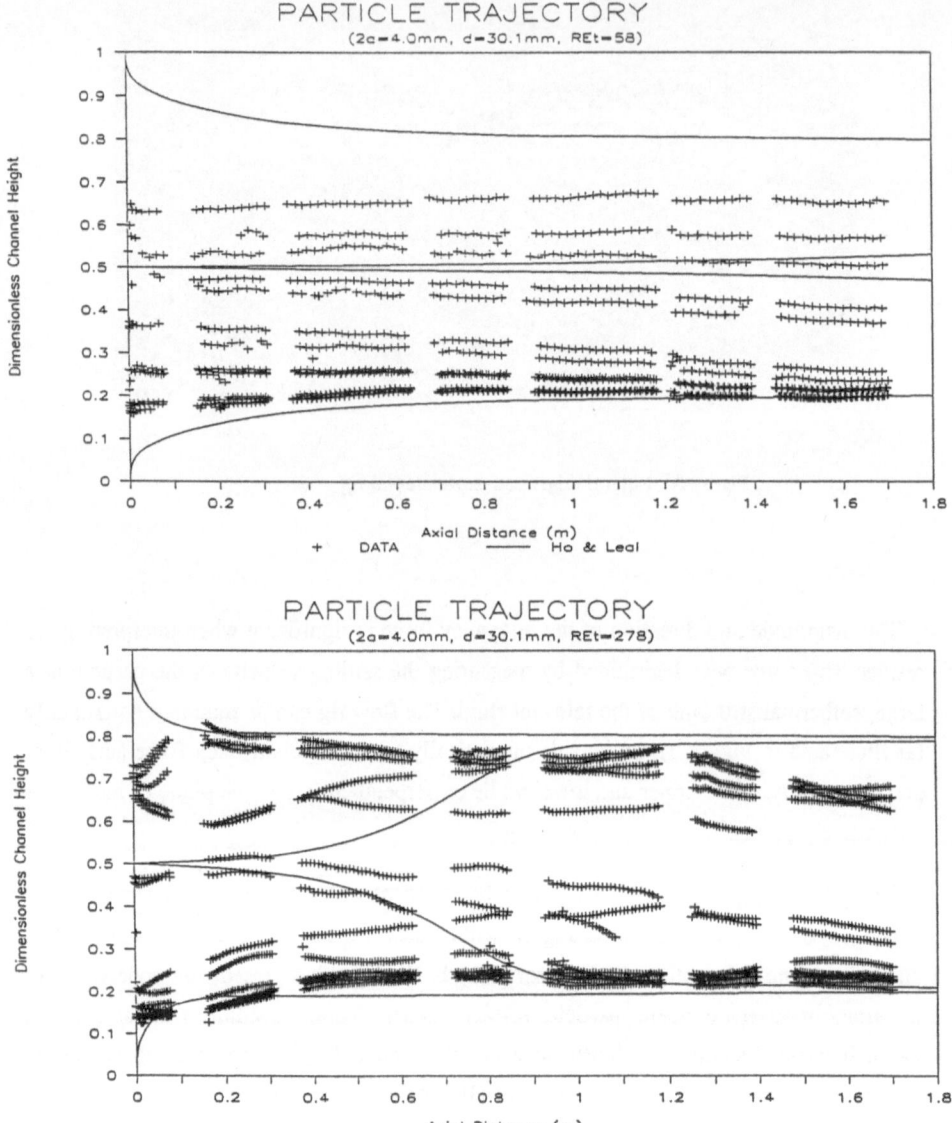

Figure 4. Trajectories of particles in 2-d Poiseuille flow

$$\text{Axial Distance to equilibrium} \approx \frac{1.13\, d^4}{Re_T\, a^3} \tag{10}$$

Since laminar Poiseuille flow is being used, the maximum shear rate, β_{max}, within the channel can be determined by:

$$\beta_{max} = 8\, U_{max}\left(\frac{a}{d}\right) \tag{11}$$

The actual shear experienced by an entrained particle will be less than this dependent on its relative velocity.

The results given here represent a worst case analysis intended to provide a first order of magnitude estimate for predicting the particle shear and size of channel necessary to achieve a separation. This paper presents no evidence to quantify the effect of particle concentration as a limitation to the technique, however there is evidence (15, 18) to believe that lateral migration can significantly concentrate dilute suspensions. If the channel is operated vertically buoyancy forces cause the relative velocity of the particle to be greatly altered from that due to just the skin and form drag of the particle. This can be used to greatly enhance the rate of migration. Similarly if the particle is not rigid, eg liquid or elastic, and/or not spherical then migration may again be greatly improved. Thus initial results suggest that a large number of biological suspensions with low settling velocities and which are shear sensitive may be amenable to continuous separation by this mechanism.

CONCLUSION

Lateral migration in 2-dimensional Poiseuille flow has been shown to provide a new mechanism for the continuous separation of dilute shear sensitive biological suspensions. The magnitude of the lateral migration force was shown to be dependent on the channel height, particle diameter, tube Reynolds number and relative density. This fluid mechanism causes particles to migrate to an equilibrium streamline from which they may then be split effecting separation of the dispersed phase from the bulk flow. Only single neutrally buoyant spheres have been considered. When the disperse phase is not neutrally buoyant, rigid, or spherical, then separation can be achieved yet more efficiently.

ACKNOWLEDGEMENT

Work described in this report was undertaken as part of the underlying research programme of AEA Technology.

NOMENCLATURE

a..........................particle radius
d...........................channel height
g..........gravitational acceleration
P.....................................pressure
Re....................Reynolds number

U....................fluid velocity
U_{max} .....centre-line velocity
V_∞..............settling velocity
μ..............................viscosity
ρ.................fluid density

REFERENCES

1. BRETHERTON, F.P; Slow viscous flow round a cylinder in a simple shear. **J. Fluid Mech.** 1962, **12** , 591-613

2. COX, R.G. and BRENNER, H; The lateral migration of solid particles in Poiseuille flow – I. **Chem. Eng. Sci.** 1968, **23** , 147-173

3. COX, R.G. and HSU, S.K; The lateral migration of solid particles in a laminar flow near a plane. **Int. J. Multiphase Flow** 1977, **3** , 201-222

4. DENSON, C.D; Particle migration in shear fields. PhD thesis. Univ. Utah, Salt Lake City, Utah, 1965.

5. GOLDMAN, A.J., COX, R.G. and BRENNER, H; Slow viscous motion of a sphere parallel to a plane wall-I Motion through a quiescent fluid. **Chem. Eng. Sci.** 1967, **22** , 637-651

6. GOLDSMITH, H.L. and MASON, S.G; The flow of suspensions through tubes. **J. Colloid Sci.** 1962, **17** , 448-476

7. HALOW, J.S. Radial migration of solid spheres in Couette systems. PhD. thesis, Virginia Polytechnic Inst., Blacksburg, Va., 1967

8. HAPPEL, J. and BRENNER, H; **Low Reynolds number hydrodynamics.** Noordhoff Pub., 1973.

9. HILLER, W. and KOWALEWSKI, T.A; An experimental study of the lateral migration of a droplet in a creeping flow **Exp. in Fluids** 1987, **5** , 43-48

10. HO, B.P. and LEAL, L.G; Inertial migration of rigid spheres in two-dimensional unidirectional flows. **J. Fluid Mech.** 1974, **65** , 2, 365-400

11. JEFFREY, R.C. and PEARSON, J.R.A; Particle motion in laminar vertical tube flow. **J. Fluid Mech.** 1965, **22** , 4, 721-735

12. KARNIS, A. and MASON, S.G; Particle motions in sheared suspensions. XIII . Wall migration of fluid drops. **J. Colloid Interface Sci.** 1967, **24** , 164-169

13. LEAL, L.G; Particle motions in a viscous fluid. **Ann. Rev. Fluid Mech.** 1980, **12** , 435-476

14. MAJUMBAR, A. and GRAHAM, A.L; Experimental study on the solid particle dynamics in shear flow. **Powder Tech.** 1987, **49** , 217-226

15. McTIGUE, D.F, GIVLER, R.C. and NUNZIATO, J.W; **J. Rheology** 1986, **30** , 5, 1053-1076

16. OLIVER, D.R; Influence of particle rotation on radial migration in the Poiseuille flow of suspensions. **Nature** 1962, **194** , 1269-1271, 1962

17. PURDOM, G.J.; Lateral migration of single rigid buoyant and neutrally buoyant in two dimension Poiseuille flow. PhD thesis. Imperial College, London, 1990

18. RAKOW, A.L. and CHAPPELL, M.L; Axial migration of *spirulina* microalgae in laminar tube flow. **Biorheology** 1987, **24** , 763-768

19. REPETTI, R.V. and LEONARD, E.F; Segre-Silberberg annulus formation: a possible explanation. **Nature** 1964, **203** , 1346-1348

20. SEGRE, G. and SILBERBERG, A; Behaviour of macroscopic rigid spheres in Poiseuille flow. Part2. Experimental results and interpretation. **J. Fluid Mech.** 1962, **18** , 312-317

21. SMALL, H; Hydrodynamic chromatography: A technique for size analysis of colloidal particles. **J. Colloid Interface Sci.** 1974, **48** , 1, 147-161

22. TACHIBANA, M; On the behaviour of a sphere in the laminar tube flows. **Rheol. Acta** 1973, **12** , 58-69

23. VASSEUR, P. and COX, R.G; The lateral migration of a spherical particle in two-dimensional shear flows. **J. Fluid Mech.** 1976, **78** , 2, 385-413

24. YANIZESKI, G.M; Phenomenological characteristics of the laminar flow of neutrally buoyant particles in a rectangular of high aspect ratio. PhD thesis. Carnegie-Mellon Univ., Michigan, 1968

THE DEWATERING OF BIOLOGICAL SUSPENSIONS
IN INDUSTRIAL CENTRIFUGES

P.N. WARD* and M. HOARE **
*ICI Biological Products, Billingham, Cleveland, TS23 1LB.
**Department of Chemical & Biochemical Engineering
University College London, Torrington Place
London, WC1E 7JE.

ABSTRACT

For optimal design and operation of solid-liquid separation processes a knowledge both of the characteristics of the separating machine and of the suspension which affect clarification and dewatering is required. In this paper we examine those characteristics which determine the extent of dewatering achievable and those which control sediment discharge from continuous centrifuges. Suspensions of *Methylophilus methylotropus*, *Fusarium graminearum* and borax flocculated *Saccharomyces cerevisiae* (bakers yeast) debris are studied.

SOLID-LIQUID SEPARATION TECHNOLOGY

Throughout biotechnology centrifugal separation is often a key step in primary product recovery. The efficient operation of such a step not only requires attention be paid to the removal or recovery of the particles from the liquid phase, but also to the dewatering step so as to prevent undesirable entrainment of liquid in the sediments. This latter objective is increasingly important with the trend towards feed suspensions of high solids content and the need to avoid excess dilution of the soluble phase, for example in a wash step. In this paper we examine how the characterisation and classification of suspension types and centrifuge configurations can aid process development.

There are a number of basic machine types, each of which have different characteristics and are each therefore suited to suspensions with different types of processing properties. For efficient operation all solids must settle out of the flowing liquid layer in the finite time taken for liquid to flow from the machine inlet to the exit, ie clarification. The basic parameters governing clarification are particle size, shape, density and the suspension viscosity, as well as any treatment leading to particle aggregation. This area of separation technology is well understood, and procedures for optimising particle characteristics and treatment, ultimately leading to minimisation of product loss, has been well established. Further investigation of clarification and machine characteristics has also led to the establishment of scale-up procedures for machine rating such as the sigma concept [1].

Factors governing dewatering are generally given less attention in process development than those affecting clarification. One of the basic characteristics of suspensions governing dewatering is the rheology of the concentrated suspension and the importance of this in dewatering can be appreciated from examination of a scroll decanter centrifuge (Figure 1). In this machine sedimented solids are scrolled (scraped) up a 'beach' to discharge, whilst the liquid is decanted off over a weir. Clearly for this type of machine it is important that solids can be scrolled to discharge against considerable centrifugal force.

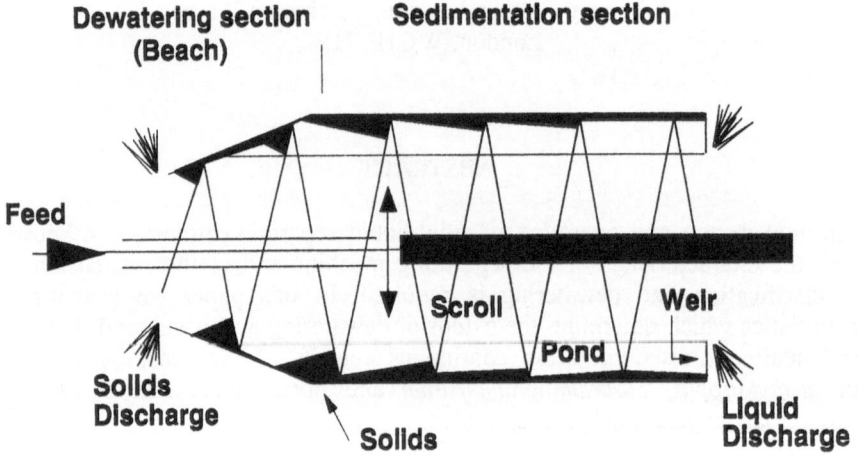

Figure 1. Scroll decanter centrifuge cross-section.

RHEOLOGICAL CHARACTERISATION

There are three major aspect of rheology that are of importance in determining dewaterability and flow of sediments. These are continuous shear rheology, visco-elasticity and network property characterisation. Continuous shear rheology gives an insight into the flow characteristics of sediments within the centrifuge when subject to continuous shearing forces, as in the discharge from a disk stack machine. The visco-elastic properties give an insight into flow characteristics at high solids concentrations, and along with network characteristics, an indication of suspension compressibility.

Visco-Elasticity

Concentrated suspensions can exhibit significant viscoelastic characteristics, that is solid as well as liquid-like properties. This has been illustrated with studies on a wide range of biological and non-biological suspensions [2-5]. As with shear rheology the basics of visco-elastic characterisation, and interpretation of the findings is well documented in text books [2,6]. However the key aspect here is to apply this powerful technique to suspension dewatering technology.

The basic principles of visco-elastic characterisation involve measurement of the relative effects of shear rate dependent stresses (viscous) and shear magnitude dependent stresses (elastic) on a sample exposed to oscillatory shear. Techniques for measuring visco-elastic characteristics involve the use of conventional rheometers with the ability to expose samples to oscillatory shear. Typically the results of such characterisations are expressed as shear moduli (cf Youngs modulus for extensional strength) and the loss angle as a function of oscillation frequency. The loss angle is a measure of the proportion of suspension strength attributable to interactions of a viscous nature and an elastic nature.

A further means of measuring visco-elastic characteristics involves exposing the sample to a single frequency pulsed shear wave. This technique only gives a measure of the material 'strength'. The principal technique for measuring characteristics by this means involves use of the pulse shearometer device [3,7,8]. Some visco-elastic characteristics of a flocculated and a non-flocculated cell debris suspension as a function of frequency are illustrated in Figure 2.

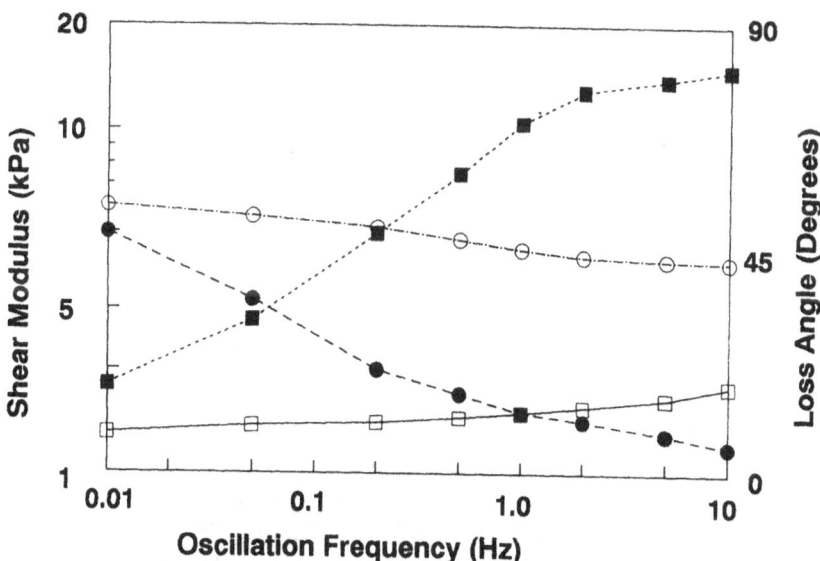

Figure 2. Visco-elastic spectra for borax flocculated (closed symbols) and nonflocculated (open symbols) cell debris. Loss angle, O, ● ; Shear modulus, □, ■.

Although the relationship between measurement frequency and shear rate is not fully established, the characteristics at high frequencies (> 1 Hz) can be taken as relevant to rapid or 'forced' flow, whilst those at low frequencies (< 1 Hz) can be taken as relevant to slow or 'creeping' flow. Furthermore a material where the solid-like properties dominate (loss angle close to zero) is likely to fracture or break upon deformation, whilst one in which liquid-like properties dominate (high loss angle) is likely to flow. An indication of the resistance to deformation or flow is given by the magnitude of the modulus. The consequence of these properties in centrifuge recovery is discussed later.

Compressional Network Strength (or Network Modulus)

Many of the visco-elastic characterisation techniques described above have been primarily aimed at measuring the strength of networks, and the resulting suspension resistance to flow. Also of interest is the strength of the network in compression as it will influence the solids content obtainable in compressional dewatering processes. Any relationship between shear and compressional strength is also of interest as in centrifuges such as scroll decanters the sediment is continuously sheared.

When subjected to compressive pressure, P, the solids network in many flocculated suspensions do not collapse to give a cake having randomly close packed solids. Usually the solids network compresses to an equilibrium value where it is sufficiently strong to resist the pressure. Any increase in the pressure will cause further network compaction to a new equilibrium value, and so on. Therefore varying the pressure will allow relationships between the equilibrium sediment volume and applied pressure to be obtained for any given suspension.

The concept of the network modulus, K, has been defined as given in equation 1, where P is the applied compressive pressure, ϕ is the solids volume fraction and ϕ_o is the minimum solids fraction at which a network is formed. It has been used in a number of studies to establish characterisation procedures for predicting the compressional properties of suspensions.

$$K = \phi \frac{dP}{d\phi} \Rightarrow P = \int_{\phi_o}^{\phi} \frac{K(\phi)\, d\phi}{\phi} \tag{1}$$

The network modulus may be measured in a compression cell [9], but more recently a centrifugation test has been established [10] and further developed [11].

A number of studies [9-11] have investigated the relationship between network strength in compression and the shear modulus. From these it has been shown that the network modulus is approximately numerically equal to the shear modulus measured at high frequencies (200 Hz) and low strains (10^{-4}) using shear wave propagation. However shifts between shear and network moduli do occur and these are probably due to the nature of the particles and interactions. Since the shear modulus is simpler than the network modulus to measure in many cases [12] the former has been used to characterise the strength of suspension network. This is generally because the network modulus measured by the centrifuge technique takes many hours, since the equilibrium conditions for each centrifuge speed have to be attained, whilst for shear modulus measurement there is no need to wait for an equilibrium to be reached [12]. The similarity between the magnitude of the shear and network moduli is illustrated in Figure 3 for suspensions of *F. graminearum* [13].

USE OF RHEOLOGICAL CHARACTERISATIONS IN PRACTICE

It is now possible to apply such rheological characterisation to concentrated suspensions so as to aid process development. One of the first areas to benefit from rheological characterisation was the development of Pruteen (a single cell protein, SCP, product). In this process suspensions of *Methylophilus methylotropus* was recovered using disk stack separators.

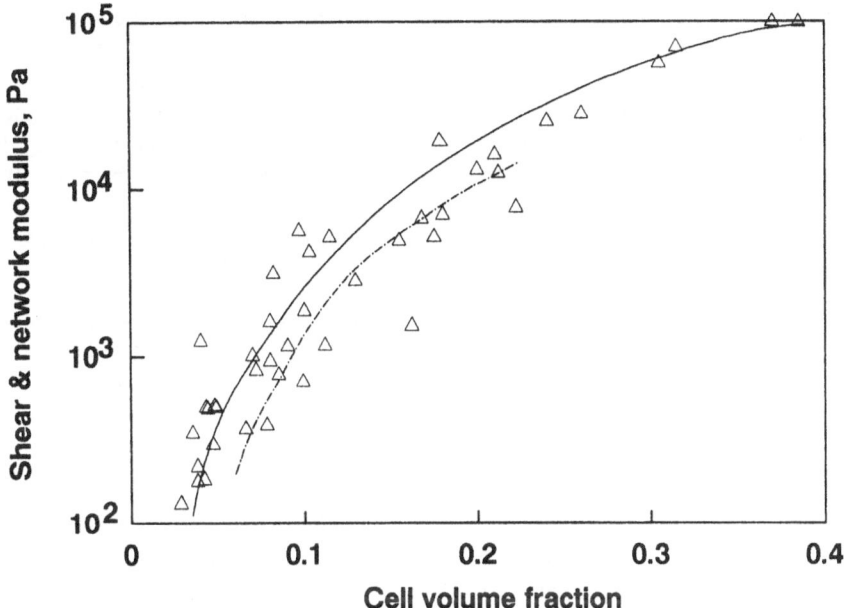

Figure 3   Shear and network moduli for *F. graminearunm* broth.   Shear modulus measurements, —△— ; calculated network modulus from centrifugation compression tests, — · — .

Pruteen *
The   development of the Pruteen* process required optimal recovery of bacterial cells followed by product drying.   For effective clarification solids aggregation by heat and pH treatment was required.   This gave a very wide range of possible treatment conditions and the prime concern was to select the best conditions for efficient process operation. To this end the shear and network rheological characteristics were measured as a function of treatment conditions and used to

* Pruteen in an ICI Trademark

predict the performance of a continuous disc nozzle centrifuge for the dewatering or 'thickening' of the suspension [12].

Figure 4 illustrates the influence of pH on the shear modulus, with lower pH values leading to a stronger bacterial network structure. From such data it was possible to evaluate the effect of the maximum centrifugal compressive pressure on the dewatering achieved. This assumes that the residence time in the sedimentation zone is sufficient for the dewatering to approach new equilibrium and that the shear modulus relates to the network modulus (equation 1). For example it was possible to predict that effective dewatering could still be achieved for the stronger network structures formed at low pH values. The use of such low pH values was advantageous is assisting degassing of the feed to the centrifuge.

Flocculation and Dewatering of Cell Debris Suspensions

The use of borax ions for the selective flocculation of yeast cell debris in the recovery of intracellular soluble proteins has been demonstrated in the laboratory [14] and on the pilot plant scale [15]. The formation of complexes involving the 1,2 cis diols of the mannan groups of yeast cell debris and the borax ions results in large flocs which are readily recoverable using low centrifugal forces. For such processes the feed stream is of relatively high solids content and it is necessary to either wash the sediment or to achieve good dewatering in order to maintain the required yield of soluble product. The former option both increases the required centrifuge capacity and yields a dilute product stream. For the latter it is necessary to examine the mechanism of dewatering and solids discharge in a centrifuge.

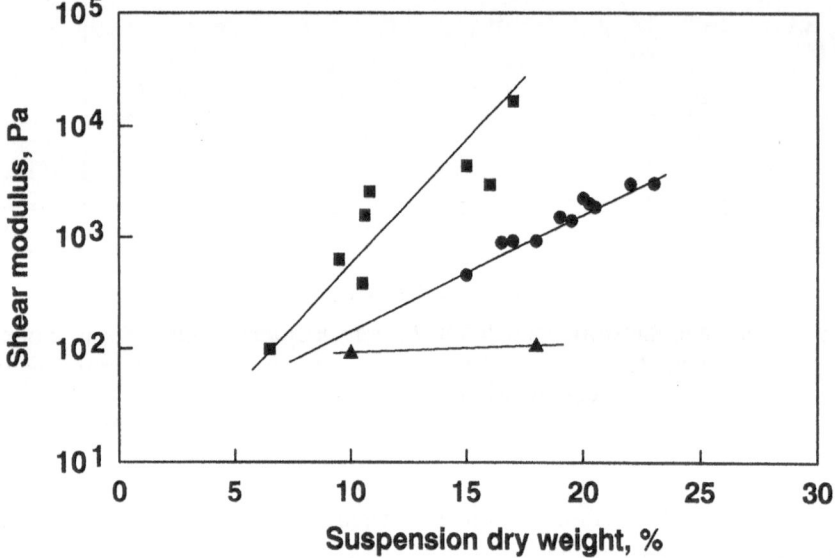

Figure 4. Network moduli characteristics for Pruteen; flocculation by treatment at pH4, ■ ; pH5, ● ; pH6, ▲.

One suitable centrifuge for processing feeds of a high solids content is the scroll decanter (Figure 1). Efficient clarification of non-aggregated debris is not possible in commercial scroll decanter machines. In addition to the small size of the particles the concentrated cake properties are such that the sedimented solids are easily resuspended and fluidised by turbulence within the machine. Furthermore, and more importantly to the discussion, it is not possible to scroll the sediment up the beach due to the flow of sediment, under centrifugal force, back into the settling pond. Such a phenomenon is observed with a centrifuge operating with a dry beach (as shown in Figure 1). A flooded beach (achieved by operating at an increased weir height) reduces much of the centrifugal pressure on the sediment allowing discharge, albeit in a wet (less dewatered) state.

From the visco-elastic spectra for flocculated and non-flocculated cell debris (Figure 2) it is evident that non-flocculated sediment with 24% dry weight has very little shear strength (a modulus of only a few kPa), with dominant viscous properties. As a result of aggregation using borax the modulus rises significantly (greater than 10 fold) and solid-like properties become dominant at frequencies corresponding to rapid deformations (> 1Hz). This indicates a change in the suspension characteristics from those of a fluid to a more solid-like nature. This rigidity indicates that in a decanter centrifuge the sediment should not readily flow, that is it should not work against the discharge mechanism of a decanter. Armed with this data pilot plant trials on the dewatering of borax aggregated debris using a decanter centrifuge were carried out using both flooded and dry beach configurations.

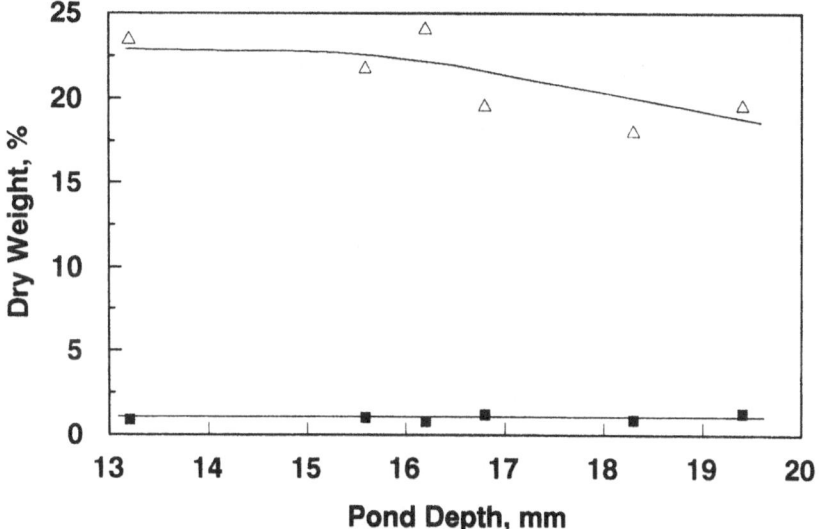

Figure 5.   The effect of decanter beach confugurations on flocculated cell debris suspension dewatering. Pond depth 13mm - dry beach; pond depth 19.3 mm -   flooded beach. Sediment dry weight, △ ; suspended solids in liquid discharge, ■.

From Figure 5 it is apparent that a decrease in pond depth gives an increase in sediment dry weight with no loss in clarification. For dry beach operation the solids are discharged at a concentration in the region of 24%, as would be expected from the network modulus characterisations [15]. As the pond depth is increased the solids are scraped against centrifugal force over a shorter length once clear of the liquid layer and there is less chance of sediment dewatering by drainage and greater chance of liquid carry over. For material where viscous properties dominate, ie the non-flocculated debris, there is no sediment discharge even though a considerable proportion of the solids size distribution is such that they should settle out within the centrifuge.

CONCLUSION

The basis has been established for using standard rheological characterisation techniques in the laboratory as an aid to the selection of the operating conditions of solid-liquid separator and suspension treatment conditions. This will help to reduce the number of expensive and time consuming pilot plant trials needed throughout process development. Further work is underway to use measurements of the rheological and structural properties of sediments to gain an improved insight into the interaction between the process stream and the centrifuge dewatering mechanism.

REFERENCES

1. Ambler, C.M., *Ind. Eng. Chem.*, 1961, 53(6), 430-433.

2. Whorlow, R.W., *Rheological Techniques*, Ellis Horwood, Chichester 1980.

3. Buscall, R.,Stewart, R.F and Sutton, *D. Filtrn. Sepn.*, May/June 1984, 183-188.

4. Buscall, R., Mills, P.D.A, and Yates, G.E. *Colloids and Surfaces*, 1986, 18, 314-358.
5. Buscall, R., McGowan, I.J., Mills, P.D.A., Stewart, R.F., Sutton, D. and Yates, G.E. *J. Non-Newtonian. Fluid Mechan.*, 1987, 24, 183-202

6. Ferry, J.D., *Visco-elastic Properties of Polymers*. 3rd Ed. John Willey, New York, 1980.

7. Van Olphen, H., Proc. *4th Nat. Conf. Clay and Clay Mining* (National Acadamy of Science, USA) 1956, 202-224.

8. Goodwin, J.W. and Khidhar, A.M. *In Colloid and Interface Science vol. 4* Ed Kirker,M.,1976.

9. Callaghan, I.C. and Ottewill, R.H. *Disc. Faraday Soc.*, 1974, 57, 110-118.

10. Buscall, R., *Colloids and Surfaces*, 1982, 5, 269-283.

11. Buscall, R. and White, L.R., *J. Chem. Soc., Faraday Transactions 1*, 1987, 83, 873-891.

12. Mills, P.D.A., Stageman, J.F. and Stewart, R.F. *5th Int. Conf. Surf. Coll. Sci.,* Potsdam, New York, June 24-27, 28, 1985.

13. Ward, P.N., PhD Thesis, University of London, 1989.

14. Bonnerjea, J, Jackson, J.C.., Hoare, M and Dunnill, P. *Enzyme and Microbiol. Tech*, 1988, 10, 357 - 365.

15. Bentham, A.C., Bonnerjea, J., Osborn, C.B., Ward, P.N. and Hoare, M. *Biotech.Bioeng.* Accepted for publication.

Part 2

Concentration Processes

Part 2

Concentration Processes

CONCENTRATION AND SELECTIVITY IN BIOSEPARATIONS

D.L.Pyle
Biotechnology Group,
Department of Food Science and Technology,
University of Reading, RG6 2AP, UK.

ABSTRACT

This brief chapter is a personal reflection on some of the aspects involved in attempting to reconcile simultaneously the objectives of concentration and selective purification of bioproducts. The review is written from a process engineering perspective, and it attempts to provide a context for the many papers in this volume addressing these problems, and also to identify problems and targets which are worthy of research and development.

INTRODUCTION

In trying to resolve the problems peculiar to bioseparations it is natural to seek solutions either in the traditional areas of chemical processing or to base them on recently developed analytical techniques. Whilst both may be helpful, there are many situations in which they lead only to a dead end. Unlike most chemical processes, bioseparation starts with a dilute or very dilute culture fluid, best described as a thin soup. The concentration of some products may be up to around 10 wt%, but recombinant proteins are more likely to be below 0.5 wt% (and much lower with animal cell culture systems). Clearly, a first objective must be concentration. However, thermal, solvent-based or other severe techniques may be excluded because of the need to conserve biological activity. Other techniques may be unsuitable because of the propensity of biomolecules to aggregate and form fouling layers. Also, the increasing interest in injectable therapeutic products makes demands on product purity which are unknown in the chemical industry. Thus bioseparations pose unique challenges if these objectives and constraints are to be reconciled. It is tempting to look towards analytical techniques as an alternative source of separations technology; whilst some techniques open exciting possibilities, others are unlikely ever to work their way into process technology for reasons such as cost, over-complexity, inappropriateness, or the sheer difficulties of translation to efficient process technology. By definition, we are concerned with many processes which are immature in their state of development for process scale operations in comparison with technologies such as distillation and absorption.

PREREQUISITES FOR A SUCCESSFUL TECHNOLOGY

Economics.

Only in the earliest stages of market penetration are economics relatively unimportant. Downstream processing costs are invariably a large fraction of total costs in biotechnology, but the ratio of recovery costs to fermentation culture costs is much higher for rDNA products and enzymes than for antibiotics or other traditional small molecular weight products : one estimate of these ratios puts them in the order: enzymes - 2:1; penicillin -1:1; ethanol - 0.16:1 [1]. In r-protein production chromatographic separations alone typically account for 25-50 % of the total processing costs. Cost reduction must be a significant objective whether one is concerned with chromatographic or any other process. For example, one reason why aqueous two phase systems have not yet realised their potential lies in their operating cost - mainly due to the cost of the polymers currently available and proven. One key R&D objective therefore must be to reduce these costs-either by developing new polymer systems or by designing better integrated processes.

Simplicity.

Given a choice, the process engineer will seek simplicity rather than complexity. This has many implications - sometimes not very palatable to the purist - for technology choice. Chromatographic systems developed for multiproduct detection are not the best candidates for single product recovery. Buffer exchanges will be avoided where possible. Stepwise elution will be preferred to gradient methods. Proven, mature technologies will be preferred to novel, apparently risky ones. Equilibrium stage processes will probably be preferred to rate-driven separations.

SELECTED CONCENTRATION AND PURIFICATION PROCESSES

The discussion which follows considers only post-harvesting techniques; crystallisation and precipitation methods are not included here since they fall within the compass of another review at this meeting. However we note the advantages of replacing isolation and concentration techniques - where these involve copurification - by ones with a degree of selectivity.

Equilibrium versus rate processes.

It is sometimes convenient to divide separation techniques into these two categories. In the former solute concentration or selective separation depend on the attainment of a favourable equilibrium state. Examples include liquid/liquid extraction and related partitioning techniques, many chromatographic processes (ion exchange, HIC, etc) etc. In these processes the primary determinant of yield and selectivity resides in the equilibrium law; it is convenient to define an equilibrium or partition coefficient for component i as the ratio between its concentrations in the two equilibrated phases:

$$K_i = y_i/x_i$$

Then selectivity, as measured by relative enrichment, is given by:

$$S_{i,j} = y_i/x_i \, /(y_j/x_j) = K_i/K_j$$

Good processes will be ones in which K- and S- values are both high, so that equilibrium strongly favours partitioning into one phase over the other, and in which the property differences between solutes result in markedly different equilibrium ratios or partition coefficients. Although such processes are not rate driven, kinetics are still very important, since the shorter the time to equilibrate, the more compact or intensive will be the process. There are other implications which we return to below.

Rate processes - such as electrophoresis - fractionate different components on the basis of their response to some imposed field. Mobility and other similar properties determine the feasibility of this type of operation, and successful processes will be ones where solutes have markedly different mobilities.

The design of any of these processes from first principles relies on an understanding of the principal features governing partitioning or mobility. Unfortunately in the majority of bioseparation processes this fundamental underpinning is deficient. Although partition coefficients for small molecules in common solvents can be predicted, the situation is far more complex for large biopolymers, or where mixtures of solvents are used, or in the complex environments typical of aqueous two phase systems. However a redeeming feature of liquid/liquid systems is that over the range of concentrations met equilibrium relationships are usually linear (that is partition coefficients don't change with concentration) and solute/solute interactions can usually be neglected. Solid-based adsorptive processes, however, often have nonlinear equilibria, and can readily saturate; moreover intermolecular interactions (eg protein-protein competition and interactions in ion exchangers) and exclusion phenomena are important in determining equilibrium capacities and, thus, selectivity. No general theory yet exists to cope with this. If the prediction of equilibrium behaviour is difficult in some situations, the analysis of rate processes involving biopolymers lags further behind. For example, the art of electrophoresis is well developed, but there is as yet no convincing theory to predict the behaviour of flexible polymeric particles, with changing conformation, in their electrically-driven passage through a gel which hinders diffusive motion. This is not a very satisfactory basis for rational design. We now consider some selected bioprocesses.

Extractions

Traditional methods for concentrating and separating polar and non-polar biomolecules typically use successive stagewise solvent extractions. For example a first stage extraction with methylene chloride will extract such nonpolar species as the tetracyclines; subsequent extraction of the remaining polar phase against n-butanol will separate highly (eg acids)and moderately polar species (such as penicillins, cephalosporins). Where these processes are appropriate (that is, denaturation is not a constraint) they have many significant features. Concentration and selectivity can be enhanced by a variety of strategies. The key is to find ways of increasing K. Since at equilibrium the chemical potential of the solute in the two phases is equal, this implies reducing the standard chemical potential of the extract phase. There are various strategies for this: by the choice of solvent or solvent(s) (there are still many opportunities for developments in this field - see for example, the paper by Schugerl's group in this volume [2]); by ion pairing; changing some key property of the solute:for example, changing the pH, as exploited in separation of penicillins [3], can have a crucial influence on the extraction of weak acids and bases and in processes using ion-pairing. Another strategy is to relax the thermodynamic constraint by using a reagent in the stripping phase [4]. This idea is readily exploited in another variation on the same theme, in the use of liquid membranes, which are particularly suitable for dilute feeds. The major potential advantages of LM systems are in process intensification (extraction and stripping are carried out simultaneously), in improving transfer kinetics, and in the case of supported systems, in solvent retention. The effect of a stripping reagent can be easily seen by considering the distribution of a solute between three phases, with distribution coefficients $K_{1,2}$ and $K_{2,3}$. Then at equilibrium:

$$x_2 = K_{1,2} \, x_1 \; ; \; x_3 = K_{1,2} \, x_2 = K_{1,2}{*}K_{2,3} \, x_1$$

where x_1 etc are the equilibrium concentrations in phases 1, etc. Maximising the two partition coefficients will maximise transfer, but will always leave some residual solute in phases 1 and 2. An irreversible reaction in phase 3, however, corresponds to an infinite parition coefficent between phases 2 and 3, so that at equilibrium - or the end of a batch extraction -no solute remains in phases 1 and 2. In these processes the principal costs are usually in solvent recovery and make-up. With solvent extractions the major problem to be overcome is to increase selectivity-the process technology is mature ; LM systems need much more work both on system design for selectivity and on process design aspects,[4],[5].

New and exciting possibilities for the use of modified solvent systems have opened up in parallel with the discovery of the potential of solvent-based, low-water enzymatic processes. There have been a number of papers demonstrating the use of reversed micelles for protein solubilisation and purification [6]. Whilst it is clear that a key determinant of solubility is the charge distribution on the protein and the stabilising surfactant, there is a need for much more work. The number of protein systems studied so far is small; the limits of the process (presumably determined by the reverse micelle dimension, as well as surfactant charge) are not known. There are few reports on transfer kinetics.

It is often not fully appreciated that in these and other extraction processes, purification and selectivity can be significantly improved by multistage operation; yield and selectivity depend on the number of equilibrium stages, n, and the extraction factor SK_i/F where S is the solvent flow, F is the feedrate, and K_i the partition coefficient for the solute i. When the feed enters some way down the chain of contactors, enrichment and extraction are both realised, thus meeting the dual requirements of concentration and selective enrichment. In the limit, that is when the effective number of stages is high, such processes can be run (countercurrently) in a column as a differential contactor, in which case mass transfer becomes crucial. Chromatographic separations fall into this category: they are not rate processes.

The several papers on aqueous two phase systems (ATPs) in this volume illustrate the growing interest in this technique, which is already being applied on a process scale. The main advantages of ATPs are that reasonable and selective purifications (purification factors of up to 15 are readily achievable) are possible, although there remains a need to develop and test well-founded equilibrium theory. The low interfacial tension permits the formation of good dispersions with fast interphase transfer and low denaturing; since the technology is closely related to other liquid extraction processes, scale up should not be too much of a problem. It will be necessary to investigate transfer aspects more comprehensively if the logical development to differential contactors is to be pursued. Another great advantage is that concentrates can then be fed directly to a chromatographic separator, which is often not the case with streams from centrifuges or membrane concentrators. Apart from developing a sound theoretical and empirical base for K-factors, the main problem to be overcome is economic. ATP systems are limited by high material (polymer) costs -especially where dextran is a preferred component. New phase components can thus be expected to give better technical and economic performance.

Foam and froth separations

In contrast to extraction processes which have a significant niche both in biotechnology and in other areas of process technology, foam separations are almost unheard of for product separation in biotechnology (Nisin production being an exception) whilst being very important in other sectors such as minerals processing. On the contrary, foams are perceived to be a nuisance. However, there is a strong argument for a more realistic assessment of the possibilities of exploiting the surface active nature of many biomolecules in concentration and purification. There is no doubt about the possibility of the selective concentration of proteins and peptides, and that significant changes are found by changing pH; what need to be studied are the conditions under which proteins and other active biomolecules will collect or be sequestered in foams, and how the processing conditions affect conformation and denaturation [7].

Membrane processes

It must be recognised that there still remain problems to be overcome in the development of solid membrane systems. Protein fouling is still a problem, and several papers in this conference address this issue, including ways of reducing the effects of polarisation (eg the paper by Cooney et al [8]). A major challenge is to incorporate selectivity into the membrane action by suitable choice of matrix, manipulation of the pore size distribution where this is amenable to control, the use of ion exchange and affinity ligands, and the use of supported solvents and carriers (including reverse micelles).

Adsorption and chromatography

This discussion will be very restricted in scope. Only a small range of processes will be discussed: affinity methods are discussed elsewhere [9]. Since elution chromatography has the process drawback that, whilst it may isolate a solute, it also involves dilution by the eluting solvent, the tension between concentration and resolution is a live issue. A limited set

of scientific and engineering issues will be considered. We can identify
the following apparent areas of actual or potential concern with adsorptive
processes:

* Relatively limited capacity
* Non-linear equilibria (affecting capacity, dynamics etc)
* Solute/solute interactions influencing response and equilibria
* Possible mass transfer limitations
* Process engineering considerations:
    pressure drop constraints
    pellet compressibility
    packing and flow distribution
    generally non-steady behaviour
    scale-up limitations

The interactions between some of these aspects produce several
particular problems. A constant theme is the conflict between capacity,
resolution and throughput on scaling up. From an engineering perspective
the pressure drop across a fixed bed is very important both in relation to
energy use and to materials integrity. Under the usual conditions of low
particle Reynolds number, pressure drop increases linearly with bed length,
superficial velocity, and with the inverse square of support particle
diameter, d; moreover as particle sizes approach the micron range it
becomes increasingly hard to maintain uniform flow conditions. The conflict
arises because mass transfer and capacity considerations favour smaller
particles. External mass transfer may limit with very dilute feeds.
Internal mass transfer is a potential problem because of the need to
generate high internal areas for adsorption. Both capacity and resolution
are affected if mass transfer is a problem; both suffer because solute
molecules elute from the column before having time to equilibrate, so that
the breakthrough curve or elution band will be diffuse rather than sharp.
When in-pore diffusion controls, the effective kinetics have a pseudo first
order rate constant which is approximately proportional to d^{-1};
bandspreading is then proportional to Ud^2/L .When the effects of physical
dispersion are taken into account the spread of the elution front
deteriorates both with increasing particle size and flow velocity, U. Thus
high take-up of potential adsorptive capacity can only be realised at low
loadings, which is unattractive. If improvements in support particles (eg
to reduce internal diffusional constraints) are not possible, the result is
short, low aspect ratio, columns or multi-stage units. However, it may be
possible to reduce diffusional limitations by using membranes with shorter
controlled pore lengths or support particles with a bimodal pore size
distribution: the recent report by Afeyan et al [10] describes particles
with a tree-like structure of large diameter pores in which convective flow
may be possible and diffusion is unhindered, connected to a network of
short small diameter pores. Thus a large internal area for adsorption can
be achieved, without imposing severe mass transfer limitations even at high
velocities.

It is clear then that there are many possibilities for developing
adsorptive methods which can handle process scale throughputs, and achieve
acceptable levels of capacity and resolution. Apart from developing rigid
support materials compatible with biopolymers, a key objective must be to
minimise mass transfer limitations, since these cause many problems. An
ideal design is one in which equilibrium is achieved on a time scale
significantly below the scales of competing processes.

Rate processes

Finally, a few comments on the potential applications of rate-driven separations. The development of analytical scale electrically-driven processes such as electro-phoresis, iso-electric focussing and electro-dialysis is well-advanced in biotechnology. These are all methods which potentially could be engineered for selective separations, and which provide significant intellectual challenges; typical problems are to minimise mixing and thermally driven convective effects, and to provide for heat removal and temperature control. It is interesting to speculate why there is so little evidence of work in these areas.

CONCLUSIONS

Whilst there have been remarkable developments in bioseparations over the last few years, many significant and worthwhile challenges remain both for basic scientists and for engineers. I have tried to identify a few such problems which are relevant to the early stages of downstream processing. There are many more waiting to be solved.

REFERENCES

[1] Datar, R. Economics of primary separation steps in relation to fermentation and genetic engineering. Proc. Biochem. 1986 (Feb),19-26.
[2] Schugerl, K., Degner, W., Frieling P v.,Handojo, L. and Kirigios,I. Separation of low-molecular organic compounds from complex aqueous mixtures by extraction. This volume. 1990.
[3] Belter, P.A., Cussler, E.L. and Hu, W-S. Bioseparations. 1988. Wiley & Sons,N.Y.
[4] Chaudhuri, J.B. and Pyle, D.L. A model for liquid emulsion membrane extraction of organic acids. This volume. 1990.
[5] Sirman, T.,Grandison, A.S. and Pyle, D.L. Extraction of organic acids using a supported liquid membrane. This volume. 1990.
[6] eg Abbott, N.L. and Hatton, T.A. 1988. Chem. Eng. Progress. 31-41.
[7] Zlokarnik, M. Trends and needs in bioprocess engineering. Chem.Engng. Progr. 1990. 86(4), 62-67.
[8] Cooney C.L., Holechovsky, U. and Agarwal G. Vortex flow filtration for ultrafiltration of protein solutions. This volume. 1990.
[9] Chase, H.A. The future of high resolution purification. This volume. 1990.
[10] Afeyan N.B., Fulton, S.P., Gordon N.F., Mazsaroff, I., Varady, L. and Regnier F.E. Perfusion Chromatography : an approach to purifying biomolecules. 1990. Bio/Technology, 8, 203-206.

A MODEL FOR EMULSION LIQUID MEMBRANE EXTRACTION OF ORGANIC ACIDS

J.B. CHAUDHURI[*] and D.L.PYLE,
Biotechnology Group, Department of Food Science and Technology
University of Reading, Whiteknights, PO Box 226, Reading RG6 2AP

[*]Currently: Centre for Biochemical Engineering,
University of Birmingham, PO Box 363, Birmingham B15 2TT.

ABSTRACT

Studies on the batch extraction of lactic acid using an emulsion liquid membrane system are reported; the effects of the surfactant, Span 80, and tertiary amine carrier, Alamine 336, on the system stability, kinetics and swelling are examined. A new quantitative model is developed and tested to describe the facilitated extraction of lactic acid using Alamine 336. The stoichiometry of the facilitated reaction is elucidated. This model also accounts for the convective solute transport which results from emulsion swelling. Very good agreement is observed between the model and experimental results; the variation in extraction with carrier concentration is successfully predicted.

INTRODUCTION

A liquid membrane is an insoluble liquid, usually an organic solvent which is selective for a solute, separating two other liquid phases. Extraction occurs by transport of solutes through the membrane. This paper is concerned with emulsion liquid membrane extraction (figure 1) which is potentially of great interest for downstream processing in biotechnology: the specific surface areas for extraction are large; separation and concentration can be achieved in a single step; process scale-up may be based on the existing, well-defined systems used in conventional solvent extraction.

Research on liquid membrane extraction has focussed mainly on the recovery of metal ions. Where facilitated extraction systems have been used in the recovery of biochemicals [1] there are no quantitative models for the extraction kinetics. Also, there are no quantitative models of extraction kinetics which incorporate the effect of swelling which results from water transport across the organic film, thus diluting the recovered product, destabilising the system, and transporting water-soluble solutes into the emulsion. This paper reports a kinetic model for the kinetics of the batch extraction of lactic acid by both facilitated and convective transport.

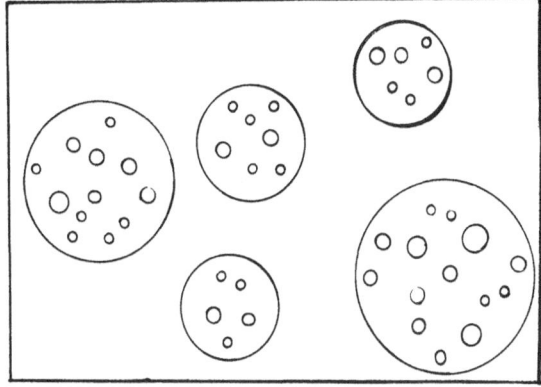

Figure 1. Emulsion liquid membrane.

PRINCIPLES OF EXTRACTION AND TRANSPORT MECHANISMS

There have been several recent reviews on the fundamentals and applications of liquid membrane systems since Li's studies [2, 3]. Lorbach and Marr [4] summarise the principles of transfer; some quantitative models are reviewed by Chan and Lee [5]. Frankenfeld and Li [6] cover fundamental principles and applications; many aspects are discussed by Noble and Way [7].

In general, liquid membrane extraction systems can be divided into two types, classified by the dominant transport mechanism. The simplest is unfacilitated transport where solutes transfer into the membrane and stripping phase because of solubility effects. If there is a reagent in the stripping phase to react with the extracted solute, this removes the thermodynamic constraint so that all the solute can be removed from the feed. Unfacilitated transport is generally used for the extraction of small un-ionised molecules. Facilitated or carrier-mediated transport involves the formation of a carrier-solute complex, which diffuses through the membrane phase. The carrier should be specific for the solute, react reversibly with it, and be insoluble in both aqueous phases. Larson *et al*. [8] found that a tertiary amine carrier significantly increased the rate of acetic acid extraction. Using the same carriers, Boey *et al* [9] and Chaudhuri and Pyle [10] have reported studies on the extraction of citric acid and lactic acid respectively. Extraction is improved by a high reaction equilibrium constant and high concentrations of carrier; its rate may be controlled by the reaction rate or the diffusion of the carrier complex. This paper is concerned with the facilitated extraction of lactic acid.

DEVELOPMENT OF A KINETIC MODEL FOR THE BATCH EXTRACTION OF LACTIC ACID

Extraction Chemistry of Lactic Acid by Alamine 336

A reliable model for the kinetics of extraction must incorporate the correct stoichiometry and equilibrium relationships for the extraction and the stripping reactions.

It is known that tertiary amines are effective in extracting organic acids including lactic acid [11, 12]. Alamine 336 is a commercial product (Henkel Corporation) which consists of approximately 95% trioctylamine [11]. Studies on the chemistry of tertiary amines have shown that these species require protonation before they can participate in ion-exchange reactions [13]. Hence the proposed reaction for lactic acid extraction by Alamine 336 is

$$H^+_{aq} + La^-_{aq} + A\ R_3N_{org} \longleftrightarrow (R_3N)_A H^+ La^-_{org} \tag{1}$$

where H^+ is the hydrogen ion, La^- and R_3N represent lactate and the tertiary amine respectively, $(R_3N)_A H^+ La^-$ is the quaternary amine salt formed in the complexing reaction, A is the stoichiometric coefficient of the tertiary amine, and the subscripts aq and org refer to the aqueous and organic phases respectively. It is assumed that the quaternary amine salt is insoluble in the aqueous phase. It has been suggested [11,13] that the acid reacts with the amine via an acid/base reaction to form an ion pair complex. For a monocarboxylic acid such as lactic acid, this would suggest that $A = 1$; this is assumed in the development of the facilitated transport models here; equilibrium studies [14] confirm this assumption.

The equilibrium constant K_1 for the above reaction is therefore defined as

$$K_1 = \frac{[(R_3N)H^+La^-]_{org}}{[H^+]_{aq}\ [La^-]_{aq}\ [R_3N]_{org}} \tag{2}$$

Lactic acid is also assumed to be in equilibrium with its dissociated ions in the aqueous phase; K_A is its dissociation constant.

The proposed overall extraction mechanism is shown below:

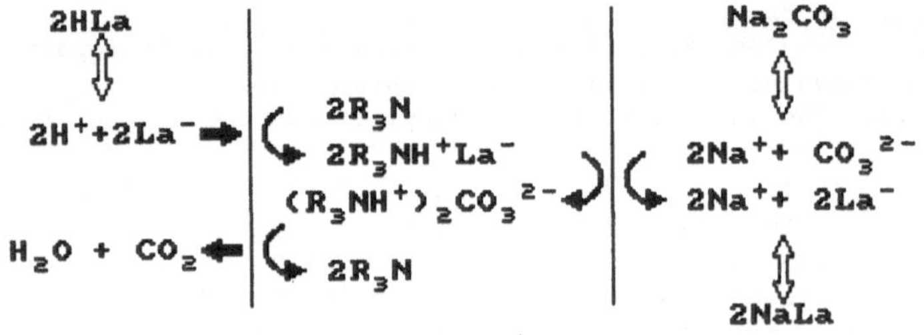

FIGURE 2 : FACILITATED EXTRACTION OF
LACTIC ACID BY ALAMINE 336.

The reaction between the amine salt and sodium carbonate at the membrane / internal phase interface can also be characterised by the equilibrium constant K_2:

$$K_2 = \frac{[(R_3NH^+)_2CO_3^{--}]_{org}\,[Na^+]^2[La^-]^2_{aq}}{[Na^+]^2_{aq}[CO_3^{--}]_{aq}\,[R_3NH^+La^-]^2_{org}} \tag{3}$$

where $(R_3NH)_2CO_3$ is the amine carbonate formed by the reaction between the quaternary amine salt and sodium carbonate

Kinetics of Facilitated Extraction

It is assumed that the extraction reaction is confined to the interface, and is fast in comparison to the rate of diffusion of the amine salt through the membrane phase. Thus eqn. (1) relates the concentrations at the interface. The flux of lactic acid from the bulk external phase to the interface with the membrane phase is given by:

$$j_{HLa} = k([HLa] - [HLa]_i) \tag{4}$$

where k is the external phase mass transfer coefficient, and $[HLa]_i$ is the interfacial lactic acid concentration. The flux of the species in the membrane phase (Alamine 336 and amine salt) to and from the reaction front with sodium carbonate can be described by diffusion through a sphere [14] viz, for Alamine 336:

$$j_{R3N} = \frac{r_c D_{R3N}}{RL}([R_3N]_L - [R_3N]_i) \tag{5}$$

where D_{R3N} is the diffusion coefficient of Alamine 336 in the membrane phase, r_c is the position of the stripping reaction in the emulsion globule, R is the globule radius, and L is the diffusion distance to the reaction front $(= R - r_c)$. In this study a shrinking core model for the movement of the reaction front is found appropriate. It is also assumed that the diffusion coefficient of the amine salt is equal to that of Alamine 336. Assuming steady state at the emulsion globule interface:

$$j = j_{HLa} = j_{R3N} = j_{R3NHLa} \tag{6}$$

By combining Eqns (2)-(6) the interfacial concentration of lactic acid can be expressed:

$$[HLa]_i = 0.5\left\{[HLa] - \frac{r_c D_{R3N}\,[R_3N]_0}{RkL} - \frac{1}{K_1\,K_A}\right\} - $$
$$- \left\{0.25\,([HLa] - \frac{r_c D_{R3N}\,[R_3N]_0}{RkL} - \frac{1}{K_1\,K_A})^2 + \right.$$
$$\left. + \frac{[HLa]}{K_1\,K_A} + \frac{r_c D_{R3N}\,[R_3NHLa]_L}{RkLK_1K_A}\right\}^{0.5} \tag{7}$$

where $[R_3N]_0$ is the initial concentration of Alamine 336, which equals the average carrier concentration in the membrane phase [14, 15]; the solute diffusion distance (L) is estimated using a shrinking core model based on the depletion of the internal phase reagent [10, 14]. Substituting Equation 7 into Equation 4 and solving the differential equation resulting from a mass balance on lactic acid gives the lactic acid flux versus time.

Water Transport

Under certain conditions water transfers into the stripping phase, and is likely to cause convective transfer of water-soluble lactic acid. Thus solute extraction may occur by both facilitated and convective transport; the development of a reliable kinetic model must accomodate both these effects.

The convective flux of lactic acid will be equal to the product of the volumetric water flux (giving rise to swelling) and the interfacial lactic acid concentration:

$$j'_{HLa} = j_W [HLa]_i \tag{8}$$

where j'_{HLa} is the convective lactic acid flux, and j_W is the volumetric water flux. Assuming that the membrane phase solvent has negligible water solubility, the volumetric flowrate of water (Q_W) into the emulsion will equal the rate of increase in emulsion volume V_S:

$$Q_W = A j_W = dV_S/dt \tag{9}$$

where A, the area for water transport, is the emulsion globule surface area. A quantitative model has been developed to describe water transport driven by the osmotic pressure difference across the membrane phase [14]; this work will be reported elsewhere.

Assuming that the facilitated and convective fluxes are additive, the net flux of lactic acid is equal to $j_{HLa} + j'_{HLa}$. From a mass balance on the external phase, the rate of lactic acid loss is

$$d(V_E [HLa])/dt = - A \{k([HLa] - [HLa]_i) + j_W [HLa]_i\} \tag{10}$$

By a volume balance the rate of increase in the emulsion volume (dV_S/dt) must equal the rate of decrease in the external phase volume ($-dV_E/dt$). Substituting this into Equation 10 and rearranging gives :

$$\frac{d([HLa])}{dt} = - \frac{1}{V_E} \{kA([HLa] - [HLa]_i + \frac{dV_S}{dt}([HLa]_i - [HLa])\} \tag{11}$$

The solution of this equation with Equations (2),(3),(6),and (7) gives the concentration-time profile of lactic acid in the external phase resulting from facilitated and water transport.

MATERIALS AND METHODS

n-heptane and light paraffin (the membrane phase) were obtained from B.D.H. Chemicals (Poole, Dorset). The surfactant Span 80 (sorbitan monooleate) used to stabilise the emulsion was obtained from Sigma. The "carrier" species used was Alamine 336, a tertiary amine which was kindly supplied by the Henkel Corporation (Cork, Eire). Lactic acid was obtained from Fisons, and the stripping phase reagent, sodium carbonate, was obtained from B.D.H. Chemicals. Commercial grade Alamine 336 was used as supplied; all other chemicals were reagent grade or better.

Equilibrium studies were carried out by adding lactic acid (0.024-0.23 M) and the organic phase (solvent containing (2-20%v/v) Alamine 336) to a stoppered conical flask. The flask was agitated in a rotary shaker at 100-120 rpm at 25°C for 24 hours. The phases were settled for 24 hr. The lactic acid concentration in the aqueous phase was then measured.

The membrane phase consisted of Span 80 (1-6% by volume of the membrane phase) and Alamine 336 (2-10%v/v) dissolved in the solvent. The solvent was 70%(v/v) n-heptane and 30%(v/v) paraffin. To form the emulsion liquid membrane, typically 50 ml of sodium carbonate solution was added dropwise to 50 ml of the membrane phase during emulsification. This was carried out using a compressed-air driven high-speed stirrer (Silverson L2-Air) operating at 5100 rpm. Emulsification lasted 10 min and was carried out in an ice bath. Batch extractions were carried out in a 500 ml beaker fitted with four baffles to prevent vortexing. The emulsion was dispersed by an eight-bladed turbine at 285 rpm. 200 ml of the feed (external) phase were poured into the beaker, then the emulsion was poured slowly into the beaker to form a separate top layer (it was found that negligible mass transfer occurred during this procedure), and then stirring was started. Extractions were carried out at 20-22°C. At intervals samples were removed by pipette, transferred into a test-tube and settled. The settled emulsion was broken in a water-bath at 80°C. Both the internal and external phases were analysed for lactic acid. Lactic acid concentrations were measured enzymatically (Boehringer Mannheim GmbH); Alamine 336 concentrations were determined using the method of Fritz [16].

RESULTS AND DISCUSSION

Extraction Chemistry

The value of K_1 was found from equilibrium data using a method based on work by Kertes and King [17]. The stiochiometry of eqn. 1 was confirmed with Alamine 336 and lactic acid with $A = 1$ and $K_1 = 5414$ $l^2 mol^{-2}$.

Lactic Acid Extraction Kinetics

Figure 3 shows the results from the batch extraction of 0.19 M lactic acid using a membrane phase containing Span 80 concentrations at 1-6 % (by volume) of the membrane phase, and 5% Alamine 336. The internal phase consisted of 0.57 M sodium carbonate which is 50% above that needed to neutralise all the acid. The results are presented in the form of the dimensionless concentration of lactic acid in the external phase. In all four cases, there was an initial fast reduction in the lactic acid concentration. The initial extraction rates, (over the first minute) in mmol cm^{-3} (total system volume) min^{-1} were respectively 0.109 with 1% Span 80, 0.112 with 2%, 0.116 with 4%, and 0.128 with 6% Span 80. It was also apparent that after about 3-4 minutes contact time, the lactic acid concentration in the external phase increased at the lower Span 80 concentrations (1 and 2%).

During extraction the system became increasingly viscous, and at longer contact times it became hard to disperse the emulsion. Based on the measurements of the solute concentration in the internal and external phases, emulsion swelling was quantified by a mass balance. It was experimentally verified that there was no significant solubility of water of lactic acid in the membrane phase. Significant swelling occurred, and it was found that the Span 80 concentration had a marked effect: the internal phase volume increased from 50 ml to around 90 ml with 1% Span 80, whilst with 6% Span 80 the increase was more than 3-fold in around 15 minutes.

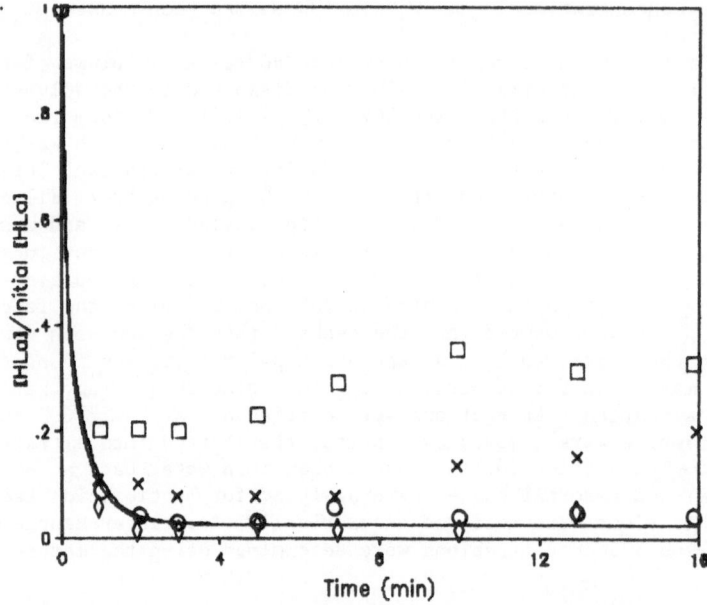

Figure 3. Effect of surfactant concentration on extraction kinetics.
(Span 80: ▫ = 1%; x = 2%; o = 4%; ◊ = 6%)

The swelling results clearly indicate that the surfactant is implicated in water transport. Colinart et al. [18] suggest that this occurs via hydration of the surfactant headgroup, and so the rate and extent of swelling should increase with the surfactant concentration; this is consistent with these results. The concentration of Span 80 also significantly affects emulsion stability. The increase in concentration of lactic acid in the later stages of extraction, most noticeably with 1 and 2% Span 80, is ascribed to solute leakage from the internal phase.

Figure 4 (x = 2%; o = 5%; # = 10%) shows the experimental and predicted results for the effect of Alamine 336 on extraction. Alamine 336 concentrations from 2-10% were used in the membrane phase, with a Span 80 concentration of 4%. Increasing Alamine concentrations result in faster initial extraction rates, as seen by the following values (in mmol cm^{-3} min^{-1}): 0.107 at 2% alamine, 0.112 at 5%, 0.117 at 10% respectively. This is consistent with the role of Alamine 336 as a facilitated carrier.

Comparison of Experimental Results With Kinetic Model

The mass transfer coefficient required for the model was determined using a correlation for dispersions [19], and was estimated to be $4*10^{-3}$ cm s^{-1}. The diffusion coefficient of the quaternary amine salt in the membrane phase was estimated as the diffusion coefficient of lactic acid in n-heptane using the Wilke-Chang correlation [20], and was found to be $3.6* 10^{-5}$ cm s^{-1}. The initial emulsion surface area was estimated from the emulsion volume by assuming a monodisperse, spherical distribution of droplets, and using an initial (observed) globule diameter of 0.04 cm [14]. The value of K_2 was unknown. To calculate the convective transport of lactic acid the change in emulsion volume with time was fitted by an empirical expression of the form

$$V_S = 100 + A_1 t /(B + t) \qquad (12)$$

where V_S is the emulsion volume in ml, and t is the contact time in seconds, with the following values: for 1% Span 80, A_1 = 34.0 ml, B = 36.3 s; 2%, A_1 = 55.8 ml, B= 42.0 s; 4%, A_1 = 91.7 ml, B = 70.7 s; 6%, A_1 = 93.9 ml, B = 53.3 s. The rate of water transport to the emulsion (or the rate of emulsion volume increase) is given by differentiating eqn (12).

   The equations for the kinetic model were solved using a Runge-Kutta algorithm: the theoretical model was fitted to the experimental data by adjusting the value of K_2; the value which fitted the data was 5×10^6. The large value of K_2 indicates that the formation of the stripping products is favoured. This behaviour is expected from a weak acid-strong base reaction. Figure 4 shows very good agreement between the model and experiment for a range of carrier concentrations.

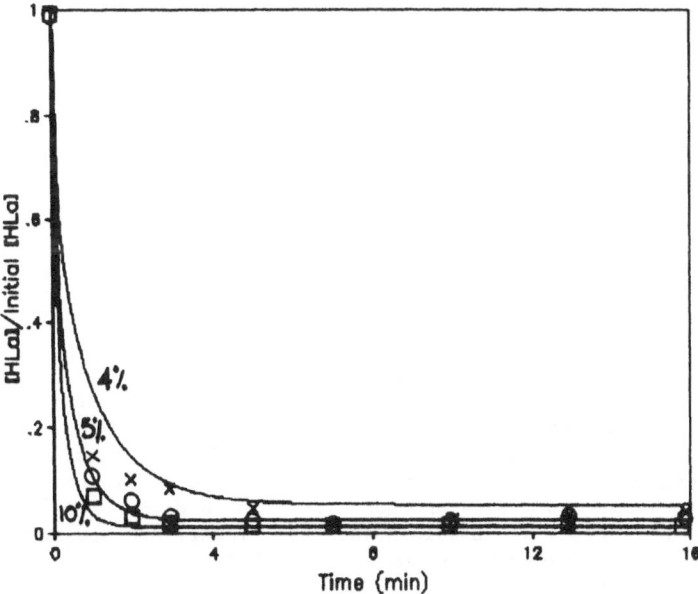

Figure 4. Effect of alamine concentration on extraction kinetics.

From Figure 3 it is seen that there is good agreement between experiment and theory for 4 and 6% Span 80. The theory predicts no effect of Span 80 concentration. In particular, since breakage of the droplets or globules is neglected, the model does not predict the observed increase in external phase lactic acid concentration at lower surfactant concentrations.

The rate limiting process can be identified by considering the relative size of the external and membrane phase mass transfer resistances. From the model this is approximately represented by the ratio of $k:r_cD_{R3N}[R_3N]_O/RL$. At the onset of extraction the diffusion distance L, is zero and the external phase mass transfer is rate limiting. For the conditions in this study when the diffusion distance is equal to $9.4*10^{-4}$ cm the resistances in the two phases are equal; this condition occurs for a C/C_O value around 0.8, after about 0.016 min. After this time lactic acid extraction is dominated by membrane phase diffusion.

Finally, consider the effect of swelling on facilitated transport. The greatest swelling occurred with 6% Span 80: for this run the initial extraction rate of lactic acid was 0.128 mmol cm^{-3} min^{-1}, and the initial rate of water transport was 0.16 cm^3 cm^{-3} min^{-1}. The maximum possible rate of lactic acid extraction by convection may be estimated by assuming that the acid carried by the water flux was at its initial concentration (0.19 M), giving a rate of extraction of 0.03 mmol cm^{-3} min^{-1}. This represents 24% of the initial extraction rate. Although this is a significant proportion of the overall solute transport, it cannot account for the high initial extraction rates shown in Figures 3 and 4. In practice, lactic acid transported by the water flux is carried at its interfacial concentration which is lower than the concentration used here. It is concluded therefore that the independent transfer of lactic acid by facilitated transport dominates the initial extraction.

REFERENCES

1. Thien, M.P., Hatton, T.A. and Wang, D.I.C., Separation and concentration of amino acids using liquid emulsion membranes. Biotech. Bioeng., 1988,32, 604-615.
2. Li, N., Permeation through liquid surfactant membranes. A.I.Ch.E.J., 1971, 17, 459-463.
3. Li, N., Separation of hydrocarbons by liquid membrane permeation. Ind.Eng. Chem. Proc. Des. Dev., 1971, 10, 215-221.
4. Lorbach, D. and Marr, R., Emulsion liquid membranes. Part 2: Modelling mass transfer of zinc with bis(2-ethylhexyl)dithiophosphoric acid. Chem. Eng. Proc., 1987, 21, 83-93.
5. Chan, C.C. and Lee, C.J., Mechanistic models of mass transfer across a liquid membrane. J. Memb. Sci., 1984, 20, 1-24.
6. Frankenfeld, J.W. and Li, N.N., Recent advances in liquid membrane Technology. In Handbook of Separation Process Technology, ed.R.W. Rousseau, John Wiley and Sons, 1987, pp.840-861
7. Noble, R.D. and Way, J.D., Liquid Membranes - Theory and Applications, American Chemical Society Symposium Series, 1987, 347.
8. Larson, K.M., Hanna, G., Hanson, S. and Way, J.D., Carrier enhanced acetic acid extraction with emulsion liquid membranes. American Institute of Chemical Engineers Annual Meeting, San Francisco, Nov. 1984. Cited by Thien et al. (1988).
9. Boey, S.C., del Cerro, M.C. and Pyle, D.L., Extraction of citric acid

by liquid membrane extraction. Chem. Eng. Res. Des., 1987, **65**, 218-223.

10. Chaudhuri, J. and Pyle, D.L., Liquid membrane extraction. *In* Separations for Biotechnology, eds. M.S. Verrall and M.J. Hudson, SCI/Ellis Horwood Ltd., Chichester, 1987, pp. 241-259.

11. Reschke, M. and Schugerl, K., Reactive extraction of penicillin: Stability of penicillin G in the presence of carriers and relationships for distribution coefficients and degrees of extraction. Chem. Eng. J., 1984, **28**, B1-B9.

12. Wennersten, R., The extraction of citric acid from fermentation broth using a solution of a tertiary amine. J. Chem. Tech. Biotech., 1983, **33B**,85-94.

13. Cox, M. and Flett, D.S., Metal extractant chemistry. *In* Handbook of solvent Extraction, eds. T.C. Lo, M.H.I. Baird and C. Hanson, John Wiley and Sons, 1983, pp. 53-89.

14. Chaudhuri, J.B., PhD Thesis, University of Reading, 1990.

15. Cussler, E.L., Diffusion, Mass Transfer in Fluid Systems, Cambridge University Press, New York, 1986.

16. Fritz, J.S. Titration of bases in nonaqueous solvents. Anal. Chem., 1950, **22**, 1028-1029.

17. Kertes, A.S. and King, C.J. Extraction chemistry of fermentation product carboxylic acids. Biotech. Bioeng., 1986, **28**, 269-282.

18. Colinart, P., Delepine, S., Trouve, G. and Renon, H., Water transfer in emulsified liquid membrane processes. J. Memb. Sci., **20**, 167-187.

19. Calderbank, P.H. and Moo-Young, M.B., The continuous phase heat and mass-transfer properties of dispersions. Chem. Eng. Sci., **16**, 39-54.,

20. Wilke, C.R. and Chang, P., Correlation of diffusion coefficients in dilute solutions. A.I.Ch.E.J., 1, 264-270.

VORTEX FLOW FILTRATION FOR ULTRAFILTRATION OF PROTEIN SOLUTIONS

CHARLES L. COONEY, ULRICH HOLESCHOVSKY AND GOPAL AGARWAL
Biotechnology Process Engineering Center and
Department of Chemical Engineering
Massachusetts Institute of Technology
Cambridge, MA 02139 USA

ABSTRACT

The induction of secondary flow by Taylor vortices during ultrafiltration of protein solutions provides a means to enhance this unit operation. Vortex flow filtration is readily implemented by placing a membrane on a high speed rotating cylinder in a narrow annular gap. The resulting increase in mass transfer coefficient makes the filtration dependent on membrane properties such that filtration flux is controlled by membrane permeability and protein transmission is enhanced.

INTRODUCTION

The major problem in membrane filtration is concentration polarization leading to the formation of a concentrated protein layer of macromolecules on the surface of the membrane. This concentration polarization (CP) layer becomes the primary barrier *e.g.* resistance and molecular sieve, to filtration. As a consequence, performance of membrane filters becomes relatively insensitive to the type of membrane. For these reasons, much of the research on membrane filtration has focused on approaches to minimize the CP layer.

Cross flow filtration (CFF) has evolved as the most commonly used method to minimize CP; by increasing the velocity across the membrane surface, it is possible to reduce CP and enhance filtration flux. CFF is readily achieved in either plate and frame or hollow fiber configurations [1] The problem is that to achieve high velocity, it is necessary to increase the pressure. This leads to greater convective transport of macromolecules to the membrane surface and enhancement of CP; the result is a steady state between convective transport of macromolecules to the membrane and back diffusion to the bulk solution. When this occurs, the system performance is independent of pressure and operates at the maximum flux. The formation of the CP layer, not only restricts flow through the membrane, but also acts as a molecular sieve controling molecular selectivity of the membrane. The result is that a membrane with a nominal 100,000 Da cutoff becomes much "tighter" and will not easily pass molecules of much lower molecular weight.

An alternative approach to minimize CP employs the use of secondary flow to increase the mass transfer coefficient at the membrane surface to enhance removal of macromolecules transported to the membrane surface by convective flow. Vortex flow filtration (VFF) has emerged as an attractive means to minimize formation of a CP layer [2,3,4]. Secondary flow in VFF is established by rapid rotation of a cylinder in a narrow annular gap; a membrane may be mounted on one or both of the rotating or stationary cylinder surfaces. The result is formation of stable counterrotating Taylor vortices in the gap [5]. Vortex formation is independent of pressure and relatively insensitive to axial flow. As a result, it is possible to increase the mass transfer coefficient at the membrane surface without increasing the pressure. This raises two interesting questions: Can pressure be used to increase flux beyond the values for onset of the pressure independent region in CFF, and by minimizing CP layer formation, does the membrane play a greater role in controlling selectivity in molecular transmission?

MATERIALS AND METHODS

A Benchmark™ filtration system (Membrex, Garfield NJ) was used for this work. The system was adapted to allow temperature control of the stationary cylinder with a jacket. Figure 1

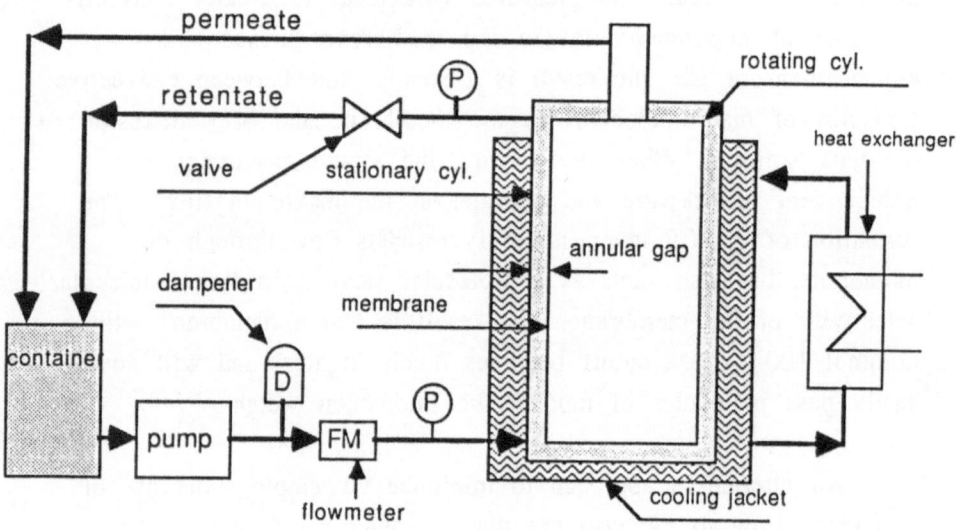

Figure 1. The experimental set-up for VFF using a Membrex Benchmark system.

depicts the experimental unit. The feed is pumped into the annular gap between the two concentric cylinders. The inner cylinder rotates, while the outer one is stationary. The pressure was measured at the entrance and exit of the filtration unit by means of test gauges. The pressure drop along the membrane was negligible (< 0.2 bar). The feed rate was monitored by a flowmeter while the permeate flux was measured manually, by measuring the time necessary to collect a volume of permeate. Both permeate and retentate were recycled in order to maintain the bulk concentration constant. As a further condition for a differential operating mode, the ratio between feed and permeate had to be large enough not to change the bulk concentration in the axial direction of the cylindrical membrane. The ratio between feed and permeate was chosen to be at least 25. The characteristics of the system were as follows: radius of rotating cylinder, 2.23 cm; length of membrane cartridge, 17 cm; gap width 2.15 mm. The operating conditions were: transmembrane pressure as indicated in the text; temperature of filtered solution: 20°C. The experiments were performed with Membrex Ultrafilic MX-10 and MX-100 membranes with nominal molecular weight cutoff values of 10k Da and 100k Da.

These studies were done with a defined single protein system to examine the effect of operating parameters on flux and molecular transmission during VFF. Transmission is the ratio of protein concentration in the permeate over the feed. Bovine serum albumin (69,000 Dalton) was used as a model protein for membrane filtration studies and provides a frame of reference for comparison with other work in the literature. Bovine serum albumin (BSA) at 98-99% purity was purchases from Sigma (A 7906) and suspended at pH = 7.4, in 0.15 M NaCl, 0.01 M phosphate buffer at concentrations indicated in the text. Protein concentration was monitored by absorbance at 280 n m.

RESULTS

Secondary flow in VFF is caused by the rotation of a cylinder in a narrow annular gap, thus a primary operating parameter is the rotation speed, N. This was varied and its effect on filtration flux examined as shown in Figure 2. The experiment was done by increasing the rotation rate at a fixed operating pressure. The results show that an increase in rotation rate permits one to achieve higher flux.

The effect of protein concentration on performance is seen in a separate experiment in which BSA concentration was increased up to 180 g/liter. The results are seen in Figure 3. In this case a MX-10 membrane was used. Because the viscosity increased with protein concentration, the rotation rate was adjusted to maintain the tangential Reynolds number constant. The flux performance is quite high for concentrated protein solutions and decreases as the protein

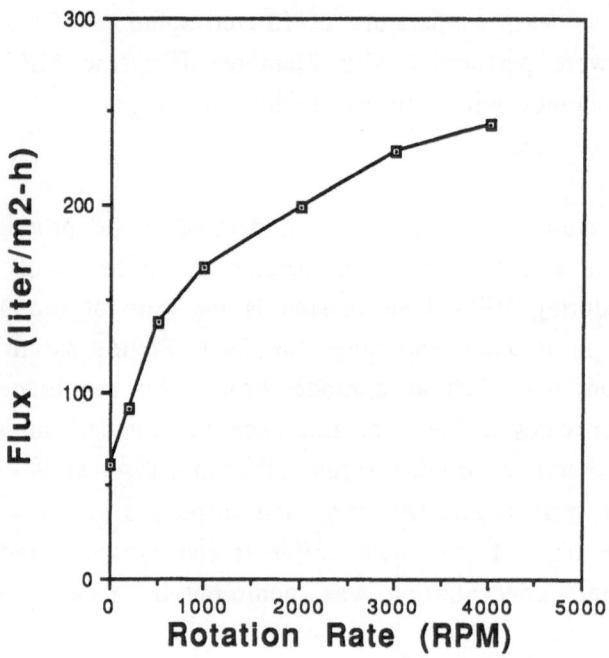

Figure 2. The effect of rotation rate on filtration flux with 1 g/liter BSA solution at 0.67 bar with a MX-100 membrane.

concentration is increased. Furthermore, one can see the earlier onset of the pressure independent region as the concentration is increased. In these experiments, both permeate and retentate were recycled so as not to change the protein concentration during the experiment.

The high flux values in VFF are believed due to reduction of the CP layer, thus, it is interesting to ask: is the CP layer reduced sufficiently such that molecular transmission is controlled by the membrane rather than the CP layer?   This question was addressed by a series of experiments examining the effect of pressure and rotation rate on transmission of BSA through a MX100 membrane. The results are presented in Figure 4. As pressure increased, BSA transmission also increased

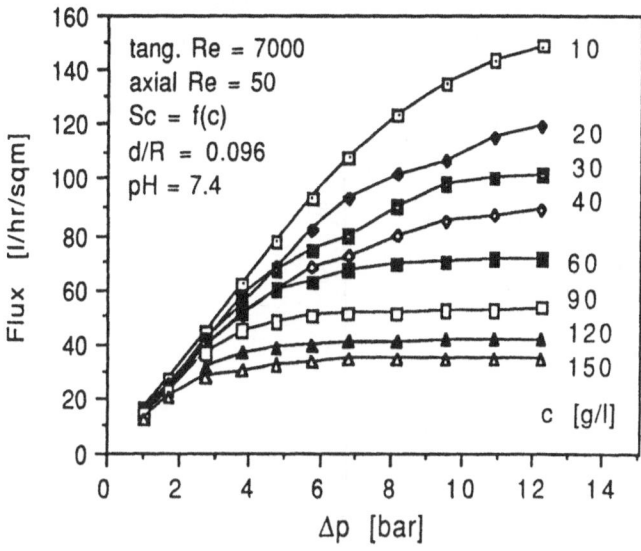

Figure 3. The effect of increasing BSA concentration on filtration flux with a MX-10 membrane (see text for further details)

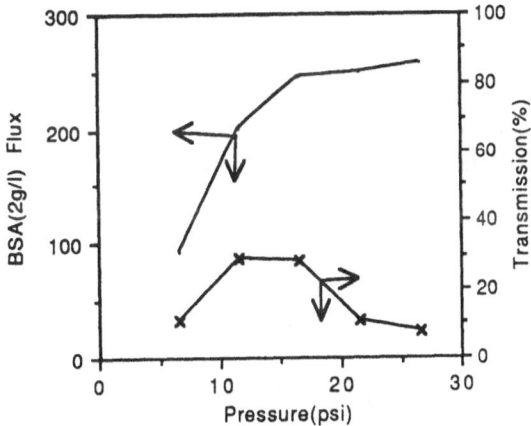

Figure 4. The effect of increasing pressure with a rotation rate of 2000 rpm on filtration flux (liter/m^2-h) and protein transmission during ultrafiltration of 2 g/liter of BSA.

as long as the system performance was linearly dependent on pressure; when the flux began to fall with increasing pressure the transmission also declined. As seen in Figure 5, the maximum transmission is dependent on the rotation rate; when rotation is increased to 3000 rpm, the maximum transmission increases to 44%.

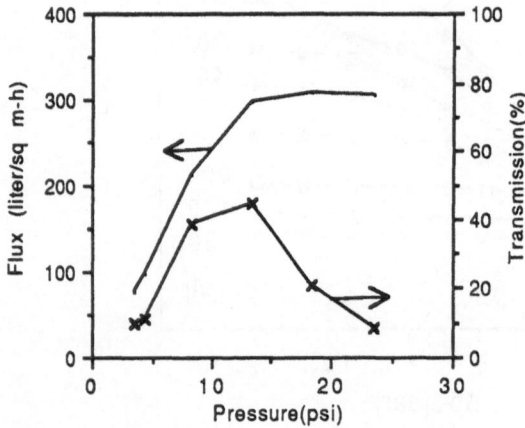

Figure 5 The effect of increasing pressure with a rotation rate of 3000 rpm on filtration flux and protein transmission during ultrafiltration of 2 g/liter of BSA.

DISCUSSION

The benefits of VFF illustrated above are the result of an increased mass transfer coefficient at the membrane surface caused by the Taylor vortices. Holeschovsky and Cooney [6] have shown that the mass transfer coefficient can be correlated with operating and design parameters with the following equation:

$$Sh = 0.75\ Re_t^{0.5}\ (2d/R)^{0.42}\ Sc^{0.33}$$

Where $Sh = k2d/D$ is the Sherwood number, $Re_t = vR2d/v$ is the tangential Reynolds number and $Sc = v/D$ is the Schmidt number. From this relationship, it is seen that the primary operating parameter is rotation rate as expressed in the tangential Reynolds number. It is important to note that when working with viscous solutions, as in the case shown in Fig. 3, to keep the tangential Reynolds number constant the rotation rate must be altered. There is little impact of axial flow rate; as a consequence one of the benefits of VFF is that it is not necessary to use high velocities across the membrane. This reduces the need for pumping and the likelihood of shear mediated damage.

Perhaps the most surprising finding is the observation that molecular transmission is dependent on the membrane at pressures where flux is dependent on pressure. There is a maximum value for the transmission. As the pressure is increased, convective transport of protein to the surface increases; at a point when accumulation of protein in the concentration polarization layer becomes significant, this layer is the primary barrier and less protein reaches the surface of the membrane. Further increase in protein leads to increased concentration polarization and less transmission.

By minimizing CP, filtration performance becomes more dependent on the membrane, its surface properties and permeability. Surface properties are important in preventing the initiation of fouling and loss of product through adsorption. Furthermore, membranes with high permeability and not restricted by CP are likely to provide high sustained flux values during ultrafiltration for concentration and diafiltration. The control of transmission selectivity by the membrane rather than by the CP layer should improve their effectiveness in diafiltration.

CONCLUSIONS

Secondary flow induced by rotating a cylinder inside a stationary cylinder increases mass transfer at the surface of a membrane mounted on the rotating cylinder; the result makes filtration performance more dependent on the membrane properties and allows the full potential of the membrane to be exploited. For these reasons, VFF provides and effective means for enhancement of ultrafiltration leading to improved filtration flux and protein transmission.

ACKNOWLEDGEMENTS

We acknowledge support of the National Science Foundation through the Biotechnology Process Engineering Center and Membrex, Inc. , Garfield, NJ. We are also grateful Dr. J. Hildebrandt and Dr. P. Rolchigo of Membrex, Inc for their insightful discussions during this work.

REFERENCES

1. Cheryan, M., Ultrafiltration Handbook, Technomic Publishing Co., Lancaster, PA, 1986.

2. Kroner, K. H. and H. Hustedt, Separation of enzymes from cell homogenates by means of a dynamic filtration. Dechema Biotechnology Conferences 1 - VCH Verlagsgesellschaff 1988, pp 477-483.

3. Hildebrandt, J. R. and Saxton, J. B., The use of Taylor vortices in protein processing to enhance membrane filtration performance, Bioprocess Engineering Colloquium Ed. Dean, R. C. and Nerem, R. M., American Society of Mechanical Engineers, New York, pp. 93-96, 1987.

4. Rushton, A. and Zhang, G. S. , Rotary Microporous Filtration of Organic and Inorganic Particulates. I. Chem. E. Series No. 113, European Federation of Chemical Engineering Publication N. 72, 1989.

5. Lopez-Leiva, M., Ultrafiltration at low degrees of concentration polarization: technical possibilities, Desalination, vol. 35, 115-128, 1980.

6. Holeschovsky, U. and Cooney, C. L. ,Quantitative Description of Ultrafiltration in Rotating Filtration Device. <u>Am. Inst. Ch. Engr. Journal</u>, (submitted).

NOMENCLATURE

d   gap width (m)

D   Diffusivity (m^2/s)

k   mass transfer coefficient (m/s)

R   inner cylinder radius (m)

v   axial fluid velocity (m/s)

v   kinematic viscosity (m^2/s)

MEMBRANE FILTRATION:
A *PRIORI* PREDICTIONS OF THE IDEAL LIMITING FLUX

Robert Field
School of Chemical Engineering
University of Bath
Bath BA2 7AY

ABSTRACT

The present theoretical work explores the effect that the modification of fluid properties (principally viscosity) at the membrane surface plays in determining flux. The first achievement is to show that a limiting flux exists independent of any supposed gelation or osmotic pressure effects.

Secondly, the flux, viscosity, pressure drop equation has been developed by distinguishing between the pressure drop required to drive fluid into the pore entrance and that required to drive fluid through the pore itself. The resulting equations predict the same limiting flux as that referred to above. In addition, *a priori* predictions of the flux-transmembrane pressure drop relationship can be obtained. Comparisons with experimental data are in progress.

INTRODUCTION

Many separation processes based on artificial membranes suffer from the consequences of a particular phenomenon. As with some heat exchangers, a minute fraction of the fluid (the boundary layer) has transport properties (diffusivity, viscosity, density) very different from those of the bulk. In particular there may be a large viscosity gradient. Even though the hydrodynamics are principally determined by bulk conditions, radial heat or mass transfer is dependent upon conditions in the boundary layer. When these conditions differ significantly from bulk conditions the classical chemical engineering correlations must be altered to account for this situation.

The aim of this paper is two-fold. Firstly, it is to summarise a method to account for variations in mass transfer coefficients in ultrafiltration due to variations in the transport properties in the boundary layer. With certain assumptions about the dependency of viscosity on concentration, it follows that a limiting flux exists independently of any supposed gelation effects. Secondly a new flux-pressure drop relationship which is consistent with the limiting flux predictions of earlier work is developed.

BACKGROUND

In 1984, Clifton *et al*[1] improved a previous integral model by Leung and Probstein[2]. They introduced the concentration dependence of the viscosity in the boundary layer. More recently, Aimar *et al*[3] have shown that the variation in mass transfer coefficient for proteins was mostly due to the viscosity variation of proteins with concentration: the concentration and thus the viscosity being greater at the wall. Their approach was based on using the properties at the wall to calculate the dimensionless numbers.

Recently, Gekas and Hallström[4] and Gekas and Olund[5] extensively reviewed the mass transfer and heat transfer corrections and suggested a correction factor of $(Sc/Sc_w)^{0.11}$ for the evaluation of mass transfer coefficients in membrane units subject to turbulent flow. However, this is highly questionable because it is based not on a heat transfer analogue for cooling conditions where $\mu_w > \mu_b$ but on a factor developed for heating conditions where $\mu_w < \mu_b$. Obviously this does not parallel the conditions in membrane units where $\mu_w \gg \mu_b$. Gnielinski[6] work in which a single equation incorporating both a correction factor for the entry length and the term $(Pr/Pr_w)^{0.11}$ was used to justify the Schmidt number ratio. However the term $(Pr/Pr_w)^{0.11}$ had been introduced *inter alia* by Hufschmidt and Burck[7] and by Yakovlev[8] who were concerned with heating duties. Indeed the former authors quote a range of $1 < Pr/Pr_w < 28$. This restriction to heating duties was ignored by Gnielinski.

A critical reading of the literature suggests that there is no reason to alter the conclusions of Petukhov[9] that n = 0.11 for heating duties ($\mu_w/\mu_b < 1$) and n = 0.25 for cooling duties ($\mu_w/\mu_b > 1$). In support of the latter mention can be made of Mikheev's[10] recommendation that the term $(Pr_b/Pr_w)^{0.25}$ be used. Furthermore recent work by Field[11] has provided a theoretical basis for a correction factor that is applicable for $\mu_w > \mu_b$. The approximate form is:

$$Nu/Nu_o = (\mu_b/\mu_w)^{0.27} \tag{1}$$

and with regard to heat transfer, it is in good agreement with previous experimental work. The predicted exponent depends slightly on the viscosity ratio, and for $\mu_w \gg \mu_b$ the following more complex expression of the correction factor is applicable:

$$F = \left\{ \frac{\mu_b}{\mu_w} \frac{4(6e^\alpha - \alpha^3 - 3\alpha^2 - 6\alpha - 6)}{\alpha^4} \right\}^{\frac{1}{3}} \tag{2}$$

where α reflects the change of viscosity with respect to temperature.

Naturally these equations are also relevant to mass transfer problems. The relatively large size of the correction factor has been shown to be significant in determining the theoretical limiting flux.

The approach adopted by Aimar and Field[12] for the calculation of limiting flux is similar to that proposed by Aimar and Sanchez, but it is no longer referred to the osmotic model. The dimensionless groups are calculated using the bulk properties, with a correction factor for the mass transfer coefficient, that is theoretically justified, being introduced.

PREVIOUS THEORETICAL DEVELOPMENTS CONCERNING LIMITING FLUX

The key work may not be available when this paper is first read and so a short summary is included.

In the following analysis no particular relationship between mass transfer coefficient and viscosity is initially assumed. Nevertheless it is shown that, given certain assumptions about the dependency of viscosity on concentration, a limiting flux exists independently of any supposed gelation effects.

Now given that film theory is valid

$$J_v = k \ln \left\{ \frac{C_w}{C_b} \right\} \tag{3}$$

Experimentally J_v has been found to be a function of membrane (wall) concentration, C_w. It follows that a limiting flux will correspond to a zero of the derivative dJ/dC_w. Differentiation of (3) for a given value of C_b gives:

$$\frac{dJ_v}{dC_w} = \frac{dk}{dC_w} \ln \frac{C_w}{C_b} + \frac{k}{C_w} \tag{4}$$

The theoretical condition for a limiting flux is given by $dJ_v/dC_w = 0$, ie

$$\frac{1}{k} \cdot \frac{dk}{dC_w} = - \frac{1}{C_w \ln \frac{C_w}{C_b}} \tag{5}$$

Substituting for k using an equation of the form

$$k = k_0 (\mu_b/\mu_w)^z \tag{6}$$

gives a practical version of the limiting condition (ie equation (5)):

$$- \frac{z}{\mu_w} \left\{ \frac{d\mu}{dC} \right\}_w = - \frac{1}{C_w \ln \frac{C_w}{C_b}} \tag{7}$$

The wall concentration that arises from this equation depends on the bulk concentration, C_b, on the wall viscosity, μ_w, and on the sensitivity of the mass transfer coefficient to the viscosity gradient (ie value of index z). Such an equation can be solved numerically for each case, provided that the variations in viscosity with respect to concentration are known.

In certain macromolecules, Aimar et al[3] have shown that simple relationships of the following form are applicable.

$$\mu = a(C)^v \tag{8}$$

For the conditions at the membrane surface, this relationship gives,

$$\frac{1}{\mu_w} \left\{ \frac{d\mu}{dC} \right\}_w = v/C_w \tag{9}$$

Substituting for $d\mu/\mu_w$ in (7) gives:

$$\frac{zv}{C_w} = \frac{1}{C_w \ln \frac{C_w}{C_b}} \tag{10}$$

The value of C_w from equation (8) corresponds to the wall concentration when the limiting flux is reached. This is as follows:

$$C_{w,lim} = C_b \exp(1/vz) \tag{11}$$

On substituting for C_w in the film equation, a value for the limiting flux is obtained. This is:

$$J_{lim} = k/vz \tag{12}$$

Using equations (6) and (8) the following is true for limiting conditions: $k = k_o(C_b/C_{w,lim})^{vz}$. Then, by use of equation (11), equation (12) can be further developed to give:

$$J_{lim} = k_o/\{vz \exp(1)\} \tag{13}$$

As mentioned above, recent theoretical developments suggest that for membrane filtration (where $\mu_w > \mu_b$) z should be taken to be 0.27.

NEW THEORETICAL DEVELOPMENTS

A number of workers have modelled the flux-transmembrane pressure relationship in a manner which ignores the variation of viscosity with concentration. The following is due to Fell and co-workers[13]:

$$J_v = \frac{1}{\mu} \frac{(P-\Pi)}{R_m + R_{ret}} \tag{14}$$

In the following analysis a distinction is made between the pressure drop through the pore, p_{por}, its associated resistance, R_{por} (for which the appropriate viscosity is μ), and the pressure drop in the boundary layer/pore entrance region, p_{sur}. The associated resistance is R_{sur} and the viscosity in this region is μ_w. The transmembrane pressure is the sum of p_{por}, p_{sur} and the osmotic pressure Π.

$$p = \Pi + p_{por} + p_{sur} \tag{15}$$

The flux through the surface boundary layer and the pores is given by:

$$J_v = \frac{1}{\mu_w} \frac{p_{sur}}{R_{sur}}$$

$$= \frac{1}{\mu_w} \cdot (p - \Pi - p_{por})/R_{sur} \tag{16}$$

Also $\qquad J_v = \frac{1}{\mu} \cdot p_{por}/R_{por} \tag{17}$

Thus $\qquad J_v = \frac{1}{\mu_w} \cdot (p - \Pi - \mu J_v R_{por})/R_{sur}$

$$= \frac{p - \Pi}{\mu_w R_{sur} + \mu R_{por}} \tag{18}$$

The flux may be limited by the dependency of μ_w upon the wall concentration. This will increase as p, and hence J_v, increase. The limiting condition is found by differentiation.

$$\frac{dJ_v}{dp} = \frac{1}{\mu_w R_{sur} + \mu R_{por}} - \frac{(p-\Pi) R_{sur}}{(\mu_w R_{sur} + \mu R_{por})^2} \frac{d \mu_w}{dp}$$

$$= \frac{1}{\mu_w R_{sur} + \mu R_{por}} \left\{ 1 - \frac{(p-\Pi)}{\mu_w R_{sur} + \mu R_{por}} \frac{d \mu_w}{dp} \right\} \tag{19}$$

A relationship between viscosity and concentration is required in order to obtain a practical form of equation (19). Using

$$\mu_w = a \, C_w^v, \qquad \frac{d \mu_w}{\mu_w} = \frac{v}{C_w} d \, C_w$$

and noting that

$$dC_w/dp = \frac{d \, C_w}{d \, J_v} \cdot \frac{d \, J_v}{dp}$$

it follows that

$$\frac{d \, J_v}{dp} = \frac{1}{\mu_w R_{sur} + \mu R_{por}} \left\{ 1 - \frac{(p-\Pi) \, v \, \mu_w}{(\mu_w R_{sur} + \mu R_{por}) C_w} \frac{dC_w}{dJ_v} \cdot \frac{dJ_v}{dp} \right\} \tag{20}$$

Defining $R = R_{sur} + \mu R_{por}/\mu_w$, it follows that

$$\frac{d\ J_v}{dp} = \frac{1}{\mu_w\ R}\ \left\{1 - \frac{(p-\Pi)\ v}{R\ C_w}\ \frac{dC_w}{dJ_v}\cdot\frac{dJ_v}{dp}\right\} \tag{21}$$

Recalling (18):
$$J_v = \frac{p - \Pi}{\mu_w\ R_{sur} + \mu\ R_{por}} = \frac{p - \Pi}{\mu_w\ R}$$

and combining with equation (21)

$$\frac{d\ J_v}{dp} = \frac{1}{\mu_w\ R} - \frac{J_v\ v}{C_w}\ \frac{dC_w}{dJ_v}\ \frac{dJ_v}{dp}$$

$$= \{1/\mu_w\ R\}\ /\ \left\{1 + \frac{v\ J_v}{C_w}\ \frac{d\ C_w}{dJ_v}\right\} \tag{22}$$

Again film theory is assumed to be valid and equation (3) is used to eliminate the differial term on the right-hand side of equation (22).

From $J_v = k\ \ell n\ (C_w/C_b)$ one obtains:

$$\frac{d\ J_v}{d\ C_w} = \frac{dk}{d\ C_w}\ \ell n\ (C_w/C_b) + \frac{k}{C_w} \tag{23}$$

Following previous analysis:

$$k = k_o\ \left\{\frac{\mu_b}{\mu_w}\right\}^z = k_o\ \left\{\frac{C_b}{C_w}\right\}^{vz}$$

Hence

$$\frac{dk}{d\ C_w} = -\ \frac{vz\ k}{C_w} \tag{24}$$

and

$$\frac{d\ J_v}{d\ C_w} = \frac{k}{C_w}\ \left\{1 - vz\ \ell n\ \frac{C_w}{C_b}\right\}$$

$$= (k - J_v\ vz)/C_w \tag{25}$$

$$\frac{d\ C_w}{dJ_v} = \frac{C_w}{k - J_v\ vz} \tag{26}$$

Substitution into equation (22) yields an equation of practical importance.

$$\frac{d\ J_v}{dp} = \{1/\mu_w\ R\}\ /\ \left\{1 + \frac{v\ J_v}{k - J_v\ vz}\right\}$$

$$= \frac{1}{\mu_w \ R} \ \frac{k - vz \ J_v}{k + v \ J_v \ (1-z)} \qquad (27)$$

The limiting flux obtained from equation (27) is consistent with that suggested by the earlier work *ie*

$$\frac{d \ J_v}{dp} = 0 \quad if \quad J_v = k/vz$$

However, the new development also provides a model for the pre-limiting flux region.

DISCUSSION

From the seminal work of Porter[14] onwards the ultrafiltration flux from solutions of macromolecules has been found to be of the same order as the mass transfer coefficient (when expressed in consistent units), although this is rarely remarked upon. The mass transfer coefficient is obtained from the Lévêque or Dittus-Boelter relationships. These relationships permit one to predict k_o, and so the only unknown in equation (13) is v. Unfortunately the variation of viscosity with concentration has not been extensively studied. Very approximately, one can expect v to approach unity for dilute solutions, in which case (given z = 0.27), J_{lim} is predicted to approach 1.4 k_o, whilst for concentration solutions, v tends to 3.0, which gives a theoretical limiting flux of 0.45 k_o. These predictions are consistent with the trends reported by Porter[14] for albumin and carbowax, but the absence of viscosity- concentration data prevents a quantitative comparison.

In order to obtain numerical results for the effect of transmembrane pressure, integration of equation (27) requires knowledge of R (*ie* R_{por} and R_{sur}), and information on v, z and k. R_{por} will be a function of pore radius and length, and the number of pores whilst R_{sur} will be a function of pore spacing and pore radius.

For water, $R = R_{sur} + R_{por}$, because $\mu_w = \mu$. Thus for water:

$$J_v = \frac{P}{\mu \ R}$$

Hence a value of $R_{por} + R_{sur}$ can be obtained from knowledge of the clean water flux. This is the true membrane resistance.

Knowledge of R for a given membrane-fluid system at a particular value of μ_w will enable R_{por} and R_{sur} to be obtained separately. Once obtained, they should prove to be independent of the fluid system and solely a function of the membrane type and size. Comparisons with experimental data are in hand.

An approximate integration of equation (27) has been

Table 1   Dependency of dimensionless flux and transmembrane pressure upon concentration polarisation for $z = 0.27$, $v = 3.0$

$\dfrac{C_w}{C_b}$	$\dfrac{J_v}{k_o}$	$\dfrac{dJ_v}{dp} \cdot \mu_w R$	$p/(\mu_w R k_o)$	$p/(\mu_b R k_o)$
1.005	0.005	0.985	0.0025	0.0025
1.01	0.010	0.971	0.005	0.005
1.05	0.047	0.868	0.045	0.052
1.1	0.088	0.764	0.096	0.128
1.15	0.125	0.679	0.147	0.224
1.2	0.157	0.609	0.190	0.328
1.4	0.256	0.419	0.383	1.05
1.7	0.345	0.264	0.644	3.16
2.0	0.395	0.174	0.872	6.98
2.4	0.431	0.100	1.135	15.7
2.8	0.447	0.051	1.347	29.6
3.2	0.453	0.016	1.526	50.0
3.436	0.454	0.000	1.776	72.0

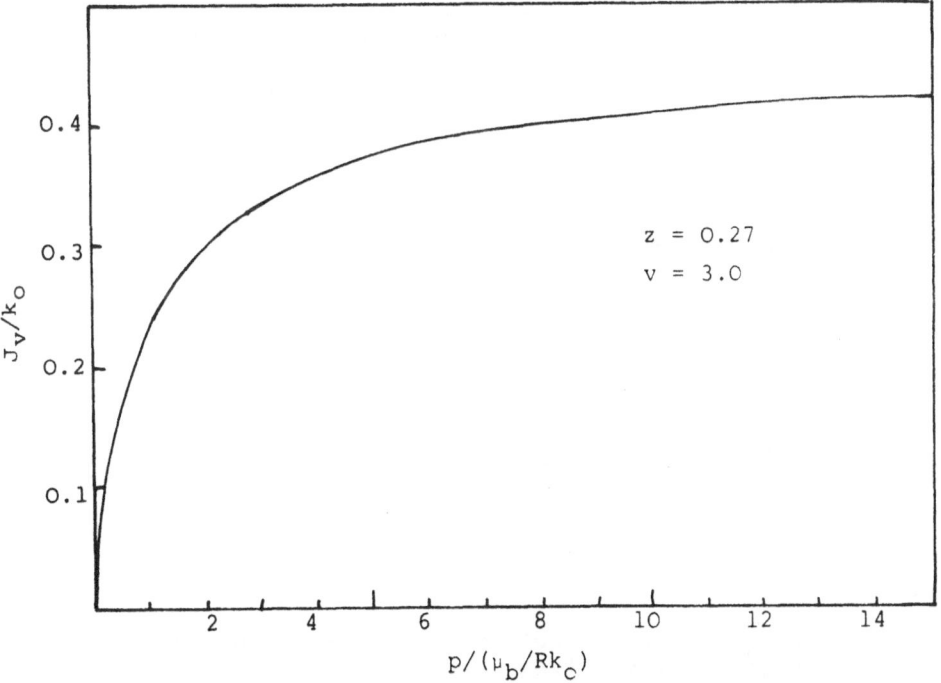

Figure 1   Plot of dimensionless flux against dimensionless pressure

completed for z = 0.27 and v = 3.0, the latter value being suggested by experimental data for Dextran. The results are displayed in Table 1 and Figure 1. The effect of v upon the results is currently being explored and will form part of the poster display.

CONCLUSIONS

1 The modification of viscosity at the membrane surface as a result of concentration polarisation has a significant effect upon mass transfer. The strength of the effect is such that a limiting flux exists independent of any supposed gelation or osmotic pressure effects.

2 By distinguishing between the pressure drop required to drive a fluid into a pore and that required to drive the fluid through the pore itself, the full role of the wall viscosity, μ_w, can be revealed. An equation for the pre-limiting flux region has been derived. With further development this should lead to useful predictive equations.

NOMENCLATURE

C	concentration
F	correction factor
J_v	volumetric flux through membrane
k	mass transfer coefficient
k_o	mass transfer coefficient at bulk conditions
Nu	Nusselt number
n	index on viscosity correction factor
P	transmembrane pressure difference
p	pressure difference
Pr	Prandtl number
R	hydraulic resistance
Sc	Schmidt number
v	index in equation (8)
z	index in equation (6)

Subscripts

b	bulk
lim	limiting
m	membrane
por	pore
ret	retentate
sur	surface
w	wall

Greek

α	constant in thermal or concentration boundary layer equation reflecting rate of change of viscosity with temperature or concentration
μ	viscosity
Π	osmotic pressure

REFERENCES

1   Clifton, M J, Abidine, N, Aptel, P and Sanchez, V,
    "Growth of the polarization layer in ultrafiltration with
    hollow-fibre membranes", *J Memb Sci*, 1984, <u>21</u>, 233-246

2   Leung, W F anmd Probstein, R F, "Low polarization in
    laminar ultrafiltration of macrocolecular solutions", *Ind
    Eng Chem Fund*, 1979, <u>18</u>, 274-

3   Aimar, P, Baklouti, S and Sanchez, V, "Membrane-solute
    interactions: influence on pure solvent transfer during
    ultrafiltration", *J Memb Sci*, 1986, <u>29</u>, 207-224

4   Gekas, V and Hallström, B, "Mass transfer in the membrane
    concentration polarization layer under turbulent cross
    flow: Part I", *J Memb Sci*, 1987, <u>30</u>, 153-170

5   Gekas, V and Olund, K, "Mass transfer in the membrane
    concentration polarization layer under turbulent
    cross-flow: Part II", *J Memb Sci*, 1988, <u>37</u>, 145-163

6   Gnielinski, V, "New equations for heat and mass transfer
    in turbulent pipe and channel flow", *Ind Chem Eng*, 1976,
    <u>16</u>, No 2, 359-368

7   Hufschmidt, W and Burck, E, "The effect of temperature-
    dependent properties of materials on heat transfer in the
    turbulent flow in liquids in tubes at high heat fluxes
    and Prandtl Numbers" (in German), *Int J Heat Mass
    Transfer*, 1968, <u>11</u>, No 6, pp1041-1048

8   Yakovlev, V V, "Local and average heat transfer in the
    turbulent flow of non-boiling water and high heat
    loadings" (in German), *Kernenergie*, 1960, <u>3</u>, No 10/11,
    pp1098-1099

9   Petukhov, B S, "Heat transfer and friction in turbulent
    pipe flow", in Advances in Heat Transfer, 1970, <u>6</u>, Ed
    Hartnett, J P and Irvine, T F Jr, 503-564

10 Mikheev, M A, Izv Aka Nauk SSSR Otd Tekhn Nauk, 1952,
    No 10

11 Field, R W, "A theoretical correction factor for heat
    transfer and friction in pipe flow", *Chem Eng Sci*, 1990
    (in press)

12 Aimar P and Field, R W, "Viscosity: some influences on
    mass transfer and limiting flux in membrane separations",
    Chem Eng Sci, 1990 (submitted for publication)

13 Suki, A, Fane, A G and Fell, C J D, "Flux decline in
    protein ultrafiltration", *J Memb Sci*, 1984, <u>21</u>, 269-283

14 Porter, M C, "Concentration polarisation with membrane
    ultrafiltration", *Ind Eng Chem Prod Res Develop*, 1972,
    <u>11</u>, 3, 235-248

CLARIFICATION OF TISSUE CULTURE FLUID AND
CELL LYSATES USING BIOCRYL® BIOPROCESSING AIDS

KATHLEEN FLETCHER AND SHERYL DELEY
TosoHaas
Independence Mall West, Philadelphia, PA 19105

ROBERT J. FLEISCHAKER, JR.
Vista Biologicals Corporation
2120-C Las Palmas Drive, Carlsbad, CA 92009-1523

IAN T. FORRESTER[1], ANTHONY C. GRABSKI AND W. NICK STRICKLAND
Protein Purification Facility
University of Wisconsin Biotechnology Center
1710 University Avenue, Madison, WI 53705

ABSTRACT

Bioprocessing aids are used as direct additives during clarification (centrifugation or filtration) prior to chromatography. They improve the flocculation process, remove pyrogens, color bodies and acidic impurities (nucleic acids and acidic proteins) that can foul chromatography columns. This paper describes three examples of the uses of Biocryl bioprocessing aids: (1) a study of the clarification of pseudomonas cell lysate using bioprocessing aids as an alternative to heat treatment following high speed centrifugation; (2) a study of the effect of pre-chromatography polishing on peptide purification from a recombinant strain of yeast; and (3) DNA removal in the purification of monoclonal antibodies produced in tissue culture.

[1] On leave from the Department of Biochemistry,
University of Otaga, P.O. Box 56, Dunedin, New Zealand

INTRODUCTION

It is well known that genetic engineering and molecular biology have become enabling technologies in the development of a biochemical manufacturing industry capable of producing therapeutic and diagnostic proteins. However, with every process stream, one is faced with the need to purify a product that may be present in only minute quantities from an often intractable culture fluid or lysate. It is only through efficient downstream processing that one can scale-up to volumes that make the production and purification of these proteins commercially viable. Enhancements in the early processing steps by removal of nucleic acids, pyrogens, particulate materials and contaminating proteins will ultimately improve the high resolution and very costly separation procedures employed for production chromatography to purify proteins.

The removal of contaminants from a variety of processing streams by flocculation has been reported in the literature by a number of investigators. A review article by Bell, et al [1] describes a number of materials used for protein precipitation. Jendrisak [2] provides a thorough description of the use of polyethyleneimine (PEI), and Agerkvist, et al [3] describes the use of chitosan (deacetylated chitin) as a flocculating agent. These investigators have shown that water-soluble, cationic polymers of high molecular weight and charge density are effective for the flocculation of proteins. However, Jendrisak points out that the use of acidic polyelectrolytes requires a low operating pH which may denature many enzymes. Chitosan must be solubilized before use and held at low pH to be maintained in solution.

Jendrisak provides some examples of PEI removing only contaminants from a crude broth and leaving the target protein in the supernatant following centrifugation. However, protein fractionation with PEI often requires recovery of the desired product from a PEI pellet by extraction with buffers at higher salt concentration. This approach adds centrifugation and redispersion steps to the purification strategy and may become cumbersome in large scale applications.

This paper describes the use of highly cross-linked, cationic polymeric particles as clarification aids for pre-chromatography polishing of crude process streams. The particles (Biocryl BPA–1000) have a nominal particle size of 0.1 micron, are shear stable, and maintain their positive charge over a broad pH range (pH 2-10). A high molecular weight, water soluble polycation (Biocryl BPA–5020) acts synergistically with the BPA–1000 as an additional clarification aid. These materials can be used alone or in combination to remove contaminating nucleic acids, particulate debris, color bodies and strongly anionic proteins from tissue culture fluid, cell lysates and extracts. Given the particulate nature of the BPA–1000, it can be readily used not only to enhance centrifugation but also in conjunction with filtration for concentration and diafiltration. These processing steps produce a clear process stream suitable for direct application to a chromatography column. Three examples of the use of these agents are presented to illustrate the diversity of systems with which they are compatible.

MATERIALS

Biocryl Bioprocessing Aids (TosoHaas Philadelphia, PA)

BPA–1000 (10% W/V aqueous solution) was used as supplied without further dilution. BPA–5020 (2% W/V aqueous solution) was diluted 1:49 with water to give a 0.04% W/V solids concentration before use. The exchange capacity of BPA–1000 is 2.6 meq/gm. BPA–5020 is not directly titratable.

Reagents

BCA reagent, bovine serum albumin, bovine gamma globulin, salmon DNA, and murine IgG were purchased from Sigma Chemicals (St. Louis). Bisbenzimide H33258 was purchased from CalBiochem (San Diego). Goat anti-mouse IgG was purchased from Crystal Chemicals (Chicago).

METHODS

The determination of IgG, total protein, DNA, and turbidity (OD610) were performed using an Instrumentation Laboratory Multistat MCA III+. Those assays were adapted from existing methodologies for that purpose.

Total Protein

Total protein was determined by a slight modification of the BCA method [5]. The values of the samples and the reagents were changed slightly for use on the Multistat: 10 µl of sample is mixed with 200 µl of the active reagent (BCA reagent mixed with 4% Copper Sulfate 50:1). The reaction is quantitated at 550nm, after a 20 min reaction at 37°C. The samples are referenced against bovine serum albumin, except when measuring IgG samples which are referenced to bovine gamma globulin. The assay is linear for protein concentration between 0.1 and 1.0 mg/ml in the sample solution.

Murine IgG

IgG concentrations were determined using a nephelometric procedure. That procedure was a slight modification of the method provided by Instrumentation Laboratories in their kit for the determination of Human IgG except that goat anti-mouse IgG was substituted for the precipitating antibody and murine IgG was used for preparing the calibration standards.

DNA

The DNA concentration of the samples was determined by quantitating the fluorescence that results when the Hoechst stain 33258 chelates with DNA. Samples or DNA standards are mixed (1:20) with a Tris Buffer (0.1 M Tris, 0.1 M NaCl, 1 mM EDTA) containing 1 µg/ml H 33258 and the fluorescence (460nm) resulting from excitation at 365nm is recorded. Quinine sulfate (3.5 M) is used as reference. The assay is useful over the range of 1 to 50 µg/ml DNA in the sample solution [6].

Turbidity

The turbidity of samples following treatment with the BPA's was evaluated by quantitating the optical density at 610nm.

Cells

Pseudomonas XA (ATCC 29574) were grown in fermentation equipment ranging in size up to 7,500 liters. The cells were grown in tryptic soy broth at 28°C. Oxygen tension was controlled to between 20 and 40% of air saturation. The culture was harvested after 36 hours by centrifugation. The cell paste was stored at −70°C [4].

IgG was produced by hybridoma cells grown in DME plus 5% fetal calf serum at 37°C in a CO_2 humidified incubator.

A recombinant strain of Saccharomyces known to secrete a peptide with an approximate molecular weight of 2000 daltons, was grown by conventional fermentation procedures [7].

Treatment Optimization

A standardized, Biocryl bioprocessing aid (BPA) treatment optimization protocol [10] involved taking aliquots of broth in centrifuge tubes, adding BPA with good mixing and centrifuging. When both BPA–1000 and BPA–5020 were used, BPA–1000 was added first followed by the slow addition of BPA–5020.

Assessment of efficacy was dictated by the desired result and is discussed in the results section for each application.

RESULTS AND DISCUSSION

Clarification of Pseudomonas Cell Lysate

As the first step in preparation of an amino acid degrading enzyme from a pseudomonas cell strain, a procedure for clarifying the cell lysate was required. The original method of heat treatment followed by high speed centrifugation typically resulted in significant product losses of 30-45%. Furthermore, this method imposed a number of restrictions on the scale-up of this process from benchtop to pilot scale. Both diatomaceous earth and Cell Debris Remover (Whatman) were tested and found to be ineffective in producing clarification.

Using a combination of 5000 ppm Biocryl BPA–1000 and 67 ppm Biocryl BPA–5020 as described in Materials and Methods, it was possible to flocculate the lysate. The resultant flocs were then readily removed by either low speed centrifugation or by cross-flow microfiltration. By either method a clear supernatant or permeate was produced for final purification by chromatography. Enzyme losses in this one-step clarification were reduced to an acceptable level of between 5% and 10%. Figure 1 demonstrates the degree of clarification that was achieved.

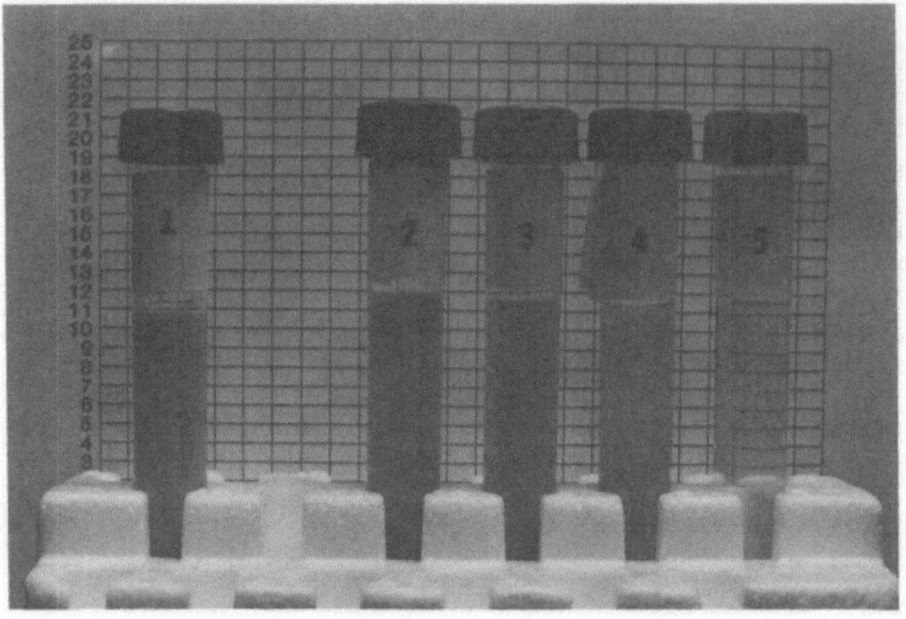

Figure 1.　　Pilot 1.　Starting Material
　　　　　　　Pilot 2.　2000 ppm BPA–1000:　　26.6 ppm BPA–5020
　　　　　　　Pilot 3.　3000 ppm BPA–1000:　　　40 ppm BPA–5020
　　　　　　　Pilot 4.　4000 ppm BPA–1000:　　53.4 ppm BPA–5020
　　　　　　　Pilot 5.　5000 ppm BPA–1000:　　66.6 ppm BPA–5020

Peptide Purification

A pale-brown colored culture medium was separated from the yeast cell mass by centrifugation followed by ultrafiltration through a 2,000 nominal molecular weight cut-off (NMWCO) membrane (flowchart Figure 2.). The retentate (R1) was discarded and the permeate (P1) concentrated by a second ultrafiltration [8] using a 500 NMWCO membrane. The retentate (R2), 50 ml., pH 6.7, derived from this second ultrafiltration procedure was dark brown colored (due to the substantial concentration) and contained the active peptide. In these studies, active peptide was quantitated by reverse phase high performance liquid chromatography (HPLC) [9] using the system detailed in Figure 3. Analysis of R2 by this HPLC system resolved the active peptide components into an elution region characterized by a cluster of peaks, with elution times between 16.5 min. and 19.6 min. (see line #1, Figure 3). The integrated area of these peaks was nominally defined as peptide units. Using this method of quantitation, it was established that R2 contained a total of 19,068 active peptide units.

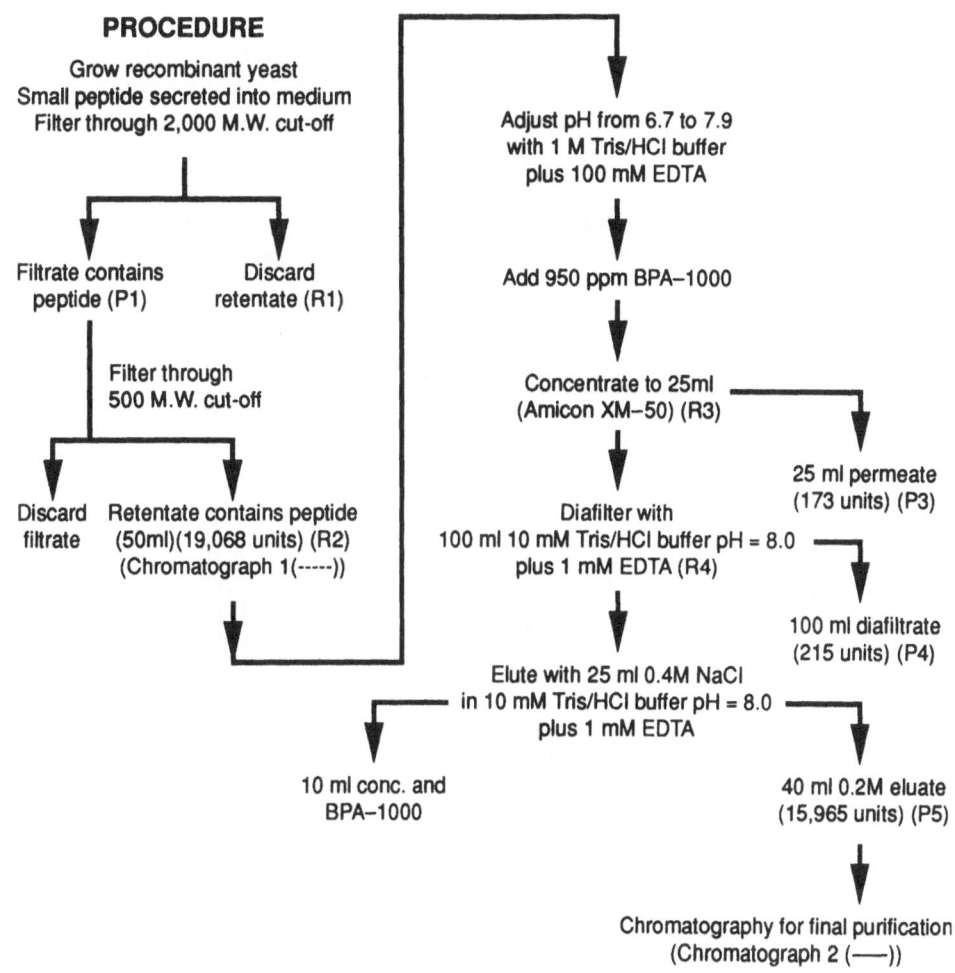

Figure 2. Peptide Purification Strategy

In order to achieve selective binding of the peptide to BPA–1000, (but not the brown colored material), the pH of R2 was adjusted to 7.9 by the addition of 1M Tris/HCl buffer plus 100 mM EDTA. Aliquots (1.0 ml) of R2 were titrated using a standardized protocol [10] in which BPA–1000 (5-200 µl) in plastic microfuge tubes (No. 72.690 Sarastdt, W. Germany) was mixed thoroughly, incubated for 5 min. and then centrifuged at 15,000 g for 2 min. The supernatants were withdrawn and assessed for active peptide as described above. Color was qualitatively assessed by visual examination. Optimum binding of peptide was achieved at 950 ppm of BPA–1000. Consequently, sufficient BPA–1000 was then added to bring the final concentration of bulk R2 to 950 ppm BPA–1000, and the mixture was stirred for 5 min. at 4°C. The mixture was then concentrated to 25 ml. using an Amicon stirred cell with an Amicon XM–50 (NMWCO 50,000) membrane.

The permeate (P3) contained brown colored material but only 173 peptide units. The concentrated retentate (R3) was then diafiltered with 100 ml of 10 mM Tris/HC1, pH 8.0, containing 1 mM EDTA. The permeate (P4) from the diafiltration also contained brown colored material but only 215 peptide units. The peptide was then eluted from the decolorized, BPA–1000 containing retentate (R4) by the addition of 25 ml. 0.4 M NaCl in 10 mM Tris/HC1, pH 8.0 containing 1 mM EDTA and recovered in the permeate (P5) obtained after subsequent ultrafiltration with an XM–50 membrane. The eluant containing 15,965 peptide units was then chromatographed under the same conditions as above (see line #2, Figure 3).

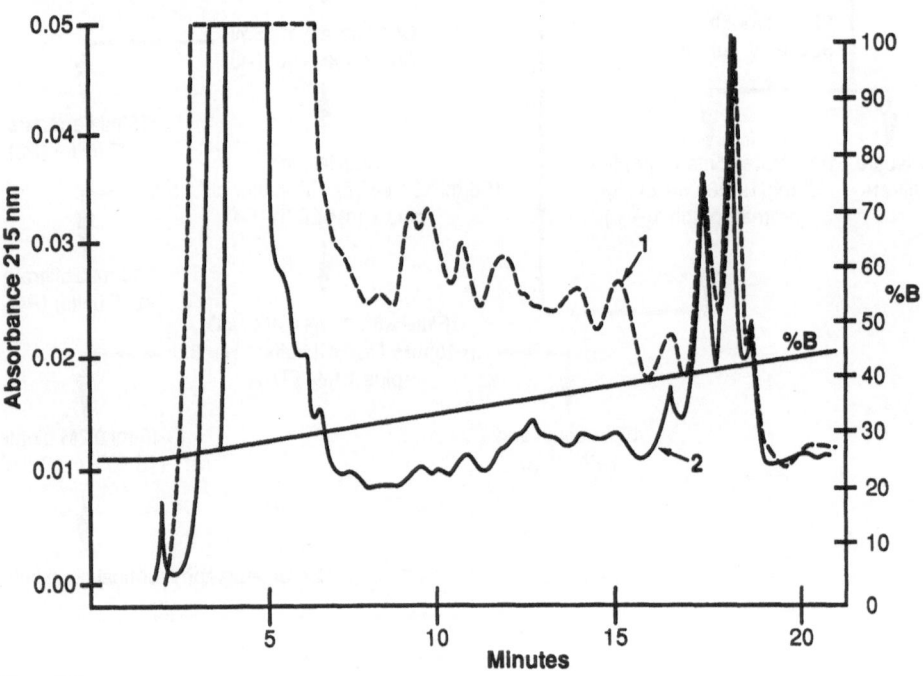

Conditions

Column: Dynamax C4 4.6 X 2.5 cm
Mobole Phase: A: 0.1% trifluoroacetic acid
B: 80% acetonitrite plus
0.09% trifluoroacetic acid
Sample Volume: 50 µl of sample indicated in flow diagram

Figure 3.

Figure 3 shows the chromatograms produced from retentate R2 (line #1) and permeate P5 (line #2) as described in Figure 2.

The absence of brown colored material in P5 is characterized by the significantly reduced absorption profile of the same reverse phase HPLC chromatogram after BPA–1000 treatment (line #2) as compared to the profile before BPA–1000 treatment (line #1). In particular, the BPA–1000 treatment has removed several discrete contaminants including the load–limiting contaminant eluting in a peak at 15.5 minutes.

DNA Removal From Tissue Culture Fluid

In the purification of monoclonal antibodies (MAbs) from tissue culture fluid, the use of cation exchangers and mixed ion exchange resins are becoming increasingly attractive. The MAbs are typically loaded onto the column under acidic conditions. However, one drawback is that upon acidification the crude sample often becomes turbid from the precipitation of DNA and proteins. This can greatly reduce column life and performance.

Pretreatment of the crude tissue culture fluid with BPA–1000 can greatly reduce or eliminate the turbidity that develops when the sample is acidified. The example shown in figure 4 shows the results of a standard titration of increasing amounts of BPA–1000 in crude tissue culture fluid to determine optimum pretreatment for the antibodies. At different doses of BPA–1000 (0-500 ppm active ingredient) the concentration of IgG, total protein and DNA were determined. OD610 was monitored as an indication of turbidity.

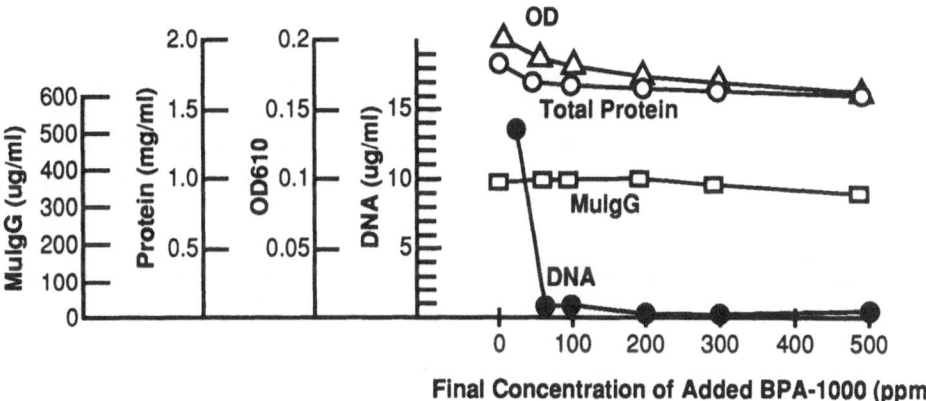

Figure 4. The BPA–1000 treatment optimization pilot study.

These results indicate that at low concentrations of BPA–1000 (50-200 ppm) there is selective removal of DNA and presumably other proteins with a strong negative charge. At higher concentrations, the BPA–1000 begins to show non-specific removal of total protein and with it IgG. It is possible with these results to target a treatment level at the highest level tolerable before the IgG losses become significant (>5%). In this case, BPA–1000 at a 200 ppm level removed 98% of the DNA with less than 4% loss of IgG.

CONCLUSIONS

It can be seen from the examples presented here and others in the literature [9–13] that Biocryl bioprocessing aids are effective reagents that enhance unit operations in downstream processing, producing polished streams for chromatography. Improvements achieved in the efficiency of centrifugation and filtration and the removal of potential foulants ultimately decrease yield losses during processing and improve the performance of the final chromatography steps for protein purification.

The potential value to an overall processing and purification strategy of utilizing a clarification aid comes from close examination of the benefits obtained as a consequence of decreasing the number of processing steps or increasing the efficiency of the mandatory procedures.

The pseudomonas cell lysate clarification can be seen as a generic example where cell debris, extraneous protein, nucleic acids and color bodies posed serious processing problems. Yield losses from the standard procedure of heat treatment were unacceptable. However, the use of a combination of BPA–1000 and BPA–5020 was effective in reducing the yield loss during clarification to an acceptable 5 to 10%. In addition to that obvious benefit, the need for cumbersome large scale heat treatment and high speed centrifugation were eliminated.

Peptide purification from a recombinant strain of yeast presented a unique challenge. The established process that generated a purified product was in operation. However, the efficiency of the reverse phase chromatography step was compromised by the presence of color bodies and a column load-limiting contaminant. BPA–1000 was shown to be effective in removing that contaminant and it is expected that the column loading can be increased 100-fold in scale-up.

Finally, the production of monoclonal antibodies in tissue culture is widely accepted. However, the inevitable presence of contaminating nucleic acids in cell-free, concentrated tissue culture fluid places in jeopardy any chromatographic media, be it for hydrophobic interaction, ion exchange or affinity chromatography. BPA–1000 at an optimized treatment level was shown to effectively remove 98% of the DNA while removing less than 4% of the IgG. The added lifetime of chromatographic media by the elimination of severe fouling by nucleic acid is significant.

Biocryl is a registered trademark of Rohm and Haas Company.

REFERENCES

1. Bell, D.J., M. Hoare and P. Dunnill, "The formation of protein precipitates and their centrifugal recovery" in <u>Advances in Biochemical Engineering and Biotechnology</u>, ed. A. Fiechter, Springer-Verlag, New York, 1982, p.1.

2. Jendrisak, J., "The use of polyethyleneimine in protein purification. In Protein Purification: Micro to Macro, Alan R. Liss, Inc., 1987, pp. 76-97.

3. Agerkvist, I. and S-O. Enfors, "Flocculation as a Separation Tool in Primary Recovery Stages: 32nd IUPAC Congress, Stockholm, Sweden, August 2-7, 1989.

4. Roberts, J. and Rosenfeld, H., Isolation, Crystallization, and properties of Indolyn-3-alkane alpha Hydroxylase, J. Biol. Chem. 255 (8), pp. 2640-2647 (1977).

5. Brown, R., Jarvis, K.L., Hyland, K.L. Protein Measurement Using Bicinchoninic Acid: Elimination of Interfering Substances, Anal. Biochem. 180, pp. 136-139 (1989).

6. Labarca, C. and K. Paigen, A simple, rapid and sensitive DNA assay procedure. Analytical Biochemistry, 1980, Volume 102, p. 344.

7. Corbett, K., "Design, preparation and sterilization of fermentation media" In Comprehensive Biotechnology, ed. A.T. Bull and H. Dalton, Pergamon Press, Oxford, 1985, Volume 1, pp. 127-139.

8. Le, M.S. and J.A. Howell, "Ultrafiltration". In Comprehensive Biotechnology, ed. C.L. Cooney and A.E. Humphrey, Pergamon Press, Oxford, 1985, Volume 2, pp. 383-438.

9. Hancock, W.S., "Introduction to high performance liquid chromatography". In Handbook of HPLC for the separation of amino acids, peptides and proteins, ed.l W.S. Hancock, CRC Press, Boca Raton, 1985, Volume 1, pp. 3-11.

10. Forrester, I.T., A.C. Grabski and W.N. Strickland, Bioprocessing aids in the purification of proteins. In Fermentation Technologies: Industrial Applications, ed., I.S. Maddox, Elsevier, Barking, England, 1989, in press.

11. Kim, C.W. and C.K. Rha, Application of submicron-sized polymeric particles in bioseparation processes. Enzyme Microb. Technol. 1987, 9, 57-59.

12. Cabral, J.M.S., E.M. Robinson and C.L. Cooney, Membrane filtration of cell culture media with charged particles, Patent number 4,830,753, May 16, 1989.

13. Forrester, I.T., A.C. Grabski, G. Mishra, B.D. Kelley, W.N. Strickland, G.F. Leatham and R.R. Burgess, Characteristics and N-terminal amino acid sequence of a manganese peroxidase purified Lentinula edodes culture grown on a commercial wood substrate. Applied Microbiology and Biotechnology, 1989, in press.

ECONOMIC FABRICATION AND UTILISATION OF AGAR COMPOSITES IN BIOSELECTIVE RECOVERY IN FIXED AND FLUIDISED BEDS.

Nigel B Gibson and Andrew Lyddiatt
Biochemical Recovery Group,
School of Chemical Engineering,
University of Birmingham, Birmingham B15 2TT, UK.

ABSTRACT

The assembly and characterisation of beaded forms of agar (100-500 μm) formed in simple, economic emulsification procedures is described. Such particles have been chemically cross-linked to enhance physical, temperature and pH stabilities, and have been subsequently derivatised with ion-exchange groups, textile dyes and blood antigens to form selective adsorbents. Materials have been characterised in single-step, fixed bed concentrations and/or purifications of albumin from solution and monoclonal antibodies from animal cell culture broths. The introduction of inert, dense particles during bead emulsification has yielded material suited to similar product recoveries from biological suspensions in fluidised bed contactors. Direct comparison has been made with conventional commercial chromatographic materials in terms of performance efficiency and cost effectiveness in realistic biorecoveries.

INTRODUCTION

Contemporary development of solid phases for affinity chromatography has concentrated upon improved hydrodynamic performance in fixed beds. Spherical, mono-dispersed beads, characterised by low flow resistance and dispersion, high rigidity and physical strength, and low non-specific binding, are commonly available [1]. Such have been developed to meet specifications judged relevant by the standards of classical chromatography. The manufacture of agarose, acrylamide or silica materials in homogeneous or composite fabricates has yielded costly solid phases largely restricted to fixed bed contactors and beyond the process budget of all but the highest value products.

Techniques of affinity chromatography may be readily described in terms of bioselective adsorption since chromatographic qualities of solid phases are rarely involved in most practical processes [2]. In the ideal state, products bind to reactive sites on solid phases to the exclusion of all impurities, and may be subsequently desorbed in a simple front of

available capacity. Conventional solute partitioning between mobile and
stationary phases, characteristic of ion-exchange, reversed phase and
other adsorptive chromatographies, is largely absent. The simplicity of
bioselective adsorption, and the stability of exploited interactions in
the face of extremes of pH and ionic strength, offer a diminution of
solid-phase specification in respect of refined, chromatographic geometry
and stringent chemical treatment which minimise non-specific interaction.
Such 'de-refinement' may offer sufficient cost-advantage to solid-phase
manufacture and assembly sufficient to justify the application of the
unique powers of bioselective adsorption to lower value products. The
relaxation of geometric specification also permits a move away from fixed
bed to fluidised bed contactors on a cost basis, wherein the tolerance of
suspended particles permits ready adsorptive recoveries from whole broths
or cell homogenates [3,4].

EXPERIMENTAL METHODS

Preparation of cross-linked agar and agar composite beads

Particles were fabricated with adaptions of reported methods [5,6,7]. A
4% solution of high gel strength agar (GPR grade, BDH) was obtained by
heating pre-swollen material in an autoclave for 45 minutes at $121^{\circ}C$. One
litre of hot agar ($90-95^{\circ}C$) was agitated with 1 litre of hot commercial
vegetable oil ($115-120^{\circ}C$) for 5 minutes. The preparation of composite agar
particles required the addition of 4% (w/v) TiO_2 (GPR grade, BDH) to the
hot agar with a 5 minute mixing period prior to addition of oil. The
system was cooled to $20^{\circ}C$ using an ice bath and the oil was displaced by a
series of wash steps. Thus: twice with ether (500ml), twice with acetone
(500ml) and twice with water (1l). Cross-linking exploited an
epichlorohydrin method [8]. The resultant beads were resuspended in 10mM
tris buffer pH7.5 (buffer A) containing 0.01% azide and stored at $4^{\circ}C$.

Physical characterisation of agar and agar composite beads

The diameters of spherical particles of agar were estimated using a
Malvern 3600E analyser. Sepharose CL-4B (Pharmacia, Sweden) facilitated a
comparison with a commercial product. The flow resistance of both types of
particles in fixed bed (10mm x 75 to 85mm) operations was estimated using
a pressure transducer ('Mediamate', Data instruments) connected upstream
of the column inlet. Record was made of back pressures applied by the
adsorbents and bed fittings in response to changes in operational flow
rates. Comparisons were made with Sepharose CL-4B and Macrosorb K4AX
(160-1000μm range). The latter, a 1:1 composite of kieselguhr and agarose,
was formerly manufactured by Sterling Organics. Comparisons of fluidised
bed expansion under various linear flow rates were estimated for the
composite agar and Macrosorb K4AX at various solid concentrations (5-20%
v/v) using a glass tower (2.15 x 50 cm; 180ml volume). A mercury manometer
was connected across the bed to measure pressure differences.

Weight determinations were made of the composite materials to assess
water and solid content. Samples of known weight (0.1-1.5g suction dried
weight) were heated to $120^{\circ}C$ for 2 hours, cooled and re-weighed. The beads
were then heated to $900^{\circ}C$ for 2 hours to remove all organic material,
cooled and re-weighed. Density determinations of the composite material
were estimated by a method of difference.

Assembly of adsorbents

Preparation of Cibacron blue F3GA (Sigma, England) derivatives of agar or
Sepharose exploited a method descibed by Dean *et al.* [6]. The amount of
immobilised Cibacron blue was determined as described by Chambers [9].
Immunoadsorbents were assembled by the covalent immobilisation of human
immunoglobulin (huIgG) to agar and Sepharose CL-4B, activated by
1,1'-carbonyldiimidazole (CDI) [10]. The huIgG was purified by anion
exchange fractionation [11] of outdated blood obtained from the National
Blood Transfusion Service, Birmingham. Activated adsorbents (60ml) were
washed and equilibrated in PBS, and mixed with 150ml huIgG (1.9 mg/ml) in
PBS (pH 7.5). The slurry was agitated at 4°C for 12 hours. After
displacement of unbound IgG with PBS, un-reacted groups were capped with
one volume of 0.3M ethanolamine and subsequently equilibrated for use in
PBS. Preparation of DEAE agar composites used the method of Peterson *et
al.*[12] and the small ion capacity was determined by titration with
aqueous 0.1M HCl.

Biochemical characterisation of adsorbents

Effective dissociation constants (Kd) and maximum binding capacity (qm)
were determined for the adsorption of purified preparations of human serum
albumin (HSA; 0.3-2.6mg/ml, buffer A) to dye adsorbents, bovine serum
albumin (BSA; 0.5-6mg/ml, buffer A) to DEAE adsorbents, and antihuIgG-fc
monoclonal antibodies (MAB; 5-55ug/ml,PBS) to huIgG adsorbents [13]. The
buffer systems were respectively buffer A pH 7.5 in the presence or
absence of 1M NaCl, buffer A, and PBS of varying sodium chloride
molarities. Equilibrium concentrations of HSA and BSA were estimated by UV
absorbance at 280 nm (HSA 1mg/ml=0.58 [14], BSA 1mg/ml=0.67 [15]) after 4
hours reacting at 25°C. MAB concentrations were determined by an ELISA
method [16] after 12 hour reaction at 4°C. The equilibrium binding
capacity (q^*) was estimated from the initial (Co) and equilibrium (C^*)
concentrations according to the method of Chase [17], assuming Langmuir
adsorption [18]. Values of Kd and qm were estimated from plots of C^*/q^*
against C^* according to the equation :

$$C^*/q^* = (C^* + Kd)/qm \qquad [17]$$

where the intercept on the x-axis gives Kd and the slope 1/qm.

Equilibrium binding capacities (qmx) were also estimated in fixed
beds (10 x 100 mm) challenged with excess protein. Adsorbents (5-6ml) were
equilibrated with the appropriate buffer (see above) at a flow rate of
1-1.5 ml/min. The dye adsorbents were challenged with 25ml HSA (2.5mg/ml)
or BSA (2.1mg/ml) and the huIgG adsorbents were challenged with 25ml
filtered animal cell culture medium containing MAB (50 µg/ml). After
washing all adsorbents were eluted with 2 column volumes of aqueous 3M
KSCN. HSA and BSA were dialysed overnight and the amount of protein
determined by UV absorbance at 280 nm. The MAB was de-salted by gel
filtration on a Sephadex G-15 column (2.6 x 40cm) equilibrated in PBS and
quantified by ELISA. Pre-equilibrated (buffer A) anionic exchangers (20ml)
were challenged under recirculating fluidised bed conditions with BSA
(4.5mg/ml; 500ml) at various flow rates for 4 hours at 25°C. The beds were
collapsed by flow reversal, washed with buffer A and protein eluted with
1M NaCl in buffer A. Recovery were estimated spectrophotometry at 280 nm.

Estimations of non-specific binding (NSB) were made in fixed and fluidised beds with appropriate adsorbents. Underivatised adsorbents (2-3 ml) were washed with buffer A containing NaCl (0, 0.05, 0.5 or 1M). Adsorbents were challenged with either BSA (2 mg/ml) or lysozyme (1.5 mg/ml) in buffer A at selected salt concentrations. Fluidised systems were subsequently collapsed and treated as fixed beds. Both system types were washed with buffer and eluted with 3M KSCN. After dialysis, the recovered protein was determined by UV absorbance at 280 nm (lysozyme 1mg/ml=2.55 [19]).

Composite anion exchangers were tested in fluidised beds for the adsorptive recovery of a bulk protein fraction from baker's yeast. Yeast slurry (500ml; 40% w/v, in 0.5M tris buffer pH 7.5) was disrupted by wet milling (Dynomill, KDL), adjusted to pH7.5 with 0.5M HCl and recycled for 2 hours at 50 ml/min through a single stage fluidised bed (2.15x50 cm) containing pre-equilibrated adsorbent (20ml). Fluidised beds were washed with buffer A to yield clear wash, before collapse by flow reversal and transfer to a fixed bed. After a further wash (5 column volumns), the bound material was eluted using a 0-0.75M NaCl linear gradient in buffer A. Collected fractions were analysed for protein and RNA by the Bradford [20] and orcinol [21] assays respectively.

RESULTS AND DISCUSSION

Preparation and application of adsorbents

Preliminary investigations serve to emphasise that refined chromatographic media are not mandatory requirements in successful biorecovery processes. Crude inexpensive agar particles have been produced by simple bead emulsification procedures which permit reproducible fabrication of cheap, agar solid phases characterised by prescribed physical and biochemical specifications. The agar particles were stable to a wide range of pH (pH 1-13), temperature (0-85°C) and solvent (acetone and 1,1 dioxane) environments used during their preparation and subsequent derivatisations.

The coupling of various ion exchange and affinity ligands to agar, whether by aqueous 'salting' methods for the dye, or by CDI activation and coupling, produced affinity adsorbents with lower ligand densities compared to Sepharose CL-4B (40-160 μm). Large particle diameter (350 μm) required longer reaction times to achieve equilibrium during ligand coupling, washing and adsorption procedures. In both cases, the presence of charged impurities on agar [22] may hinder, by repulsion or by shielding, the access of dye or CDI molecules to the available covalent binding sites at coupling. In the case of the composite agar particles, the degree to which DEAE reacted was lower than on Macrosorb K6AX by virtue of a higher percentage of agarose content in the keiselguhr-agarose. It is interesting that efforts to prepare Macrosorb K4AX by the same method as for 4% agar proved unsuccessful, as the rigid keislguhr backbone fractured.

Tests involving the productivity of albumin-dye adsorbent interaction in fixed beds, shown in Table 1, indicates that albumin recovery was a function of the ligand density and not the adsorbent material. While for similar tests for the MAB-IgG interaction there appears to be a slight increase for Sepharose over agar. Recoveries of BSA by composite ion-exchangers show that the degree of binding is inversely proportional to the flow rate.

The operation of anion exchange adsorbents in fluidised beds to recover yeast protein from milled yeast homogenates poses a severe test of fractional properties. Extracts contain protein, nucleic acid as well as charged cell debris. The protein and RNA profiles for fixed bed elution of previously fluidised beds are shown in Figures 1a and 1b. The agar-TiO_2 composite produced a clear separation of protein and RNA. In contrast the elution profile of Macrosorb K6AX was compromised by severe fouling with cell debris at adsorption which promoted coalescence of adsorbent particles. The shape of the protein peak eluted from agar-TiO_2 comfirms a narrower particle size range than for Macrosorb K6AX.

TABLE 1

Binding characteristics of agar and agarose adsorbents

Adsorbent material	Ligand (mg/ml$)	qmx (mg/ml$)	Productivity (Molar)	Total challenge (mg)	% recovery
Cibacron blue-HSA					
4% Agar	3.14	5.20	1.9×10^{-2}	62.5	41.6
4% Agar (1M NaCl)	3.14	1.89	6.9×10^{-3}	62.5	15.1
Sepharose	6.01	12.30	2.4×10^{-2}	62.5	59.0
Sepharose (1M NaCl)	6.01	4.67	8.9×10^{-3}	62.5	37.4
Cibacron blue-BSA					
4% Agar	3.14	2.56	1.0×10^{-2}	52.5	24.4
4% Agar (1M NaCl)	3.14	0.10	3.7×10^{-4}	52.5	1.0
Sepharose	6.01	8.06	1.1×10^{-2}	52.5	46.1
Sepharose (1m NaCl)	6.01	0.31	5.9×10^{-4}	52.5	4.8
huIgG-MAB					
4% Agar					
PBS (0.15M NaCl)	0.63	0.12	0.19	1.25	48.2
PBS (1.15M NaCl)	0.63	0.18	0.29	1.25	72.5
Sepharose					
PBS (0.15M NaCl)	1.09	0.15	0.14	1.25	60.3
PBS (1.15M NaCl)	1.09	0.19	0.17	1.25	76.4
DEAE BSA					
Agar composite					
50ml/min	0.05*	54.1		2250	48.1
70ml/min	0.05*	34.4		2250	30.6
Macrosorb K6AX					
50ml/min	0.13*	68.0		2250	60.4
70ml/min	0.13*	35.1		2250	31.2

($ in terms of settled adsorbent
  * in terms of milliequivalents/g(wet weight)

Agar is the impure source of agarose (arabinose chain), having an increased degree of charged groups. The magnitude of impurity is governed by the increased presence of amylopectin (branched arabinose chain with higher degree of charged substituents) in agar, and is reduced during refining which yields commercially available agarose [24]. In order to reduce NSB to the mono-dispersed particles, a range of ionic environments were imposed. At comparible flow rates the degree of protein interaction decreased with an increase in the salt content of the loading environment as shown in Table 3. The results indicate that the degree of NSB is higher for agar as compared to Sepharose at intermediate salt concentrations, whilst a uniform limit is reached in 1M NaCl. In the case of the fluidised system, the increase in flow rate has an advantageous effect on the reduction of NSB (data not shown). Decreased residence times in contactors clearly deminishes both specific and non-specific interactions. We conclude that the degree of NSB to all particle types can be reduced by manipulation of operating environments and may facilitate the use of less refined materials in bioseparations.

TABLE 2

Batch binding characteristics of agar and agarose adsorbents

Adsorbent material	ligand density (mg/ml$)	qm (mg/ml$)	Kd (M)	Molar binding ratio (M(adsorbate)/M(ligand))
Cibacron blue-HSA				
4% agar	1.76	1.08	1.2×10^{-5}	7.2×10^{-3}
4% agar (1M NaCl)	1.99	0.39	6.6×10^{-6}	2.2×10^{-3}
Sepharose	6.01	1.76	4.5×10^{-6}	3.4×10^{-3}
Sepharose (1M NaCl)	6.01	1.37	9.3×10^{-6}	2.6×10^{-3}
huIgG adsorbents				
4% agar				
PBS (0.15M NaCl)	0.63	0.09	4.37×10^{-8}	0.157
PBS (1.15M NaCl)	0.63	0.21	9.78×10^{-8}	0.340
Sepharose				
PBS (0.15M NaCl)	1.09	0.23	8.61×10^{-8}	0.207
PBS (1.15M NaCl)	1.09	0.09	8.99×10^{-8}	0.078
DEAE adsorbents				
agar composite	0.05*	96.4	1.66×10^{-6}	
Macrosorb K6AX	0.13*	140.1	6.06×10^{-6}	

($ in terms of settled adsorbent

* in terms of milliequivalents/g(wet weight)

Resistance to fluid flow was estimated under fixed bed conditions and typical response curves are shown in Figure 2. The mono-disperse agar was able to with stand greater linear flow rates than Sepharose, resulting from the larger bead size of agar and the compressability of the narrow ranged Sepharose. Comparison of the behavior of the composite materials indicate the benefit of a rigid particle. A hysteresis was noted in the spherical matrices, the extent is governed by the elastic nature of the species. The entrapment method of incorporation of TiO2 within agal

Biochemical and physical characterisation of adsorbents

Ion exchangers and immunoadsorbents were characterised by estimating the binding of BSA and HSA from pure solutions, and MAB from heterogeneous crude feedstock respectively. The maximum binding capacity (qm) and dissociation constants (Kd) were estimated and are displayed in Table 2. The values of qm estimated for huIgG adsorbents approximate to those observed in fixed bed operation (~0.1-0.2 mg/ml) for both bead types whilst variation in ionic environment had no major effect. The difference in estimated qm and qmx values for dye adsorbents infers that saturation under fixed bed conditions was more efficient than in batch mode. Multi-valent and protein-protein interactions are the cause of such increases. Total protein bound decreased for both adsorbent types with increased salt concentration of feedstocks. Leatherbarrow *et al.* [23] proposed that the dye interacts with the weak fatty acid binding site in all albumins, while HSA additionally exhibits strong dye binding at the bilirubin site. The Kd values for the albumin-dye system were observed to be two orders of magnitude lower than that found for a true immunoaffinity system. The effect of molar salt seems to indicate that the protein was bound via a stronger hydrophobic interaction. In the case of anionic exchangers, the qm values for Macrosorb K6AX were higher than for agar-TiO2 composites due to the increased site availability on the commercial adsorbent. The batch qm values were greater for both materials than the binding observed in the fluidised beds, probably due to changes in the flow characteristics in and around the beads. In fluidised systems, qmx values were reduced with an increase flow rate and decreased residence time.

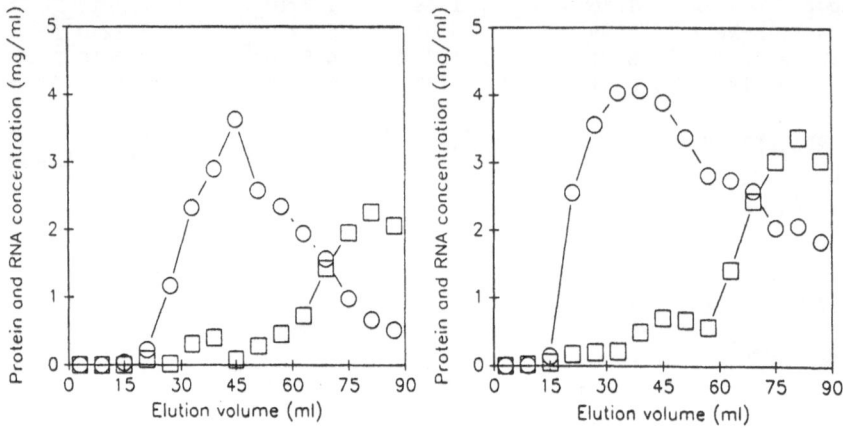

Figures 1 a and b. Elution profiles of the purification of Yeast proteins on a) agar-TiO2 and b) Macrosorb K6AX

Adsorption and elution conditions were applied as described in the text. Protein (○) and RNA (□)

In all cases the qm values for the purified agarose, whether it be as a mono dispersed or composite particle, tend to be higher. It would seem likely that contribution of impurities in agar plays a role in the hindrance of specific binding.

matrices reduces overall gel strength due to structural interferance. Attempts to incorporate more TiO2 hindered the gelation process to the extent that no bead formation occurred. Macrosorb composites by contrast are produced by incorporating agarose into a porous kieselguhr particle [25].

TABLE 3

Non-specific interaction of mono-dispersed particles at uniform flow rate

Adsorbent material	Conditions used	BSA bound (mg/ml$^\$$)	Lysozyme bound (mg/ml$^\$$)
4% agar	buffer A	0.48	1.20
	buffer A 50mM NaCl	0.21	0.30
	buffer A 0.5M NaCl	0.20	0.12
	buffer A 1M Nacl	0.07	0.03
Sepharose	buffer A	0.73	0.26
	buffer A 50mM NaCl	0.12	0.14
	buffer A 0.5M NaCl	0.09	0.07
	buffer A 1M NaCl	0.03	0.04

($ in terms of settled adsorbent)

The composite materials were tested for fluidised bed expansions. Estimated particle densities were 1.18 and 1.33 g/cm^3 for agar composite and Macrosorb resectively. Whilst it was found that Macrosorb contained 30.7% non-combustible material and agar composite was only 13.9%. The fluidisation of a particle depends on the balance of upward liquid flow and downward gravitation in relation to particle density and diameter. As expected, Macrosorb beds expanded to a lesser extent than agar composite under the similar linear flow conditions, as shown in Figure 3.

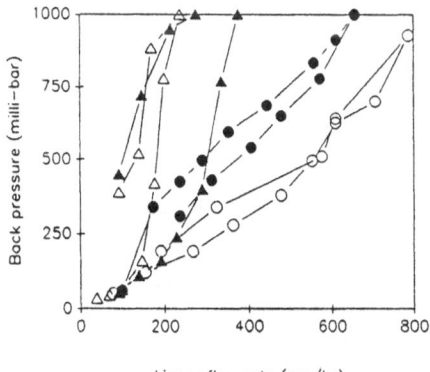

Linear flow rate (cm/hr)

FIGURE 2. Resistance to fluid flow in fixed bed adsordents

Resistance to flow as described in the text were applied to Sepharose CL4B (△–△), Agar (●–●), agar-TiO2 (▲–▲) and Macrosorb K4AX (○–○). Arrows indicate the direction of flow through fixed bed (10mm id x75 to 85mm).

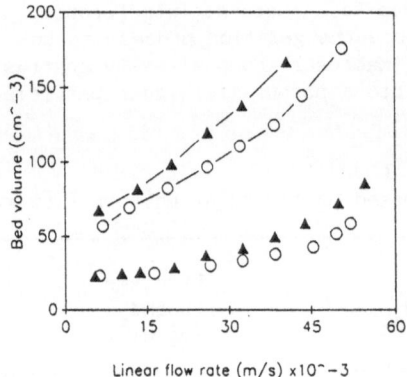

FIGURE 3. Bed expansion under linear flow in fluidised bed adsorbents.

Bed volumes of agar-TiO2 (◄ ◄ 5%v/v), ◄—◄ (20%)) and Macrosorb K4AX (○ ○ (5%), ○—○ (20%)) were measured under flow conditions as described in the text.

Conclusion

From this preliminary study we can conclude that agar and agar composites can be readily used in the construction of adsorbents for operation in fixed and fluidised beds. In comparing mono-dispersed materials, the large particle size of agar used in this study (100-500 μm) required the extension of solvent equilibrium times during selective adsorbent assembly. The presence of agarose of higher purity increased bulk capacities in both the dye and ion-exchange applications. This may be offset by the reduced cost of production and the ease by which particles of differing geometries can be fabricated to suit the requirements of the process [26].

The recovery of yeast protein by fluidised ion exchange emphasises that this type of contactor and adsorbent has not been used to its fullest potential. In particular, highly specific recovery is now possible directly downstream of the source of target materials, to the benefit of overall purification costs. We propose that further work on the design of solid phases and fluidised contactors will benefit the bioprocessing industry.

Acknowledgement

NBG acknowledges the financial support of an SERC studentship..

References

[1]Clonis Y.D., Large-scale affinity chromatography- considerations for scale-up, desirable adsorbent properties, applications. BIO/Technology. 1987,5, pp 1290-1293.
[2]Lyddiatt A., Kieselguhr-Agarose composites in adsorptive bioprocessing. 2nd Proc. Int. Conf. on Sep. Sci. Tech.. 1989,2, pp 497-504
[3] Wells C.M.,Patel K. and Lyddiatt A., Liquid fluidised bed adsorption in biochemical recovery and biological suspensions. In: Separation for biotechnol.(ed Verrall M.S.and Lyddiatt A.),Ellis-Horwood, Chichester, UK. 1987, pp 217-224.
[4] Wells C.M. PhD Thesis, University of Birmingham. (1989)
[5] Nilsson S. and Mosbach K., Entrapment of microbial and plant cells in

beaded polymers. Methods in Enzymology. 1987, 135, pp 222-230. Acadenic press, NY.

[6] Dean P., Johnson W.S and Middle F.A. In: Affinity chromatography- a practical approach . IRL press, oxford. 1985.

[7] Hjerten S., The preparation of Agarose spheres for chromatography of molecules and particles. Biochem Biophys Acta. 1964, 79, pp 393-405.

[8] Porath j., Janson J-C. and Laas T., Agar derivatives for chromatography, electrophoresis and gel-bound enzymes. J. Chromatog. 1971, 60, pp 167-177.

[9] Chambers G.K., Determination of Cibacron blue F3GA substitution in blue Sephadex and blue Dextran-Sepharose. Anal Biochem. 1977, 83, pp 551-559.

[10] Hearn M.T. and Bethell G.S., Application of 1,1'-carbonyldiimidazole-activated Agarose for the purification of proteins. J Chromatogr. 1979, 185, pp 463-470.

[11] Desai M.A. M. Phil. Thesis, University of Birmingham. (1987)

[12] Peterson E.A. and Sober H.A., Chromatography of proteins. I. Cellulose ion-exchange adsorbents. J Amer Chem Soc. 1956, 78, pp 751-753.

[13] Jefferies R., Lowe J., Ling N.R. and Servor S., Immunogenic and antigenic epitopes of huIgG. Immunology. 1982, 45, pp 71-78

[14] Hunter M.J. and McDuffie F.C., Molecular-weight studies on human serum albumin after reduction and alkylation of sulphide bond. J Amer. Chem. Soc. 1959, 81, pp 1400-1409.

[15] Everett W.W., The effects of metal ions on the isomerisation and dimerisation of bovine plasma albumin. J. Biol. Chem. 1963, 238, pp 2676-2682.

[16] Voller A., Bidwell D.Z. and Bartlett A., Enzyme immunoassays in diagnostic medicine: theory and practice. Bulletin of the World Health Organisation. 1976, 53, pp 55-65.

[17] Chase H.A., Prediction of the performance of preparative affinity chromatography. J. Chromatogr. 1984, 297, pp 179-195.

[18] Languir I., The constitution and fundamental properties of solids and liquids. J. Amer. Chem. Soc. 1916, 38, pp 2221-2235.

[19] Praissman C. and Ruply E.P., Comparison of protein structure in the crystal and in solution. Biochemistry. 1968, 7, pp 2446-2454.

[20] Bradford M.M., A rapid and sensitive method for the quantitation of microgram quantities of protein utilising the principle of protein-dye binding. Anal Biochem. 1976, 72, pp 249-254.

[21] Herbert D., Phipps D.J. and Strange D., Chemical analysis of microbial cells. In: Methods in Microbiology. 1971, 5B, pp 209-344. (eds. Morris, T.R. and Ribbons, D.W.), Academic press.

[22] Rees D.A., Structure, conformation, and mechanism in the formation of polysaccharide gels and netwoaks. In: Adv. Carbohyd. Chem. and Biochem. (eds. Wolfrom M.L. and Tipson R.S.,) Academic press, NY. 1969, 24, pp 267-295.

[23] Leatherbarrow R.J. and Dean P., Studies on the mechanism of binding of serum albumins to immobilised Cibacron blue F3GA. Biochem. J. 1980, 189, pp 27-34.

[24] Araki C., Some recent studies on the polysaccharides of Agarophytes. Proc. Int. Seaweed. Symp., 5th., 1966, pp 2-17.

[25] Bite M.G., Berezence S. and Reed F.J.S., Macrosorb kieselguhr-Agarose composite adsorbents. Applied Biochem. and Biotech. 1984, 18, pp 275-284.

[26] Porumb T., Porumb H. and Lascu I., Cibacron blue 3G-A-substituted cross-linked agar beads: an inexpensive, fast-flow affinity medium for large scale applications. J Chromatogr. 1985, 319, pp 218-221.

ELECTROMAGNETIC SEPARATION OF SMALL PARTICLES IN THE DIELECTRIC REGIME

KAREL GOOSSENS AND LEO VAN BIESEN
Department ELEC,
Vrije Universiteit Brussel VUB
Pleinlaan 2, 1050 Brussel, Belgium

ABSTRACT

The most important phenomena related to wet dielectric separation processes are resumed in this paper. From the frequency dependent polarizability it is shown that the dielectrophoretic response is sensitive to very fine differences in the electrical configuration of complex particles, such as cells. The electrophoretic behaviour of charged particles in strong divergent a.c. fields explains the repulsion of collected particles.

INTRODUCTION

Under the influence of a non-uniform electric field, a polarized particle will migrate along the field lines. The migration of the particles depends on their own effective dipole moment. The dipole force acting on the particles due to the induced charges is used in many practical applications, such as industrial filtration of liquids [1], and solid-solid separations [2]. Compared with electrophoresis, dielectrophoresis is much more sensitive to very fine differences in the electrical configuration of the particles. For this reason the dipole force opens new perspectives in cell biology as well. Dielectrophoresis appears to be a simple and useful technique which can be used to characterize, manipulate, or separate various biological components. The dipole force has already been used in different electrical techniques, such as electric field induced fusion [3],[4], or to measure the translational diffusion coefficient of polar macro-molecules [5]. However, one of the biggest potentials that can be obtained through dielectrophoresis is the ability to separate or concentrate cells [6]. One of the main problems with dielectrophoretic separation of small particles is that of removing the particles once they have been collected on one of the electrodes. Therefore, dielectric separation is mainly a laboratory technique.

We have used the dipole force to separate pollen grains from soil cores. This separation process is of practical importance in paleontological research [7]. The differences between the separation processes applied in determinative mineralogy and experimental cell research are mainly due to the dimensions and the density of the particles under investigation. The dimensions of intact pollen grains (± 50 μm) are large compared with the dimensions of cells. Therefore, dense liquids of low conductivity and permittivity are used as a separating medium to separate the matrix particles of graded samples. Dielectrophoretic experiments on cells are performed at higher frequencies and low voltages (6Vpp) compared with the experiments on pollen extraction (600 Vpp , KHz..MHz).

Our goal is to resume the most important phenomena related to wet dielectric separation processes from an electromagnetic point of view. The models given in this paper establish the basis for the study of interpretive models. This should then permit more insight into the ways in which complex particles respond to non-uniform electric fields. The frequency dependent behaviour of the total polarizability of the particles is discussed in detail. Furthermore, we will focus our attention to the electrophoretic effect of a charged and polarized particle in a strong divergent a.c. electric field. Finally, the sensitivity of the dipole force is experimentally verified under laboratory conditions.

POLARIZABILITY

In most dielectrophoretic studies a neutral, but polarized particle is examined by considering the bulk properties (permittivity ε and conductivity σ) of the particle, and of the liquid medium. Such theoretical analyses will take the bound dielectric polarization charge density σ_d, and the surface charge density σ_m, due to Maxwell-Wagner polarization, into account. These charges ($\sigma_d+\sigma_m=\sigma_{pol}$) are induced by the applied electric field, and will result in a zero net charge.

Theoretical model
In this section we consider an uncharged spherical particle, which is built up of k different shells, see figure (1). The k different regions of the sphere are characterized by the electric parameters (permittivity ε, conductivity σ), which are considered constant over the frequency domain of interest. The shells can represent the nucleus, cytoplasm, membranes, or even an ionic double layer at the particle-liquid interface.

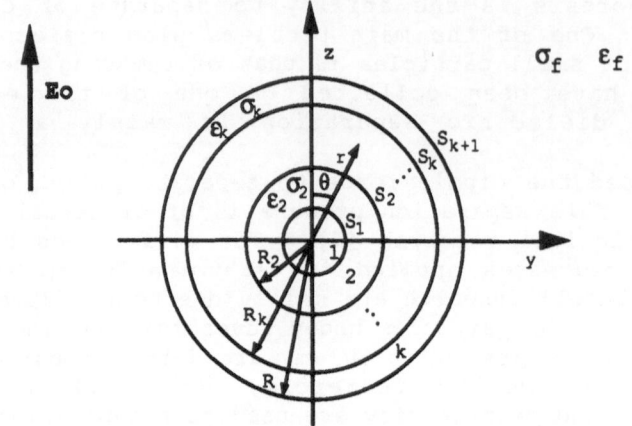

Figure 1. Spherical particle built up of k different shells.
Every region i enclosed by the surface Si is characterized by
the parameters ε_i, σ_i, and R_i.

The particle is immersed in a liquid medium (ε_f, σ_f) supporting a
uniform electric field **Eo** in the absence of the particle. This
"uniform" field approximation seems the most general approach to
the problem still suitable to an analytical treatment.   A
general way to study the time-harmonic behaviour of such system
is to solve first the well-known equivalent dielectric
electrostatic problem [8]. When the electrostatic field **Eo** is
replaced by a uniform harmonic field, the potential becomes
complex. The time-harmonic solution proceeds wholly analogously
with the derivation of the static case, provided that the
boundary conditions contain now the complex permittivity.

$$\hat{\varepsilon} = \varepsilon - j\frac{\sigma}{\omega} \tag{1}$$

The complex induced dipole moment **p**($j\omega$) for a linear polarized
particle is proportional to the uniform harmonic field.

$$\mathbf{p}(j\omega) = \frac{\varepsilon_f}{\varepsilon_o} \left\{ \sum_{i=1}^{k} \int_{Si} \hat{\sigma}_{pol_i} \, \mathbf{r} \, dSi \right\} = \alpha(j\omega) \, \hat{\mathbf{E}}\mathbf{o} \tag{2}$$

The frequency dependent instantaneous polarizability $\alpha(\omega)$ is
obtained from the complex polarizability $\alpha(j\omega)$.

$$\alpha(\omega) = \text{Re}\{ \alpha(j\omega) \} \tag{3}$$

The complexity of the mathematical treatment increases rapidly if the number of interfaces increases. It can be proved that the particle with the different shells behaves as a homogeneous particle with a complex effective permittivity. This allows us to calculate $\alpha(j\omega)$ by using equation (4).

$$\alpha(j\omega) = 4\pi R^3 \varepsilon f \left(\frac{\hat{\varepsilon}peff - \hat{\varepsilon}f}{\hat{\varepsilon}peff + 2\,\hat{\varepsilon}f} \right) \tag{4}$$

Where $\hat{\varepsilon}peff$ depends on the number of shells k.

$$k=1 : \quad \hat{\varepsilon}peff = \hat{\varepsilon}1 \tag{5}$$

$$k=2 : \quad \hat{\varepsilon}peff = \left\{ \frac{\left(\dfrac{R}{R_1}\right)^3 + 2\left(\dfrac{\hat{\varepsilon}1 - \hat{\varepsilon}2}{\hat{\varepsilon}1 + 2\,\hat{\varepsilon}2}\right)}{\left(\dfrac{R}{R_1}\right)^3 - \left(\dfrac{\hat{\varepsilon}1 - \hat{\varepsilon}2}{\hat{\varepsilon}1 + 2\,\hat{\varepsilon}2}\right)} \right\} \hat{\varepsilon}2 \tag{6}$$

$$k=3 : \quad \hat{\varepsilon}peff = \left\{ \frac{\left(\dfrac{R}{R_2}\right)^3 + 2\left(\dfrac{\hat{\varepsilon}2eff - \hat{\varepsilon}3}{\hat{\varepsilon}2eff + 2\,\hat{\varepsilon}3}\right)}{\left(\dfrac{R}{R_2}\right)^3 - \left(\dfrac{\hat{\varepsilon}2eff - \hat{\varepsilon}3}{\hat{\varepsilon}2eff + 2\,\hat{\varepsilon}3}\right)} \right\} \hat{\varepsilon}3 \tag{7}$$

Where $\hat{\varepsilon}2eff$ is given by $\hat{\varepsilon}peff$ of equation (6)

Repeated application of equations (4)...(7) can be used to obtain the effective permittivity when k>3.

Numerical results

The polarizability $\alpha(\omega)$ of a spherical particle (k=1 R=5.0 10^{-6} m) is given in figure 2. The particle is more polarizable than the liquid medium. The curves of figure 2 represent three different cases, respectively $\sigma p<\sigma f$, $\sigma p=\sigma f$ and $\sigma p>\sigma f$. These curves have been obtained by interchanging the conductivity values. It is observed from figure 2 that the dielectric regime is only obtained at frequencies > 100MHz. At frequencies < 1 MHz the system is in conductive regime, so that the migration of the particles is actually conductophoresis instead of dielectrophoresis. It is also worthwhile to note the sign reversal of $\alpha(\omega)$ in the case of $\varepsilon p>\varepsilon f$ and $\sigma p<\sigma f$.

The dielectrophoretic behaviour of intact *Micrococcus Lysodeikticus* have been studied by Inoue et al. (20 Hz to 4MHz) [9]. The parameters (ε, σ) that we have used in our previous calculation have been adopted from this interesting paper. The authors concluded from their experiments that the low frequency

behaviour of the cells could not be explained by the simple
theory of a homogeneous particle.

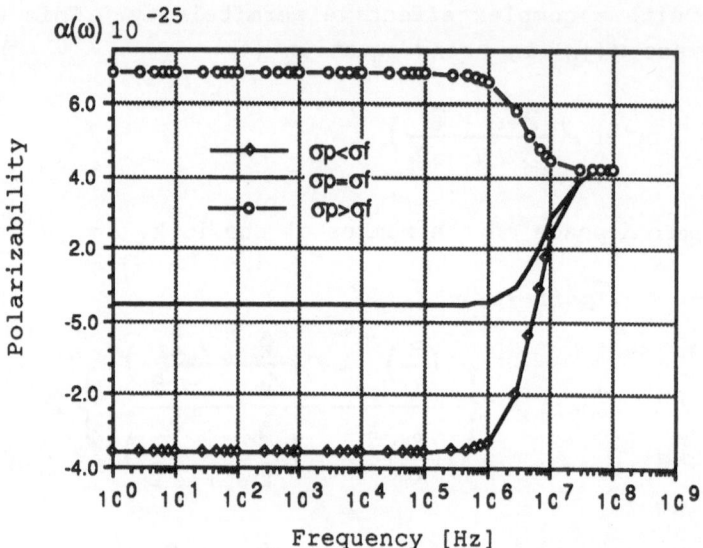

Figure 2. Polarizability versus frequency. ($\varepsilon_o=8.85 \cdot 10^{-12}$ F/m)
($R=5 \cdot 10^{-6}$ m, $\varepsilon_p=200\varepsilon_o$, $\varepsilon_f= 80\varepsilon_o$, $\sigma_p=0.05$ S/m, $\sigma_p=0.01$ S/m).

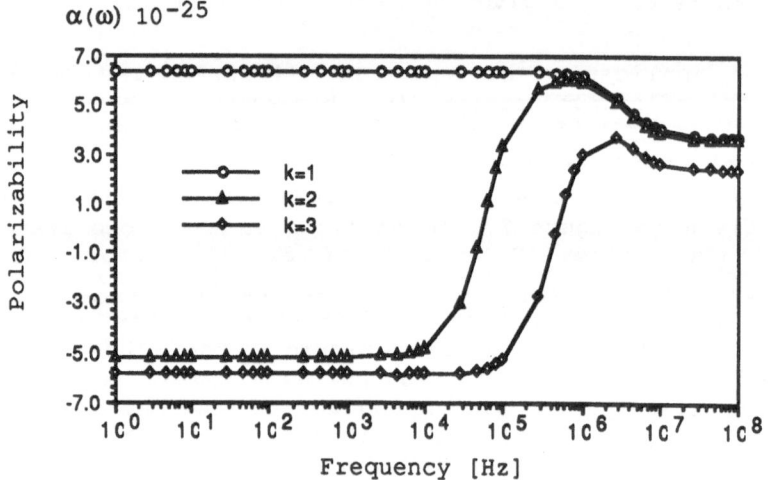

Figure 3. Polarizability versus frequency.
($\varepsilon_1=200\varepsilon_o$, $\varepsilon_2=9\varepsilon_o$, $\varepsilon_3=40\varepsilon_o$, $\varepsilon_f=80\varepsilon_o$, $\sigma_1=0.05$ S/m, $\sigma_2=10^{-6}$ S/m,
$\sigma_3=10^{-4}$ S/m, $\sigma_f=0.01$ S/m, $R_1=5.0 \cdot 10^{-6}$ m, $R_2=5.01 \cdot 10^{-6}$ m, $R=5.014$
10^{-6}, when k=1 or k=2 or k=3).

It is shown in figure (3) that the response of the particle is changed by adding extra layers to the model. Reasonable values for the parameters (ε, σ, R) were chosen for illustrative purposes [10]. In this example we consider the same cell interior and add a more insulating "cell" wall to the model (k=2). The influence of a small "ionic double" layer is represented by the curve k=3

ELECTRO-DIELECTROPHORETIC RESPONSE

It is well-known that particles suspended in a liquid of moderate conductivity acquire an electric charge [11]. Therefore, it seems worthwhile to look at the electrophoretic behaviour of a charged and polarized particle in a.c. dielectrophoresis.

Theoretical model
The electric force $\mathbf{F}$ exerted on a charged and polarized particle is given by equation (8).

$$\mathbf{F} = Q \, \mathbf{E}_0 + \frac{\alpha(\omega)}{2} \nabla \mathbf{E}_0^2 \tag{8}$$

where Q is the net charge of the particle and $\alpha(\omega)$ is the instantaneous polarizability, as given by equation (3).
Consider a cylindrical electrode system with inner radius r2 and outer radius r1. When the particle migrates through a viscous medium (η) under the influence of an applied voltage $U_p = U_m \cos \omega t$ the equation of motion is given by [12] as

$$\ddot{r}_0 = \frac{Q \, E_{om}}{m \, r_0} \cos \omega t - \frac{6 \pi \eta R}{m} \dot{r}_0 - \frac{\alpha(\omega) \, E_{om}}{m \, r_0^3} \cos^2 \omega t \tag{9}$$

$$E_{om} = \frac{U_m}{\ln(r2/r1)}$$

Where m is the mass of the particle and r_0 represents the distance between the centre of the spherical particle and the centre of the inner electrode of radius r1.

Numerical results
A numerical solution of the non-linear second order differential equation (9) is given in figure (4). The particle trajectory varies in a periodic manner with a tendency to drift towards the region of lower field intensity. Either positively or negatively charged particles show the same electrophoretic behaviour. The electrophoretic response is reduced at higher frequencies due to the mass m of the particle and the viscosity of the liquid medium.

In dielectrophoretic experiments it is often observed that collected particles at the inner electrode are suddenly repelled from this electrode. This phenomenon is observed when the electric field intensity has reached a certain level which strongly depends on the conductivity of the liquid medium and on the electrode system. Due to a local breakdown, collected particles are sometimes charged to an appreciable degree. These charged particles are, due to the electrophoretic response, repelled from the electrodes even at high frequencies. During their motion back into the liquid medium the net charge of the particle is exponentially reduced due to the conductivity of the liquid medium. When the net charge of the particle is reduced to an acceptable degree the dielectrophoretic response of the particle becomes dominant and the particle will migrate back to the inner electrode.

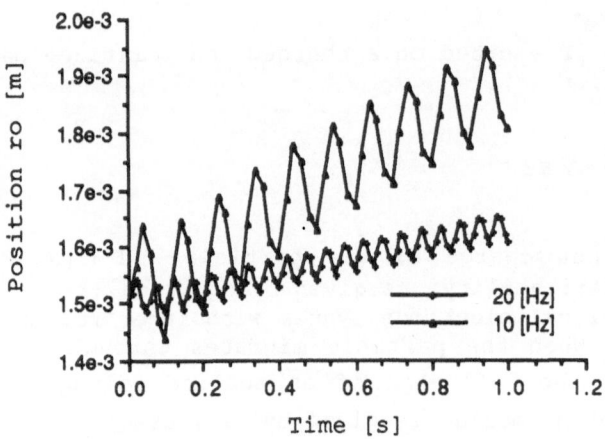

Figure 4. Particle trajectory due to electrophoresis as a function of time. (Q=5.0 10^{-15} C, Um=500 V, η=18,1 10^{-6} kg/ms, r1=50 μm, r2=10mm, R=50 μm)

SENSITIVITY AND MAGNITUDE OF THE DIPOLE FORCE IN DIELECTRIC REGIME

The sensitivity of the dipole force is experimentally verified. The test cell consists of a glass cup (ϕ=40 mm) and two diverging wires (ϕ=0.7 mm), which serve as an electrode system. The electrodes are positioned just below the liquid surface to avoid surface phenomena. As a test sample we have used a polyacetal sphere (ϕ=3 mm, ρm=1.41 gr/cm^3, ϵpr=4.0, σp=10^{-12} S/m). A liquid mixture (CCL$_4$, C2H5OH) with approximately the same density of the polyacetal sphere is used in this experiment. The permittivity of the liquid medium is changed to satisfy the condition that the particle (polyacetal sphere) is neither attracted nor repelled. To assure dielectric regime, an applied voltage (1200 Vpp) of 280 KHz was used. The permittivity of the liquids that have been used are determined by using a computer controlled measurement set-up [13].Figure 5 clearly shows how the relative permittivity ε_{fr} of the liquid medium was changed at different trials during the course of the experiments. The successive approximation based on attraction and repulsion of the sphere led us towards the correct value of the permittivity of the polyacetal sphere. The sphere ($\varepsilon_{pr}=4.0$) was clearly repelled at the fifth trial ($\varepsilon_{fr}=4.07$), and attracted at the seventh trial ($\varepsilon_{fr}=3.92$).

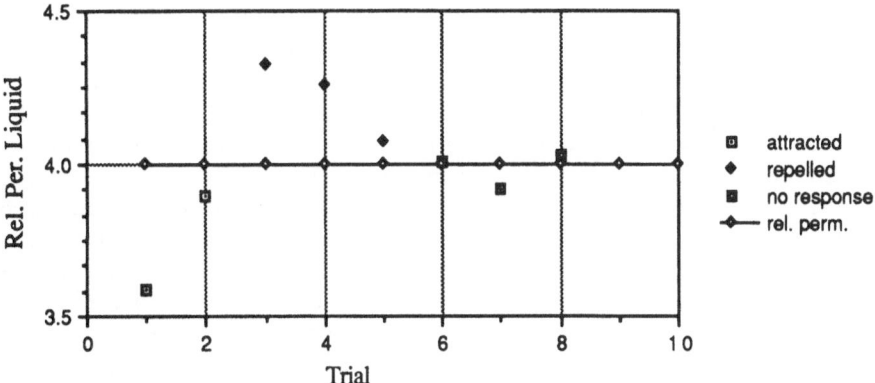

Figure 5. Relative Permittivity of the liquid medium during the course of the experiment.

With a liquid medium of relative permittivity between those two limiting values, no response of the polyacetal sphere could be observed. This experiment shows that the dipole force, under ideal laboratory conditions, is very sensitive to a small difference in permittivity of the particle and the liquid medium ($\pm 2 \Delta\%$).

In our theoretical calculations we have used the "uniform" field approximation to calculate the scalar polarizability. Considering the fact that the external electric field should be strongly non-uniform over the dimensions of the particle, the question may arise whether or not the "uniform" field approximation will yield correct results for the dipole force. A computer-aided numerical solution, which can solve 3D-electrostatic problems has been used to calculate the electric part of the Maxwell stress tensor [14]. Computational results show that when the particle size increases in relation to the characteristic length of the electric field non-uniformity, the force derived from the uniform field approximation was found too low.

CONCLUSIONS

In this paper we have discussed the most important phenomena related to dielectric separation processes. The dipole force was analysed from an electromagnetic point of view. A model to study the electro-dielectrophoretic action of a charged and polarized particle was presented in this paper. We have shown that the time-harmonic behaviour of a spherical layered particle leads to the same expression as of a homogeneous particle. Only the complex effective permittivity appears to be related to the internal structure of the particle. Numerical values for the characteristic parameters were obtained from literature. The results obtained from this model show that the dielectrophoretic spectrum was considerably changed by adding small layers to the particle. This model is used in practice to study the porosity effect in wet dielectric separations of calcareous microfossils. The model was used to explain the frequently observed repulsion of collected particles. We can conclude from experiments that the dipole force in dielectric regime is very sensitive to the permittivity differential of the particle and the liquid medium. We have also mentioned that the dipole force, as derived from the uniform field approximation, was too low. More detailed models which can handle non-spherical particles and variable parameters will be available in the near future.

REFERENCES

1. Fritsche, G.R., Electrostatic separator removes FCC catalyst fines from decanted oil. The Oil and Gas J., 1977, 28, pp. 73-81.

2. Ralstone, C., Electrostatic Separation of Mixed Granular Solids, Elsevier Press, New York, 1961, pp.105.

3. Vienken, J. and Zimmerman, V., Electric field-induced fusion. Febs Letters, 1982, 137 (1), pp.11-13.

4. Masuda, S., Washizu, M. and Nanba, T., Novel method of cell fusion in field constriction area in fluid integrated circuit. IEEE Trans. Industry Appl., 1989, 25, pp.732-737.

5. Eisenstadt, M. and Scheinberg, H., Dielectrophoresis of macromolecules. Biopolymers, 1973, 12, pp.2491-2512.

6. Craene, J. and Pohl, H., A study of living and dead yeast cells using dielectrophoresis. J. Electrochem. Soc., 1968, 115, pp.584-586.

7. Van Overloop, E., An adapted preparation method for clayey and "poor" pollen samples. Bulletin de la Société Belge de Géologie, 1984, 93 (3), pp.271-273.

8. Böttcher, C. and Bordewijk, P., Theory of Electric Polarization. Volume 2 : Dielectrics in time-dependent fields, Elsevier, New York, 1978, pp.23-24.

9. Inoue, T., Pething, R., Al-Ameen, Bust, J. and Price, J., Dielectrophoretic behaviour of micrococcus lysodeikticus and its protoplast. J. Electrostatics, 1988, 21, pp.215-223.

10. Schwan, H., Electrical properties of tissues and cells suspensions. Adv. Biol.Med.Phys., 1957, 4, pp.147-209.

11. Shaw, J., Introduction to Colloid and Surface Chemistry, Butterworths, London, 1980, pp.148.

12. Goossens, K.and Van Biesen, L., Electrophoretic effects in a.c.dielectrophoretic solid-solid separations. Sep Sci. Technol.,1989, 24 (1), pp.51-62.

13. Goossens, K. and Van Biesen, L., Measurements of time-harmonic behaviour and transient response of conductive liquids. Electromagnetic Metrology ISEM89, Pergamon Press IAP, 1989, pp. 123-127.

14. Goossens, K. and Van Biesen, L., A computer-aided analysis of dielectrophoretic force calculations. IEE Conference Publication, 1988, 289, pp.131-134.

REACTIVE SOLVENT EXTRACTION OF BETA-LACTAM ANTIBIOTICS

Tim A.J. Harris, Simi Khan, Bryan G. Reuben * and Tina Shokoya
Department of Chemical Engineering, South Bank Polytechnic, London SE1 OAA
and Michael S. Verrall Beecham Pharmaceuticals, Brockham Park, Surrey

ABSTRACT

The reactive solvent extraction of olivanic acid MM13902, other beta-lactams and phenoxyacetic acid has been investigated. Extractants were quaternary nitrogen compounds or amine salts, for example tetrabutyl ammonium hydrogen sulphate, Amberlite LA2 hydrochloride, Alamine 336 hydrochloride and Aliquat 336. Solvents included dichloromethane and butyl acetate.

Under appropriate conditions, extractions in excess of 90% could be achieved with all compounds except amoxycillin. Back extraction could be brought about by nitrate and iodide ions, the latter being more effective but more expensive. With nitrate, several extractions were sometimes required to give satisfactory yields.

Small differences in lipophilicity and hence extracting power, for example between Alamine 336 hydrochloride and Aliquat 336, suggest the possibility of selective extraction.

INTRODUCTION

Solvent extraction of beta-lactam antibiotics is hindered by their being ionic at high pH and unstable at low pH. Penicillin G and V are sufficiently stable to be solvent-extracted in centrifugal separators at low pH, but other beta-lactams require elaborate and expensive processing involving low capacity techniques such as preparative chromatography.

The overall aim of the present project is to develop an extended fermentation technique where a beta-lactam containing broth is continuously separated from mycelium by crossflow flitration, and the beta-lactam extracted either by ion-pair extraction or liquid-liquid ion exchange. Because these take place in neutral solution, they overcome the principal problem associated with conventional solvent extraction.

---

* Principal author for future correspondence

We have already reported work on the fermentation and the filtration [1],[2]. This paper reports exploratory work on the extraction. Reactive solvent extraction of penicillin has been studied by Reschke and Schügerl [3],[4], and some work on olivanic acids has been reported by Butterworth et al. [5],[6]. As phenoxyacetic acid is added as a side chain precursor to the penicillin V fermentation mixture, its extraction has also been studied to see how selective a penicillin ion-pair extraction might be.

MATERIALS AND METHODS

Beta-lactams were analysed by HPLC with a reverse phase ODS-2 column (C_{18}, 5 microns) using a buffered aqueous mobile phase containing acetonitrile.
Phenoxyacetic acid was estimated by u-v absorption. Solvent extraction experiments were performed in separating funnels and the phases were afterwards centrifuged. The volume of organic phase was always identical with the volume of aqueous phase. Back extraction was evaluated by the reverse process, the loaded organic phase being shaken with an equal volume of 0.1M sodium nitrate solution. The HPLC system did not respond satisfactorily to organic solutions and most organic layer concentrations were deduced from aqueous layer measurements by subtraction. All experiments were performed at room temperature.

Technical grades of chemicals were used without further purification. Amines were converted to their hydrochlorides by repeated treatment with concentrated hydrochloric acid.

Streptomyces olivaceus produces three olivanic acids [5] with the structures shown in Figure 1. In this work, MM13902 was used. Potassium clavulanate was used in the form of Augmentin * tablets (Beechams). The fermentation broth was prepared as described by Harris et al.[1].

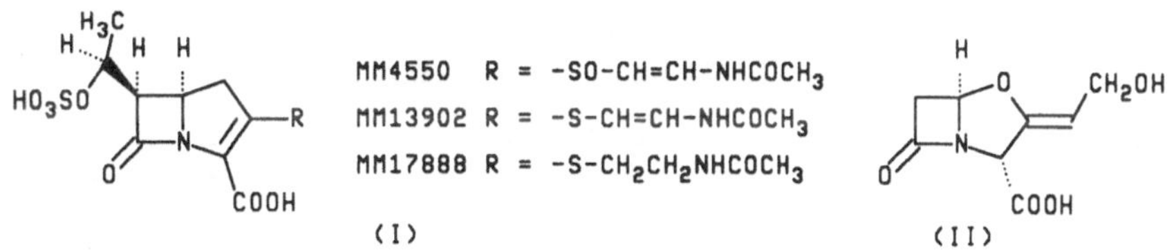

MM4550 R = -SO-CH=CH-NHCOCH₃
MM13902 R = -S-CH=CH-NHCOCH₃
MM17888 R = -S-CH₂CH₂NHCOCH₃

(I)   (II)

Figure 1. Olivanic acids from *Streptomyces olivaceus*(I), and clavulanic acid (II).

RESULTS AND DISCUSSION

Dilute solutions were used throughout and activities equated with concentrations. Results of extractions are expressed as % extractions, or as distribution coefficients given by:

$$D = [P_{org}]/[P_{aq}]$$

*Registered trademark of Beecham Group

where $[P_{org}]$ and $[P_{aq}]$ are the concentrations of material being extracted in the organic and aqueous phases. Extraction coefficients have also been calculated. In the case of water-soluble extractants such as tetrabutyammonium hydrogen sulphate (TBAHSO$_4$), the ammonium ion combines with the ion to be extracted and the two give an ion pair in the organic layer:

$$TBA^+_{aq} + P^-_{aq} \rightleftharpoons TBAP_{org} \tag{1}$$

whence the extraction coefficient is given by:

$$E = [TBAP_{org}]/[TBA^+_{aq}][P^-_{aq}]$$

In the case of liquid-liquid ion exchange with a water-insoluble amine hydrochloride (QCl), the process becomes:

$$QCl_{org} + P^-_{aq} \rightleftharpoons QP_{org} + Cl^-_{aq} \tag{2}$$

and $\qquad E = [QP_{org}][Cl^-_{aq}]/[QCl_{org}][P^-_{aq}]$

Olivanic acid

Data were gathered for extraction of MM13902 into dichloromethane with Amberlite LA2 hydrochloride (a long chain secondary amine hydrochloride, Rohm & Haas), Alamine 336 hydrochloride (tri-C_8-C_{10}-amine hydrochloride, Henkel) and Aliquat 336 (methyltri-C_8-C_{10}-ammonium chloride, Henkel). Some MM13902 was extracted in the absence of extractant. The results are shown in Table 1 together with extractions for different concentrations of MM13902 and extractant. E is not constant and increases with concentration. Back extraction results with 0.1M sodium nitrate and overall extraction yields are also shown.

Two measurements were made on extraction from filtered fermentation broth, and these are the last two lines of the table. The low yield with the lower extractant concentration probably reflects the fact that more lipophilic compounds than MM13902 are extracted first and that the beta-lactam is extracted when stoichiometry permits. This is confirmed by the identical levels of back extraction in both cases.

TABLE 1
Extraction of MM13902 into dichloromethane with various extractants

	MM13902 conc. (mM)	Extractant conc. (mM)	% ext.	Back extraction MM13902 (mM)	% ext.	Overall % ext.
No extractant	5	0	5.7			
Amberlite LA2/HCl	5	5	5.7	0.286	0	0
Alamine 336/HCl	5	5	10.6	0.53	10.1	1.1
Aliquat 336	5	5	98.6	4.93	86.5	85.3
	0.025	0.025	8.0	0.002	94.0	7.5
	0.25	0.25	38.0	0.095	86.0	32.7
	0.025	0.25	93.0	0.0232	89.0	82.8
broth extraction {	0.021	0.25	16.9	0.0036	90.1	15.2
	0.013	0.5	84.0	0.011	90.0	75.6

175

Extraction with Amberlite LA2 hydrochloride appears no more effective than solvent alone. Aliquat 336 is much more effective than Alamine 336 hydrochloride and so is the back extraction. Quaternary ammonium compounds are generally more effective than tertiary ammonium compounds in ion-pair extraction but this difference is somewhat surprising in view of the similarity between the extractants and the fact that only ion exchange is involved.

The effect of different solvents is shown in Table 2. Butyl acetate and dichloromethane are the most effective solvents. Butyl acetate is less effective than hexane, underlining the absence of a role for hydrogen bonding in the dissolving of ion pairs.

TABLE 2

Effect of solvent on MM13902 extraction

Solvent	% extraction	Back extraction MM13902 conc. (mM)	% ext.	Overall % ext.
Butan-1-ol	49	0.01225	87	42.6
Tributyl phosphate	60	0.015	88	52.8
Hexane	79	0.1975	69	54.5
Diethyl ether	85	0.02125	58	49.3
Dichloromethane	93	0.02325	89	82.8
n-Butyl acetate	98	0.0245	86	84.3

Initial concentration of MM13902 was 0.025mM; of extractant was 0.025mM. Back extraction was with 0.1M sodium nitrate.

Penicillin V

The ion-pair extraction of the penicillin V anion with TBAHSO$_4$ (Bofors) was studied as a function of solvent and concentration of penicillin and extractant. The pH was maintained at 6. With 0.75mM penicillin and 7.5mM extractant, the yields were 33.0% with hexane, 76.7% with amyl acetate, 77.7% with octan-1-ol and 81.6% with dichloromethane. Thus dichloromethane was the most effective but only hexane was substantially worse.

Extractions into dichloromethane at different concentrations were measured. Figure 2 is a plot of [TBAP$_{org}$] vs [TBA$^+_{aq}$][P$^-_{aq}$] for the low concentration runs. A reasonable straight line is obtained with a small positive intercept, probably due to finite extraction at pH6 in the absence of extractant. The gradient of the line leads to a value for E of 800 dm^{3}mol^{-1}. At higher concentrations, over 90% extraction is obtained. Penicillin V is more lipophilic than the olivanic acids, hence back extraction is more difficult. Typically, a single back extraction with 0.1M sodium nitrate recovered 20% of the ion-paired penicillin V; four back extractions raised this to about 60%.

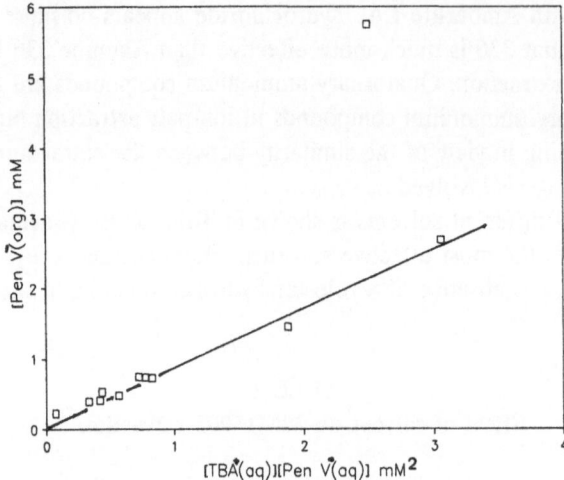

Figure 2. Ion pair extraction of penicillin V with TBAHSO₄.

Extractions with Amberlite LA2 hydrochloride into butyl acetate were very effective in contrast to the ineffectiveness of the same extractant with olivanic acids. The yields were all over 98% but they varied with concentration of starting solutions, whereas the process would be expected to follow equation 3 and be independent of concentration. The phenomenon was also noted with olivanic acids. Figure 3 shows organic phase penicillin concentration vs. the square of the aqueous phase concentrations. The approximate linearity corresponds to an ion-pairing process of the type in equation 1 rather than an ion exchange process as in equation 2.

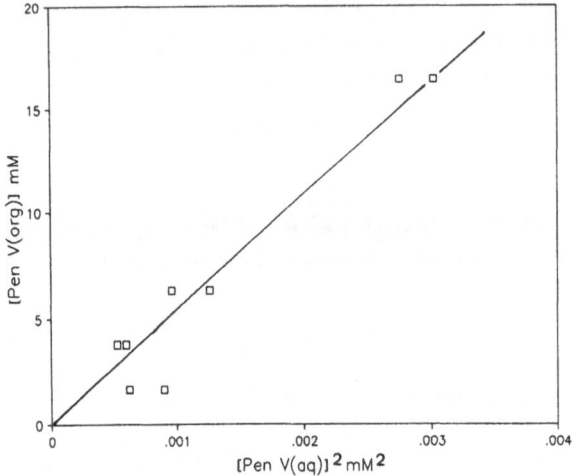

Figure 3. Extraction of penicillin V with Amberlite LA2 hydrochloride.

Clavulanic acid and amoxycillin

Potassium clavulanate (125mg) and amoxycillin trihydrate (250mg) are marketed in tablet form as Augmentin. The tablets give a solution of pH6.4. Extraction with TBAHSO$_4$, and Amberlite LA2 and Alamine 336 hydrochlorides gave little if any extraction of either beta-lactam. Aliquat 336 in a range of solvents, however, extracted substantial amounts of clavulanate ions (Figure 4). The straight lines are reasonable but points lie above the line at the upper end of the graph where large Aliquat:clavulanate ratios were used. Negligible concentrations of amoxycillin ions were extracted even when the quaternary was in large excess, showing that the amoxycillin ion is less lipophilic than the olivanate and chloride ions. In butyl acetate, extraction could be measured sufficiently accurately for a value of E ~ 0.03 to be derived.

Phenoxyacetic acid

Identical solutions of phenoxyacetic acid (1.974mM;pK$_a$ = 7.6 x 10^{-6}) were extracted into dichloromethane at a range of pH values in the presence and absence of TBAHSO$_4$ of the same concentration.

The results are shown in Table 3. At low pH, where the phenoxyacetic acid exists entirely as molecules, there is no ion-pair extraction and about 77% of the phenoxyacetic acid is extracted by conventional solvent extraction. At high pH, where the phenoxyacetic acid exists entirely as ions, there is no conventional extraction and about 10% ion-pair extraction.

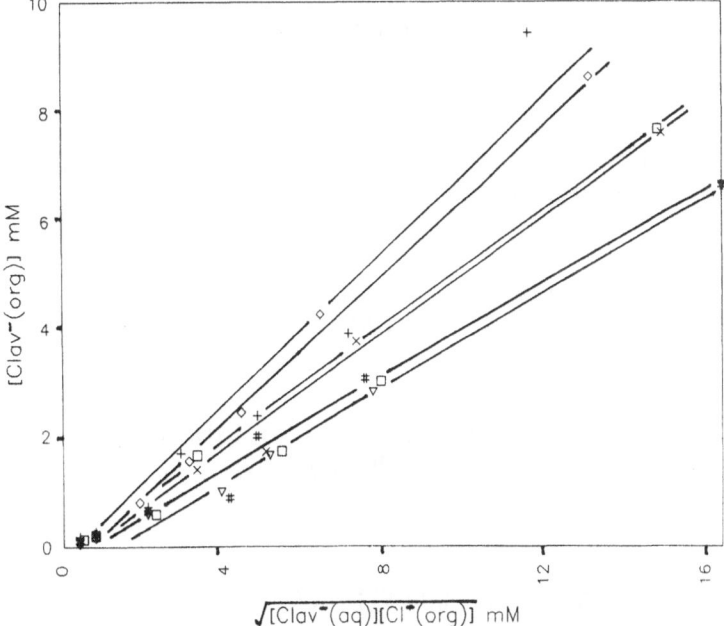

Figure 4. Extraction of the clavulanate ion. (+ butyl acetate, E = 0.65; □ octan-1-ol, E = 0.27; × octan-2-ol, E = 0.28; ▽ dichloromethane, E = 0.17; # trichloromethane, E = 0.18; ◇ tetrachloromethane, E = 0.48).

TABLE 3
Extraction of phenoxyacetic acid into dichloromethane in the presence and absence of TBAHSO$_4$

pH	% extraction	
	with extractant	without extractant
0.96	76.5	78.0
3	69.6	75.1
4	68.2	51.9
5	56.9	31.9
7	10.4	0.35
9	11.2	0.20
10	10.1	0.15

The region around pH6 is of interest in this work, and under these conditions there are competing equilibria. These have been calculated by Reschke and Schügerl [3] but the following derivation is simpler and more suited to the purposes here. If HPAA represents the phenoxyacetic acid molecule and PAA- the ion, then these may be written:

$$HPAA_{aq} \underset{K_a}{\rightleftharpoons} H^+_{aq} + PAA^-_{aq}$$

$$\updownarrow D \qquad\qquad E \updownarrow TBA^+_{aq}$$

$$HPAA_{org} \qquad\qquad TBAPAA^-_{org}$$

Let the initial molar concentrations of HPAA and TBAHSO$_4$ be a.
TBAHSO$_4$ is overwhelmingly retained in the aqueous layer at low concentrations as is the hydrogen ion. Hence, the concentrations of the relevant species at equilibrium can be written:

$[HPAA_{aq}] = a(1-x-y)$ $\qquad$ $[HPAA_{org}] = ay$
$[PAA^-_{aq}] = ax(1-z)$ $\qquad$ $[TBAPAA_{org}] = axz$
$[TBA^+_{aq}] = a(1-xz)$ $\qquad$ $[H^+]$ is defined by the pH

K_a (the acid dissociation constant) $= [H^+][PAA^-_{aq}]/[HPAA_{aq}]$
$$= [H^+]x(1-z)/(1-x-y) \qquad (3)$$

$$D = [HPAA_{aq}]/[HPAA_{org}] = y/(1-x-y) \qquad (4)$$

$$E = [TBAPAA_{org}]/[TBA^+_{aq}][PAA^-_{aq}] = z/a(1-z)(1-xz) \qquad (5)$$

From (4), $\quad y = D(1-x)/(1+D) \qquad (6)$

Substituting (6) in (1): $K_a = [H^+]x(1-z)(1+D)/(1-x)$ (7)

If $F = K_a/[H^+](1+D)$, then from (7),
$F(1-x) = x(1-z)$ and $x = F/(1-z+F)$ (8)

Substituting (8) in (5) gives:

$$E = \frac{z}{a(1-z)\{1 - Fz/(1-z+F)\}}$$ (9)

k_a for phenoxyacetic acid $= 7.6 \times 10^{-6}$. Data at very high and low pH give $D = 3.26$ and $E = 65.4$. (9) can then be solved for z by trial and error ($0<z<1$) and the various concentrations and total extraction determined. The predicted and experimental pH dependence of the extraction are shown in Figure 5.

Variation of distribution coefficient with concentration leads to a value for E of 38 litre mol^{-1}.

Comparison of these data with those for penicillin indicates that, if ion pair extraction with the reagents employed here were to be used for a mixture of penicillin and phenoxyacetic acid, the process would discriminate more efficiently against phenoxyacetic acid at pH7 than at pH6. Penicillin V is only fractionally more unstable at the higher pH.

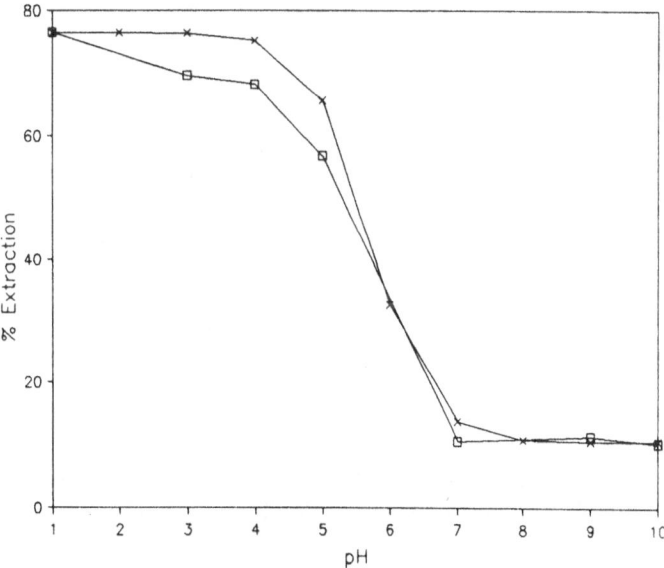

Figure 5. Experimental x and theoretical □ extraction of phenoxyacetic acid at different pH values.

CONCLUSIONS

Penicillin V and phenoxyacetic acid can be extracted into organic solvents as ion pairs with tetrabutylammonium, the E value for penicillin being 800 and for phenoxyacetic acid 47 dm^3mol^{-1}. Variation of phenoxyacetic acid extraction with pH follows theoretical predictions and indicates that extraction of both materials at pH7 will discriminate against phenoxyacetic acid.

The other beta-lactams studied are more effectively treated by liquid-liquid ion exchange. Olivanic acid MM13902 is best extracted with Aliquat 336 into dichloromethane or butyl acetate.

Clavulanic acid may also be extracted with Aliquat 336, the E values varying from 0.65 to 0.17 for different solvents. It is thus slightly less lipophilic than the chloride anion of the Aliquat.

The sensitivities of the extractions to small changes in environment suggests the possibility of developing selective extraction processes for beta-lactams at pH values where they are the most stable.

REFERENCES

1.   Harris, T.A.J., Reuben, B.G., Cox, D.J., Vaid, A.K. and Carvell J., The cross-flow filtration of an unstable beta-lactam antibiotic fermentation broth. J. Chem. Tech. Biotechnol., 1988, 42, 19-30.

2.   Curbishley, P., Harris, T.A.J. and Reuben B.G., Cross-flow filtration of an unstable beta-lactam fermentation broth - the effect of antifoam. SCI Symposium on Membrane Reactors in Biotechnology, London, November 1989.

3.   Reschke, M., and Schugerl, K., Reactive extraction of penicillin, I. Chem. Eng. J. 1984, 28, B1-B9.

4.   Part II, ibid, pages B11-B20 and other papers in this series.

5.   Butterworth, D., Cole, M., Hanscomb, G. and Rolinson, G., Olivanic acids, a family of beta-lactam antibiotics with beta-lactamase inhibitory properties produced by Streptomyces species. J. Antibiotics, 1979, 32, 287-294.

6.   Butterworth, D., Hood, J.D. and Verrall, M.S., The new beta-lactam antibiotics. In Advances in Biotechnological Processes, Vol 1,eds. Mizrahi A. and Van Wezel A.L., A.R. Liss, New York, 1983, pp. 251-292.

Aqueous Two-phase Fractionation:
Practical evaluation for productive biorecovery.

J.G. Huddleston, K.W. Ottomar, D. Ngonyani, J.A. Flanagan and A. Lyddiatt.
Biochemical Recovery Group, School of Chemical Engineering,
University of Birmingham, Edgbaston, Birmingham B15 2TT.

ABSTRACT

The recovery, by partitioning in aqueous two phase systems, of a range of potential bioproducts is considered. These include food functional protein and microbial enzymes fractionated from complex feedstocks and required in various states of purity. The influence of the basic parameters of aqueous two-phase systems: tie-line length, molecular weight and pH, upon partition of macromolecules is discussed in relation to molecular characteristics such as molecular weight, charge and hydrophobicity. The degree of purity attainable in limited numbers of discrete partitioning stages is addressed. Enhancement of resolution by adoption of approaches involving bioaffinity interactions is considered. Critical appraisal enables the current status of partitioning in aqueous two phase systems as applied to productive biotechnology to be considered.

INTRODUCTION.

Aqueous two phase systems have been widely recommended in recent years for the entire range of biochemical separations problems posed by cells, organelles, membrane preparations and macromolecules. [1,2]. In general, but particularly at laboratory scale this has involved partition in aqueous solutions of poly(ethylene glycol) (PEG) and dextran. Economic, and physical constraints (polymer costs, phase densities and viscosities) have favoured the pilot scale development of only PEG-salt systems deployed in a limited number of discrete contacting stages [3]. Use of the term PEG-salt implies reference to the extensive literature on PEG-potassium phosphate buffered systems [4], however, recent work implies that many other PEG-salt systems are potentially available [5, and see Table 1]. Current practice in the application of such systems indicates that large scale and laboratory scale multi-stage, chromatographic processes have largely yet to be realised. Published processes at productive scale utilise one to three (or more rarely four) discrete partitioning steps [1].

The incompatibility of pairs and multiples of polymers, and of polymers and salts, in aqueous solution is characterised by a phase

TABLE 1.

Some polymer-salt aqueous two-phase systems of potential use in partition.

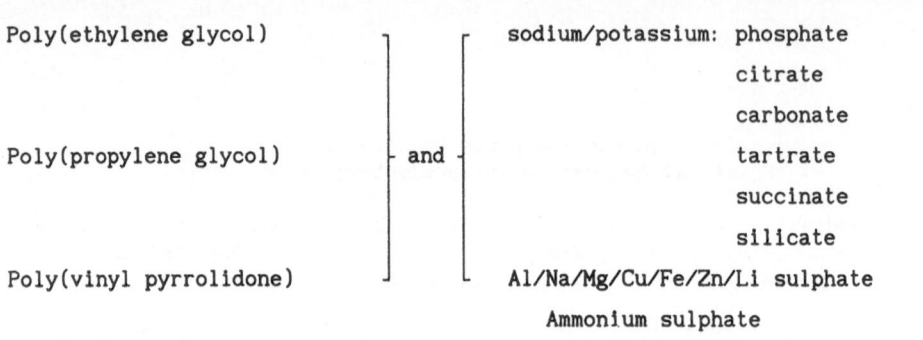

Poly(ethylene glycol)		sodium/potassium: phosphate
		citrate
		carbonate
Poly(propylene glycol)	and	tartrate
		succinate
		silicate
Poly(vinyl pyrrolidone)		Al/Na/Mg/Cu/Fe/Zn/Li sulphate
		Ammonium sulphate

diagram (see Figure 1). At all points above the binodal curve TXB, biphasic systems form. All systems on a line TB have identical compositions of top and bottom phases, but differing in phase ratio. To account for this the relative compositions of the phases of a given system are defined by the tie-line length [6].

Phase separation in polymer-polymer systems occurs as a result of the varied hydrogen bonding activities of polymers with water resulting in a repulsive "hydration pressure" [7]. PEG-salt systems form as a result of interactions between cations with the ether oxygen atoms of the polymer molecule. These interactions are mediated by the size and valency of associated anions such that large polyvalent anions effect exclusion of salt from the polymer surface region [4]. Such salt effects are also important factors determining partition in PEG-dextran systems arising from salts added to influence pH and ionic strength [8].

The present report focuses on factors of importance in the design of aqueous two-phase partitioning procedures for the isolation of a variety of macromolecular products. These range from a bulk protein fraction (of interest as a source of food functional protein or as a feedstock for multi-enzyme isolation), through specific enzyme preparations, to the production of monoclonal antibodies. Against this varied technical portfolio, the current status of the technique in modern productive biotechnology is discussed.

METHODS.

Detailed description of the methods used in the preparation of aqueous two-phase systems, microbial homogenates and the analysis of partitioning species by specific assay and electrophoretic analysis may be found elsewhere [9-12].

RESULTS.

Experimental observation here, and by other authors, confirms that the effects observed in the partitioning of proteins from yeast are not species specific but depend upon individual properties of macromolecules. Thus partition of macromolecules from a wide variety of sources, whilst

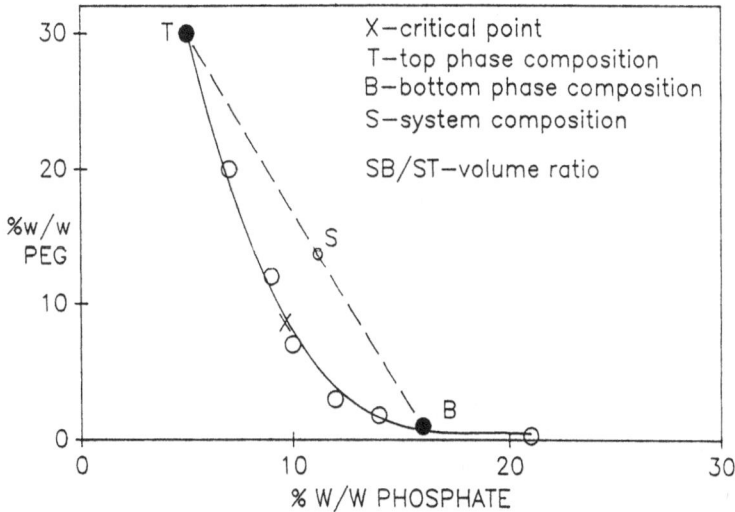

FIGURE 1. Phase diagram of PEG 3350 - phosphate aqueous two-phase system.

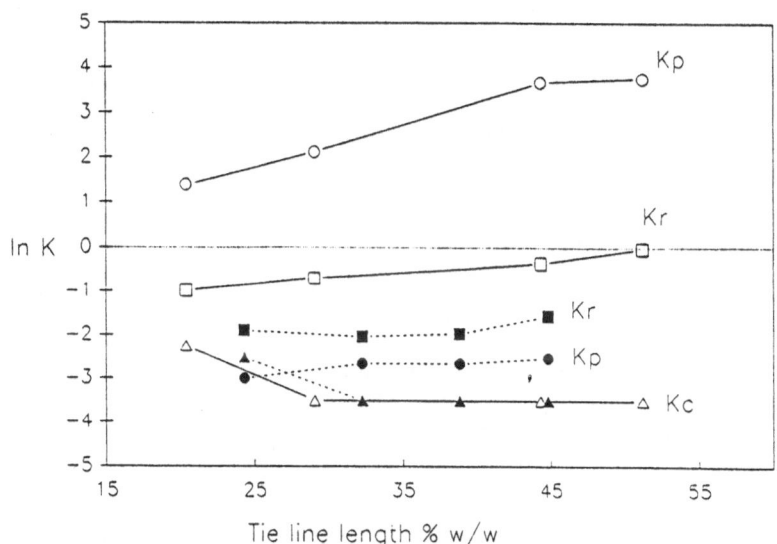

FIGURE 2. Variation of partition coefficient of products derived from yeast homogenate (40% w/v) with tie line length and molecular weight of PEG. Kp, Kr and Kc denote partition coefficients of protein, nucleic acid and carbohydrate respectively. Continuous lines indicate low molecular weight PEG (700 Daltons composed of 50% PEG400: 50% PEG1000). Broken lines indicate high molecular weight PEG (3350 Daltons). (Sigma Chemicals, Poole, Dorset)

differing in particular detail, may be expected to broadly parallel the present results.

Analysis of the results of the partition of intracellular proteins from wet milled bakers yeast in PEG-phosphate aqueous two-phase systems reveals that, for the majority of protein species, the useful molecular weight range of PEG extends from below 1000 Daltons to as high as 8000 Daltons. However, 3000 Daltons represents a more common upper limit. In addition increase in molecular weight of PEG brings about a general reduction of protein partition coefficients (see Figure 2). Only very large or very hydrophobic species, such as β-galactosidase or interferon-β, partition to the top, PEG-rich, phase in systems composed of high molecular weight PEG. It is evident that proteins having such unusual surface properties (which, for the examples given, are typically purified by hydrophobic interaction and reversed phase chromatography respectively) may be expected to be recoverable at relatively high degrees of purity from aqueous two-phase systems. The majority of water soluble intracellular proteins show changes in phase preference when partitioned in systems composed of PEG with average molecular weight between 1000 and 2000 Daltons. In such cases, resolution of particular protein species from the bulk proteins present, based upon differences in response to PEG molecular weight, may be expected to yield much lower degrees of purity than β-galactosidase and interferon-β.

In PEG-dextran systems, increasing the molecular weight of one of the phase forming polymers promotes a shift in partition of proteins to the opposite phase. The useful molecular weight ranges of PEG and dextran have been found to lie between 4000 Daltons and 40,000 Daltons and between 24,000 Daltons and 500,000 Daltons respectively. It has been shown that there is little or no relationship between protein molecular size and partition coefficient [15]. However, change in partition coefficient due to alteration of polymer molecular weight has been shown to be proportional to protein molecular size [15].

Increasing tie line length in PEG-salt aqueous two phase systems is associated with increase in protein partition coefficients (see Figure 2). Partition coefficients of nucleic acids also increase but to a lesser degree, whilst that of carbohydrate, after initial decrease, remains relatively constant. Among intracellular proteins derived from wet milled yeast, electrophoretic analysis (data not shown) revealed a small number of proteins having unusual properties in showing little tendency to partition to upper phases in response to manipulation of tie line length and PEG molecular weight [9]. This may reflect high carbohydrate content or other unusual surface properties..

The increase in partition coefficient of proteins in response to increase in tie line length may be attributed to changes in the relative composition of the phases as the tie line length is increased (see Figure 3). The concentration of phosphate remains relatively constant in the upper phase, where phosphate is at the limit of its solubility in the PEG solution despite the increase in tie line length. As a result phosphate concentration in the lower phase increases rapidly with increase in tie line length. The tendency for protein partition coefficients to increase is a consequence of the salting out effects of potassium phosphate mediated by the excluded volume effects of PEG. The partition coefficient of proteins is dependent upon their relative solubilities in each phase

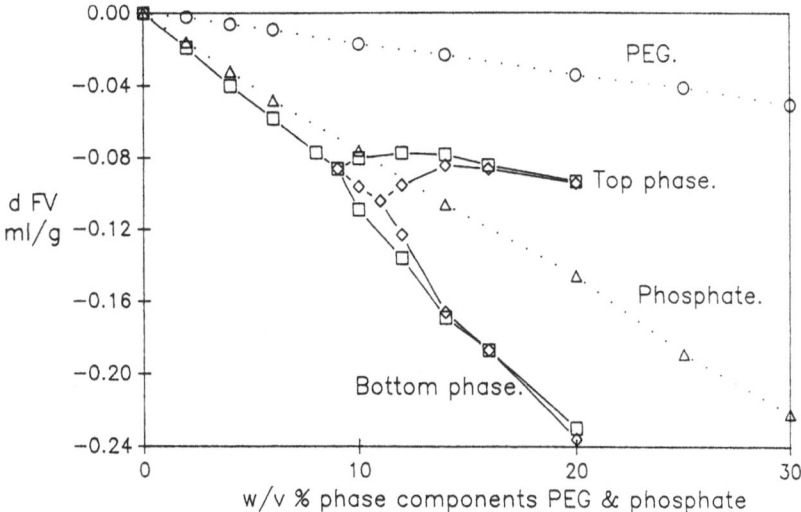

FIGURE 3. Change in free volume (dFV) with composition of PEG-phosphate aqueous two-phase systems. dFV is the specific volume of solvent and solutes compared to that of pure solvent.
    Molecular weight of PEG indicated by :- square - 1450, diamond - 3350.

and as tie line length increases proteins will ultimately be precipitated as a solid phase typically at the interface of top and bottom phase. This ultimate tendency to precipitate has been modelled [16] by combining modifications of Sinangolu's solvophobic theory with excluded volume models of PEG. In PEG-dextran systems the partition coefficient of proteins decreases with increase in tie-line length [17] and may be attributed to relative solubilities in the different polymer solutions.

    Addition of buffering salts to PEG-dextran systems critically affects the partitioning of macromolecules in relation to charge above and below their isoelectric points. Such effects have generally been attributed to differential partition of ions promoting an electrostatic potential difference between phases [4]. Other workers have stressed that the measured potential differences may simply reflect the differing hydration properties of the phases [8]. Hydrogen bonding ability and hydrostatic repulsion arising from the ability of some ions to associate with the pseudo-ionic structure of PEG and causing others to be rejected, may be expected to be important in allowing solvation of macromolecules in PEG phases associated with their surface charge. Proteins show changes in partition coefficient in relation to net charge above and below their isoelectric points, an effect which increases with tie line length. [18]. Proteins having net positive charge display increases in partition coefficient dependent upon salt type; perchlorate being more effective than chloride. Salts such as sulphates and phosphates reduce the partition coefficient of positively charged species. Proteins bearing net negative charge display opposite behaviour under these conditions.

    In PEG-potassium phosphate systems, increase in partition coefficient is seen above protein isoelectric points, and such effects are most pronounced in intermediate molecular weights of PEG. The effect is reduced or lost in higher molecular weights of PEG due to excluded volume effects

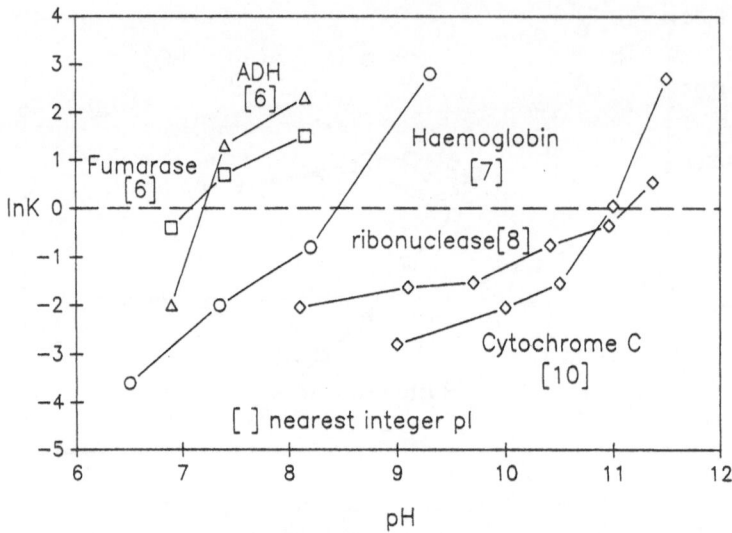

FIGURE 4. Variation of partition coefficient with system pH in PEG 1540 phosphate aqueous two-phase systems.

and is less pronounced where partition coefficients are already high, for example in lower molecular weights of PEG (see Figure 4).

Partition of proteins in PEG-salt systems has also been influenced by the addition of neutral salt, up to 1 mole NaCl/ Kg of system, when protein partition coefficients are initially lowered and then raised [2,19]. This may be analogous to the replacement of phosphate by chloride in PEG-dextran systems. The effect has been observed for positively charged particles where Q Sepharose changes preference from bottom to top phase in low molecular weights of PEG in PEG-phosphate systems on addition of NaCl [19]. The phenomenon seems not to have been rigorously investigated and is not seen in all cases.

Consideration of the overall properties of soluble intracellular proteins reveals them to be generally hydrophilic and acidic with native molecular weights in the range 60,000 Daltons to 200,000 Daltons. Only a minority have distinctively different properties such as high hydrophobicity, basic isoelectric point or very large size. Given the properties of aqueous two-phase systems, large and highly hydrophobic species may be partitioned to top phase of relatively high PEG molecular weight systems and high degrees of purity attained in a few steps. Smaller more hydrophilic species will require systems composed of PEG of lower molecular weight and so be subject to greater co-purification of proteins having similar characteristics. Similar considerations apply in relation to protein purification by manipulation of pH in aqueous two phase systems. Only basic proteins may be recovered to high degrees of purity by manipulation of pH in two stage extraction schemes as has been shown by reference to a simple model [4]. Addition of salts may in some cases be effective as indicated above. Ability to operate at lower pH in for example PEG-citrate systems may ameliorate but cannot solve this problem of resolution. This is because it is related to the similarity of surface properties of target species and total protein. Resolution of closely similar species cannot be expected in a few discrete equilibrium stages.

TABLE 2.

Single stage extraction of bulk yeast protein.						
	Top phase			Bottom phase		
	Protein	RNA	CHO	Protein	RNA	CHO
concentration mg/ml	20.6	1.08	0.88	5.5	2.49	5.96
partition coefficient	3.75	0.43	0.15			
fractional yield	0.74	0.25	0.10	0.26	0.25	0.90

RNA and CHO denote nucleic acid and carbohydrate respectively
Fractional yields are given for optimum resolution (4).

TABLE 3.

Two stage purification of cytochrome c and fumarase from waste yeast.				
cytochrome c			fumarase	
System I 17% PEG1540 7% K_3PO_4 pH 11.7			17% PEG1540 8% K_2HPO_4 pH 8.9	
System II 10% PEG1540 15% K_3PO_4 pH 10.8			70% Top phase I 7% K_2HPO_4 pH 6	
	System I	System II	System I	System II
K protein	14	5	0.2	0.61
K cytochrome	2.2	0.2	6.3	0.12
specific activity pM/mg or U/mg	247	2100	0.63	1.01
Purification Factor	2.2	19	4	6
Overall Recovery per cent	83	65	88	56

Other chromatographic techniques take advantage of increased numbers of theoretical stages, or displacement development, to acheive resolution.

In technological application, advantage is found in partitioning in PEG-salt aqueous two-phase systems in the presence of cell debris thus replacing a difficult solid/liquid separation step with a more favourable liquid/liquid separation. It has been reported that conditions can usually be found in which debris partitions in favour of the lower phase [2]. However, it appears that it is more usual for debris to form a third, solid phase at the interface which will contain some upper phase in such systems [13,19]. In this event, efficiency is lost in removing debris with the interface and restrictions on the choice of phase ratio are imposed.

The efficiency and degree of separation attainable in the isolation of proteins by partition in aqueous two-phase systems may be illustrated from the results of applying the technique to specific separations. Table 2 shows the preparation of a general bulk protein fraction from waste yeast in a single stage extraction using reduced molecular weight PEG-phosphate at relatively high pH. Partition of bulk protein to top phase is maximised and resolution from nucleic acids optimised by manipulation of the phase ratio.

Table 3 illustrates the extraction in two discrete stages of two proteins from yeast to varying degrees of purity by manipulation of system pH at each stage. In the first stage extraction (system I), high pH is utilised to partition cytochrome c to the top phase. However, under these conditions almost all proteins are above their isoelectric points (IEP) and, subject to mediation by excluded volume effects of PEG, will tend to partition to the same phase. As a consequence, the purification factor for this stage is low. In the first stage extraction of fumarase a lower pH is used and although some additional material, including cytochrome c, remains in the lower phase the purification factor remains low. In the second stage extraction (system II), in which the first stage top phase is used to form a new system, the pH is lowered in both cases to repartition the target protein to the lower phase. For cytochrome c this leaves the operating pH of the stage above the IEP of the bulk of the protein (4-6) which thus remains largely in the top phase. With fumarase, the required operating condition (pH 6) leads to considerable protein co-purification and the purification factor of the stage is much lower than for cytochrome c. Thus degrees of purity attainable in limited stage extractions are critically dependent on the degree to which targetted species differ in surface properties (principally hydrophobicity and charge) from the average properties of impurities in source materials - a feature self evident in all purification procedures. In some cases PEG-salt aqueous two phase systems may be unable to provide sufficient resolution from contaminating proteins without the increase of discriminatory power offered by exploitation of affinity interactions. Attempts to purify monoclonal antibodies from murine hybridoma cell culture supernatant was hindered by precipitation of the product. This was overcome by reduction in PEG molecular weight in the systems used, but this resulted in significant copurification of the major contaminant serum albumin. Under these circumstances, resolution was achieved by the inclusion of two solid phases, respectively derivatised with antibodies specific to monoclonal antibody product and serum albumin contaminant, and displaying partition coefficients favouring opposite phases. In this way fractionation of three principle components was achieved at a single step [12].

CONCLUSION.

We conclude from experimentation that the partition of biological products in aqueous two-phase systems is strongly influenced by the molecular mass and concentration of phase forming components, the operating pH of systems, and the molecular properties of phase and feedstock solutes. Surface charge and hydrophobicity of protein macromolecules, as mediated by system conditions of pH and salt concentration, are measureable parameters against which sytem productivity can be assessed. Simple descriptive models may account for observed partition behaviour. However, no mathematical models, of practical use in process design, describe aqueous two-phase systems in the context of these, or other molecular characters acting in concert to influence partition. In their absence, development and optimisation of aqueous two-phase processes must presently remain heuristic. This contrasts with operations such as adsorption and partition chromatography whose quantifiable dependance upon principal, bulk variables (charge, bioaffinity, mass etc) enables ready development into optimised, production processes. Experimentation also confirms the limited extent of purification possible for many types of biological product in a limited number of discrete partition stages, and the likely high inventory costs of such processes unless phase recycling can be efficiently implemented. We conclude that developments in molecular characterisation, enhanced selectivity of systems, and multi-stage processing are urgently required if the technology of aqueous two-phase partition is to successfully compete with extant unit operations of downstream processing.

REFERENCES.

1. Albertsson. P-Å., _Partition of cell particles and macromolecules_, Wiley, New York, 1986.

2. Kula. M-R., Liquid-liquid extraction of biopolymers. In _Comprehensive Biotechnology_, Vol. 2, Ed, Moo-Young. M., Pergamon, New York, 1986, pp 451-473.

3. Kroner. K.H., Hustedt. H. and Kula. M-R., Extractive enzyme recovery: economic considerations. _Proc. Biochem._, 1984, 19, pp 170-179.

4. Huddleston. J.G. and Lyddiatt. A., Aqueous two-phase systems in biochemical recovery: systematic analysis, design and implementation of practical processes for the recovery of proteins. _Arch. Biochem. Biotechnol_, 1990, In press.

5. Ananthapadmanabhan. K.P. and Goddard. E.D., A correlation between clouding and aqueous biphase formation in poly(ethylene oxide) inorganic salt systems, _J. Coll. Interface Sci._, 1986, 113, pp 294-296.

6. Bamberger. S., Brooks. D.E., Sharp. K.A., Van Alstine. J.M. and Webber. T.J., Preparation of phase systems and measurement of their physico-chemical properties, In _Partitioning in Aqueous Two-phase Systems_, Eds. Walter. H., Brooks. D.E. and Fisher. D., Academic Press, New York, 1985, pp 85-130.

7. van Oss. C.J., Chaudhury. M.K. and Good. R.J., The mechanism of partition in aqueous media, Sep. Sci. Technol.1987, 22 pp85-130.

8. Zaslavsky. B.Yu., Miheeva. L.M., Aleschko-Ozhevskii. Yu.P., Mahmudov. A.V., Bagirov. T.O. and Garaevaraev. E.S., Distribution of inorganic salts between the co-existing phases of aqueous polymer two-phase systems, J. Chrom., 1988, 439, pp 267-281.

9. Ottomar K.W., Evaluation of PEG-salt systems in the recovery of intracellular protein fractions from waste brewers yeast. MSc. Examination Thesis, University of Birmingham, 1988.

10. Ngonyani D., Characterisation of aqueous two-phase fractionation for the recovery of intracellular protein., MSc. Examination Thesis, University Birmingham. 1989

11. Huddleston. J.G. and Lyddiatt A., Aqueous two-phase separations in biochemical recovery. In Proc. I. Chem. E. Conf. on Adv. in Biochem. Eng., Newcastle upon Tyne, 1989, pp93-104.

12. Huddleston J.G. and Lyddiatt A., Aqueous two-phase partition in biochemical recovery from mammalian cell culture. In Separations using aqueous two-phase systems., Eds. Fisher D. and Sutherland I.A. Plenum, 1989, pp 309-317.

13. Köhler. K., von Bonsdorff-Lindeberg. L. and Enfors. S.O., Influence of disrupted biomass on the partitioning of β-galactosidase fused Protein-A., Enz. Microb. Technol., 1989, 11, pp 730-735.

14. Menge. U., Morr. M., Mayr. and Kula. M-R., Purification of human fibroblast interferon by extraction in aqueous two-phase systems. J. Appl. Biochem., 1983, 5, pp 75-90.

15. Albertsson. P-Å., Cajaravill. A., Brooks. D.E. and Tjerneld. F., Partition of proteins in aqueous polymer two-phase systems and the effect of molecular weight of the polymer, Biochim. et Biophys. Acta., 1987, 926, pp 87-93.

16. Kim. C.W., Interfacial condensation of biologicals in aqueous two phase systems, PhD Thesis, M.I.T., 1986.

17. Johansson. G. and Andersson. M., Liquid-liquid extraction of glycolytic enzymes from baker's yeast using triazine dye ligands, J. Chrom. 1984, 291, pp 175-180.

18. Johannson. G., Partitioning of proteins., In Partitioning in aqueous two-phase systems., Eds. Walter. H., Brooks. D.E. and Fisher. D., Academic Press, New York, 1985, pp 161-226.

19. Flanagan. J., Huddleston. J.G. and Lyddiatt. A., unpublished results.

UREASE IMMOBILIZED ON AN ACRYLIC CARRIER FOR ENZYMATIC REMOVAL OF UREA

BARBARA KRAJEWSKA, MACIEJ LESZKO, WIESŁAWA ZABORSKA
Faculty of Chemistry, Jagiellonian University
30-060 Kraków, Karasia 3, Poland

ABSTRACT

Urease was covalently immobilized on an acrylic enzyme carrier: aminated butyl acrylate-ethylenedimethacrylate copolymer using bifunctional agent glutaraldehyde. The enzyme retained 56 % of its original activity. Compared with free urease, the immobilized enzyme exhibited: 1) higher Michaelis constant, 2) insensitivity to pH changes in the range 5-7, and similar sensitivity at pH > 7, 3) considerably improved storage stability, 4) reusability. The urease-acrylic carrier system combined with an ammonium ion exchanger in a column operation proved efficient in removing urea from aqueous solution.

INTRODUCTION

In recent years interest in the application of enzyme reactors in industry, biotechnology, biomedicine etc. has markedly increased as a result of the preparation of immobilized enzymes with high lifetimes and good productivity (1). One such system, an immobilized urease, can be used for removal of urea from aqueous solutions via enzymatic hydrolysis to ammonium carbonate e.g. in purification of ground waters or in the treatment of uraemia.

Enzymes can be immobilized by physical and chemical techniques (2). Chemical bonding of enzymes is usually irreversible and the products are more stable than those obtained by physical techniques, but it often leads to profound changes in the physico-chemical properties of enzymes. It is essential to know these changes because they can be favourable or unfavourable for the intended application. Among different kinds of carriers used for covalent enzyme immobilization, synthetic acrylic carriers constitute a large group (Eupergit, Spheron, Enzacryl, Acrylex, Bio-Gel) (3).

In this study urease was covalently immobilized on an acrylic carrier, and the selected operational properties of free and immobilized urease were determined and compared. Additionally, the potential of the immobilized enzyme for practical application was tested in a column operation.

MATERIALS AND METHODS

The acrylic carrier under study was aminated butyl acrylate-ethylenedimethacrylate copolymer in the form of fine beads (4). The following are its basic physico-chemical properties (4): surface area - 30 m^2/g, porosity - 39 %, pore volume - 0.49 cm^3/g, amount of -NH_2 groups - 0.73 mmoles/g, water sorption - 1.36 g/g.

The urease was Sigma type IX of specific activity 53 units/mg protein. One unit is the amount of enzyme that liberates 1.0 µmol NH_3 from urea per minute at pH 7 and 25°C.

Glutaraldehyde (25 % w/v solution in water) was from BDH Chemicals Ltd, Poole, England.

Urea (analar grade) and all other chemicals (analar grade) were from POCh, Gliwice, Poland.

Urease and urea solutions were prepared as needed in phosphate buffers, ranging from pH 5.3 to 8.2, 22 mM containing 1 mM EDTA and NaCl added in such amounts that ionic strength of each of the buffers was I 0.07. Redistilled water was used throughout.

Urease was immobilized on the carrier according to the following procedure (5): 1 g of the carrier was treated with 5 cm^3 of 5 % w/v solution of glutaraldehyde in the phosphate buffer pH 7.0 for 1 h, then it was washed and reacted with 5 cm^3 of 1 % w/v urease solution in the phosphate buffer pH 7.0 for 3 h at room temperature and overnight at 4°C. Next the protein solution was filtered and the carrier was washed with the phosphate buffer pH 7.0 until the washings were free of urease.

The amount of bound enzyme protein was determined from the decrease in the protein content in the enzyme solution after immobilization by the method of Lowry et al. (6).

Activity of both free and immobilized urease was determined by measuring the amount of ammonia liberated from the urease-catalyzed hydrolysis of urea per minute, and was expressed in µmol NH_3/min mg protein for the free enzyme, and in µmol NH_3/min mg bound protein for the immobilized enzyme.

Unless otherwise stated, the reaction was carried out in the phosphate buffer pH 7.0 at 25°C, with urea concentration 10 g/dm³ i.e. 10 times higher than the estimated Michaelis constants. The concentration of ammonia was determined by the phenol-hypochlorite method (7).

RESULTS AND DISCUSSION

The measured activity of the urease-acrylic carrier system is equal to 1191.7 µmol NH_3/min g dry carrier, the determined amount of protein bound to the carrier is equal to 44.8 mg protein/g dry carrier, which means that the activity of the immobilized urease is equal to 26.6 µmol NH_3/min mg bound protein, and that the urease retained 56 % of its original activity.

Physico-chemical properties of free and immobilized urease

Kinetic parameters: Michaelis constant, K_M, maximum reaction rate, v_{max}
The kinetic parameters of both the ureases were determined by estimating the amount of ammonia liberated from the reactions at different urea concentrations. The results are presented in Fig. 1 in the form of Lineweaver-Burk plots, 1/v versus 1/(S), where v denotes reaction rate, and (S) substrate concentration.

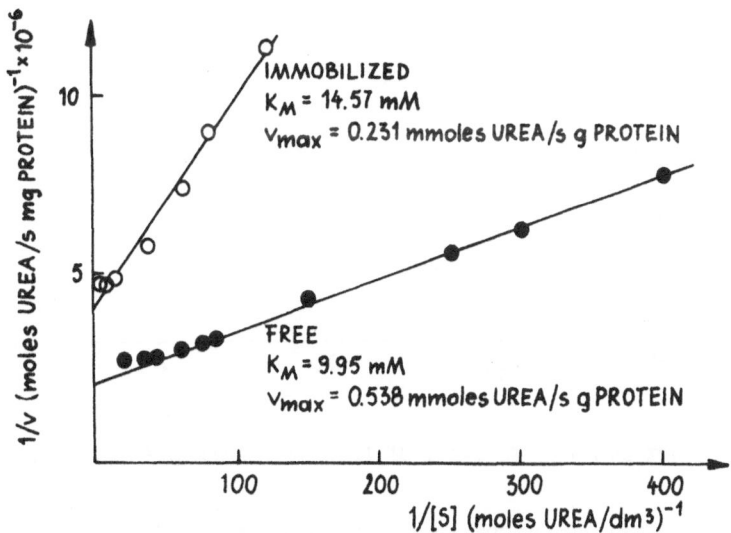

Figure 1. Lineweaver-Burk plots, 1/v versus 1/(S), for free and immobilized urease

As shown in Fig. 1 immobilization leads to an increase in the Michaelis constant from 9.95 mM for the free enzyme to 14.57 mM for the immobilized enzyme, and to a decrease in maximum reaction rate from 0.538 to 0.231 mmol urea/s g protein, respectively. The above effects result from changes introduced to the enzyme and to the enzymatic reaction by immobilization. These include: alteration of the enzyme active site, steric hindrance of the approach of substrate to the active site, alteration of the three-dimensional structure of the protein, diffusional limitations. The first two changes are usually responsible for the increase in K_M, whereas the latter two for the decrease in v_{max}. Both the increase in K_M and the decrease in v_{max} are responsible for the reduced activity of immobilized enzymes as compared to the corresponding free enzymes (8).

Effect of pH on activity

The effect of pH on activity of the ureases was studied in phosphate buffers in the pH range 5.3 - 8.2. The results are shown in Fig. 2.

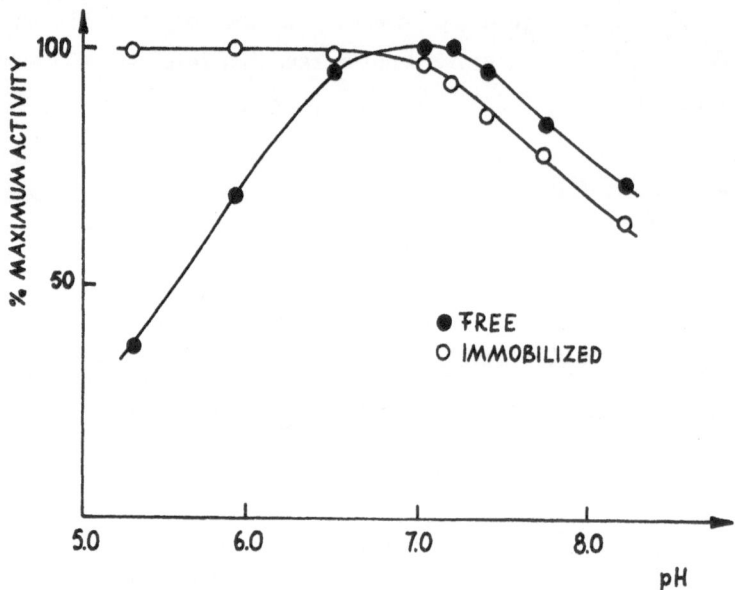

Figure 2. Effect of pH on activity of free and immobilized urease.

In contrast to free urease the immobilized urease is not sensitive to pH changes within the range of pH 5 - 7. At pH >7 both the enzymes are approximately equally sensitive to changes in pH. These results are in good agree-

ment with a general observation that positively charged supports displace
pH-activity curves of enzymes attached to them toward lower pH values. This
is usually explained as a microenvironment effect: at pH $<$ 7 $-NH_2$ groups of
the support are protonated which is why the local concentration of hydrogen
ions in the vicinity of the enzyme active site is lower than that in the
bulk solution, due to which the enzyme is stabilized (8).

Storage stability at 4 and 25^{0}C

A solution of the free urease in the phosphate buffer pH 7.0 (10 mg/100 cm³)
and several samples of the carrier with immobilized urease in the phosphate
buffer pH 7.0 were stored at 4 and 25^{0}C. The activities of consecutive
fresh samples taken from the store were determined periodically at 25^{0}C.
The results are presented in Fig. 3.

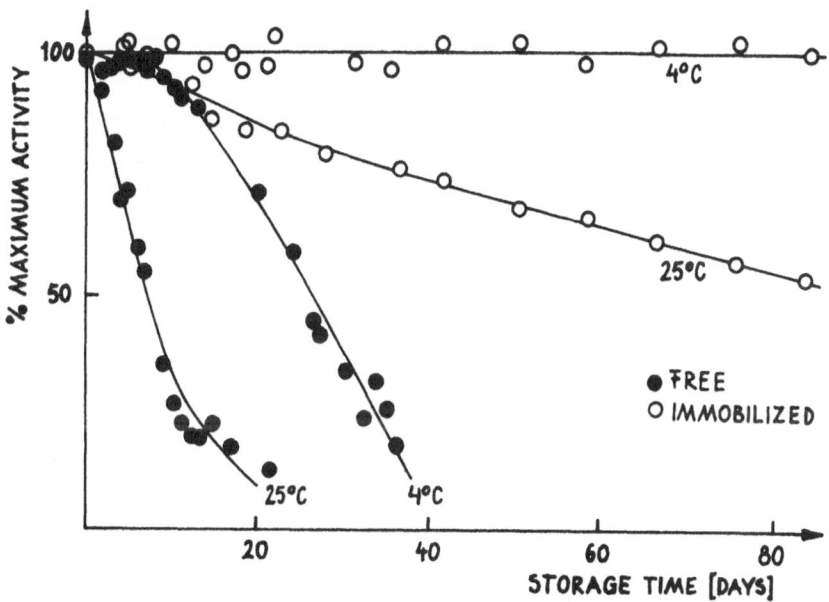

Figure 3. Storage stability of free and immobilized urease at 4 and 25^{0}C

Immobilization considerably stabilizes the enzyme during storage. The half-
times of activity decay of the free urease are equal to 25 and 7 days at 4
and 25^{0}C, respectively. The immobilized urease stored at 4^{0}C retained 100 %
of its original activity within the period 85 days, and stored at 25^{0}C lost
50 % of its original activity within the same period of time. The urease-

-acrylic carrier system, however, has to be stored in the wet state. Drying
of the system leads to an abrupt loss of enzymatic activity.

Reusability

Two columns, 1.0 cm i.d., were filled with the enzymatic carrier to a height
of 1.5 cm and with the phosphate buffer pH 7.0. One column was stored at
$4^{O}C$, the other one at $25^{O}C$. The activities of the columns were determined
periodically over the time of 200 days. The results are shown in Fig. 4.

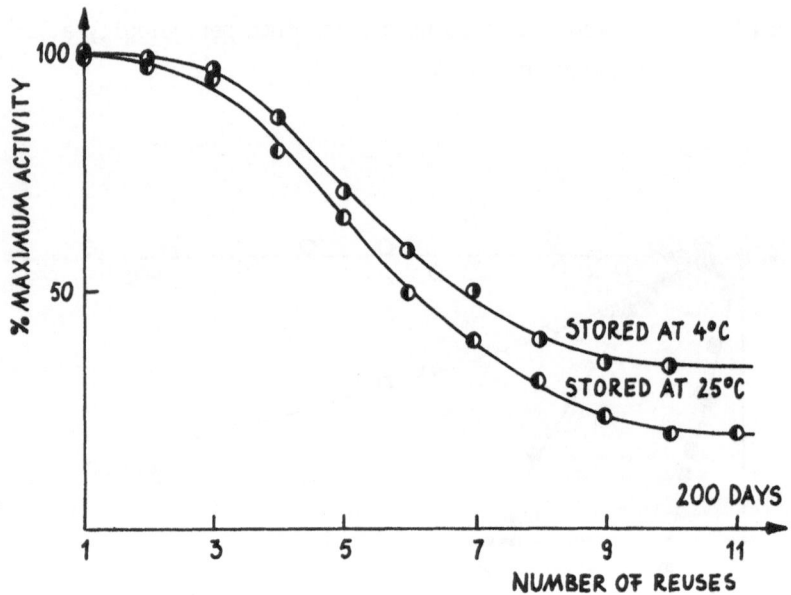

Figure 4. Reusability of immobilized urease

The activity of the column stored at $4^{O}C$ decreased to 50 % of its original
activity after 7 reuses performed within 30 days, whereas the activity of
the column stored at $25^{O}C$ decreased to the same level after 6 reuses within
60 days.

Column operation

A column, 2.0 cm i.d., was loaded with zeolite of erionite type (10 g, Na-
-K form, fraction of grade 0.16 - 0.49 mm) (9) to a height of 7 cm, and with
the previously prepared urease-acrylic carrier (2.83 g of moist preparation)
to a height of 1.4 cm (Fig. 5). The feed solution was a solution of urea,

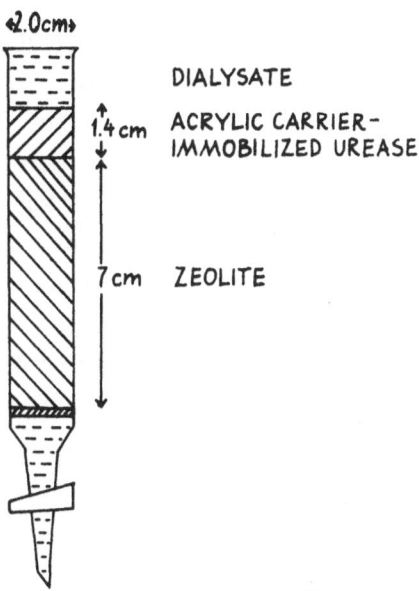

Figure 5. Schematic representation of the column used for removal of urea.

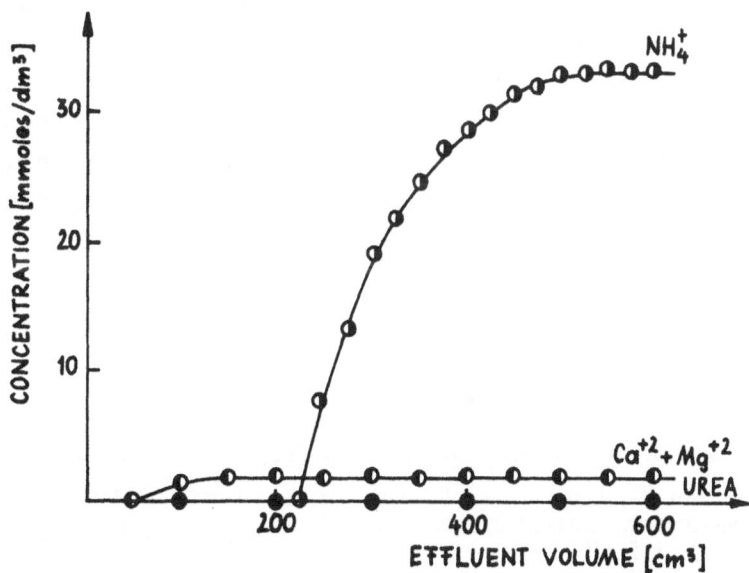

Figure 6. Removal of urea from dialysate solution (16.67 mmol urea/dm³) in the column presented in Fig. 5. Experimental details are described in the text.

similar to the dialysate solution flowing out of the artificial kidney. The following was its composition: urea - 16.67 mmol/dm^3, Ca^{+2} - 1.25 mmol/dm^3, Mg^{+2} - 0.75 mmol/dm^3, K$^+$ - 2.0 mmol/dm^3, Na$^+$ - 133 mmol/dm^3. The solution was passed through the column at a flow rate of 50 cm^3/h. The process was carried out continuously for 12 h at 25°C. The effluent was sampled period-ically and analyzed for urea, NH$_4^+$, and Ca^{+2}+ Mg^{+2}. As shown in Fig. 6 the concentration of urea was zero throughout the whole period of the process. Ammonium ion appeared in the effluent after 250 cm^3 of the feed solution had been passed i.e. after 5 h of the process. The total amount of urea hydrolized by the urease-acrylic carrier layer during 12 h of the process was 10 mmoles. The activity of urease decreased to 44 % of the initial value. After another 38 h of the same process it decreased to 12 %. The total amount of NH$_4^+$ ions exchanged by the zeolite layer was 10.12 mmoles, and of Ca^{+2}+ Mg^{+2} was 0.17 mmoles.

CONCLUSIONS

The presented study shows that the acrylic enzyme carrier: aminated butyl acrylate-ethylenedimethacrylate copolymer is a promising urease carrier. The activity retention of urease after immobilization is comparatively high. The immobilization improved physico-chemical properties of urease of practi-cal importance, which confirms usefulness of the material for practical application. The urease-acrylic carrier system performed well in the column test. The efficiency of urea removal via enzymatic hydrolysis to ammonium carbonate followed by ion exchange on zeolite was limited by the ion ex-change capacity of the zeolite.

REFERENCES

1. Zaborsky, O., Immobilized Enzymes, CRC Press, Cleveland, 1973.

2. Krajewska, B., Leszko, M. and Zaborska, W., Immobilization of urease for dialysate regeneration system of artificial kidney. Post. Fiz. Med., 1988, 23, 115-130.

3. Szewczuk, A. and Rapak, A., Immobilizowane enzymy - otrzymywanie i zasto-sowanie. Wiad. Chem., 1985, 39, 31-84.

4. Kolarz, B.N., Łobarzewski, J., Trochimczuk, A. and Wojaczyńska, M., Acrylic carriers for immobilization of enzymes. Angew. Makromol. Chem., 1989, 171, 201-211.

5. Lappi, D.A., Stolzenbach, F.E., Kaplan, N.O. and Kamen, N.D., Immobilization of hydrogenase on glass beads. Biochem. Biophys. Res. Commun.,
1976, **69**, 878-884.

6. Lowry, O.H., Rosebrough, N.J., Farr, A.L. and Randall, R.J., Protein
measurement with the Folin phenol reagent. J. Biol. Chem., 1951, **193**,
265-275.

7. Weatherburn, M.W., Phenol-hypochlorite reaction for determination of
ammonia. Anal. Chem., 1967, **39**, 971-974.

8. Royer, G.P., The kinetics of immobilized enzymes. In Immobilized Enzymes, Antigens, Antibodies, and Peptides, ed. H.H. Weetall, Marcel
Dekker, New York, 1975, pp. 49-91.

9. Cichocki, A., Crystallization field of zeolite T at $100^{0}C$ for a $SiO_2/$
Al_2O_3 ratio of 28 and crystallization sequences in the $Na_2O-K_2O-SiO_2^-$
$Al_2O_3^--H_2O$ system. J. Chem. Soc., Faraday Trans., 1985, **81**, 1297-1302.

This work was supported by the State Research Programme CPBP 02.11.

SOLID PHASES FOR IMMUNOAFFINITY ADSORPTION

Physical and biochemical influences upon
preparative and analytical performance

SUDESH B MOHAN, JULIA M MALHOTRA AND ANDREW LYDDIATT
Biochemical Recovery Laboratory,
School of Chemical Engineering,
University of Birmingham,
Edgbaston, Birmingham B15 2TT UK.

ABSTRACT

The application of novel silica and polymer supports in the construction of
immunoaffinity adsorbents is described. The influence of surface chemistry,
particle geometry, ligand and product size, and stoichiometry of adsorptive
association is considered in respect of preparative and analytical
applications of immunoaffinity adsorption. Practical requirements for an
improved understanding of physical and chemical constraints upon the
optimal design of efficient and productive bioselective adsorbents are
discussed in the context of preparative and analytical applications.

INTRODUCTION

The application of affinity chromatography is frequently proposed as the
solution for the practical problems of biorecovery and purification,
particularly when products are required in states of molecular homogeneity.
Covalent immobilisation of one partner of a pair of natural interactants
(antibody-antigen, receptor-hormone, enzyme-inhibitor etc) upon a solid
phase, such that the association continues in the face of non-specific
encounters with impurities, has proved a powerful purification strategy at
many scales of operation (1,2). However, successful affinity chromatography
rarely exploits the conventional properties of partition chromatography
reflected in contemporary developments of refined mono-disperse particles
having high physical strength, fast flow and minimal diffusion resistance
in fixed beds (3). Simple reconsideration of affinity chromatography as a
process of bioselective adsorption redefines the priority qualities of a
solid phase in terms of accessibility to ligands and products, high surface
area and capacity, and suitability for different feedstock contactors (eg
fixed or fluidised beds; 4).

Agarose-based solid phases are widely used in fixed bed applications of laboratory and production-scale affinity chromatography (5). In the last decade, many new silica and polymer-based materials have been promoted to compete with the dominance of agarose. These new materials (frequently of HPLC grade) have commonly been characterised by the chromatographic refinement referred to above (3,5,6). In contrast, properties of analyte or product accessibility, essential to the exploitation of properties of high surface area, low non-specific binding and physical strength, are less frequently invoked. For example, HPLC particles having 30nm pores are sold as wide-pore solid phases suited to protein separations. This clearly conflicts with the estimated physical diameters of the proteins commonly exploited as ligands and products (10-150 k Daltons, 20-100 nm; 7) and the widespread use in liquid chromatography (LC) of 4% agarose whose carbohydrate network corresponds to a pore size of 200-300 nm (8).

The present paper reports preliminary data concerning the assembly and characterisation of immunoadsorbents based upon three novel solid phases. Two are silica-based materials designed for the preparative LC market, having average particle and pore sizes of 50 μm and 20 or 50 nm respectively. The third is a polymer material (10 μm particle diameter; 100 nm pores) targeted at HPLC applications. Conclusions are drawn concerning the influence of surface chemistry and intra-particle geometry upon performance in preparative and analytical immunoaffinity adsorption.

MATERIALS AND METHODS

Production of monoclonal antibodies

Anti-human IgG-Fc monoclonal antibodies (MAB) were produced from continuous cultures of murine hybridomas (TB/C3; 9). Cell lines were donated by Dr R Jefferis, School of Medicine, University of Birmingham. Foetal calf serum (FCS; 5%) was replaced by 1% FCS plus 4% IgG-depleted serum (Gibco and Applied Protein Products respectively). Monoclonal antibodies were directly purified from clarified culture medium by immunoaffinity adsorption to fixed beds of human-IgG Sepharose CL-4B or Macrosorb K4AX arranged downstream of the culture vessel (9-12). Clarified, unfractionated culture medium was also used directly in the characterisation of immunoadsorbents.

Preparation of immunoadsorbents

Epoxy-silica particles (1 g) were employed with diameters of 40-60 μm, and average pore sizes of 20 and 50 nm as defined by size exclusion chromatography of polystyrenes (Sorbsil C200 and C500 respectively, specified and supplied by Crosfield Chemicals; 13). Particles were derivatised with bovine serum albumin (BSA; Sigma) or purified human IgG (11,12) in 6 ml of coupling buffer (0.1 M sodium pyrophosphate, pH 8). Reaction mixtures were agitated for up to 60 h at 4°C or at room temperature, clarified by sedimentation and unbound ligand displaced by decantation washing in coupling buffer. Remaining epoxy groups were blocked with 5 volumes of 0.3 M ethanolamine (pH 8.0) for 2 h, and adsorbents equilibrated and stored in phosphate buffered saline (PBS; 0.02 M potassium phosphate, pH 7.4 containing 0.15 M NaCl) with added 15 mM sodium azide.

Sorbsil C200 and C500 (10 g) was converted to the diol derivative by hydrolysis in 0.5 l of 0.1 M HCl at 90°C for 4 h. Diol silicas were washed in water, equilibrated in dioxane and activated at room temperature for 2 h by 1,1´-carbonyldiimidazole (CDI; 14). Reacted materials were washed in water, equilibrated in 2 volumes PBS and derivatised with selected concentrations of albumin and huIgG as above. Similar methodologies were applied to the activation and derivatisation of a hydrophilic macroporous polymer support (HMPS; Polymer Laboratories, Church Stretton) having average particle and pore diameters of 10 µm and 100 nm respectively.

Characterisation of solid phases

Known masses of drained immunoadsorbents were reacted to equilibrium with a range of concentrations of affinity purified MAB in batch suspensions at 4°C for 20 h. Reaction was terminated by centrifugation and the concentration of unbound antibody (c^*) determined by spectrophotometry at 280 nm (2,4). Equilibrium concentrations of bound antibody (q^*) were determined and the method of Chase (15) employed to estimate the maximum capacity (qm) and effective dissociation constant (Kd) of immunoadsorbents, where $c^*/q^* = (c^* + Kd)/qm$. The dissociation constant represented the ratio of backward and forward reaction rates of adsorption at equilibrium. Immunoadsorbents were tested in fixed beds (6 x 10 mm and 1 x 5 cm) equilibrated and operated in PBS with standard LC or HPLC equipment respectively (Pharmacia LKB; 11,12). Charged adsorbents were washed in PBS containing 1 M NaCl, and desorbed in 3 M KSCN in PBS. Recovered protein was desalted on-line in a Sephadex G-25 column sized appropriately to service upstream fixed bed adsorbents. Sorbsil C200 and C500 silicas, derivatised with huIgG and ethanolamine, or ethanolamine alone (see above), were equilibrated in PBS and characterised in gel permeation experiments with standard mixtures of proteins and carbohydrates. Columns (1 x 15 cm cm) were loaded with 0.2 ml or 0.05 ml respectively of sample and fractions eluted in PBS were monitored by continuous UV spectrophotometry.

Protein characterisation

Concentrations of purified MAB and BSA were determined from spectrophotometric absorbance at 280 nm (16). Enzyme linked immunosorbent assay (ELISA; 17) was adapted to estimate the immunochemical activities of MAB. Polystyrene microtitre plates were coated with 5 µg/ml huIgG antigen, and blocked with 2% casein. Sheep anti-mouse IgG-horse radish peroxidase conjugate was used to quantify bound MAB (standard or unknown) in the presence of 0.08 % (w/v) o-phenylene diamine in 0.1 M citrate-phosphate buffer pH 4.3 containing 0.08 % (v/v) hydrogen peroxide. The purity of protein samples was determined by silver diamine staining after electrophoresis on 10-15 % gradient acrylamide gels under native, denatured or unreduced conditions on a Phast System (Pharmacia-LKB; 18).

RESULTS AND DISCUSSION

Derivatisation of epoxy-silicas

Various authors have reported the necessity of extended reaction times of up to 5 days for effective derivatisation of epoxy-solid phases with protein ligands (19,20). Sorbsil epoxy silicas studied herein were reacted

with various concentrations of BSA in three buffer systems (PBS; 0.1 M
sodium hydrogen carbonate, pH 9.0 containing 0.5 M NaCl; and 0.1 M sodium
pyrophosphate, pH 8.0). Buffer composition or pH had no effect on protein
uptake (data not shown). Data from variously timed reactions in
pyrophosphate at 4°C or room temperature as shown in Table 1 suggests that
the reaction with BSA was essentially complete at 1 h.

TABLE 1

Derivatisation of Sorbsil C200 and C500 with BSA

Matrix	Total protein (mg)	BSA immobilised at 22°C (mg/ml)		BSA immobilised at 4 °C (mg/ml)	
		1 h	48 h	1 h	60 h
Sorbsil C200	6.0	1.7	1.6	nd	1.3
	12.0	nd	3.4[1]	3.1	3.0
	24.0	nd	7.3	7.5	7.5
	36.0	11.6	nd	12.3	14.1
Sorbsil C500	15.0	nd	4.4	nd	4.8[2]
	30.0	9.8	10.0	nd	9.7[2]

Matrix (1g; 3 ml) was agitated in 6 ml pyrophosphate containing BSA, and
unreacted protein determined by UV spectrophotometry as described.
Immobilised protein is expressed as mg/ml adsorbent matrix.
nd = not determined; 1,2 = terminated at 20 h and 24 h respectively.

Characterisation of epoxy-silica immunoadsorbents

Sorbsil C200 was challenged with 2.5 and 5.0 mg purified huIgG per ml of
matrix (0.33 g) for 24 h at 22°C to yield immobilised concentrations of
0.96 and 1.12 mg/ml respectively. The corresponding value for Sorbsil
C500 challenged with 7.5 mg huIgG per ml matrix was 2.1 mg/ml.
Immunoadsorbent capacities of matrices were tested by saturating fixed beds
with excess challenges of purified MAB or MAB in cell culture medium. ELISA
determinations indicated low recoveries of MAB (see Table 2) which could be
ascribed to poor initial binding, incomplete desorption in KSCN, or
immunochemical inactivation. The former, exacerbated by restricted
penetration of 20 and 50 nm pores (see later), is favoured since the
immobilised huIgG-anti-huIgG-Fc MAB system has previously shown excellent
properties of desorption and immunochemical stability in KSCN (9-12,21).
Productivities (P; mg MAB yields per mg immobilised huIgG) approaching 1.7
have been observed with 4 % agarose-huIgG adsorbents (11,12) where nominal
pore sizes of 300 nm (equivalent to a 2×10^6 Dalton protein rejection; 18)
are reported (8). Bacitracin derivatives of epoxy C200 have been used
successfully to purify neutral proteases (40 k Daltons) from Bacillus,

although low adsorbent capacities equivalent to between 0.2 to 0.4 mg/ml matrix were recorded (20).

TABLE 2

Performance characteristics of epoxy-silica immunoadsorbents

Matrix	HuIgG challenge (mg/ml)	HuIgG bound (mg/ml)	Productivity with pure MAB (ELISA mg/mg)	Productivity with crude MAB (ELISA mg/mg)
Sorbsil C200	2.5	0.96	0.09	nd
	5.0	1.12	0.21	0.20
Sorbsil C500	7.5	2.10	0.2	0.16

Matrix (1g; 3 ml) was agitated in 6 ml pyrophosphate containing huIgG, and unreacted protein determined by UV spectrophotometry as described. Prepared matrices were saturated in fixed beds with 10 ml purified MAB (2mg/ml)or 200 ml clarified culture medium (15-20 µg MAB/ml) at 1 ml/minute. After extensive washing with PBS containing 1 M NaCl, bound MAB was desorbed in PBS containing 3 M KSCN and quantified after on-line desalting by ELISA. nd = not determined.

Characterisation of diol-silica immunoadsorbents

Epoxy-silicas hydrolysed to the diol form as described were activated with CDI, challenged with huIgG as ligand, and characterised in terms of their equilibrium binding, specific binding activity and productivity with purified and crude preparations of anti-huIgG MAB (see Table 3). The CDI-activated diol derivative of Sorbsil C200 reacted with between 30 and 50% of challenges at coupling to yield ligand concentrations (L) of 1.3 and 2.1 mg huIgG per ml matrix. Higher values could not be obtained by increasing the huIgG challenge. HuIgG-C-200 with ligand concentrations of 1.3 mg per ml matrix displayed productivities (P) of less than 1 mg MAB per mg immobilised huIgG when saturated with MAB from crude cell culture medium. Higher ligand concentrations yielded lower productivities. In contrast, higher efficiencies (90-95%) and concentrations (2-6 mg/ml matrix) were obtained with huIgG coupled to CDI-activated Sorbsil C500. Estimates of the maximum capacity (qm) by equilibrium batch binding (15) yielded values of 2.6 to 5.8 mg/ml whilst dissociation constants in the range $2-6 \times 10^{-7}$ M approximated to values reported for the identical immunochemical interaction exploiting huIgG-agarose (9-12). Values of specific binding activity and productivity (qm/L and P respectively in Table 3) for intermediate ligand concentrations (L = 2.5 and 2.9 mg/ml matrix) equal or better those reported for huIgG agarose and kieselguhr-agarose (11,12) and approach the theoretical stoichiometry of 2 expected for MAB interactions with the constant Fc domain of huIgG. Productivity (P) of huIgG Sorbsil

C500 was associated with MAB displaying molecular homogeneity on silver-stained SDS-polyacrylamide electrophoresis (data not shown).

TABLE 3

Performance characteristics of diol-silica immunoadsorbents

Matrix	L (mg/ml)	qm (mg/ml)	Kd (M)	SBA (qm/L)	C (mg/ml)	P (C/L)
Sorbsil C200	1.3	nd	nd	nd	1.13	0.9
	2.1	nd	nd	nd	0.65	0.3
Sorbsil C500	1.9	2.64	2.7×10^{-7}	1.4	1.98	1.0
	2.5	3.83	4.6×10^{-7}	1.5	4.72	1.9
	2.9	4.73	3.6×10^{-7}	1.6	5.83	2.0
	6.4	5.84	2.7×10^{-7}	0.9	4.63	0.7

HuIgG diol-silicas, having characteristic ligand concentrations (L mg/ml matrix), maximum capacities (qm mg/ml matrix), molar dissociation constants (Kd) and specific binding activities (SBA), were saturated in fixed beds with anti-huIgG MAB in crude cell culture broth. Saturation capacities (C) were estimated from the recovery of MAB proven pure on SDS-PAGE. Productivity (P) represents the mass of antibody recovered per unit immobilised ligand (C/L). nd = not determined.

Immunoglobulin G, having a molecular mass of 150k Daltons, have a Stoke's radius of 10 nm (7). Specific associations of one or two MAB IgG molecules with a single huIgG molecule would be significantly larger, having a maximum molecular mass in excess of 450 k Daltons. The disappointing performance of hu-IgG Sorbsil C200, with average pore dimensions of 20 nm, may be ascribed to steric hindrance of ligands at coupling and MAB during adsorption. The preliminary data in Table 3 suggests that the accommodation of immunoglobulin G (whether covalently or immunochemically associated) within Sorbsil C200 is physically limited (estimated at less than 3 mg/ml matrix). In contrast, huIgG Sorbsil C500 has characteristic maximum capacities (qm), specific binding activity (SBA = qm/L) and productivity (C/L) equivalent to agarose materials used for the identical separation (11,12). This is despite having huIgG ligand concentrations less than the maximal 10-12 mg/ml matrix achieved with agarose (12). It should be emphasised that determinations of qm and C exploited equilibrium and saturation binding respectively, where diffusional resistance would be expected to be largely overcome.

It is well documented (and confirmed in Table 3) that maximal capacities and productivities decline with increasing ligand concentrations above values characteristic of individual solid phases (11,12). However, the relationship between ligand and product sizes, solid phase porosities and surface areas, and those immobilised ligand concentrations optimal for maximal productivities is rarely discussed in the separations literature.

Pore characterisation of diol-silicas by gel permeation.

Diol silicas (Sorbsil C200 and 500), activated with CDI and reacted with huIgG and ethanolamine (or ethanolamine alone), were characterised in respect of their pore accessibility to affinity ligands and products by simple gel permeation. Figure 1 illustrates the fractionation of a mixture of blue dextran, BSA and cytochrome $\underline{c}$, or blue dextran alone on C200 and C500 diol-silicas respectively capped with ethanolamine. Blue dextran (2×10^6) Daltons elutes in the void volume of both materials, but exhibited limited inclusion in the porous volume of C-500. Albumin (68 k Daltons) appears to be just included within the pore structure of diol C200. Fractionation of blue dextran, β-galactosidase (440 k Daltons) and cytochrome $\underline{c}$ (12k Daltons) on huIgG-C200 and C500 (data not shown) emphasised the additional restriction to pore penetration imposed by the presence of the large immobilised antigen molecule (150 k Daltons). This was particularly true of C200, supporting the conclusion that adsorptive performance of such material is constrained by poor access to large molecules.

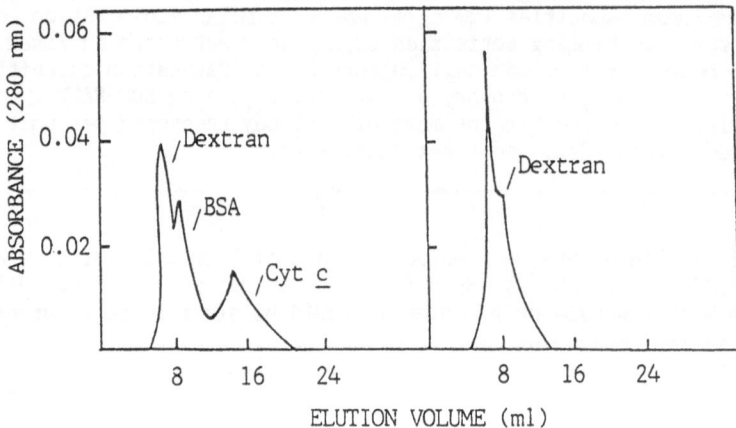

Figure 1. Gel permeation experiments on diol-Sorbsil C200 and C500

A mixture of (a) blue dextran, BSA and cytochrome $\underline{c}$, or (b) blue dextran alone was chromatographed on gel permeation columns (1 x 15 cm) of Sorbsil C200 and C500 respectively. Solid phases were activated by CDI and capped with ethanolamine. Mobile phase was PBS and column effluent monitored by continuous UV spectrophotometry.

Preliminary characterisation of polymer HPLC supports

Preliminary experiments were undertaken with hydrophilic macroporous polymer support (HMPS) materials. Quantification of CDI activation indicated (22) that the representation of reactive groups present on the surface of hydrophilic-coated polystyrene (10 um; 100nm pore size) was less than for diol-silicas. Table 4 confirms this character in respect of covalent immobilisation of huIgG (L), estimates of maximal capacity (qm) in batch binding experiments and specific binding activity (SBA). Estimates of the effective molar dissociation constant are lower than reported herein for Sorbsil silica (see Table 3), but these values were not reflected in reduced elution recoveries in 3M KSCN. Previous characterisation of the immobilised huIgG-MAB interaction (9-12,21,22), and other immunoaffinity systems (23), has suggested that the nature of the solid phase has little influence upon the strength of interaction.

TABLE 4
Characterisation of polymer HPLC supports

Matrix	L (mg/ml)	qm (mg/ml)	Kd (M)	SBA
HMPS (10 um;100 nm pore)	0.5	0.8	5×10^{-8}	1.6
	1.3	0.6	7×10^{-9}	0.5
	2.2	0.5	9×10^{-8}	0.2

HMPS was activated by CDI and coupled with huIgG to yield ligand concentrations (L mg/ml matrix). Maximum capacities (qm mg/ml) and molar dissociation constants (Kd) were estimated from batch binding experiments. SBA = specific binding activity (qm/L)

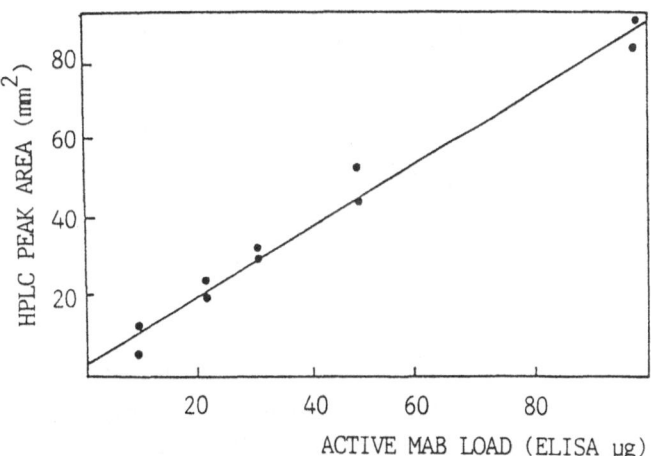

FIGURE 2. Bioquantitation of monoclonal antibodies with huIgG-HMPS

Peak areas of UV traces of pure MAB eluted from mini-preparative HPLC columns (0.3 ml) previously challenged with MAB in crude culture broth are plotted against ELISA estimations of original immunochemical activity.

Preparative applications of HMPS require further investigation in respect of increasing immunoadsorbent capacity. However, the materials assembled here were successfully utilised in miniaturised preparative systems for the bioquantitation of MAB produced in animal cell culture (21). Micro-HPLC columns (0.3 ml) were loaded with 0.2 ml MAB (10-100 ug/ml) in crude cell culture medium (2-3 mg/ml total protein) and the protein concentration of recovered pure MAB related to the immunochemical activity (ELISA) in the original sample (see Figure 2). Quantitation by immunoaffinity HPLC agreed well with ELISA data in analyses taking less than one tenth of the time. The prospect for near-online affinity HPLC assuming a role in bioquantitation and control of bioprocesses such as integrated fermentation and product recovery are good (11,21,22).

CONCLUSION

We conclude that Sorbsil LC silicas have interesting properties as bioselective adsorbents, particularly when diol chemistries are activated by agents such as CDI prior to ligand immobilisation. Comparative study of MAB adsorption to huIgG C200 and C500 emphasised the importance of selecting material with pore sizes appropriate to the dimensions of ligand and product molecules. Reference to molecular size, and simple gel permeation experiments, confirmed the suitability of Sorbsil C500 as an immunoaffinity adsorbent rivalling agarose in terms of productivity of pure MAB. Low levels of non-specific binding, high physical strength, fast flow characters and suitability for liquid fluidisation (data not shown) support that recommendation. Polymer HPLC supports possessed the purification power required of preparative media, but lacked capacity when derivatised with CDI chemistry and huIgG ligands. However, repetitive operation in miniaturised preparative systems enabled rapid immunochemical quantitation of monoclonal antibodies in crude culture feedstocks appropriate to near on-line control of integrated fermentation and recovery. This preliminary study served to emphasise the importance of relating solid phase design to the physical and biochemical nature of individual separation systems.

ACKNOWLEDGEMENT

SBM gratefully acknowledges the support of the SERC Rolling Programme in Biochemical Engineering at the University of Birmingham. JMM gratefully acknowledges the support of an SERC-CASE Studentship with Polymer Laboratories.

REFERENCES

1. Katoh, S., Scaling-up affinity chromatography, TIBTECH, 1987, 5, 281-286.
2. Hill, E.A. and Hirtenstein, M.D., Affinity Chromatography: its application to industrial scale processes. In Advances in Biotechnological Processes, Alan Liss, N.Y., 1983, 1, 31-66.
3. Clonis, Y.D., Large-scale affinity chromatography, Bio/Technology, 5, 1290-1293.
4. Lyddiatt, A., Solid phases and product contactors: new options for bioselective adsorption. In World Biotechnology Report, Online Publications. London, UK, 1988, 167-172.

5. Groman, E.V. and Wilchek, M., Recent developments in affinity chromatography supports. TIBTECH, 1987, 5, 220-224.
6. Kennedy, J.F., Rivera,, Z.S. and White, C.A., The use of HPLC in biotechnology. J. Biotechnology, 1989, 9, 83-106.
7. Porschka, M., Universal calibration of gel permeation chromatography and determination of molecular shape in solution. Anal. Biochem., 1987, 162, 42-64.
8. Attwood, T.K., Nelmes, B.J. and Sellen, D.B, Electron micrsocopy of beaded agarose gels, 1988. Biopolymers, 27, 201-212.
9. Rudge, J., Desai, M.A., Shojaosadaty, S.A. and Lyddiatt, A., Continuous culture of murine hybridomas with integrated recovery of monoclonal antibodies. In Modern Approaches to Animal Cell Technology, ed, R.E. Spier and J.B. Griffiths, Butterworth, London, 1987, pp 556-575.
10. Desai, M.A., Huddleston, J.G., Lyddiatt, A., Rudge, J. and Stevens, A.B., Biochemical and physical characterisation of a composite solid phase. In Separations for Biotechnology, ed M.S. Verrall and M.J. Hudson, Ellis-Horwood, Chichester, UK, 1987, pp 200-209.
11. Lyddiatt, A., Desai, M.A., Huddleston, J.G., Rudge, J and Shojaosadaty, S.A., Controlled assembly and operation of immunoadsorbents. J. Chem. Tech. Biotech., (1989), 45, 47-60.
12. Desai, M.A. and Lyddiatt, A., Comparative studies of agarose and kieselguhr-agarose composites for the preparation and operation of immunosorbents. Bioseparation, 1990, In press.
13. Chappell, I., Preparative media design. Laboratory Practice, 1988, 36, 61-64.
14. Bethell, G.S., Ayers, J.S., Hancock, W.S. and Hearn, M.T.W., A novel method of activation of cross-linked agaroses which gives a matrix for affinity chrommatography devoid of additional charged groups. J. Biol. Chem., 1979, 254, 2572-2574.
15. Chase, H. A., Prediction of the performance of preparative affinity chromatography. J. Chromatog., 297, 179-202.
16. Goding, J.W., Monoclonal Antobodies: Principles and Practice, Academic Press., London,. 1983, pps 98-127.
17. Voller, A., Bidwell, D.E. and Bartlett, A., Enzymatic immunoassays in diagnostic medicine: theory and practice. Bulletin of World Health Organisation, 1976, 53, 55-65.
18. Pharmacia Technical Literature, Uppsala, Sweden, 1989.
19. Larsson, P-O.,High performance affinity chromatography. In Methods in Enzymology, 1989, 18 209-220.
20 Burg, B.V.D., Eijsink, V.G.M., Stulp, B.K. and Venemam, G., One step affinity purification of Bacillus neutral proteases using Bacitracin-silica. J. Biochem. Biophys. Methods, 1989, 18, 209-220.
21. Shojaosadaty, S.A. and Lyddiatt, A. Application of affinity HPLC to recovery and monitoring operations in biotechnology, In Separations for Biotechnology, ed. M.S. Verrall and M.J. Hudson, Ellis-Horwood, Chichester, UK, 1987, pp 436-444.
22. Malhotra, J.M., Characterisation and application of a novel solid phase for bioaffinity HPLC. M. Phil. Thesis, University of Birmingham, 1989.
23. Fowell, S.L. and Chase, H.A., A comparison of some activated matrices for preparation of immunoadsorbents, J. Biotechnol., 1986, 355-368.

THE RELATIONSHIP BETWEEN PROTEIN PARTITION COEFFICIENT
AND POLYMER CONCENTRATION IN AQUEOUS TWO-PHASE SYSTEMS

GORDON W NIVEN, SUSAN J SMITH, PETER G SCURLOCK,
ANTHONY T ANDREWS
AFRC Institute of Food Research, Reading Laboratory
Reading, Berkshire RG2 9AT

ABSTRACT

We have observed that in a series of "equivalent" two-phase aqueous
systems (systems with varying polymer concentrations but constant phase
volume ratio), a direct relationship exists between the polyethylene
glycol (PEG) concentration and the reciprocal of the partition
coefficient of proteins. From a series of simple experiments a constant
can therefore be determined which describes the partition of a
particular protein at any point in the phase diagram. This is also true
for the partition of total protein in a complex mixture. This
information may simplify the optimization of the purification of
proteins using two-phase aqueous systems.

INTRODUCTION

Liquid two-phase systems, most often consisting of immiscible aqueous
and organic solvents, have been used by industry for many years for the
extraction of organic materials. However, such systems are not usually
suitable for the extraction of sensitive biological materials where the
maintenance of biological activity is required. By comparison,
two-phase systems consisting of mutually exclusive aqueous solutions of
polymers, or a polymer combined with a concentrated salt solution, are
compatible with many biological materials. These systems offer the
potential of large scale partitioning of cells, organelles or
macromolecules using relatively inexpensive equipment [1].

Various factors affect the partitioning of macromolecules in
two-phase aqueous systems. These include characteristics of the solvent

system such as polymer molecular weight and concentration, pH and salt composition; and characteristics of the molecule to be partitioned such as charge, hydrophobicity and size [2]. Although various mathematical models exist which describe partition in two-phase systems [3], the optimization of the purification of one component from a crude extract of biomass is complex and time consuming. Process optimization would be considerably simplified if it were possible to predict the partition coefficient of a solute at any point on the phase diagram using a minimum of parameters which are easily determined experimentally. Such a method has been derived by Diamond & Hsu [4] who determined that the difference in polymer concentration between the phases is directly proportional to the natural logarithm of the partition coefficient. This allows the determination of a constant which describes the behaviour of a partioned species in a particular two-phase system. The main drawback of this approach in practical terms is that this relationship only holds true for proteins of molecular weight less than 20 KDa.

We have observed that, when a series of two-phase systems of constant phase volume ratio are used, a direct relationship exists between the PEG concentration and the reciprocal of the partition coefficient. This method allows the determination of a constant which describes solute partition at any point on a phase diagram.

MATERIALS AND METHODS

All aqueous two-phase systems were 1.0 g final weight containing 0.1 g of 10 mg ml^{-1} pure protein solutions. PEG-dextran systems were buffered by the inclusion of 0.1 g of 0.1 M Tris/HCl, pH 7.5. After thorough mixing, phases were separated by centrifugation at 1000 rpm for 5 min.

Protein concentrations were determined by the method of Bradford [5].

Cell-free extracts of Steptococcus lactis NCDO 712 were prepared from cultures grown in L-M17 broth. Cells were harvested by centrifugation, washed and resuspended in 50 mM Tris/HCl at pH 7.5. After cell disruption by 5 passes through a French Press at 100 MPa, cell debris was removed by centrifugation. 0.1 g samples of crude protein preparation containing approximately 1.0 mg protein were applied to two-phase systems.

Glutamate aminopeptidase (GAP) and lysine aminopeptidase (LAP)
activities were assayed by measuring the hydrolysis of glutamic acid
α-4-nitroanilide or lysine p-nitroanilide in reaction mixtures which
contained 1 mM substrate, 50 mM Tris/HCl pH 7.5 and 0.1 ml of enzyme in a
total volume of 1.0 ml. Tubes were incubated at 50 °C (GAP) or 35 °C
(LAP) for 35 min and the reaction was terminated by the addition of 0.5 ml
30% (v/v) acetic acid. After centrifugation, p-nitroanilide concentration
was determined by measurement of the absorbance 410 nm.

RESULTS

For each of the two-phase systems studies, series of systems were used
in which the concentrations of the components were such that the volumes
of the upper and lower phases were equal (Table 1).

TABLE 1

Concentrations of components in two-phase systems

I		II		III		IV		
PEG8	Dx480	PEG4	Dx480	PEG8	K_2HPO_4	PEG8	K_2HPO_4	KH_2PO_4
				(% w/w)				
4.0	5.3	7.0	7.0	11.75	9.25	11.5	3.75	3.75
4.2	7.0	8.0	8.0	12.5	10.0	12.0	4.0	4.0
4.5	8.0	9.0	9.0	13.0	10.5	12.5	4.25	4.25
5.0	9.2	9.5	9.5	13.5	11.0	13.0	4.5	4.5
5.5	10.3	10.0	10.0			13.5	4.75	4.75
6.0	11.1							

PEG8 = polyethylene glycol, MW 8000
PEG4 = polyethylene glycol, MW 4000
Dx480 = dextran, MW 480000

The partition of samples of pure chymotrypsin and β-galactosidase
were measured in each system and plots of PEG concentration against 1/k
are shown in Figures 1-4.

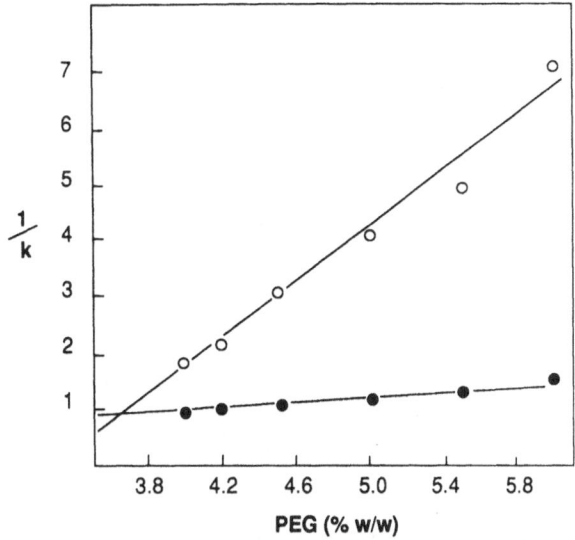

Figure 1. Plot of the reciprocal of the partition coefficients of chymotrypsin (●) and β–galactosidase (○) against PEG concentration in a PEG 8000–dextran 480 aqueous two-phase system (I)

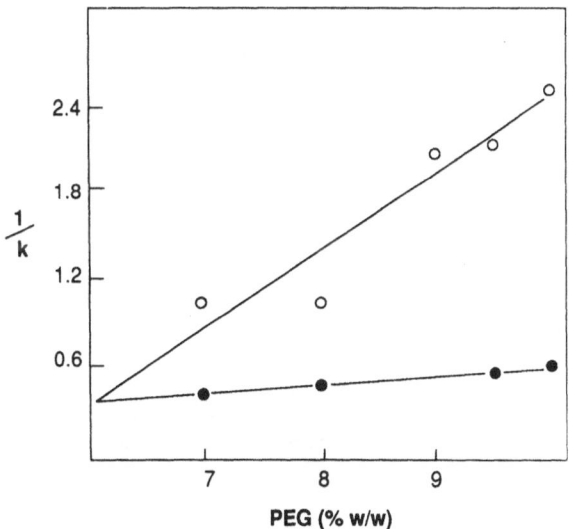

Figure 2. Plot of the reciprocal of the partition coefficients of chymotrypsin (●) and β–galactosidase (○) against PEG concentration in a PEG 4000–dextran 480 aqueous two-phase system (II)

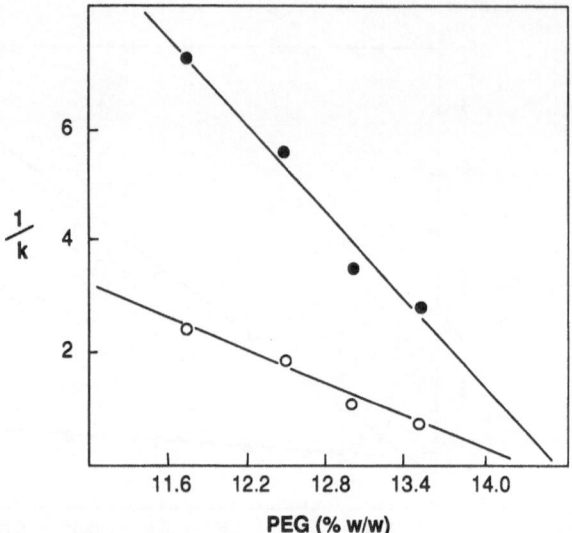

Figure 3. Plot of the reciprocal of the partition coefficients of chymotrypsin (●) and β–galactosidase (○) against PEG concentration in a PEG 8000–K_2HPO_4 aqueous two-phase system (II)

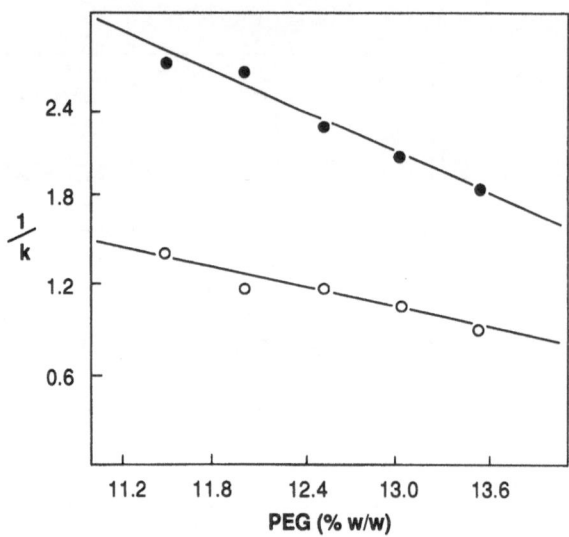

Figure 4. Plot of the reciprocal of the partition coefficients of chymotrypsin (●) and β–galactosidase (○) against PEG concentration in a PEG 8000–K_2HPO_4–KH_2PO_4 aqueous two-phase system (IV)

In each figure the linear regression coefficients were greater than 0.98 with the exception of β-galactosidase in Figure 2, the value for which was 0.95. The gradients for both proteins were positive in PEG–dextran systems and negative in PEG–phosphate systems. This implies that the separation of a mixture of two such proteins would be more effective at low PEG concentrations in PEG–phosphate systems and at high PEG concentrations in PEG–dextran systems. Of the four systems tested, the difference in gradient for the two proteins was greatest in the PEG 8000–dextran 480 system. It is therefore likely that this system would be most suitable for the separation of such proteins.

The partition of total protein, glutamate aminopeptidase (GAP) activity and lysine aminopeptidase (LAP) activity in a cell-free extract of S. lactis were compared in PEG 4000–dextran 480 two-phase systems. The gradients of plots of PEG concentration against $1/k$ for total protein, GAP and LAP were 1.28, 5.72 and 1.12, respectively (Figure 5). The similarity of the gradients for total protein and LAP suggest that this system is not suitable for the purification of LAP. In contrast, the gradient for GAP was greater which indicates a greater purification in the bottom phase. The purifications obtained at the highest PEG concentration were 1.2 fold for LAP and 2.6 fold for GAP, as assessed by specific activity.

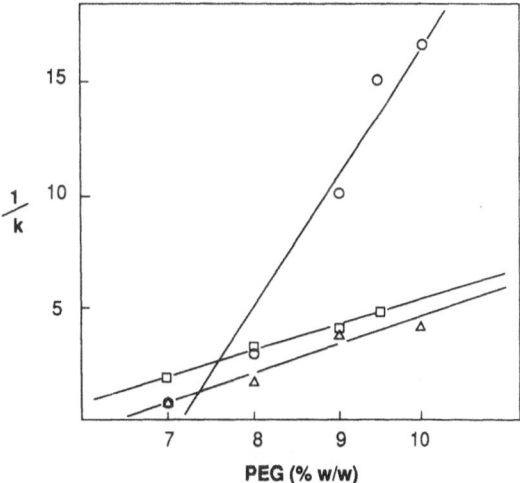

Figure 5. Plot of the reciprocal of the partition coefficients of total protein (△), GAP activity (○) and LAP activiy (□) in a cell-free extract of S. lactis against PEG concentrations in a PEG 4000–dextran 480 aqueous two-phase system (II)

CONCLUSIONS

The exact extent and limits of this relationship have not yet been fully determined. However, this method has two main potential uses. It allows a direct comparison of the partition of a compound in various different two-phase systems. For example, often in the literature data are presented showing the effect of polymer molecular weight on protein partitioning. This usually involves using the same polymer concentrations for each molecular weight which is studied. These concentrations represent different points on their respective phase diagrams and so equivalent points are not compared. A comparison of the gradients of plots of polymer concentration against $1/k$ may be more valid.

Secondly, as this method is applicable to complex mixtures of proteins, a direct comparison of the partition of a target protein and its source material may be made. This would give a direct assessment of the suitability of a particular system for the purification of a protein across the range of possible polymer concentrations. Using only a few selected polymer concentrations which result in a constant phase volume ratio, a large number of two-phase systems may be screened to determine the system which is most likely to give a satisfactory purification.

REFERENCES

1. Albertsson, P.-A., Partition of Cell Particles and Macromolecules, John Wiley & Sons, New York, 1986.

2. Johansson, G., Partitioning of proteins. In Partitioning in Aqueous Two-phase Systems, eds. H. Walker, D. E. Brooks and D. Fisher, Academic Press, London, 1985, pp. 161-226.

3. Baskir, J. N. and Hatton, T. A., An overview of theoretical developments for protein partitioning in aqueous two-phase systems. In Separations Using Aqueous Phase Systems, eds. D. Fisher and I. A. Sutherland, Plenum Press, London, 1989, pp. 217-27.

4. Diamond, A. D. and Hsu, J. T., Fundamental studies of biomolecule partitioning in aqueous two-phase systems. Biotechnol. Bioeng., 1989, 34, 1000-14.

5. Bradford, M. M., A rapid and sensitive method for the quantitation of microgram quantities of protein utilizing the principle of protein-dye binding. Anal. Biochem., 1976, 72, 248-54.

PROTEIN FOULING AND ITS IMPLICATIONS FOR SELECTION OF ULTRAFILTRATION MEMBRANES

I M REED J M SHELDON
AEA Environment and Energy, Harwell Laboratory,
Oxfordshire OX11 ORA, UK

ABSTRACT

This poster describes techniques for studying the interaction of proteins
with ultrafiltration membranes with the aim of providing guidelines on
membrane selection. It is shown that measurement of protein adsorption to
membranes under static conditions, ie. where there is no flow and minimal
hydrostatic pressure, can give a useful indication of the degree of fouling
during ultrafiltration. Electron microscope studies reported here reveal
that certain membrane materials can alter tertiary protein structure which
might lead to an irreversible loss in protein function. The implications
of these findings for membrane selection are discussed.

INTRODUCTION

The performance of ultrafiltration membranes is limited in many
applications by the interaction of the membrane with components of the feed
stream. Protein adsorption for example can cause substantial flux decline
in certain types of membrane [1,2] and, in addition, some membrane
materials may alter protein structure possibly leading to a loss in
activity. With so many types of ultrafiltration membrane available, both
organic and inorganic, it is important to make the correct choice for a
particular application. This poster describes techniques developed by
BIOSEP, the biotechnology separations club of the UK Department of Trade
and Industry, for studying solute-membrane interaction with the aim of
providing guidelines on membrane selection especially in applications
involving proteins.

Protein Adsorption Studies

Studies of protein-membrane interaction were carried out using bovine serum albumin (BSA). Several methods have been used to measure protein adsorption to membranes [3-5]. Here, quantification of adsorbed protein was achieved by labelling the BSA with the radioactive isotope iodine-125 and using γ-ray counting equipment to measure the activity of membrane samples exposed to protein solutions.

Protein adsorption to membranes was studied under both filtration and static conditions, ie. where there was no liquid flow through the membrane and minimal hydrostatic pressure.

Static tests were carried out by placing a membrane disk, separating surface uppermost, in a bottle top and putting protein solution into the bottle itself. The bottle was then inverted and placed in a thermostatically controlled incubator. After a measured period of time the membrane was removed, washed to remove unbound protein and the quantity of adsorbed protein was measured.

Protein adsorption under filtration conditions was measured using an Amicon 8010 stirred cell. Each membrane was used to filter a fixed volume of protein solution. The procedure for determining the amount of protein adsorbed was the same as that used in the static tests.

Cross-flow Filtration Studies

Cross-flow ultrafiltration experiments were carried out using an Amicon TR1C flat sheet module. This module accommodates 150mm diameter membranes and has a flow guide which directs the feed stream over the membrane in a spiral channel 14.3mm wide, 0.8mm high and 0.9m in length. Tests were carried out over runs of 6 hours duration at a Reynold's number of 3900 and a temperature of about 10°C.

Protein

The protein used in the static and stirred cell tests was sigma A0281 grade BSA. This was a highly purified, essentially fat and globulin free BSA supplied in lyophilized crystalline form. Cross-flow experiments were carried out using Fraction V BSA supplied by BDH. This was a cruder crystalline preparation containing 96-98% BSA.

Electron Microscopy

There are many examples of the use of scanning electron microscopy for examination of membranes (eg. [6, 7]). Transmission electron microscopy (TEM) has also proved a useful tool in the examination of clean and protein fouled ultrafiltration membranes [8]. One technique, freeze-fracture combined with deep-etching, has been used to obtain information on membrane surface/protein interactions. The procedure followed to obtain images using this method are outlined below.

Small pieces of the membrane were mounted in 2M sucrose on specimen holders orientated in such a way that on fracturing a cross section would be obtained. Specimens were then rapidly frozen by plunging into liquid propane and fractured at 166K (polysulphone) or 93K (regenerated cellulose) in a freeze-fracture unit. After etching of the fracture surface by subliming water from the frozen specimen replicas were made of the sample surface by rotary shadowing with carbon/platinum and replicating with carbon.

These replicas were removed from the specimen by dissolving the membrane using 1-methyl-2-pyrrolidinone in the case of polysulphone membranes of 70% H_2SO_4 for reconstituted cellulose. After washing in distilled water replicas were treated with a 70% bleach solution to remove any protein residue. They were then washed and picked up on formvar-coated grids for examination with the TEM.

RESULTS AND DISCUSSION

Static Adsorption
A range of ultrafiltration membranes were exposed to a 0.5% solution of BSA for 1 hour at pH7 and 25°C under static conditions. Table 1 presents results for membranes with 3 different molecular weight cut-offs and 3 membrane materials: polysulphone, regenerated cellulose, and cellulose acetate.

TABLE 1
BSA Adsorption under static conditions

Membrane Material	Nominal Molecular Weight cut-off	Protein Adsorption (mg/m^2)
Cellulose Acetate	5000	9.5
Cellulose Acetate	10000	15.6
Cellulose Acetate	20000	60.3
Polysulphone	10000	38.2
Regenerated Cellulose	10000	7.3

The results for the cellulose acetate membranes indicate that protein adsorption increases with increasing molecular weight cut-off of the membrane. This probably occurs because ultrafiltration membranes have a very large internal surface area. Membranes with larger pore size (ie. higher molecular weight cut-off) allow greater access of protein into the bulk and thus exhibit higher levels of protein adsorption.

From a comparison of the 10,000 molecular weight cut-off membranes it is clear that there is a substantial difference in protein binding characteristics between the various membrane materials. For instance the polysulphone membrane adsorbed over 5 times as much BSA as the regenerated cellulose membrane. This was probably due to strong interaction between hydrophobic portions of the BSA and the hydrophobic polysulphone membrane. BSA has a much lower affinity for the more hydrophilic regenerated cellulose and cellulose acetate membranes.

The Effect of Protein Adsorption of Membrane Flux

Membrane performance can be described in terms of a hydraulic resistance, which can to a certain extent compensate for variations in operating conditions. Resistance is related to flux by the following equation:

$$J = \frac{\Delta P}{\mu R_T}$$

.... (1)

where J is the flux, ΔP the transmembrane pressure drop, μ the solution viscosity and R_T the membrane resistance. As an approximation the total hydraulic resistance can be considered to be the sum of resistances in series:

$$R_T = R_m + R_{cp} + R_f$$

.... (2)

Where R_m is the resistance of the clean membrane, R_{cp} the resistance due to concentration polarisation and R_f the resistance due to fouling.

Figure 1a shows the BSA adsorption to the 10,000 molecular weight cut-off polylulphone and regenerated cellulose membranes. The resistance of these membranes before and after protein adsorption is shown in Figure 1b. The polysulphone membrane, which adsorbed the greater quantity of BSA, also showed a sharp increase in resistance due to adsorption. The regenerated cellulose membrane on the other hand showed if anything a slight reduction in resistance, ie. an increase in flux, as a result of protein adsorption. This suggests that comparison of protein adsorption under static conditions can give an indication of which membranes are most likely to sufffer from protein fouling. Nevertheless, in this particular example, despite suffering the greater flux loss (increase in resistance) due to protein adsorption the polysulphone membrane still exhibited the lower overall resistance.

Protein Adsorption During Stirred Cell Filtration

The quantity of BSA adsorbed by polysulphone and regenerated cellulose membranes during stirred cell ultrafiltration and the resultant changes in membrane resistance are shown in Figures 2a and 2b. Again the polysulphone membrane adsorbed the greater quantity of protein and exhibited the greater increase in hydraulic resistance due to adsorption.

A surprising feature of these results was that although the regenerated cellulose adsorbed more BSA during filtration than did the polysulphone membrane, under static conditions it experienced almost no increase in hydraulic resistance. This suggests that it is not only the quantity of protein adsorbed which is important in determining the effect on membrane resistance but also perhaps the location of protein within the membrane.

The Effect of Protein Adsorption During Cross-flow Ultrafiltration

Cross-flow UF experiments were performed in order to study the effect of protein adsorption on membrane performance under conditions approximating those of an actual process. Figure 3 shows the initial (R_m), final ($R_m + R_f$) and total hydraulic resistances ($R_m + R_f + R_{cp}$) for the polysulphone and regenerated cellulose membranes.

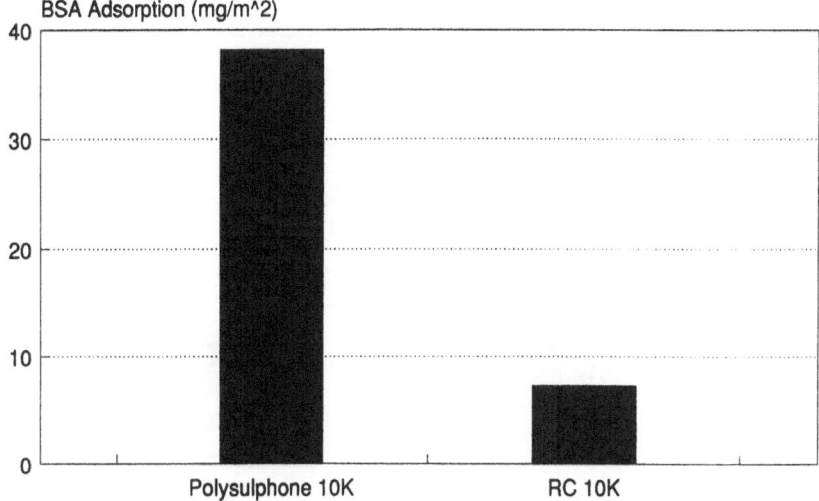

Figure 1a. BSA adsorption of polysulphone and regenerated cellulose
membranes under static conditions

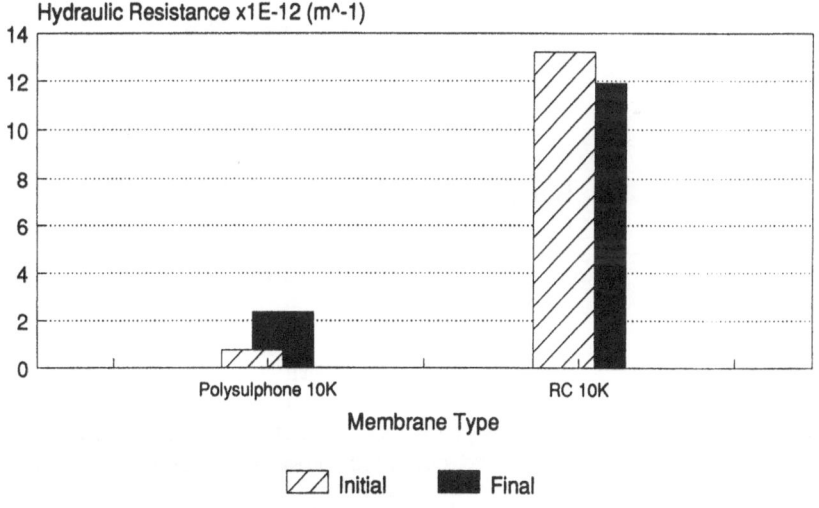

Figure 1b. Effect of BSA adsorption on hydraulic resistance of
polysulphone and regenerated cellulose membranes under static
conditions.

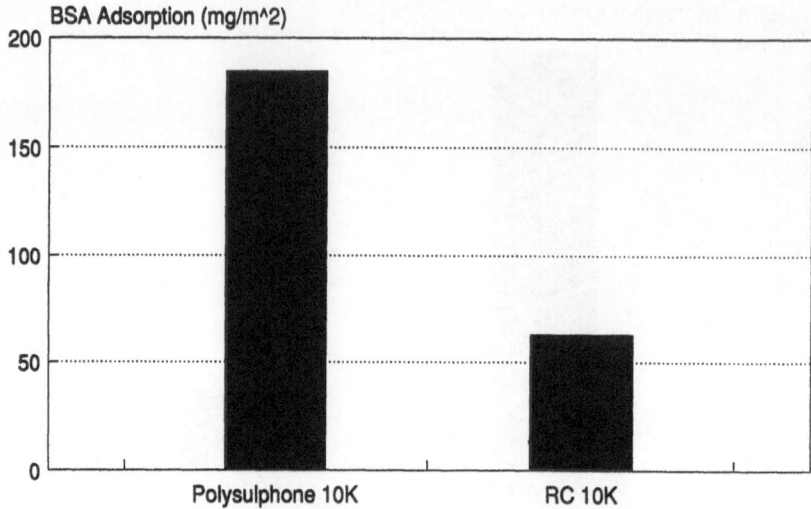

Figure 2a. BSA adsorption of polysulphone and regenerated cellulose membranes under stirred cell filtration conditions.

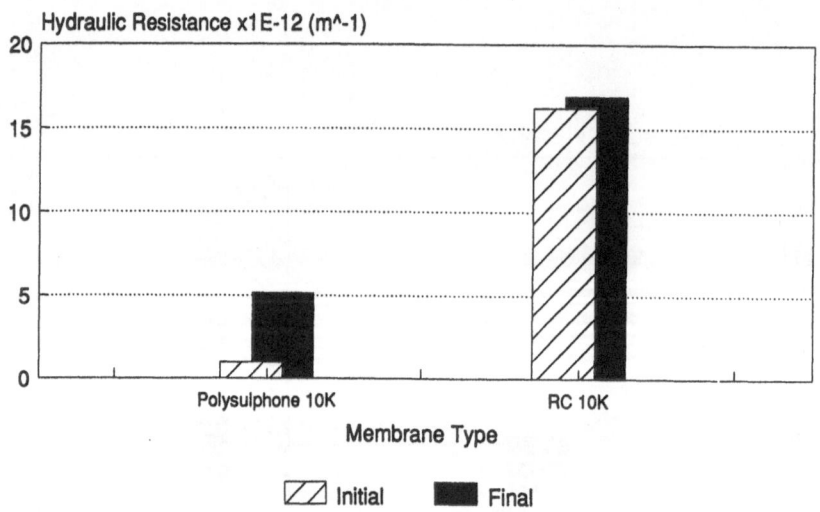

Figure 2b. Effect of BSA adsorption on hydraulic resistance of polysulphone and regenerated cellulose membranes during stirred cell filtration.

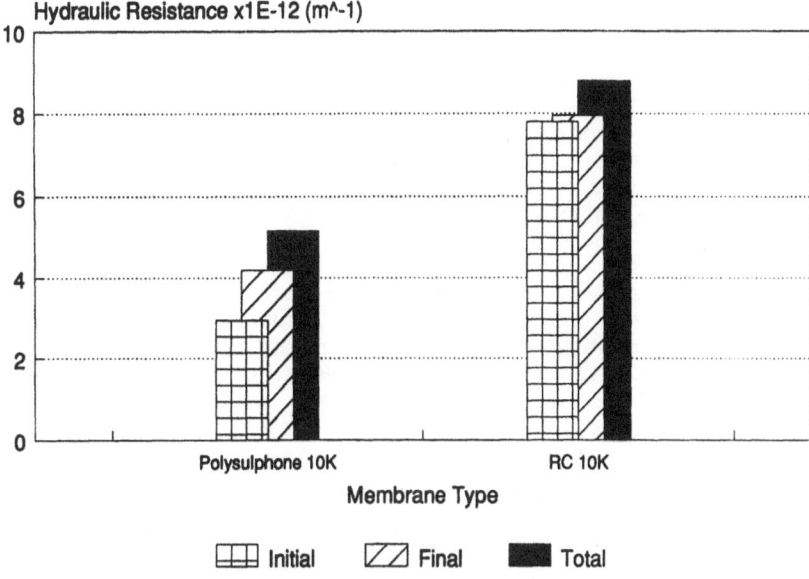

Figure 3. Effect of BSA adsorption and concentration polarisation on
hydraulic resistance of polysulphone and regenerated cellulose membranes
during cross-flow ultrafiltration

The quantities give respectively measures of the performance of the
clean membrane, after protein adsorption and during ultrafiltration. Once
again the polysulphone is seen to suffer the greater increase in
resistance due to adsorption. Nevertheless this membrane gave the lower
total hydraulic ie. greater flux in operation. In this particular
instance it is clear that the high protein binding of the polysulphone
membrane was indicative of a greater level of fouling than that
experienced by the regenerated cellulose membrane.

Thus it appears that measuring protein adsorption and the resultant
flux decline under static conditions can give a good guide to the effect
of protein adsorption under actual filtration conditions. Nevertheless
the much lower initial resistance of the polysulphone membrane still
enabled it to give the better performance in ultrafiltration operation.
Where membranes are of more similar initial permeability (resistance) a
knowledge of the protein binding characteristics might provide a good
guide as to which is likely to be more resistant to fouling.

Electron Microscopy
BSA Solution: The nature of the protein solution used in the filtration
experiments is shown by freeze-fracture and deep-etching (figure 4). BSA
is a globular protein but as the plane of fracture of the molecules varies
a number of profiles are visible. Also the BSA molecules are often
clumped.

Polysulphone Membranes: The polysulphone membranes appear to have an effect upon the tertiary structure of the BSA and the protein found at the separating surface is filamentous, rather than globular (figure 5). It is these filamentous protein molecules which are observed within the membrane itself. BSA molecules can be seen passing through the separating surface of the membrane and immediately behind this surface much protein is visible amongst the fibres of the polysulphone matrix. Filamentous protein molecules are also seen elsewhere within the membrane matrix, being found in the voids and in the micropores near the back surface of the membrane.

Regenerated Cellulose: BSA at the separating surface of the regenerated cellulose membrane appears to retain a more globular structure (figure 6) similar to the protein molecules in the BSA solution. Within the membrane matrix any protein molecules that may be present cannot be distinguished from the fibres of the membrane due to the nature of the preparation.

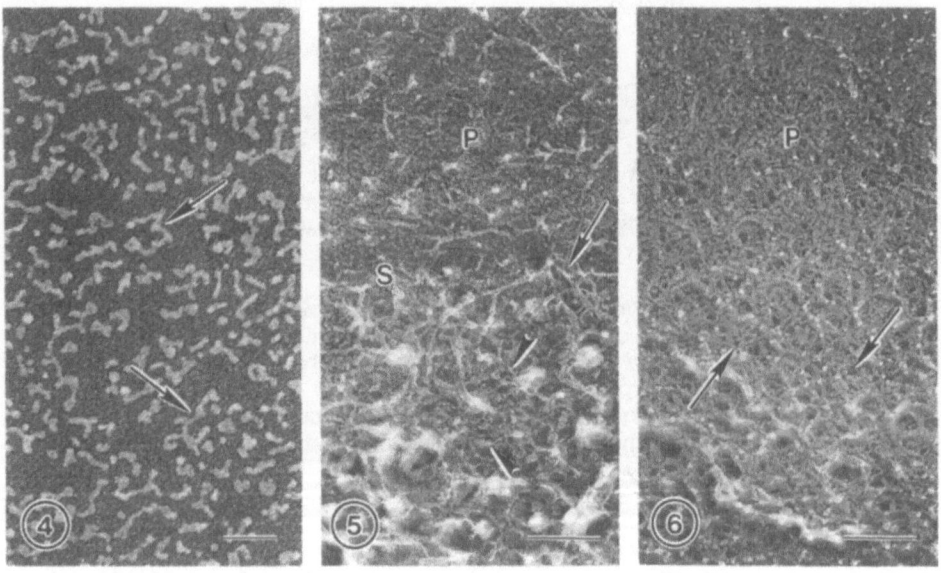

Figure 4. BSA solution after freeze-fracture and deep-etching. The protein molecules are clearly visible and are viewed from a number of angles due to the plane of fracture. In some places the molecules are clumped (arrowed). bar = 20nm.

Figure 5. Separating surface (S) of a protein fouled 10000 molecular weight cut-off polysulphone membrane after freeze-fracture and deep-etching. Note the filamentous nature of the protein (P) and the filamentous protein places the protein appears to be passing through the separating surface (arrowed). bar = 100nm.

Figure 6. As figure 5, but protein fouled 10000 molecular weight cut-off regenerated cellulose membrane. Here the protein (P) at the separating surface (arrowed) is more globular in nature. bar = 100nm.

Electron microscopy shows that certain membrane materials, for example polysulphone, can alter protein structure which might cause a loss in enzyme activity. This has a significant impact on membrane selection where small volume high value enzymes are being treated. In such cases it may be preferable to use a non-denaturing membrane material such as regenerated cellulose despite the lower fluxes. Nevertheless the much lower initial resistance of the former membrane still enabled it to give the better performance in ultrafiltration operation. Where membranes are of similar initial permeability (resistance) a knowledge of the protein binding characteristics might provide a good guide as to which is likely to be more resistant to fouling.

CONCLUSIONS

Implications for Selection of Ultrafiltration Membranes

Membrane manufacturers literature usually describes the performance of ultrafiltration membranes in terms of water fluxes and a nominal molecular weight cut-off. In actual operation membranes usually give much lower flux due to the combination of concentration polarisation, and fouling.

Concentration polarisation is an inescapable feature of the ultrafiltration process and means for reducing it are well established [9]. The factors which influence fouling and means for overcoming it are less well known. However, solute-membrane interaction is thought to play a major part. The results from the static adsorption tests shown the strength of this interaction, for a particular solute, is dependent on the membrane material. Therefore careful selection of the membrane material might enable this interaction to be minimised and to reduce protein fouling of the membrane.

Electron microscope studies reveal that the influence of protein adsorption on flux is not the only consideration when selecting materials, for example polysulphone; whereas regenerated cellulose appears to leave protein structure intact. BSA adsorbed to the surface of the polysulphone membrane appears to unfold thus losing its tertiary structure. Such behaviour has important implications when treating low volume, high value products with enzymatic activity. Thus a non-denaturing membrane material such as regenerated cellulose might be preferable despite its high resistance and low flux.

The work reported in this poster shows that protein adsorption can result in a reduction in membrane fluxes (increase in resistance). This poster describes simple techniques, such as measurement of static adsorption, which can be used to gain valuable information on the likely performance of ultrafiltration membranes in protein processing applications. In situations where high product recovery is required these simple tests can be augmented by more sophisticated electron microscope techniques.

ACKNOWLEDGEMENT

This work was funded by BIOSEP, the Biotechnology Separations Club, which is supported by the Biotechnology Unit of H.M. Governments Department of Trade and Industry.

REFERENCES

1. Tong, P.S., Barbano, D.M. and Rudan, M.A. J. Dairy Sci., 1988, **71**, p 604.

2. Nystrom, M. J. Membrane Sci., 1989, **44**, p 183.

3. Matthiasson, E. J. Membrane Sci., 1983, **16**, p 23.

4. Turker, M. and Hubble, J. J. Membrane Sci., 1987, **34**, p 267.

5. Nilsson, J.L. J. Membrane Sci., 1988, **36**, p 147.

6. Pusch, W. and Walch, A. J. Membrane Sci., 1982, **10**, 2+3, p 325.

7. Vetier, C., Bennasar, M., Tarodo de la Fuente B. and Nabias, G. Le Lait, 1986, **66**, 3, p 269.

8. Sheldon, J.M., Reed, I.M. and Hawes, C.R. The Fine structure of Ultrafiltration Membranes. ii Protein Fouled Membranes. In preparation.

9. Belfort, G. and Altena, F.W. Desalination, 1983, **47**, p 105.

PURIFICATION BY TWO-PHASE PARTITIONING OF AN HEPATITIS CORE PROTEIN-PERTUSSIS EPITOPE FUSION,EXPRESSED IN YEAST.

V. RIVEROS-MORENO and J.E. BEESLEY
Dept. of Protein Chemistry, Wellcome Biotech.
Dept. of Pharmacology, Wellcome Research Labs.
Langley Court, Beckenham, Kent.

ABSTRACT

It is known that a six amino acid epitope of p69, a membrane protein of
<u>Bordetella pertussis</u>, is involved in protection against whooping cough.
Chimeric particles were obtained when a thirty amino acid peptide
containing the epitope was genetically fused to the hepatitis B virus core
antigen and expressed in yeast. The particles were purified by extracting
the centrifuged yeast homogenate in a two-phase system composed of PEG 1450
and K phosphate. In this first step, a ten times enrichment was obtained
with a total recovery of 45% of the particles. Ninety percent of these were
found in the PEG phase. PEG was removed by diafiltration through a YM 100
Amicon membrane with simultaneous concentration of the particles. Gel
filtration through a Trisacryl GF 2000 column achieved a good purification
and an homogenous preparation as shown by electron microscopy.
Immunoaffinity as an alternative to two-phase partitioning will be
discussed as well as the advantages of the described protocol for large
scale purification of particles.

INTRODUCTION

The recognition that immunoprophylaxis through vaccination has to undergo
new developments has triggered the search for acellular vaccines that have
to perform equally well or better than the existing whole cell vaccines.
Microbial pathogens possess a complex mosaic of epitopes, not all of which
induce immunity to disease. The induction of the necessary cooperative
immune response at the level of B, T and antigen-presenting cells requires
a certain degree of antigenic complexity. The omission or deletion of any
of the essential functional epitopes would limit the effectiveness of
vaccines produced by recombinant DNA technology or peptide synthesis.
Therefore, the identification and learning about the mode of action of the
relevant antigens becomes of paramount importance. There are several
examples in the recent literature on the performance of chimeric peptide

vaccines (1, 2, 3), which in many instances have produced an immunological response far from the expected. The need for a suitable carrier protein has been recognised. The use of hepatitis core protein (Hcp) as a carrier was first reported by Clarke et al (4) for a Foot and Mouth Disease Virus (FMDV) epitope fused to Hcp and expressed in vaccinia virus. The construct rendered the FMDV epitope more immunogenic than it was when linked to other carrier proteins, probably due to the fact that Hcp, when expressed on its own or fused to peptides, is able to form particles. But most importantly, these particles have the capacity to strongly stimulate both T cell-dependent and independent antibody response (5). Bordetella pertussis, the causative agent of whooping cough, has been the centre of interest for new vaccine development for some years now, since there are conflicting reports on the side effects produced by the existing vaccine (6).
A membrane protein (69K daltons) of B. pertussis has been identified as a protective antigen against whooping cough (7). Moreover, a monoclonal antibody (BB05) against this protein was able to passively protect mice challenged by an aerosol containing virulent microorganisms (8). The epitope recognised by Mab BB05 on p69 has been identified, sequenced (9), and expressed in E. coli (unpublished results). It has been found that the expression of the BB05 epitope in conjunction with a carrier protein can protect mice (unpublished results). Here is reported the purification of chimeric protein particles consisting of a thirty amino acid peptide, containing the BB05 epitope of p69 fused to the amino terminal of Hcp, expressed in yeast. After breakage and centrifugation, the clear supernatant containing Hcp-BB05 particles was partitioned in a two-phase system formed by PEG 1450 and K phosphate; followed by gel filtration. Immunoaffinity was tested as an alternative to two-phase partitioning.

MATERIALS AND METHODS

The construction of the expression plasmids; yeast transformation and induction procedures will be published elsewhere.
Chemicals: Polyethylene glycol (PEG) Mr 4000 and Mr 1450 were obtained from BDH and Sigma respectively. TRITON X-100 normal or reduced was from Aldrich Chemicals. Bovine serum albumin (BSA) was from ICN Biomedicals code 81001. Tween 20 from Sigma. All protease inhibitors were from Boehringer.
Yeast homogenate: Pelleted cells of Saccharomyces cerevisiae were suspended in 20 mM sodium phosphate buffer pH 7 containing 0.1% TRITON X-100, 20 mM EDTA, and the following protease inhibitors: 1 mM PMSF, 0.1 mM pepstatin, leupeptin and chymostatin respectively. The complete mixture was broken in a glass bead mill, submerged in ice. The homogenate was centrifuged at 17,000 g for 30 minutes at 4 degrees C.
Sucrose Gradient Centrifugation: A linear sucrose gradient between 45 and 15 per cent in phosphate buffer saline (PBS) pH 7.2 was made. The clear yeast homogenate was layered on top and centrifuged at 20 degrees C for 3 hrs at 100,000 g. Gradient fractions were collected and analysed by DOT BLOT. The fractions containing cores were pooled and dialysed against PBS.
Dot Blot : 50 ul of each sample to be analysed was applied to wells containing 100 ul of 50 mM Tris pH 7.5, 150 mM NaCl (TBS). The nitrocellulose membrane was blocked with 3% BSA-TBS, incubated with monoclonal BB05 diluted 1/500 in the BSA buffer, washed in TBS-0.05% Tween-20, and reacted with goat anti-mouse alkaline phosphatase (Sigma A-5153) diluted 1/1000 in BSA buffer. After incubation and washing, the colour was

developed with Nitroblue tetrazolium and 5Br-4Cl-3Indolyl phosphate (both from Bio-Rad) in 0.15M Tris pH 9.5.

Diafiltration : Samples containing cores were ultrafiltered in an Amicon stirred cell with a YM 100 membrane. In order to diafilter the polymer or the sucrose, 3/4 of the Amicon cell volume was filled with PBS for at least 4 times.

Two-Phase Partitioning : The two-phase liquid-liquid systems were prepared from aqueous stock solutions of PEG (40% w/v), and 3 M potassium phosphate buffer pH 7.45 prepared according to Albertsson (10). The concentration of Hcp-BB05 was assessed by ELISA.

Gel Filtration : A Trisacryl GF2000 (LKB) column was calibrated with blue dextran and RNase. Hcp-BB05 in PBS were loaded onto the column equilibrated with PBS-0.1% reduced TRITON X-100 and 5 mM EDTA. The cores were followed by DOT BLOT.

ELISA: A sandwich ELISA specific for Hcp-BB05 was developed. Dynatech microtitre plates (code 001-010-2401) were coated with rabbit anti-Hcp IgG 1 ug/ml of 0.1 M Na carbonate buffer pH 9.5, overnight at 4 degrees C. All wells were washed 5 times with PBS-0.05% Tween 20. Doubling dilutions of Hcp-BB05 in 1% BSA-PBS-0.05% Tween 20 were made in the coated wells. The plates were incubated for 1hr at 37 degrees C, and wells washed as above. A second antibody, 1 ug/ml of Mab BB05 in the BSA-buffer, was added to all wells, and incubated for 1 hr at 37 degrees C. All wells were washed again. A conjugate antibody, goat anti-mouse-alkaline phosphatase (Sigma A5153) diluted 1/1000 in the BSA-buffer was added, and incubated for 1 hr at 37 degrees C. All wells were washed. Alkaline phosphate substrate Sigma 104 at 1 mg/ml in 0.1 M Ethanolamine pH 9.6 was added. After incubation for 30 or 60 mins at 37 degrees C, the colour reaction was read in a Titertek at 405 nm.

Protein Analysis: SDS-PAGE was performed according to Laemmli (11), the reducing agent was dithiothreitol. BCA protein assay from Pierce was used according to the makers' instructions.

Western blot: SDS-acrylamide protein gels were blotted onto nitrocellulose membrane using a Sartoblot II (Sartorius). The blot was reacted with Mab BB05 or rabbit anti-Hcp IgG basically as described for the DOT BLOT. When rabbit anti-Hcp was used, the conjugated antibody was goat anti-rabbit alkaline phosphatase (Sigma A-8025).

Coupling of Monoclonal BB05 to Control Pore Glass (CPG) and Immunoaffinity: CPG was obtained from Pierce (code no 24875), derivatised with a long alkylamine arm. The procedures employed were basically those of Nilsson et al. (12) and Nilsson and Mossbach (13) with some modifications. The IgG was coupled in 0.1M Na carbonate pH 8.3, 0.5 M Na Cl, at RT for 2 hrs. 0.87 mg of monoclonal BB05 were coupled per ml of CPG. Binding of the particles contained in the yeast homogenate or in dialysed PEG phase, was done in 50 mM phosphate, 50 mM NaCl pH 7.2 buffer by batch absorption overnight at 4 degrees C.

Electron Microscopy: Purified particles were negatively stained and also labelled using the immunogold technique as described by Beesley et al. (14). The antibody used was monoclonal BB05 diluted 1/500.

Agarose Gel Electrophoresis: The method followed was basically that of P. Serwer (15). Isogel (FMC) 0.4% in 0.05M Na phosphate pH 7.4, was used. The running buffer was the same as in the gel and recirculated at 25 mls/min. The electrophoresis was performed at 20 Volts (approx 40 m Amp) for 24 hours at RT. After such time, the gel was transferred to Gel-Bond (FMC) and dried. Finally, the gel was stained with 0.05% Coomassie blue in 10% acetic acid. Destaining was done in 10% acetic acid.

R E S U L T S

<u>Properties of Hcp-BB05 Particles</u> : The construction of the expression vector and the subsequent induction of the chimeric protein will be published elsewhere. The sequence of the p69 peptide which was fused to the amino terminal of Hepatitis B core protein (16) is shown in Fig. 1. The codon usage was optimised for yeast expression. The fusion of peptides to the amino terminal of the core protein does not affect drastically the folding of the protein since core particles can be demonstrated in the cell homogenate. Centrifugation of the homogenate in a linear sucrose gradient sedimented the particles into the middle of the gradient, as reported earlier (4). The chimeric cores can be identified by dot blot (not shown). Electron microscopy of negatively stained, sucrose purified cores, showed complete particles containing the pertussis epitope as evidenced by reaction with monoclonal BB05 (data not shown).

```
 1860            1870            1880            1890             1
   *               *               *               *
GCG CCG CCG GCG CCC AAG CCC GCG CCG CAG CCG GGT CCC CAG CCG
ala pro pro ala pro lys pro ala pro gln pro gly pro gln pro

 900            1910            1920            1930            1940
   *               *               *               *               *
CCG CAG CCG CCG CAG CCG CAG CCG GAA GCG CCG GCG CCG CAA CCG
pro gln pro pro gln pro gln pro glu ala pro ala pro gln pro
```

<u>Fig 1</u>: Sequence of p69 peptide, containing the BB05 epitope that was fused to Hcp. Before the first alanine, at the amino terminal, a methionine was added.

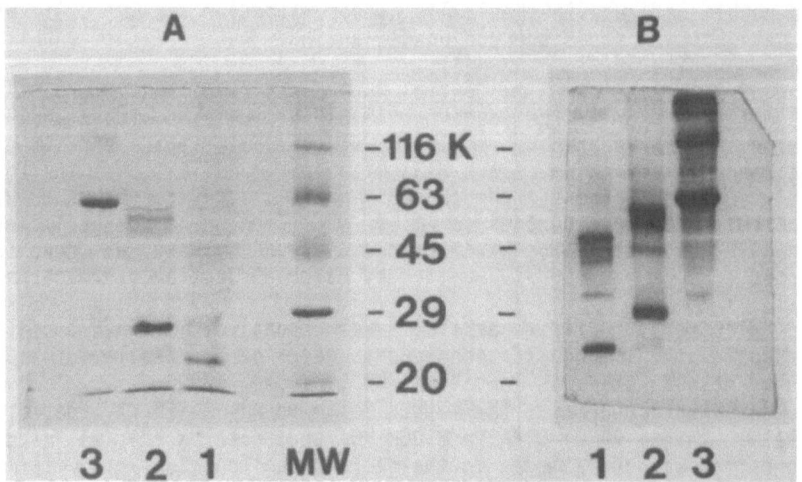

Figure 2 : Panel A: SDS-P.A.G.E. of: Track 1.core particles (Hcp); Track 2.Hcp-BB05; Track 3.Hcp-CSP. MW: molecular weight markers. Panel B : Western of the same samples as in Panel A. The antibody was rabbit anti-core IgG.

Due to the folded configuration of the core protein (17), the fused foreign epitope is believed to be located on the surface of the particle. B. pertussis epitope confirms this assumption since it can be visualised by the immunogold technique.
SDS-P.A.G.E. and further immunoblotting of different particles purified by sucrose sedimentation (Fig 2) showed very little or no degradation of the monomer.
Under reducing conditions non-fused core monomers have an apparent Mr of 23,000, Hcp-BB05 monomer 29,000 and a very large chimeric construct containing 200 amino acids of the P.falciparum sporozoite protein, Hcp-CSP, has a monomer Mr 45,000 but it runs anomalously at about 58,000, due to the three dimensional configuration adopted by the CSP part of the fusion, as documented by Wasserman et al.(18).

Quantitation of Chimeric Particles : An ELISA was developed for quantitating the Hcp-BB05 particles, as described in the Methods section. In general, the reliability of the ELISA technique when dealing with particulate samples is not very good. It was found that a sandwich ELISA gave a reasonable correlation and it is reproducible with a 20% variation as demonstrated by repeated ELISA tests in the same preparation (data not shown).

Purification of Hcp-BB05 by Two-Phase System Partitioning : Binodial curves for PEG4000/phosphate and PEG1450/phosphate were made. The systems tested with PEG4000 were: A) 19% PEG   / 0.5M K phosphate, B) 15% PEG / 0.6M K phosphate and C) 12% PEG / 0.75M K phosphate all at pH 7.45. These three PEG 4000 systems left the core particles in the bottom phosphate phase together with most of the homogenate proteins, the recovery of the cores was less than 10 per cent (data not shown). The three systems made with

TABLE 1

TWO PHASE EXTRACTION OF YEAST HOMOGENATE

	PROTEIN			Hep. Cores Pertussis (HcP)		
	mg/ml	Total mg	% Protein in System*	ug/ml	Total mg	% HcP in System*
Homogenate	68	136	—	1500	3	—
Top Phase	0.84	18.5	22.4	206	1.13	89.7
Interface	6.0	7.5	9.1	27.2**	0.072**	5.7
Bottom Phase	5.4	37.8	45.8	0.89	0.0062	0.5
Pellet	13.3	18.6	22.5			
Recovered in System		82.6	54.9		1.26	42

*   100% is the total protein or HcP recovered in the system.
**   Value for combined interface and pellet.

PEG1450/phosphate: A) 16% PEG/0.7 M K phosphate, B) 14% PEG/0.8 M K phosphate and C) 10% PEG/0.9 M K phosphate left the core particles in the top PEG phase. It was found that as the concentration of phosphate increased in the system, the recovery of particles decreased (data not shown). System A was chosen because the partition coefficient of about 200, was not very different to B or C.

Table I shows that about 50% of the total protein partitioned in the system is lost with a similar loss of particles. The appearance of a pellet and a heavy interface shows qualitatively that insolubility has occurred, both combined would account for about 30% of the total protein used for the partitioning, providing the protein assay was reliable in pellets. Also it should be noted that over 85% of the core particles in the system are taken into the PEG phase. The two-phase extraction of the homogenate produced a 10 fold purification in one step.

Particle Integrity After Two-Phase Extraction : The comparison in Fig. 3A and B between SDS-P.A.G.E. stained for proteins and a Western of the same samples shows the following:- first, the top phase contains very little protein as compared to the other phases; second, the major band reacting with monoclonal BB05 in the top phase is the monomer unit of the particle (29 k Daltons); third, very little material is reacting with the monoclonal in the bottom phase although there is a high protein concentration; fourth, a second positive band in the Western blot corresponds to the dimer size, about 58 k Daltons.

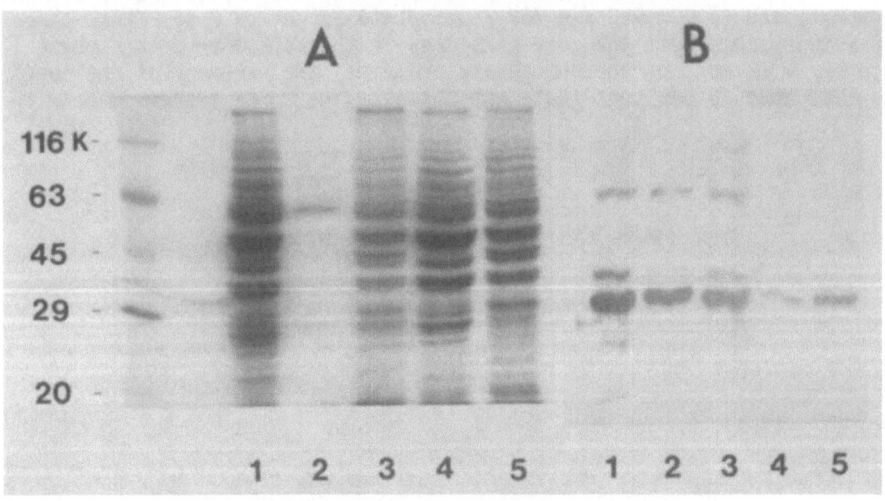

Figure 3 : Purification of Hcp-BB05 by two-phase extraction. Panel A : SDS-P.A.G.E. of: 1.yeast homogenate; 2.Top (PEG) phase; 3.Interface; 4.Bottom (phosphate) phase; 5.Pellet. Panel B : Western of a similar gel with monoclonal BB05.

There is a third reacting band at about 35 k Daltons. This could be a breakdown product of the dimer due to the homogenisation conditions. The addition of other protease inhibitors or higher concentrations did not alter the presence of this band; fifth, only in the total homogenate, are traces of degraded core monomer detected. In summary, the two phase extraction of the cores does not seem to degrade the monomer. Electron microscopy (Fig. 4) of Hcp-BB05 after two-phase extraction shows particles indistinguishable from those purified by sucrose gradient.

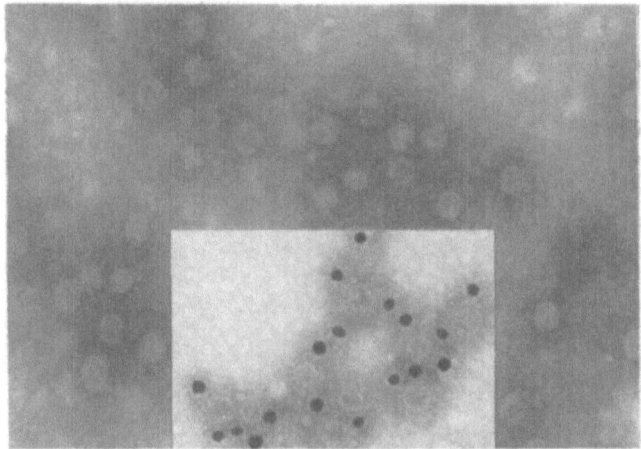

Figure 4 : Electron micrographs of Hcp-BB05 particles purified by two-phase partitioning. Outer picture negatively stained, inset immunogold with BB05 monoclonal.

Another way of assessing the presence of particles is by agarose gel electrophoresis. To obtain visible bands, the samples have to contain at least 1 mg/ml of protein, a minimum of 30 ug per band. Fig. 5 shows that the same three particles as in Fig. 2, including the cores alone (Hcp), are negatively charged and migrate towards the anode as a single band. Also a correlation between particle size and distance of migration is evident.

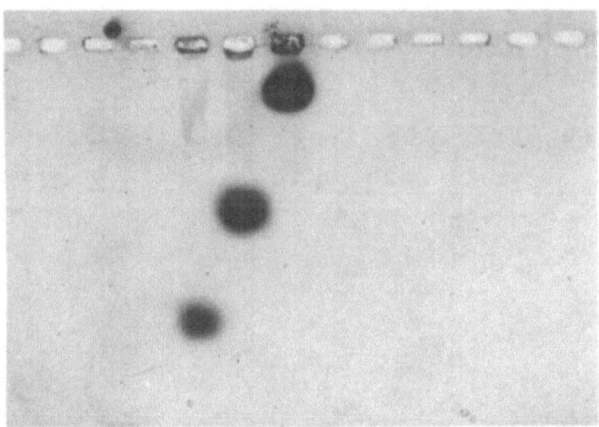

1 2 3

Figure 5 : Agarose gel electrophoresis of core particles. Track 1.Hcp.
Track 2.Hcp-BB05. Track 3.Hcp-CSP.

Gel filtration of Particles : Particles after two-phase partitioning were dialysed and applied to a Trisacryl GF2000 column that has a fractionation range between 10k-15000k Daltons. Fig. 6 shows the elution profile and the dot blot of column fractions. Most of the large peak is likely to

234

correspond to non-reduced TRITON-X 100 present in the sample.
The first peak contains the particles as shown by the dot blot and it
coincides with the void volume calibration. The recovery of the cores, as
determined by ELISA was over 90%.

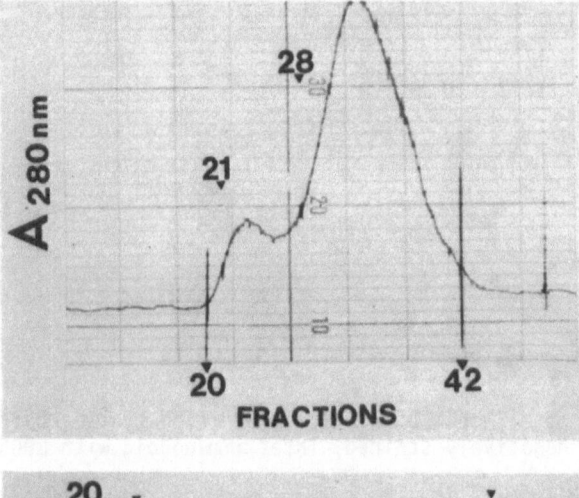

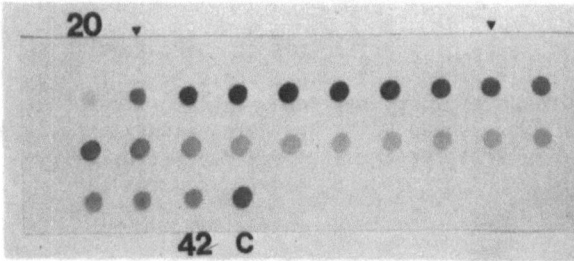

Figure 6 : Gel filtration
in Trisacryl GF2000.
Top; elution profile,
numbers correspond to
fractions.
Bottom: DOT BLOT of the
same fractions as above
reacted with monoclonal
BB05.
Arrow heads indicate pooled
samples. C: is control
Hcp-BB05.

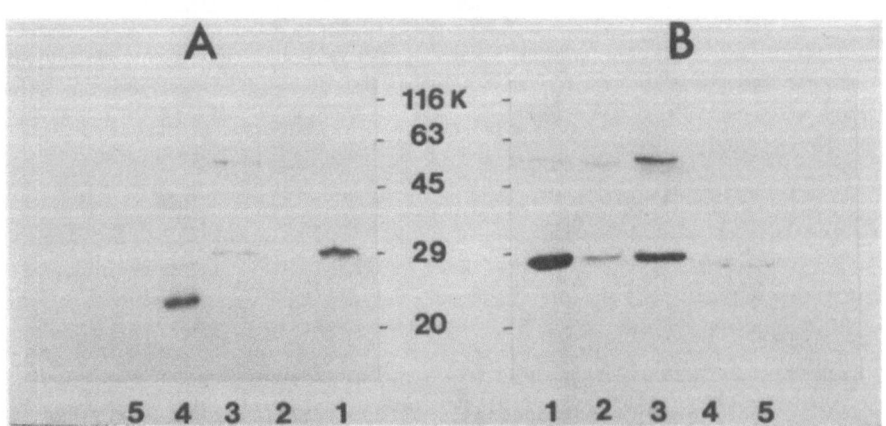

Figure 7 : Affinity purification of Hcp-BB05. Westerns of SDS-P.A.G.E.
Panel A: rabbit IgG anti-Hcp. Panel B: monoclonal BB05. Track 1: Hcp-BB05
homogenate. Track 2: absorbed homogenate to CPG-BB05. Track 3: Peak
eluted with 0.1M citrate pH 2.2. Track 4: yeast homogenate expressing Hcp
only. Track 5: control yeast homogenate.

Immunoaffinity Purification : This purification method was explored as an alternative to the two-phase extraction. The clear yeast homogenate was absorbed to CPG-BB05 batchwise at 4 degrees C. The glass beads were transferred to a column and washed extensively with 50 mM Na phosphate, 50 mM NaCl pH 7.2. The elution was first attempted with a salt gradient of up to 3M NaCl. The column was followed by dot blot, no detectable BB05 positive material came off the column. 0.1M Na citrate pH 2.2 released BB05 positive protein. Fig. 7 shows comparative Westerns of the affinity purified Hcp, obtained by reaction with monoclonal BB05 and with rabbit anti-core IgG. The usual two major bands at 29,000 and 59,000 were obtained with no traces of degradation. The cores thus purified were checked for particle integrity on an FPLC Superose 6 column. There were two included peaks that roughly corresponded to the size of the monomer and dimer in the Westerns,no excluded complete particles were detected (data not shown).

DISCUSSION

The expression of the virus like particles consisting of peptides fused to the core antigen of hepatitis B virus has been demonstrated to occur in vaccinia (4) and E. coli (19). Knistern et al.(20) and also Miyanohara et al.(21) have reported the expression of hepatitis core antigen in yeast as producing stable 28 nm particles. Beesley et al. (unpublished results) have shown the expression of chimeric particles in yeast. These particles seem to be stable and no extensive degradation beyond the monomer is detected by Western blotting under reducing conditions. Also, dot blots of sucrose gradients (reacted with Mab BB05) in which the whole yeast homogenate is sedimented, showed that only the fractions corresponding to the particle band were positive. The results obtained with agarose electrophoresis would suggest that the particle is the major component after the sucrose gradient; although small molecules would have electrophoresed out of the gel. The presence of one Coomassie blue positive band shows homogeneity at the particle level and no particle aggregates.

Sucrose gradient centrifugation, although a mild purification method, is very expensive for large scale production. Therefore, an alternative method for production has to yield intact particles and be easy to scale up. Two-phase liquid-liquid extraction appears to provide for both requisites. The BB05 particles are negatively charged as judged by their electrophoretic mobility. When PEG 4000 was used for the two-phase extraction, the particles were partitioned into the phosphate phase and due to the high salt concentration they would have been disrupted and resulted in the drastic effect observed. The PEG 1450 /phosphate system studied so far has the disadvantage of producing losses of about 50%. We are investigating the performance of the Dextran/PEG systems, on the yeast homogenate since it is known to be milder than PEG/phosphate, but more expensive to use. The particles purified by the PEG/phosphate system seem to show similar integrity to those purified in sucrose, in terms of the monomer being the lowest molecular weight band detected by Western blotting. The presence of a dimer under reducing conditions is peculiar, perhaps PEG has interacted with the particle making it resistant to full reduction. The particles purified by sucrose gradient sometimes show this band faintly. No major differences in particle appearance are detected by electron microscopy as seen in Figure 4. The ELISA developed for Hcp-BB05 was intended to quantitate the fused protein monomer assembled into particles, since the capture antibody is a polyclonal IgG raised against core particles and the monoclonal BB05 would provide the specificity for the epitope. In fact, the ELISA does not provide information on whether the particles obtained by sucrose gradient or two-phase purification are identical in their structure. One way of assessing this is to measure the presence of e antigen in core particles. In the hepatitis core protein, two immunogenicities have been identified C and e (22). Antibodies to both can be found in hepatitis B patients (23). It is also known that the e antigen has two epitopes, both near the carboxyl end of the core protein, but one e2 is buried in the centre of the assembled core particle and only available after harsh treatment (17). This analysis should give us another insight into the integrit of the particles.

Antibodies made against particles purified by both methods, sucrose gradient and two-phase partitioning, could also help in this comparative characterization. These specific antibodies would be able to quantitate differences between not only the quality of the particles but as well between the epitope density on each particle.

The results obtained with affinity purification of the particles were disappointing, since the breaking down of the particles into monomer and dimer was evident. Attempts to reconstitute the particles have not been very successful. It is possible to detect particles after higher salt concentration and TRITON X-100 treatment, but the yield is poor and consequently too laborious and risky for large scale preparation. In summary, two-phase partitioning seems to be a suitable first step for the purification of core particles, providing that a thorough characterization will show no major differences with the particles obtained after sucrose gradient. As a final purification step, gel filtration seems to be a good choice.

REFERENCES

(1) Lerner, R.A., Green, N., Alexander, H., (1981) Proc. Natl. Acad. Sci. USA 78, 3403-3407.
(2) Emini, E.A., Jameson, B.A., and Wimmer, E., (1983) Nature 304, 699-703.
(3) Zavala, F., Tam, J.P., Hollingdale, M.R., Cochrane, A.H. Quakyi, I., Nussenzweig, R.S., and Nussenzweig, V., (1985) Science 228, 1436-1440.
(4) Clarke, B.E., Newston, S.E., Carroll, A.R., Francis, M.J., Appleyard, G., Syred, A.D., Highfield, P.E., Rowlands, D.J. and Brown, F., (1987) Nature 330, 381-384.
(5) Milich, D.R., and McLachlan, A., (1986) Science 234 1398-1401.
(6) Mortimer, E.A., (1988) Vaccines 74-79 Ed. Plotkin, S.A. and Mortimer, E.A., Pub. by W.B. Saunders & Co.
(7) Novotny, P., Kobisch. M., Cownley, K. Chubb, A.P., and Montaraz, J.A., (1985) Infect. and Immun. 50, 190-198.
(8) Montaraz, J.A., Novotny, P., and Ivanyi, J., (1985) Infect. and Immun. 467, 744-751.
(9) Charles, I.G., Dougan, G., Pickard, D., Chatfield, S. Smith, M. Novotny, P., Morrisey, P. and Fairweather N.F., (1989) Proc. Natl.

Acad. Sci. 86, 3554-3558.

(10) Albertsson, P.A., (1971) Partition of Cell Particles and Macromolecules 2nd Ed. Wiley Interscience.

(11) Laemmli. U.K., (1970) Nature 227, 680-685.

(12) Nilsson, K., Norrlow, O., and Mosbach, K., (1981) Acta. Chem. Scand. B. 35, 19-27.

(13) Nilsson, K., and Mosbach, K., (1981) Biochem. Biophys. Res. Commun. 102 449-457.

(14) Beesley, J.E., Day, S.E.J., Betts, M.P. and Thorley, C.M. (1984) J. Gen. Microbiol. 130, 1481-1487.

(15) Serwer, P., (1986). Methods in Enzymology 130, 116-132. Ed. C.H.W. Hirs & S.N.T. Timasheff. Acad. Press Inc.

(16) Pasek, M., Goto, T, Gilbert, W., Zuik, B., Schaller, H., MacKay, P., Leadbetter, G., and Murray, K. (1979) Nature 282, 575-579.

(17) Argos, P., and Fuller, S.D. (1988) Embo J. 7, 819-824.

(18) Wasserman, G.F., Inacker, R., Silverman, C.C., and Rosenberg, M., (1987) Protein Purification: Micro to Macro p.337-354. Ed.Alan R. Liss, Inc.

(19) Stahl, S.J. and Murray, K., (1989) Proc. Natl. Acad. Sci. USA 86, 6283-6287.

(20) Kniskern, P.J., Hagopian, A., Montgomery, D.L., Burke, P., Dunn, N.R., Hofman, K.J., Miller, W.J., and Ellis, R.W., (1986) Gene 46 135-141.

(21) Miyanohara, A., Imamura, T., Araki, M. Sugawara, K., Ohtomo, N., and Matsubara, K., (1986) J. Virol. 59. 176-180.

(22) Milich, D.R., McLachlan, A., Stahl, S., Wingfield, P., Thornton, G.B., Hughes, J.L., and Jones, J.E. (1988). J,. Immunol. 141, 3617-3624.

(23) Okada, K., Kamiyama, I., Inomata, M., Imai, M., Miyakawa, Y., and Mayumi, M., (1976) N. Engl. J. Med. 294, 746-749.

Acknowledgements: I thank Mark Betts for the electron micrographs;K.Beesley and M.Romanos for providing the recombinant yeast cells and helpful discussions; Dr. M. J. Francis for providing the rabbit anti-core IgG; Dr. G. Allen for helpful comments.

SEPARATION OF LOW-MOLECULAR ORGANIC COMPOUNDS FROM COMPLEX AQUEOUS MIXTURES BY EXTRACTION

K. Schügerl, W. Degener, P. v. Frieling, L. Handojo,
I. Kirgios
Institut für Technisch Chemie, Universität Hannover,
Callinstr. 3, D-3000 Hannover, F.R.G.

ABSTRACT

The separation of mixtures of aliphatic carboxylic acids and amino acids by extraction with phosphorous-bonded oxygen donor (PBO) extractants as well as with quaternary ammonium extractants with long alkyl chains, with the aim of reducing their solubility in water, is investigated.

INTRODUCTION

Fermentations broths with complex medium components (yeast extract, cottonseed meal, cornsteep liquor) are highly complex mixtures that contain a large number of substrates (proteins, amino acids, carbohydrates), different primary metabolites such as alcohols, carboxylic acids (acetic acid, lactic acid, and different amino acids) and secondary metabolites (e.g., antibiotics). Protein hydrolysates also contain several amino acids of chemically similar properties. The recovery of the desired compounds from mixtures and their purification is often coupled with high expenditure, especially if the product concentration is low.

From the possible separation techniques, precipitation has a very low selectivity, and ion-exchange chromatography has a very low capacity. Extraction could be a possible separation technique, provided its selectivity is high enough.

However, extraction of organic acids with carbon-bonded oxygen donor (CBO) extractants is not selective enough.

The solvation numbers of aliphatic carboxylic acids extracted with PBO extractants sometimes indicate a stoichiometric association between an individual phosphoryl group and an individual carboxyl group. Although the high distribution ratios of the acids are due to a successful competition between the solvating phosphoryl group against the water molecules at the interface, the organic adducts with alkylphosphates are hydrated with one or more water molecules. The best results can be achieved with high-molecular-weight aliphatic amine extractants or with their combination with PBO extractants. When using aliphatic amine (secondary or tertiary-amine) extractants, proton transfer occurs during the extraction, and the acid prevails in the organic phase as an amine-acid ion pair. Therefore, they are called ion-pair (IP) extractants. They are less soluble in water than CBO- and PBO-extractants and yield the highest partition coefficients.

However, amino acids cannot be extracted with IP-extractants, but with quaternary amines, with which they form an ammonium salt.

The recovery of alcohols by extraction is not economical.

The recovery of aliphatic carboxylic acids with CBO-extractants is not economical either. PBO- and IP-extractants are preferred because of their high partition coefficients and low solubilty in aqueous solutions.

In the following, the separation of solutes with a single and more than one charged group will be discussed separately.

SEPARATION OF SOLUTES WITH A SINGLE CHARGED GROUP

The extraction of single aliphatic carboxylic acids with PBO- and IP-extractants have been investigated by several groups. An excellent review was given by Kertes and King [1]. Therefore, the extraction of single components from aqueous solutions in the absence of chemically similar components is not considered here.

The separation of different carboxylic acids in the presence of chemically similar compounds with PBO- and IP-extractants have not been considered as yet.

The stability of PBO- and IP-acid complexes is influenced by the pH-value of the solution as well as by the pK_a-value of the compounds. The separation of components is possible, based on their pK_a-value, if they are at least 1 pH-unit away from each other. Examples are: formic acid (pK_a = 3.7) and acetic acid (pK_a = 4.76). A separation of higher carboxylic acid homologues based on their pK_a-value is not possible (the pK_a-values of propanoic, butyric, pentanoic, hexanoic, heptanoic, octan-oic, etc. acids are in the range of 4.85-5.0).

The difference between, e.g., propanoic acid (pK_a = 4.85), lactic acid (pK_a = 3.86) and pyruvic acid (pK_a = 2.49) is large enough for their separation.

Also the cis- and transisomers, i.e., maleic acid and fumaric acid, can be separated by IP-extraction.

SEPARATION OF SOLUTES WITH MORE THAN ONE CHARGED GROUP WHICH FORMS ZWITTERIONS

The separation of solutes with more than one charged group, which form zwitterions in the neutral range, is a most difficult task. Typical compounds are the amino acids.

With the exception of threonine, all bifunctional amino acids have their pK_1 (COO^-) value between 2 and 3, and their pK_2 (NH_3^+) value between 9 and 10.

This holds true also for the trifunctional amino acids, with the exception of cystein with pK_1 = 1.71 and pK_2 = 10.78, and histidine with pK_1 = 1.82.

The trifunctional amino acids differ only in their pK_3- (side group) values: 3.36 (aspartic acid), 4.25 (glutamic acid), 6.0 (histidine) 8.33 (cystein), 10.07 (tyrosine), 10.53 (lysine) and 12.40 (arginine).

For the extraction of these compounds, quaternary ammonium ions, quaternary phosphonium ions, tertiary sulfonium ions, organic boride ions, organic sulfonic acids as well as dialkyl or diaryl phosphoric acids can be used.

Extraction is performed at very low or very high pH-values in order to eliminate the zwitterionic form.

The separation of amino acids of basic character (lysine, argine, histidine) from those of acidic character (aspartic acid, glutamic acid) is fairly easy; likewise, the separation of amino acids with a polar side group (glycine, serine, threonine, cystein, tyrosine, glutamine) from those with a nonpolar side group (alanine, valine, leucin, isoleucin, proline, phenylalanine, tryptophan, methionine).

The separation of amino acids of the same characters (polar and/or nonpolar side groups, the same basicity and/or acid-ity) needs a fine gradation of the pH-value, and the solubilisation usually is influenced by the ion strength.

An example of the separation of amino acids with a nonpolar side group and with a positively charged side group (basic character) is shown in Table 1.

Table 1

---

aqueous solution
leucine, isoleucine (bifunctional, neutral)
lysine (trifunctional, basic)

---

extracted by 100 mM TOMAC (trioctyl-methylammonium choride,
anion exchanger) 6% decanol in methylcyclohexane
at pH 10-11

---

org. phase	aqu. phase

---

leucine, isoleucine	lysine
back-extraction	
with HCl- or NaCl-solution	
into aqueous phase	

---

extracted by 0.2 M aliquat 336 (trialkyl-methyl ammonium
chloride, anion exchanger) 4 M NaCl, in cylcohexane

---

org. phase	aqu. phase

---

leucine	isoleucine

Another example is given in Table 2.

Table 2 [2]

---

aqueous solution
tryptophan (highly soluble in water),
tyrosine (low solubility in water)

---

extracted by 15 mM TOMAC in xylene at pH 10-11

---

org. phase	aqu. phase

---

tryptophan, tyrosine	----

---

back-extraction with 1.5 M NaCl-solution

---

org. phase	aqu. phase

---

tryptophan	tyrosine

By increasing the NaCl concentration, the degrees of extraction of both Trp and Tyr improve, but at 0.15 M NaCl, the
selectivity difference is the highest (Fig. 1).

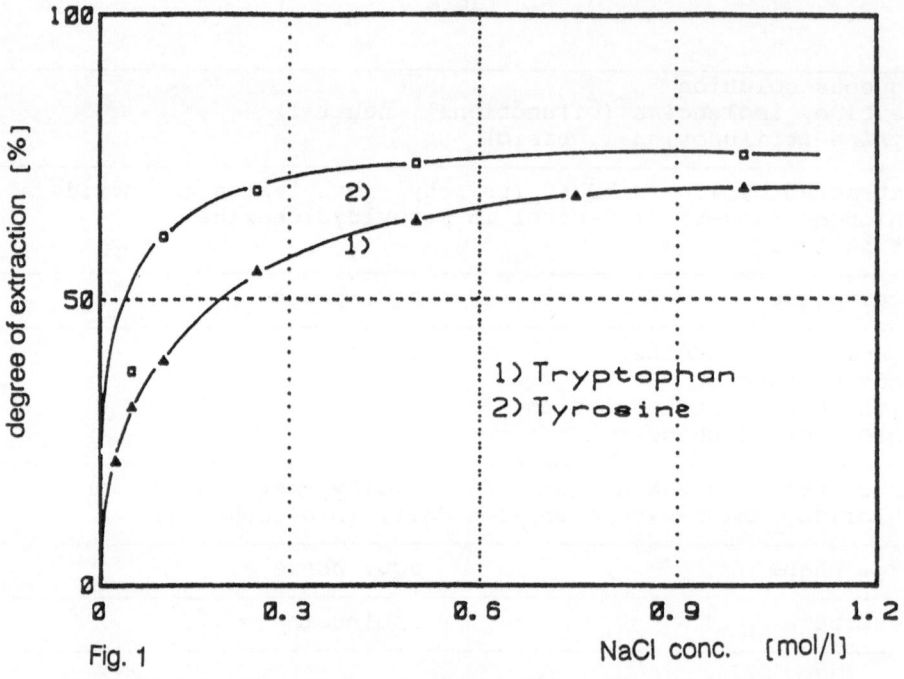

Fig. 1

Re-extraction of tryptophan and tyrosine from xylene by NaCl solution

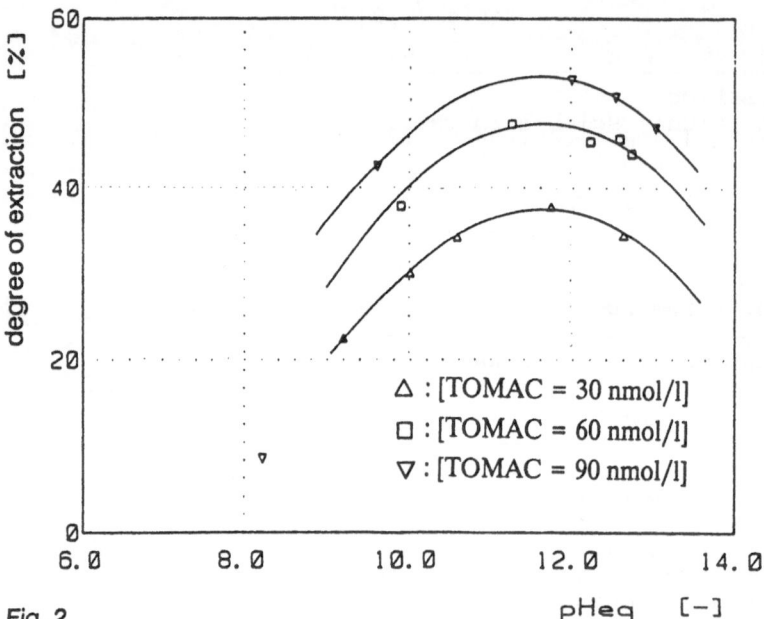

Fig. 2

Extraction of aspartic acid with TOMAC as a function of the equilibrium pH value
$C_0 = 10\,mM$

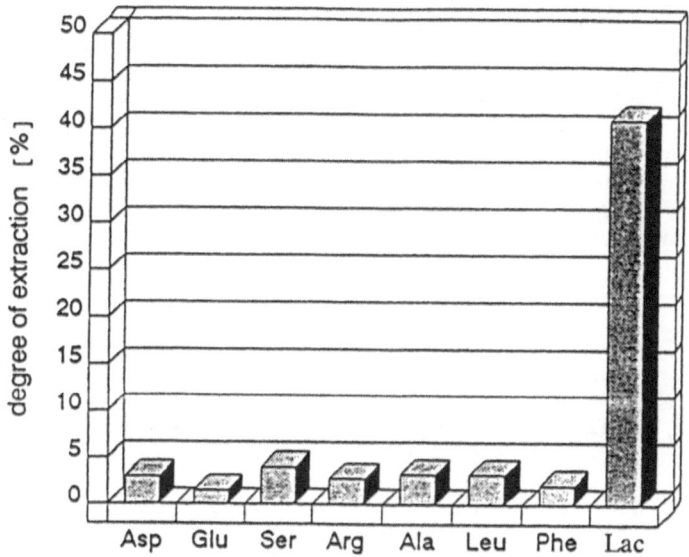

Fig. 3

Selectivity of lactic acid extraction in presence of amino acids. Extraction with Hostarex A 327 in xylene at pH 1.8 and re-extraction with 1 M NaOH.

When treating an aqueous solution of Arg and Lys, which are amino acids of basic character, as well as Asp, which is an amino acid of acidic character, with 0.3 M TOMAC in xylene at pH 12, aspartic acid is extracted into the organic phase, Arg and Lys remain in the aqueous phase. The degree of extraction of aspartic acid passes through a maximum wit at pH 12 due to the coextraction of OH^--anions at higher pH values (Fig. 2) [3].

The separation of lactic acid (pK_a = 3.86) from amino acids (pK_1 = 2-3) is possible based on the difference of their pK_a-values. For example, when treating an aqueous solution of lactic acid and amino acids with 40% HOSTAREX A 327 (1:1 tri-n-octyl/tri-n-decylamine-mixture) in the presence of 10% isodecanol in 50% kerosene at pH 1.8, only the lactic acid is extracted into the organic phase (Fig. 3). After back-extraction with 1 M NaOH, the pure lactate is recovered [4]. When using TOPO (trioctyl-phosphinoxide) as extractant, similar results can be obtained.

In a patent (Ref [5]) the separation of various amino acids are considered. This extractive recovery uses the differences in the polarity of their side groups, their solubility in water, the very low pK_a-value of histidine, the very high pK_3-value of the side group of arginine and their different solvatations in NaCl and/or $NaSO_4$-solutions [6].

The main problem with these separation processes is that the influence of the type of salts, their position in the Hoffmeister-ion-series, their influence (conservation or de- struction) on the water structure and their concentration on the solvatation of the PBO-acid complex and on the quaternary amine-acid salt is not known as yet. New physical methods should be devel-oped for the quantitative evaluation of the salt effects on the size and composition of association com- plexes, their aggregate shapes and sizes, to make the solvation calculable.

REFERENCES

1 Kertes, A.S., King,C.J., Biotechnol. Bioeng. 28 (1986) 269

2 Kirgios, I., Rhein, H.B., Haensel, R., Schügerl, K., Chem. Ing. Techn. 58 (1986) 908

3 Handojo, J., Dissertation, University of Hannover, 1988

4 v. Frieling, P., Master thesis, University of Hannover, 1988

5 Bitar, M., Sabot, J.L., Aviron-Violett, P., E.Pat. P 02512 852 C07C99/02 (12.06.1987) (Rhone-Poulenc Chimie)

6 Schügerl, K., Degener, W., Chem. Ing. Techn. 61 (1989) 796-804

EXTRACTION OF CITRIC ACID USING A SUPPORTED LIQUID MEMBRANE

T. Sirman, D.L. Pyle and A.S.Grandison
Biotechnology Group, Dept of Food Science and Technology, University of
Reading, Whiteknights, Reading RG6 2AP.

ABSTRACT

Citric acid was extracted from aqueous solutions using a supported liquid membrane. Alamine 336 diluted in heptane or xylene was used as carrier and various concentrations of sodium carbonate in aqueous solution were used as the stripping agent. It was found that systems with heptane as diluent were unstable, whereas those with xylene gave acceptable performance. Increasing concentrations of carbonate and citric acid gave rise to higher extraction fluxes, in agreement with qualitative predictions. Doubling the membrane thickness halved the flux, confirming that in this case the rate limiting process is transmembrane diffusion. At a given carbonate concentration, the extraction rate first increased then decreased with increasing carrier concentration. This can be explained by an increase in membrane phase viscosity and the resulting reduction in the diffusion coefficient of the complex.

INTRODUCTION

The use of supported liquid membranes containing a carrier to facilitate extraction has been suggested for the selective separation and extraction of metals and biochemical fermentation products from dilute aqueous solutions. Whereas in emulsion liquid membranes [6-9] the internal stripping phase is encapsulated by emulsification in the organic solvent, supported liquid membranes (SLMs) consist of a thin porous support retaining the solvent, which may also contain a carrier. In this latter case extraction and selectivity can be enhanced by effectively changing the solute partition coefficient; with a stripping reagent present the solute can be transported across the membrane against the apparent concentration gradient. SLMs offer the potential advantages of high separation factors, low capital and operating costs, higher fluxes with respect to conventional solid membranes, ease of scale-up and the possibility of continuous extraction [2,13].

There have been very few reported studies on the application of SLMs to bioseparations: this paper describes preliminary studies on the extraction of citric acid. The work is concerned with the effect of system components on stability and kinetics and with exploring the mechanisms of transfer using Alamine 336 (a tertiary amine) as carrier in xylene.

THEORY

For facilitated transport to proceed the solute (citric acid) must first be transferred from the bulk external phase to the interface with the membrane. The citric acid then complexes reversibly with the amine: tertiary amines such as Alamine have been shown to possess great potential for extracting citric and other organic acids [14 - 16]. The citric acid/amine complex diffuses across the supported membrane to the interface with the internal phase, where it is removed in an irreversible reaction with the stripping agent (here sodium carbonate) to form the product - sodium citrate. It is important to note that, apart from being insoluble in water, the carrier is highly viscous and that the solutions in xylene become increasingly viscous with increasing carrier concentrations. Boey et al [7] proposed the mechanism shown in figure 1 for citric acid transport based on studies of an emulsion liquid membrane; it was assumed that the reactions are fast and confined to the two interfaces.

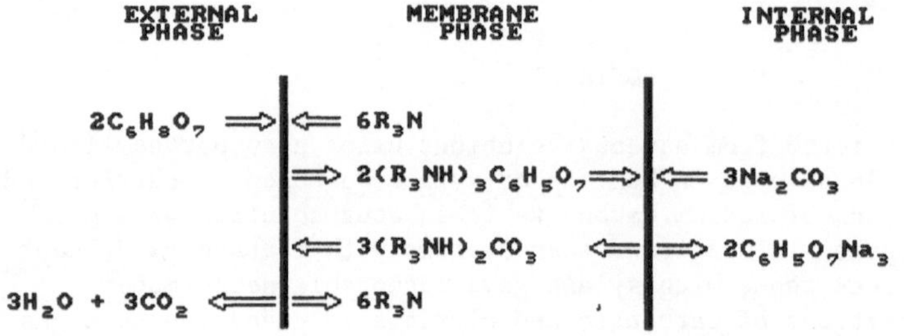

<pre>
  EXTERNAL   MEMBRANE   INTERNAL
   PHASE   PHASE   PHASE

2C₆H₈O₇ ⟹⟸ 6R₃N

             ⟹ 2(R₃NH)₃C₆H₅O₇ ⟹ ⟸ 3Na₂CO₃

             ⟸ 3(R₃NH)₂CO₃   ⟸ ⟹ 2C₆H₅O₇Na₃

3H₂O + 3CO₂ ⟸ ⟹ 6R₃N
</pre>

FIGURE 1 : PROPOSED TRANSFER MECHANISM (Boey et al)

Boey's results show that the rate of citric acid transfer falls as the pH of the feed solution rises above the first pK value (≈ 3.13); similar results were also found in this work. Below this first pK value the feed solution comprises essentially water (and hydroxyl ions and protons), undissociated citric acid, protons and $C_6H_7O_7^-$ ions, which suggests that the complexing may occur through ion-pairing in the same way as inferred by Chaudhuri [20] for lactic acid extraction. This would suggest an alternative mechanism and stoichiometry for the complex formation to that shown in figure 1 viz:

$$H^+ \;+\; C_6H_7O_7^- \;\;+R_3N \;\;\; \text{-->} \;\;\; R_3NH^+C_6H_7O_7^-$$

Other stoichiometries are possible and the clarification of the mechanisms involved in complexing are the subject of current work. However the effect of different stoichiometries on the predictions outlined below is likely to be quantitative rather than qualitative. Since the pH on the stripping side is high the end product is $Na_3C_6H_5O_7$, with likely **overall** stoichiometry:

$$2C_6H_8O_7 \;+\; 3Na_2CO_3 \;\;\text{--->}\;\; 2Na_3C_6H_5O_7 \;+\; 3H_2O \;+\; 3CO_2$$

During extraction the rate of mass transfer of citric acid will be determined by three factors:

1.   The rate of transfer of species to and from the membrane through the aqueous boundary layers at the interfaces with the external and internal

(strippping) phases;
2. The rate of diffusion of the complex across the membrane phase;
3. The rates of reaction in the membrane and internal phases.

The first factor may be neglected as a controlling factor provided the agitation (or Reynolds number) in the two bulk phases is sufficiently high to keep the resistances in the boundary layers to a minimum. In the stirred cell experiments here this is a reasonable assumption since transfer across the membrane occurs by diffusion alone and will normally be significantly slower than external transfer. Further simplification may be introduced by assuming that any reactions are fast and confined to the interfaces between the membrane and the bulk (Chaudhuri (20)). With these assumptions the rate limiting process is transfer across the membrane, thickness L, and the flux is:

$$ j = (D/L)(c_1 - c_2) \tag{1} $$

where c_1 and c_2 are the concentrations of the diffusing species at the membrane boundaries. D is the **effective** diffusion coefficient of this species in the membrane. We can identify several possible situations:

a) Unfacilitated transport : no carrier; no reagent in stripping phase. Then the diffusing species is citric acid and $c_1 = mc_f$, $c_2 = mc_s$ where m is the distribution coefficient between the membrane and the aqueous phases; c_f and c_s are the concentrations of citric acid in the feed and stripping solutions. Thus

$$ j = (D/L)m(c_f - c_s) \tag{2} $$

b) Unfacilitated transport, reagent (base) in stripping phase. Then c_2 will depend on the extent of reaction between the acid and base. Often a good assumption will be that this reaction goes rapidly to completion and $c_2 = 0$, so that

$$ j = (D/L)mc_f \tag{3} $$

c) Facilitated transport with carrier and stripping reagent. Here c_1 and c_2 are determined by the equilibrium between species at the interface. With a tertiary amine R_3N as carrier the simplest overall description of the equilibrium reactions with citric acid (= AH) at the first interface is, on the basis of the discussion above:

$$ AH \quad <\text{===}> \quad A^- + H^+ $$

where, for convenience, A^- represents $C_6H_7O_7^-$, and where the equilibrium constant is:

$$ K_1 = [A-][H+] / [AH] \tag{4} $$

where all concentrations are in the aqueous phase. The complexing reaction is assumed to be:

$$ A^- + H^+ + R_3N \quad <\text{==}> \quad R_3NH^+A- $$

with,

$$ K_2 = [R_3NH^+A^-]_{mem} /([A^-]_{aq}[H^+]_{aq}[R_3N]_{mem}) \tag{5} $$

Given the initial concentrations of citric acid and Alamine, and the two equilibrium coefficients, the equilibrium concentrations of all species can be found from equations 4 and 5. Note that, at equilibrium, the interfacial concentration of the transported species (the complex) c_1 is related to the other equilibrium concentrations by:

$$c_1 - [R_3NH^+A^-]_{mem} - K_1K_2 [AH]_{aq} [R_3N]_{mem} \qquad (6)$$

Very approximately, therefore, we may expect the complex concentration c_1 to increase with increasing initial citric acid and Alamine concentrations. Provided there is sufficient excess of base (stripping agent) it may also be assumed that c_2 is close to zero. With these assumptions and the stoichiometry suggested the flux of citric acid is:

$$j - (D^*/L) c_1 \qquad (7)$$

where D^* is the effective diffusion coefficient of the **complex**.

Transfer between two stirred cells

We consider three cases of batch transfer corresponding to the three transport situations identified earlier.

a) Consider first the batch transfer of citric acid, concentration c_f, from a well-stirred feed volume V to an equal volume of stripping solution, with citric acid concentration c_s, separated by a membrane with cross sectional area A and thickness L. Transfer is controlled by transmembrane diffusion. Then in the absence of carrier and stripping agent:

$$V \, dc_f/dt - - (DA/L)m (c_f - c_s) - - Vdc_s/dt$$
and, $$c_f + c_s - c_o$$

where c_o is the initial feed side concentration (and $c_s - 0$ at $t - 0$). Thus:

$$c_f(t) - 0.5c_o (1 + exp-(2mDAt/VL)) \qquad (8)$$

b) The second situation is comparable to (a) above, except that there is a fast reaction on the strip side and we assume that $c_2 - 0$; then:

$$V \, dc_f/dt - - (DA/L)m \, c_f$$
so that: $$c_f(t) - c_o exp - (mDAt/VL) \qquad (9)$$

c) Finally consider facilitated transport with a stripping reaction. Normally, c_2 must be obtained from the reaction equilibria at this interface; if the stripping agent is in large excess and the reaction is fast, $c_2 - 0$. Then a mass balance yields:

$$V \, dc_f/dt - - (D^*A/L) c_1$$

where D^* is the effective diffusion coefficient of the carrier complex and c_1 is the complex (ie $R_3NH^+A^-$) concentration at the feed/membrane interface, see eqn 6, which is given by the solution to eqns 4 & 5. Therefore

$$V \, dc_f/dt - - (D^*A/L) K \, c_f \qquad (10)$$

where $K - K1K2 [R_3N]$; when Alamine is present in substantial excess $[R_3N]$ - initial Alamine concentration. Assuming K remains constant:

$$c_f = c_o \exp - (D^*A/VL)Kt \qquad (11)$$

The greater the value of K in comparison to the distribution coefficient, m, the greater is the facilitating effect of the carrier.

MATERIALS AND METHODS

Materials

The solid support membrane was a microporous hydrophobic membrane ("Durapore") made of polyvinylidene difluoride (PVDF), Millipore Ltd (UK). The membrane had reported a porosity of 75% and thickness and pore size of 125 and 0.22 micrometers respectively. Citric acid (Analar) and other chemicals (GPR grade) were obtained from BDH chemicals Ltd (Poole, UK). Alamine 336 (trioctylamine) was a gift from Henkel Corporation (Cork, Eire). [1,5- ^{14}C] citric acid (specific activity 10 mCi/mmol) was from Amersham Radiochemicals (Amersham, UK). Pseudocumene-based liquid scintillator (Supersolve 'X') was supplied by Koch-light Ltd., Haverhill, UK.

Apparatus

The apparatus used in the kinetic studies consisted of two compartments each of working volume 80 ml and separated by the PVDF membrane of cross sectional area 13.8 cm^2, sandwiched in the vertical plane between two Teflon gaskets. The equipment was constructed in plexiglass. The aqueous solutions in the two compartments were stirred synchronously and constantly by impellers driven by a single motor.

Methods

To prepare the SLM the PVDF membrane was immersed in a mixture of defined proportions of carrier (Alamine 336) and diluent (n-heptane or xylene) for one hour at 25°C. After rapid draining the membrane was either weighed or quickly sandwiched between the two compartments of the kinetic cell. The fractional extent of pore filling by the solvent/carrier mixture was calculated from the weighed values and the membrane specifications. These fractions were 0.17 and 0.89 for 5 v/v% Alamine in n-heptane and 20 v/v% Alamine in xylene respectively.

For each extraction experiment 80 mls each of an aqueous solution of citric acid of determined composition and sodium carbonate (up to 20 w/v%) were placed in the adjacent feed and stripping compartments of the cell. (Note: a 5 w/v% solution of citric acid is 0.24 M; the corresponding stoichiometric sodium carbonate solution (0.36 M) is 4 w/v%). Upon commencement of stirring 10 ul of ^{14}C-citric acid (50 uCi/ml) was added to the feed solution. Starting immediately, samples (0.5 ml) were taken from both compartments at regular intervals over a typical run time of 8 hrs. Each sample was placed in a scintillation vial containing 10ml of "Supersolve X" liquid scintillator; counting was carried out in a Rackbeta 1215 LKB liquid scintillation counter; results were obtained as counts/minute/ml (cpm/ml). Citric acid fluxes were calculated from the (slowly-changing) concentration time profiles. Feed and strip side pH values were monitored using a digital pH Meter (Corning model 240). All experiments were carried out at room temperature.

RESULTS AND DISCUSSION

(a) Effect of stirrer speed.

The effect of stirrer speed on the rate of mass

transfer of citric acid through the SLM was studied over the range 50 -800
rpm : there was no significant difference in the extraction rates,
implying that the resistance to mass transfer across the two external
boundary layers is not rate-limiting.

(b) Extraction using n-heptane as diluent.

Citric acid (5w/v%) was extracted
using an SLM containing 5% alamine in heptane, using different initial
concentrations of sodium carbonate. Contrary to expectations [7,20] the
extraction rate increased with decreasing carbonate concentration, and
high rates of extraction were achieved with deionised water on the
stripping side. Furthermore there was evidence of back transfer into the
citric acid on the feed side. Together with the measured fractional
pore-fill, these results indicate that the system was unstable and that
the liquid membrane was not continuous. Rapid evaporation of heptane
during assembly of the membrane was a severe problem. Similar erratic
results have been observed by Komasawa et al [21]. Alamine concentrations
above 5% led to a complete breakdown of the liquid membrane.

(c) Extraction of citric acid using xylene as diluent.

The principal part of this study used xylene as the membrane solvent.
Batch extraction studies have been carried out using a range of Alamine
(0-20%), citric acid and sodium carbonate concentrations.

Unfacilitated transport.

Experiments with no Alamine carrier showed no
citric acid transfer: the distribution coefficient m is effectively zero.

Membrane thickness.

The theory above predicts that under a defined set of
conditions the instantaneous flux should be inversely proportional to the
membrane thickness. To test this hypothesis, experiments were carried out
with a single membrane (thickness 125 microns) and a sandwich of two
membranes (combined thickness 250 microns). In these experiments the
initial concentrations were: citric acid 0.24M; Alamine 336 20% and sodium
carbonate 20%. The initial fluxes were 1.4 x 10^{-5} mmol/(cm^{2}s) and 6.5 x
10^{-6} mmol/(cm^{2}s). The flux ratio of 2.18 confirms the hypothesis.

Sodium carbonate concentration.

Figure 2 (citric acid 0.24M and Alamine 20
w/v) shows that increasing carbonate concentrations resulted in increased
extraction rates of citric acid. In all experiments overall flux rates
were low and low extraction yields were achieved in the time allowed for
each run. The increase in flux with increasing base concentration reflects
the effect of the complex/base stripping reaction on the transmembrane
driving force: c_2 will be highest at low base concentrations; at zero base
concentration the complex dissociation reaction will be the reverse of
that occurring at the feed/membrane interface, allowing however for the
different pH (which will start at pH7). The results show that as the
sodium carbonate concentration approaches the stoichiometric requirement
the flux reaches a maximum, constant value, corresponding to $c_2 = 0$ in the
flux equation (eqn7). There is some evidence at low carbonate
concentrations that the complex/carbonate reaction may influence the
overall rate. The results confirm that the overall transfer rate is
controlled by trans-membrane diffusion.

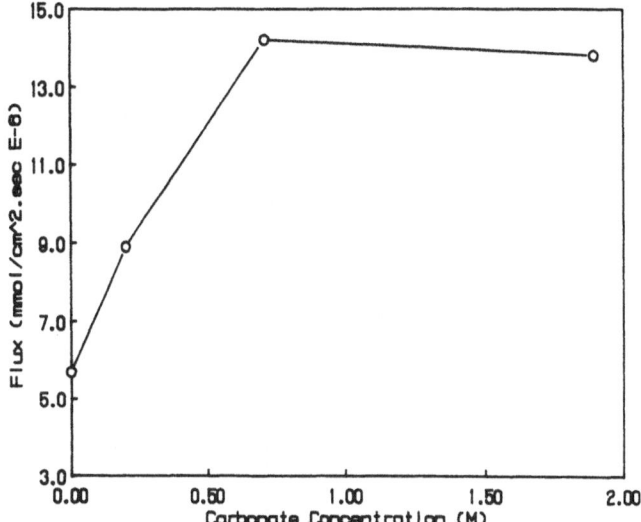

Figure 2 : Effect of ᴄarbonate concentration on flux.
(Flux measured in mmol/cm^{2}sec x 10^{-6})

Citric acid concentration.

          Figure 3 shows that, as expected from the theory
(eqn 7), the initial flux increases with increasing citric acid
concentration in the initial feed. These results correspond to Alamine and
sodium carbonate concentrations of 20% - thus there is a large excess of
base. As expected the flux is only approximately proportional to the
citric acid concentration.

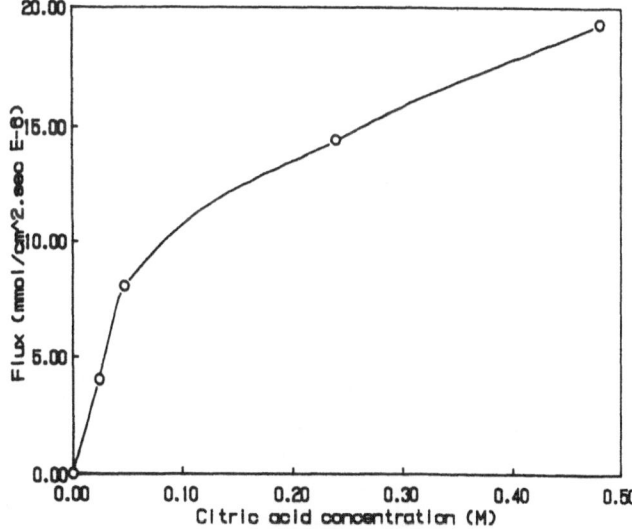

Figure 3: Effect of citric acid concentration on flux
(Flux in mmol/cm^{2}sec x 10^{-6})

As figure 4 illustrates the concentration-time profile follows the
predicted form (eqn 11); however, as extraction proceeds, the
concentration of citric acid on the feed side declines and this, together

with the accumulation of dissolved carbon dioxide, leads to a gradual
increase in pH. The assumption of constant 'K' in eqn 10 is no longer
valid and a more complete model is currently being developed to deal with
this situation. As the pH approaches and increases beyond pK_1 (≈ 3.13) the
citric acid flux falls off; we believe that this confirms our hypothesis
of complex formation by ion pairing between the tertiary amine, protons
and $C_6H_7O_7^-$ ions. Until direct measurements of the parameters in eqn 10 are
available direct comparison of theory and experiment is not possible.

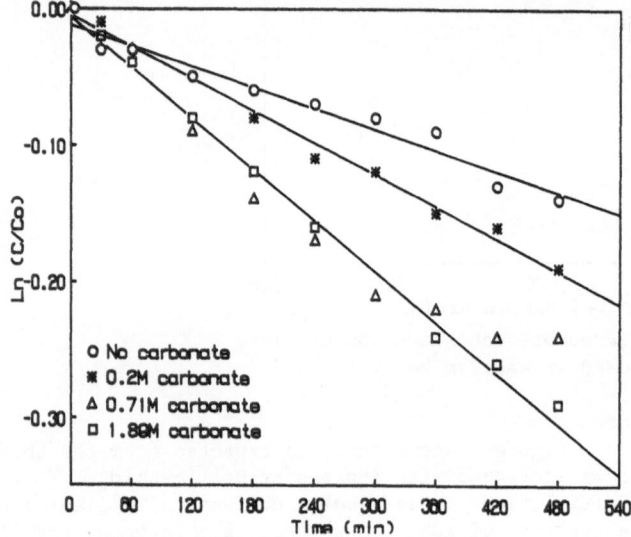

Figure 4 : Time profile of batch extraction.

Alamine concentration

Figure 5 (initial citric acid 0.24 M; sodium
carbonate 20%) shows the apparently surprising result that the alamine
concentration does not have a monotonic effect on flux rates, as might be
expected from eqns 7 and 10. At low Alamine concentrations the flux is
proportional to the concentration of tertiary amine, as suggested by the
theory. However, there is a clear maximum in the average flux at an
Alamine concentration of around 20%. This can be explained qualitatively
from the theory. As we have seen increasing alamine concentrations lead to
a higher effective driving force across the membrane ; however, as the
alamine concentration increases the viscosity of the alamine/xylene
mixture increases sharply. Thus the diffusion coefficient of the complex
in the membrane will decrease (for example the Stokes-Einstein equation or
the Wilke-Chang correlation [22] predict an approximately inverse
relationship between diffusion coefficient and viscosity). Measured values
of the viscosities suggest that this effect can explain the observed
maximum. Further experimental and theoretical studies are in progress to
explore these effects.

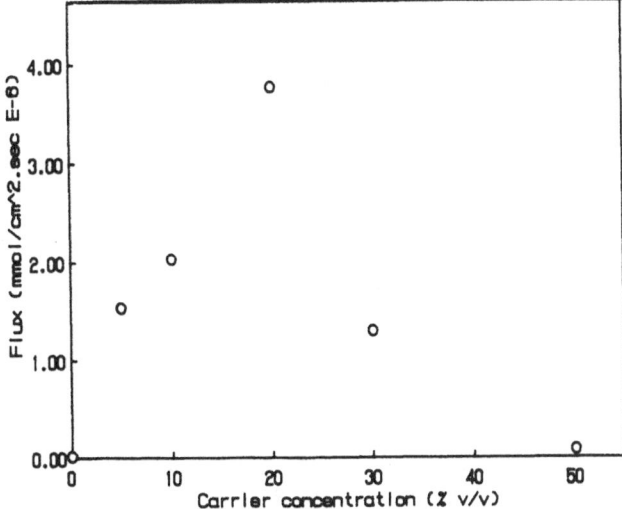

Figure 5 : Effect of carrier concentration on flux
(Flux in mmol/cm^{2}sec x 10^{-6})

CONCLUSIONS

The experimental results presented here show that the choice of system (in particular, the diluent) and concentrations of carrier and stripping solution are crucial to the extraction rate and yield of citric acid extraction using a SLM. A stable SLM system was obtained using xylene as diluent with Alamine 336 as carrier. Theory is developed for certain limiting cases, and experimental results on citric acid fluxes under a range of conditions show qualitative agreement with simple theory. In particular, the theory correctly predicts that fluxes will increase with citric acid concentration and, at low concentrations, with the base. At higher sodium carbonate concentrations the concentration at the membrane/strip side interface is effectively zero and fluxes are then independent of increasing carbonate concentration. The theory also predicts that fluxes should increase with increasing carrier concentration; however it is also shown that it is necessary to allow for the reduction in the effective diffusion coefficient of the complex which occurs with increasing Alamine concentration because of the significant increase in membrane viscosity. Experiments with two different membrane thicknesses also provide confirmation of the theory which assumes that in this situation the rate limiting process is molecular diffusion across the membrane. Current work is aimed at measuring some of the key physico-chemical parameters in the models and to developing rigorous theory for transfer kinetics.

REFERENCES

1. Danesi, P.R. Separation of metal species by supported liquid membranes. Separation Sci. Tech., 1984, 19, 857-894.
2. Loiacano, O., Drioli, E. and Molinari, R. Metal ion separation and concentration with supported liquid membranes. J Memb. Sci., 1986, 28, 123-138.
3. Kuo, Y. and Gregor, H.P. Acetic acid extraction by solvent membrane.

Separation Sci. Tech., 1988, **18**, 421-440.

4. Kyung-He, L., Evans, D.F. and Cussler, L. Selective copper recovery with two types of liquid membranes. AIChEJ, 1988, **24**, 860-868.

5. Deblay, P., Minier, M. and Renon, H. Separation of L-valine from fermentation broths using a supported liquid membrane. Biotech. Bioeng., 1990, **33**, 123-131.

6. Draxler, J. and Marr, R. Emulsion liquid membranes. Part I: phenomenon and industrial applications. Chem. Eng. Proces., 1986, **20**, 319-329.

7. Boey, S.C., Garcia del Cerro, M.C. and Pyle, D.L. Extraction of citric acid by liquid membrane extraction. Chem. Eng. Res. Design, 1987, **65**, 218-223.

8. Chaudhuri, J. and Pyle, D.L. Liquid membrane extraction. In: "Separations for Biotechnology", Verrall, M.S. and Hudson, M.J. (Eds), Ellis Horwood, Chichester, UK, 1987, 241-159.

9. Thien, M.P., Hatton, T.A. and Wang, D.I.C. Separation and concentration of amino acids using emulsion liquid membranes. Biotech. Bioeng.,1988, **32**, 604-615.

10. Babcock, W.C., Baker, R.W., Lachapelle, D. and Smith, K.L. Coupled transport membranes III: the rate limiting step in uranium transport with a tertiary amine. J. Membr. Sci., 1980, **7**, 89-100.

11. Nishiki, T. and Bautista, R.G. Platinum (IV) extraction with supported liquid membrane containing trioctylamine carrier. AIChEJ., 1985, **31**, 2093-2095.

12. Nuchnoi, P., Yano, T., Nishio, N. and Nagai, S. Extraction of volatile fatty acids from diluted aqueous solution using a supported liquid membrane. J.Ferm. Tech., 1987, **65**, 301-310.

13. Danesi, P.R., Chiarizia, R., Rickert, P. and Horwitz, E.P. Separation of actinids and lanthanides from acidic nuclear wastes by supported liquid membranes. Solv. Extract. Ion Exch., 1985, **3**, 111-147.

14. Wennersten, R. The extraction of citric acid from fermentation broth using a solution of a tertiary amine. J. Chem. Tech. Biotech.,1983, **33B**, 85-94.

15. Kertes, A.S. and King, C.J. Extraction chemistry of fermentation product carboxylic acids. Biotech. Bioeng., 1985, **28**, 269-282.

16. Vanura, P. and Kuca, L. Extraction of citric acid by the toluene solutions of trilaurylamine. Coll. Czech. Chem. Commun., 1987, **41**, 2857-2877.

17. Sengupta, A., Basu, R. and Sirkar, K.K. Separation of solutions by contained liquid membranes. AIChEJ., 1988, **34**, 1698-1708.

18. Babcock, W.C., Baker, R.W.,Lachapelle, D. and Smith K.L. Coupled transport membranes II: The mechanism of uranium transport with tertiary amines. J. Membr. Sci., 1980, **7**, 71-87.

19. Huang, T.C. and Juang, R.S. Transport of zinc through a supported liquid membrane using di(2-ethyl hexyl) phosphoric acid as a mobile carrier. J. Membr. Sci.,1987, **31**, 209-226.

20. Chaudhuri J.B. Kinetic studies on the emulsion liquid membrane extraction of lactic acid. PhD Thesis, 1990, University of Reading.

21. Komasawa, I., Otake, T. and Yamashita, T. Mechanism and kinetics of copper permeation through a supported liquid membrane containing a hydroxyoxime as a carrier. Ind. Eng. Chem. Fund., 1983, **22**, 127-131.

22. Wilke, C.R. and Chang, A. Correlation of diffusion coefficients in dilute solutions. AIChEJ., 1955, **1**, 264-270.

Part 3

High Resolution Separations

THE FUTURE OF HIGH RESOLUTION PURIFICATION

HOWARD A. CHASE
Department of Chemical Engineering, University of Cambridge
Pembroke Street, Cambridge, CB2 3RA, U.K.

ABSTRACT

This chapter presents the author's personal view of the fundamental mechanisms underlying high resolution separations and of the areas where further effort is needed in the development of new and improved techniques. The main emphasis is on the various types of partitioning methods, involving solid, liquid, or precipitated phases, used for high resolution separation. These separation processes can be classified into highly selective or multiple stage separations and recent trends in the development of these techniques are discussed. Additional sections consider the possibilities for optimisation by process design and control and the development of high resolution separations for small molecular weight species. It is shown that there are opportunities for important contributions from genetic engineers, chemical engineers, chemists and biochemists.

THE PRESENT SITUATION - MECHANISMS OF ACHIEVING HIGH RESOLUTION

The definition that will be taken of a high resolution purification step is one which usually occurs near the end of a purification flow-sheet and normally involve the separation of the bioproduct of interest from similar molecules. Sometimes it is necessary to separate a molecule from a variant of its own structure. As progress is made on analysing the fine structure of complex biomolecules, there is an increasing need to be able to remove variants differing in amino acid sequence, level and type of glycosylation (or other post-transcriptional modifications) or even three-dimensional conformation Such high resolution steps are vital in order that compounds can be purified to the levels required for their use in applications such as therapeutic agents. It can be expected that a high resolution step can eliminate the need for a series of steps each with lower resolution. Hence, if practicable, the inclusion of a high resolution step at an early stage in down-stream processing can substantially reduce the complexity of the overall process flow-sheet. The theme that will taken is a discussion of whether future progress in the development of new and more powerful high resolution steps will be governed by chemical engineering, genetic engineering or chemistry.

All separation methods work by exploiting some difference between the molecules to be separated. In principle any separation mechanism can form the basis of a high resolution step provided it can produce a compound in a highly purified form. One major technique for high resolution purification involves some form of partitioning of molecules between two phases. One of those phases is usually the aqueous phase in which the molecules are originally present whilst the other is a solid (for adsorption and chromatographic techniques) or a liquid (for liquid extraction techniques). The situation is a little different with aqueous two-phase systems. Although it is not normal to treat liquid and solid phase separations as similar as a result of the different types of equipment used, there are in fact a number of underlying similarities arising from the similar nature of the partitioning between the two phases. There is even some argument that precipitation can be thought of as a system in which partitioning occurs, in this case with molecules partitioning between being in a soluble or precipitated form.

Whatever the nature of the partitioning, there are essentially two approaches to achieving high resolution with such techniques:-

(1). The highly selective approach. The partitioning between the phases is very selective such that only a single compound partitions to one of the phases. This is the basis of most highly selective techniques. Often the high selectivity is the result of the derivatisation of an "unpopular" phase with a ligand that specifically interacts with (i.e. has a high affinity for) the compound of interest. The ligand may interact with this compound in a manner similar to interactions involved in the biological functions of the ligand and ligate. However, it must be stressed that this is not always the case, and that high selectivity can result from interactions which rely on traditional chemical principles.

(2). The multiple stage approach. Here resolution can be achieved by using a partitioning system where there is not much selectivity between the two phases. However, high resolution, i.e. the complete separation of one compound from others can be achieved if the partitioning process can be carried out many times in succession. In such a situation, the small degree of separation that can be achieved in a single stage can be multiplied to achieve much greater separations. This is essentially the basis of most chromatographic techniques and in particular, the high resolution achieved in HPLC techniques.

There are of course other separation principles that rely on techniques other than the partitioning between two phases. Electrophoretic separations and certain types of membrane methods can be considered not to fall into partitioning methods.

THE FUTURE OF THE HIGHLY SELECTIVE APPROACH

It is often argued that the highly selective approach is the desired approach to high resolution separations as once a separation system has been established, it is usually easy to scale-up and operate on the large scale. However, the development of the initial highly selective system is often complex, expensive and results in a method that may not be as robust as other non-selective methods. Hence, there remains much interest in the development of new systems and there are a number of aspects that can be investigated as described below.

New ligands
Highly selective ligands that interact with biomolecules by mimicking their *in vivo* functions include antibodies, Protein A and other proteins. However, many of these ligands are expensive and of high relative molecular mass. The latter property is undesirable on account of the likelihood that such molecules when associated with an otherwise inert phase may result in the creation of sites for non-specific adsorption. Other smaller ligands have been used and although the costs of ligand are reduced and the robustness of the system is increased, it is often not possible to achieve such high degrees of specificity. The future development of smaller sections of monoclonal antibodies that retain their highly selective affinity for antigens may

prove to be important in extending the scope of such techniques. If is desired to use a highly selective step as early in the process flow-sheet as possible, it is essential to ensure that the ligand/matrix system chosen is robust enough to withstand the rigours associated with operation in the presence of crude materials.

High resolution methods have been developed based on separations involving some of the more fundamental properties of the molecules such as size and gross surface properties (charge, hydrophobicity). These essentially constitute examples of less selective, multiple stage methods except in extreme cases where the compound of interest has either naturally or as a result of genetic intervention been endowed with exaggerated properties. The use of affinity tails involved repeated occurrences of charged amino acids giving rise to unusually strong interactions with ion-exchangers is an example of this approach and is discussed further below.

The technique of affinity precipitation of multimeric proteins using bi-functional affinity ligands removes the need to have any form of "phase-forming" material and therefore is an attractive procedure. Care must be taken to distinguish this technique from another affinity precipitation technique in which ligands are attached to polymers whose solubility can be altered in a cyclic manner by changes to the composition of the liquid phase. Molecules become associated with the polymer as a result of affinity interactions with the attached ligands and then can subsequently be "transferred" to a precipitated phase by flocculation of the polymers.

Modified ligates

The techniques of genetic engineering have enabled a different approach to be taken to the development of highly selective separations. Instead of attempting to tailor the separation procedure to the characteristics of the compound of interest, the approach taken has been to alter the compound such that it interacts more selectively with a particular separation method. Hence for improved protein separations, additional sequences of amino acids, either oligopeptides or whole proteins, have been added onto an end of the native molecule by additions to the nucleic acid sequence coding for it. These so-called tails enable the molecule to interact highly selectively in a two-phase system (normally, always an adsorptive or chromatographic separation) such that high resolution can be achieved in a process involving a single equilibrium stage. There are many variations on the theme, but perhaps the most attractive ones involve cheap and robust metal chelate adsorbents which interact very specifically with the "tailed" protein. Other examples, involving macro-molecular affinity ligands will continue to suffer problems associated with the expense and lack of robustness of such ligands. There are some systems where the affinity tail added to the protein is longer than the native protein and the tail may represent a complete protein in its own right such as β-galactosidase or β-lactamase. Although this may be desirable in order to exploit a convenient affinity interaction between the tail protein and a cheap affinity ligand, the method is must be considered as being undesirable on stoichiometric grounds as the majority of hybrid protein synthesized by the cell does not end up as a constituent of final product.

This approach to the creation of highly selective separation has a number of attractions; firstly, it is likely that gene splicing techniques will have been used in preparing the genetic information for the product. Hence the extra stages involved in adding an affinity tail are not likely to influence greatly the overall cost or complexity of the process. Secondly there is little need to make the two-phase system highly specific by immobilising a large ligand thus reducing both expense and any problems associated with leakage of the ligand or non-specific adsorption onto it. However, it is necessary to remove the tail by chemical or enzymic cleavage and part of the art of the method is the development of a specific cleavage system that results in good yields of the required product. There is still plenty of scope for contributions from genetic engineers, molecular biologists and biochemists in refining and extending these procedures which may have a radical influence on the future of high resolution separations.

New solid phases

The ideal requirements of solid phases have been the focus of much discussion and most materials fail to meet all the desired criteria. However, progress is still being made on the development of new features. One of the problems of the highly selective approach is the synthesis or formulation of a truly inert phase that has no affinity towards any molecule until derivatised with a ligand which suddenly confers it with a selective affinity towards a particular compound or group of compounds. The problem is confounded by the fact that the ligand itself has to be confined to this phase in a selective manner, normally as a result of some form of chemical immobilisation *via* reaction with the support. One major limitation on the use of highly selective separations is the possibility that the attached ligand may leach into the product during the elution phase. This problem often has to be overcome by including a subsequent separation technique designed to resolve the product from the leached ligand. The further development of inert solid phases and the successful immobilisation of ligands is essentially a problem for the chemist. Despite the introduction of numerous new materials and immobilisation chemistries over the last 20 years, there have been only comparatively modest improvements to the systems available.

Attention has been focussed on speeding up the cycle-time for each round (cycle) of the separation procedure. One strategy for overcoming mass transfer limitations arising from pore diffusional resistances in porous particles is to use small (5-10 µm) particles and these have for many years formed the basis of so-called HPLC separations. A similar strategy is the use of totally non-porous particles in which these diffusional resistances are eliminated. However, unless the particles are very small, the available surface area for ligand immobilisation and adsorption is very limited. The use of very small particles results in the development of very high pressure drops across packed beds of such particles. Hence considerable effort has been expended on the development of new materials or modifications of existing ones capable of withstanding such pressures whilst retaining all the other desirable properties of a solid phase material.

A different solution to this problem is to use the pores in a membrane as the surface for adsorption. This is a equivalent to using a very wide bed of length equal to the thickness of the membrane. Convective flux of molecules through the membrane provides transport to the adsorption sites and allows separations to be carried out rapidly. The kinetics of membrane affinity separations are much faster than those achieved with conventional porous solids as the governing mass transfer resistance becomes either film diffusion or the intrinsic chemical kinetics of the adsorption/desorption reactions. Hence the overall cycle time can be greatly reduced although care has to be taken to prevent fouling and blockage of the membrane pores.

Another approach is the development of porous media containing two sizes of pore :- "through-pores" large enough to allow some convective flux through the particles and smaller "diffusive" pores lining the through-pores to provide high adsorption area. Convection through the through-pores allows transport of adsorbate into the particle at greater rates than can be achieved by passive pore diffusion. POROS (PerSeptive Biosystems Inc) is a new solid phase material based on such a principle in which linear flow velocities of up to 3,600 cm/h can be used without diminution of the chromatographic separation achieved. These beads are composed of polystyrene/divinyl-benzene polymers coated with a hydrophilic, non-fouling surface.

New liquid phases

Most liquid separation techniques involve partitioning of a species as a result of selective *dissolution* in the two liquid phases. Hence it is always going to be difficult to achieve high resolution in a single stage system. This is because it is difficult to find a two phases system with a liquid phase in which all species are insoluble bar the one of interest, even if that phase contains ligands with a highly selective affinity for the solute. Indeed it is even difficult to ensure that the affinity ligand itself is confined to only one of the two phases. Hence, the use of conventional two-phase liquid extraction techniques is probably not best suited for high selective separations. However, by analogy with less selective solid-phase chromatographic separations, high resolution with liquid extraction methods can be achieved by using multiple stages in the separation process and this is discussed later.

An alternative approach to liquid extractions is to use a liquid phase in which all solutes are *insoluble* and where selective interactions occur at the *surfaces* of droplets in a stabilised emulsion droplet of this liquid phase. The principle is identical to using non-porous solid phases, but with a number of advantages. The surfaces of perfluorocarbon liquid droplets can be coated with affinity ligands derivatised with perfluoroalkyl groups in order that selective interactions with the adsorbent surface can be created. Droplets can be stabilised in an emulsion with the use of perfluorocarbon surfactants or polymers that are cross-linked *in situ* to coat the droplets. The use of a high-density liquid phase adsorbent permits the adoption of countercurrent, continuous adsorption techniques thus providing the potential for new unit operations in high resolution systems. The development and exploitation of these ideas will require close co-operation between chemists and chemical engineers.

THE FUTURE OF THE MULTIPLE STAGE APPROACH

Comparison with highly selective methods
The success and rapid growth of HPLC techniques particularly on the small scale suggests that the resolution obtained is sufficiently high for analytical purposes. The emphasis during analytical separations is to obtain sufficient resolution of components so that their absolute or relative amounts in the original sample can be unambiguously assessed by non-discriminatory techniques such as flow spectrophotometry. Scaling up analytical separations in order to prepare larger amounts of compound is primarily the business of the chemical engineer. The skills involved are in the physical construction of the chromatographic apparatus to withstand the operating pressures and to ensure that separation performance is not lost during scale up; problems arising from column packing and flow distribution across the bed can easily diminish the quality of the separation.

In preparative techniques, the emphasis is on obtaining a single compound in a pure form either so that its properties can be investigated without complications arising from the presence of other components, or to be put to some final use that requires the compound to be pure (e.g. use as a therapeutic agent). Scaled-up analytical methods do not necessarily result in the most efficient preparative procedures. The throughput in an analytical system is normally rather low as the techniques were initially developed to resolve more than one component using the smallest amount of sample. However, most HPLC separations are based around a very short cycle time, which when repeated frequently can result in high productivities. However, if highly selective systems are used then a higher proportion of the adsorption capacity of the bed is being used at any one time for purification. Hence when highly selective methods are used with short cycle times they still can be more productive than their less selective counterparts.

The principle attraction of multiple stage techniques is that there is no need to use any form of highly selective agent (e.g. the ligand in partitioning methods). This makes the approach intrinsically more general and avoids problems of expense, toxicity and lack of robustness that may be associated with the use of such agents.

New ligands and phase systems
In solid-phase separations, it is unlikely that there will be much progress in the development of new non-selective ligands as most mechanisms of interaction with molecules have been exploited. However, the composition of the liquid phase can greatly effect the relative strengths of interactions with different compounds. There is still plenty of scope for the development of new solid phases for use in these methods and this is essentially in the hands of chemists. The

requirements of these phases are often very similar to those for high selective methods, although the requirement of using multiple-stage separations is an additional constraint .

It has already been stated that the use of conventional two-phase liquid extraction techniques is probably not best suited for high selective separations. However, by analogy with less selective solid-phase chromatographic separations, high resolution with liquid extraction methods can be achieved by using multiple stages in the separation process. Multiple stage liquid extraction techniques involve the use of a series of mixer/settler units, centrifugal separators or countercurrent column methods. However the move to multiple stage separations will require the continued development of laboratory scale equipment in which such separations can be investigated. Inevitably such equipment will be more complex than the multi-stage chromatographic column that is the basis of HPLC separations and is likely to limit this approach to high resolution separation. The technique of electrically-enhanced extraction in which very small droplets with high interfacial area are created as a result of electrostatic charge on the droplets may have considerable application in this area.

One philosophy for the separation of very similar species involves the partitioning of those species between two phases that only differ very slightly from one another. In the absence of any mechanism to selectively discriminate between the two species, the use of greater differences between the phases will result in the two compounds partitioning in an essentially identical manner. Two phases with minimal difference between them can be achieved by operating close to the binodial of a liquid two-phase system. However, the small difference in partitioning that will be achieved in such systems will inevitably result in the need for multiple-stage system.

OPTIMISATION BY PROCESS DESIGN AND CONTROL

It is sometimes possible to increase the performance of high resolution separations by understanding more fully the principles on which they work. This approach leads to the development of computer-based simulations which can be used both for initial process design and also for optimisation of existing procedures. The main hold-up in this approach is the lack of physical data describing the interactions between solutes and phases. In particular, most separations are carried out in the presence of multiple components. Especially when less selective methods are used, it would be necessary to obtain large amounts of information about many of these components before accurate modelling could proceed. Such information would probably be obtained from small scale experiments, although more general approaches to predicting the necessary parameters would be highly desirable. It may prove possible in some cases to reduce the problem to that of a two-component separation, the two components being the compound of interest and the contaminant which is the most difficult to resolve it from by the particular technique under investigation.

A somewhat different approach is the use of artificial intelligence systems. The creation of expert systems may allow the heuristic, rule of thumb approach to be applied in a more extensive manner. No longer will the limited experience of those those involved in developing a high resolution process be the governing step to process optimisation. The ability to combine the experience of a large number of practitioners should enable the most appropriate purification system to be chosen in the first instance in addition to its subsequent optimisation.

The development of on-line assay systems for particular compounds raises the possibility that existing separations can be further improved by the introduction of process control. Such systems may allow throughputs to be raised to higher levels as the separation system becomes able to respond interactively to variations in feed-stocks etc.

OTHER METHODS OF HIGH RESOLUTION SEPARATION

The above discussions have been centred around high resolution separation methods based around some form of partitioning between two phases. However, it is worth considering whether other separation principles have the potential of achieving high resolution. Electrophoretic separations have been frustrated on the large scale by the presence of dispersive convection of liquid arising from heating effects due to the electric current. Attention has been turned to separations in space under micro-gravity (where density differences between hot and cold liquids disappear no longer give rise to convection) and fluid mechanical solutions such as the use of Taylor vortices in couette flow of liquid in an annulus to stabilise the pattern of flow.

Separations based on size and shape will depend primarily on the ability to create pores, either in membranes or molecular exclusion media that can accurately discriminate between molecules of similar sizes and shapes. Molecular exclusion with membranes is essentially a single stage process with the result that resolution will be poor unless discrimination is extremely accurate. Molecular exclusion chromatography, on the other hand is a multi-stage process thus enabling resolution with a broad distribution of pore sizes. A possible future direction for membrane technology involves further development of "gated" pore systems by which particular molecules are selectively allowed to enter pores and consequently pass through the membrane in a selective manner. The future development of this analogy with cell membranes presents a challenge for the chemist and biochemist.

An alternative approach to the improvement of high resolution bioseparations is to eliminate the need to have such systems. This philosophy implies that the product can be produced in a pure form. Perhaps the nearest approach to this ideal is the production of genetically-engineered proteins in inclusion bodies. Although these inclusion bodies can be isolated from the bulk of contaminating material comparatively easily, problems have arisen in subsequently obtaining the product in its bioactive re-natured form. Further development of the techniques of protein re-folding are needed to enable this route of product formation and isolation to by-pass the need for the inclusion of a high resolution step. However, unless complete conversion to the desired bio-active form can be guaranteed, there will remain a need for a high resolution step to separate undesirable forms of the product.

HIGH RESOLUTION SEPARATIONS FOR SMALL BIOPRODUCTS

It has been comparatively easy to develop new high resolution separations for macro-molecules. However, the main methods for small molecules are still based on solvent extraction, re-crystallisation although adsorption and HPLC methods are being used increasingly. The major challenge remains the development of highly selective methods for these small molecules. The only *general* approach to the use of highly selective ligands requires the use of high molecular weight ligands such as antibodies. Although the costs of using immobilised antibodies for recovering small amounts of therapeutic proteins may be justified considering the value of the product, the same cannot be said for smaller compounds of lower value. In additional, the use of large, expensive ligands to purify small products is also not favoured on stoichiometric grounds as the mass of product purified per cycle of operation is proportional to the molecular weight of the product. However, much cheaper adsorbents (such as XAD resins) are sometimes found to show unexpectedly selective interactions with certain compounds. The origins of these fortuitous interactions are often not known, but their elucidation, so that similar effects can be exploited on a rational basis, is a challenge for the chemist.

The use of template-imprinted polymers may revolutionise this area. The ability to be able to create a solid-phase with cavities that are specific for particular molecules may result in a completely new approach to the development of highly selective separations

There is often a need to be able to resolve different isomers of a small molecular weight compound particularly when it has been synthesized *via* synthetic organic chemistry. This is an example of a system in which an extremely high degree of resolution is required to separate two isomers, particularly when enantiomers are involved. There has yet to be established a rational protocol for selecting between two enantiomers. Approaches adopted have involved the use of chiral ligands or chiral additions to the mobile phase. Further knowledge of the three-dimensional nature of interactions between molecules may enable new separation systems to be devised.

In liquid extraction techniques, use has been made of conventional solvents for small molecules and the selective partitioning has been altered by techniques such as ion-pair formation and manipulation of physical parameters. Greater knowledge of the the physical chemistry of such systems may enable further improvements to be achieved.

CONCLUSIONS

It has been shown that there is plenty of scope for input to the development and improvement of high resolution separations from disciplines in addition to the obvious one of biochemical engineering. Bioseparation has not traditionally been thought of as one of the exciting areas of biotechnology. It is clear however, that it is genuinely a multi-disciplinary activity requiring the input of many fundamental skills. The challenge that remains is to persuade those with the requisite skills to turn their talents to this task. Clever, successful bioseparations are the key to the eventual success of many bioproducts and this fact must continue to be stressed to encourage people into this area. High resolution separation will continue to play a major part in the overall process flow-sheet.

ACKNOWLEDGMENT

I would like to thank Dr M.K. Turner for stimulating conversations which have formed the basis of some of the ideas expressed in this chapter.

STABLE SYNTHETIC AFFINITY LIGANDS FOR USE IN PROTEIN FRACTIONATION

STEVEN J. BURTON, JAMES C. PEARSON AND DAVID L. STEWART
Affinity Chromatography Ltd,
307 Huntingdon Rd, Cambridge CB3 OJX, U.K.

ABSTRACT

Stable synthetic affinity ligands such as reactive dyes have many advantages over biological ligands such as antibodies or cofactors when used in industrial protein fractionation processes. The properties of conventional reactive textile dyes can be improved by synthesising dye ligands with structures better suited to selective protein binding. Affinity adsorbents based on biomimetic dye ligands hold great promise for the biotechnology industry.

INTRODUCTION

Affinity Chromatography has traditionally exploited the natural bio-recognition phenomena between proteins and ligands as a means of effecting selective protein separations. Compounds such as immobilised antibodies, lectins and nucleotide cofactors are frequently used for such purposes [1]. Whilst affinity adsorbents based on these compounds have proved to be highly successful and often invaluable research tools, their use in industrial protein fractionation has been somewhat limited.

In addition to possessing selective protein binding properties, adsorbents used in downstream processing must be chemically and biologically inert and inexpensive. Ability to withstand treatment with caustic alkalies is particularly important for industrial applications since sodium hydroxide provides an economic treatment for cleaning, sterilising and removing pyrogens from chromatography columns in-situ without leaving toxic residues. Consequently, relatively few 'natural' ligands are used in industrial processes. Of the affinity adsorbents that are used on a large scale, the majority utilise stable commodity/synthetic chemicals. Examples include textile dyes, metal chelators and amino acids [2]. Unfortunately, in particular circumstances these compounds can suffer from lack of selectivity which may limit their use.

Progress in our understanding of the binding interactions between proteins and compounds such as synthetic textile dyes enables the prediction and de-novo synthesis of novel synthetic affinity ligands with improved selectivity. A number of examples of synthetic "biomimetic" ligands have appeared in the scientific literature in recent years [3-6] and the first generation of such products are now available commercially. This present paper reports on the interaction of synthetic dyes with proteins and the design and properties of improved biomimetic dyes. In particular the stability and general utility of these materials is compared to the current needs of the biotechnology industry.

USE OF REACTIVE DYES IN AFFINITY CHROMATOGRAPHY

The use of reactive dyes as affinity ligands as opposed to natural biological molecules offers many advantages. Reactive dyes are comparatively inexpensive and may be attached to chromatographic support matrices relatively easily by direct immobilisation. The resulting adsorbents have broad specificity and may be used to purify a wide range of frequently unrelated proteins [7]. However, since both protein adsorption and desorption can be performed in a selective manner, the degree of purification afforded by reactive dye adsorbents is often much greater than the levels of purity attained using conventional non-selective techniques such as ion-exchange, gel permeation and hydrophobic interaction chromatography [8].

To understand the basis of the interaction of reactive dyes with proteins, one needs to consider the structure of the dyes in question. C.I. Reactive Blue 2 (figure 1) is probably the most widely used dye in affinity chromatography [7]. In common with most reactive dyes, C.I. Reactive Blue 2 possesses an abundance of hydrophobic aromatic rings which are connected by relatively inert secondary amine linkages. Strongly anionic sulphonic acid residues are also incorporated into the molecule to confer water solubility. A reactive chlorotriazine ring enables direct covalent attachment of the dye to nucleophiles such as primary hydroxyl or amine groups. Colour is imparted by the diaminoanthraquinone chromophore. Other commonly of dye powders. As protein binding is often very sensitive to dye structure [4], the presence of varying proportions of isomeric components or contaminating dye species in immobilised dye preparations gives real cause for concern.

Although the alkoxy-triazine bond formed when triazine dyes are reacted with hydroxyl groups is sufficiently stable for dying purposes, it is insufficiently stable for chromatographic applications. Hydrolysis of the alkoxy-triazine bond is particularly marked at extremes of pH where dye leakage from affinity matrices becomes highly visible [7]. Dye leakage may also be observed during elution of bound proteins. The presence of dye leakage products in protein preparations is highly undesirable, particularly if the protein in question is destined for therapeutic use.

In order to identify a dye which exhibits selective binding properties towards a particular protein from a particular source, some form of dye screening exercise is generally performed [11]. Success largely depends on having a suitably diverse range of dye structures with which to perform the screening experiment. Unfortunately reactive dye structures are rarely published and a suitable range of dye samples can be difficult to acquire. Consequently, it is often necessary to synthesise many tens of adsorbent samples (some of which inevitably have very similar structures) prior to evaluation of their protein binding properties [7,11]. This is usually a very time consuming process. The need to screen dye ligand affinity adsorbents would be simplified, if not eliminated entirely, by improving dye specificity. This would have the added benefit of improving the potential degree of purification.

Most of the failings of reactive dye affinity adsorbents stem from the use of commercial textile dyes. By synthesising dyes specifically for use in affinity chromatography, most of the disadvantages outlined above can be overcome. In particular, by use of rational design strategies, biomimetic dye ligands have now been synthesised which exhibit considerably improved specificities relative to conventional textile dyes [3-6,12-14].

RATIONAL DESIGN OF BIOMIMETIC DYE LIGANDS

Biomimetic dyes may be defined as dye-like molecules which have been synthesised to resemble the structure and function of biological ligands, or mimic their interaction with the binding sites of proteins. Retention of the basic polyaromatic dye

encountered chromophores include azo and phthalocyanine groups
which give rise to yellow-red and green-blue shades
respectively [7].

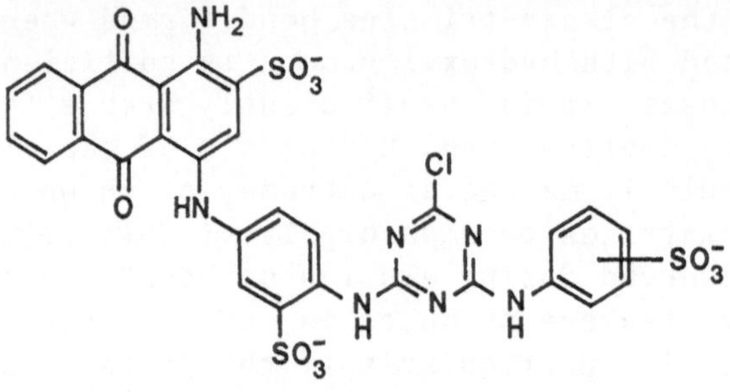

Figure 1. Structure of C.I. Reactive Blue 2.

It is the abundance of hydrophobic, hydrogen bonding and
charge groups which reactive dyes possess that promotes
interaction with the amino acid side chains of proteins.
Furthermore, the dyes are a similar shape and size to many
natural ligands, especially nucleotides and cofactors [9].
Consequently, reactive dyes are ideally suited to binding a
great many proteins and are thus considered to be general
affinity ligands. The specificity of the dye-protein
interaction is determined by the degree of compatibility
between the structure of the dye and the protein binding site,
and the conditions selected for binding and elution.
    As protein binding to reactive dyes is largely independent
of isoelectric point or molecular weight [7], purification
procedures based on dye-ligand affinity chromatography may be
developed without in-depth knowledge on the protein to be
purified. This is a considerable advantage when isolating
previously unknown or uncharacterised proteins. Furthermore,
compared to other immobilised affinity ligands, reactive dye
affinity adsorbents often have high capacities and immobilised
ligand utilisation can exceed 30%.
    Immobilised reactive dyes would provide ideal affinity
adsorbents were it not for several undesirable features. Most
commercially available reactive textile dyes are heterogeneous
in that they are frequently composed of isomeric mixtures of
dyes and are invariably contaminated with reaction
intermediates and hydrolytic by-products [10]. In addition,
diluents, preservatives and anti-dusting agents are added
deliberately to improve the shelf life and handling properties

structure is important since the chemistry required to produce such compounds is relatively simple and well understood, and the resulting molecules are very stable and resistant to chemical and biological degradation. Furthermore, since the raw materials required to produce biomimetic dyes are frequently readily available commodity chemicals, the cost of synthesis is relatively low. Three basic approaches to the synthesis of biomimetic dyes have been utilised to date: modification of existing textile dyes to improve their specificity, incorporation of the active binding groups of natural ligands onto a dye backbone, and complete de-novo synthesis of entirely new structures by utilisation of computer-aided molecular modelling.

Examples of the first category of biomimetic dyes are provided from work performed on the interaction of C.I. reactive Blue 2 analogues with the enzyme horse liver alcohol dehydrogenase [3,4,12,13]. The interaction of C.I. Reactive Blue 2 and related textile dyes with horse liver alcohol dehydrogenase has been investigated by a number of techniques including affinity labelling [15] and X-ray crystallography [9]. The dye interacts with the enzyme at the NAD$^+$ binding site where it adopts a similar conformation to bound NAD$^+$(H). The anthraquinone ring, aminobenzenesulphonate bridging ring and triazine ring of the dye mimic the binding of the adenine, adenyl-ribose and pyrophosphate moieties of NAD$^+$ respectively. However, the terminal aminobenzenesulphonate ring of the bound dye is located in a position not normally occupied by NAD$^+$. By modifying the structure of the terminal aminobenzenesulphonate ring of C.I. Reactive Blue 2 to incorporate a meta-orientated carboxyl group as opposed to a meta or para orientated sulphonate group, the affinity of the dye for the enzyme was increased by two orders of magnitude [4]. The decrease in dissociation constant from 1.6 μM for a meta orientated sulphonate group to 0.06 μM for a meta orientated carboxyl can be explained in terms of improved steric and electrostatic interactions between the terminal ring of the dye and its anomalous binding site.

Introduction of a diaminohexane spacer arm between the support matrix and the triazine ring of C.I. Reactive Blue 2 has been shown to increase the degree of purification of horse liver alcohol dehydrogenase from a crude liver extract [12]. The ortho-sulphonate isomer of C.I. Reactive Blue 2 immobilised to agarose by the conventional direct coupling regime gave a 4.3-fold purification, whereas the diaminohexane spacer dye matrix gave a 7.3-fold purification under identical conditions.

However, X-ray crystallographic studies show that the anthraquinone 1-amino group protrudes into the exterior solvent when C.I. Reactive Blue 2 is bound to the enzyme, in a manner reminiscent of the adenyl-N^6 amino group of bound NAD$^+$ [9]. In contrast, the triazine ring of the dye is bound within the interior of the protein and so provides a less favourable point of attachment. By immobilising the dye via a spacer arm attached to the anthraquinone 1-amino group, the degree of purification was increased to 10.3-fold when horse liver alcohol dehydrogenase was isolated from a crude liver extract under identical conditions to the experiments described above [12]. Thus binding specificity can be influenced by the way in which the dye is immobilised.

In instances where detailed information on the interaction of the desired protein with reactive dyes is unavailable, highly effective affinity ligands can be developed by grafting the active parts of known ligands, including enzyme substrates and inhibitors, onto a dye backbone. In one instance a phosphonic acid group, a known inhibitor of alkaline phosphatase, was incorporated into the terminal aminobenzene ring of C.I. Reactive Blue 2 [5]. When immobilised to agarose the resulting adsorbent was very effective in binding calf intestinal alkaline phosphatase. A 330-fold purification was achieved in a single step when the enzyme was purified from a crude intestinal extract. The resulting enzyme was equivalent to commercially available 'high purity' preparations. Similarly, para-aminobenzamidine, a known inhibitor of trypsin like enzymes, has been incorporated into a simple azo dye structure [6]. The immobilised dye provides an inexpensive adsorbent which afforded an 11-fold purification of trypsin from crude pancreatic extract and was able to resolve trypsin from chymotrypsin.

By the use of computer aided molecular modelling it is possible to tailor dye structures to suit particular proteins. This can be achieved by mimicking the structures of bound ligands or by synthesis of entirely new ligands using known structural features on the protein as a template. As an example of the former, the structure of a novel anthraquinone based dye was derived from the known structure of the NAD$^+$ binding site of horse liver alcohol dehydrogenase, and the conformation of NAD$^+$ when bound to the enzyme [3,13]. The new dye was of equivalent size to NAD$^+$ and had a similar conformational mobility and distribution of hydrophobic and ionic groups. When immobilised to agarose, the new dye gave a 64-fold purification of the enzyme in one step from a very crude liver extract. The resulting enzyme preparation was of acceptable purity for use in industrial applications.

More recently, a cationic ligand with a dye-like structure was synthesised for the purification of kallikrein [14]. The ligand, which contained a benzamidine group, was designed using computer graphics to interact with both the primary arginine binding site and the secondary phenylalanine binding site of the enzyme. The immobilised ligand displayed high affinity for kallikrein but did not bind trypsin, an enzyme which normally binds to benzamidine agarose and is a common contaminant of kallikrein preparations.

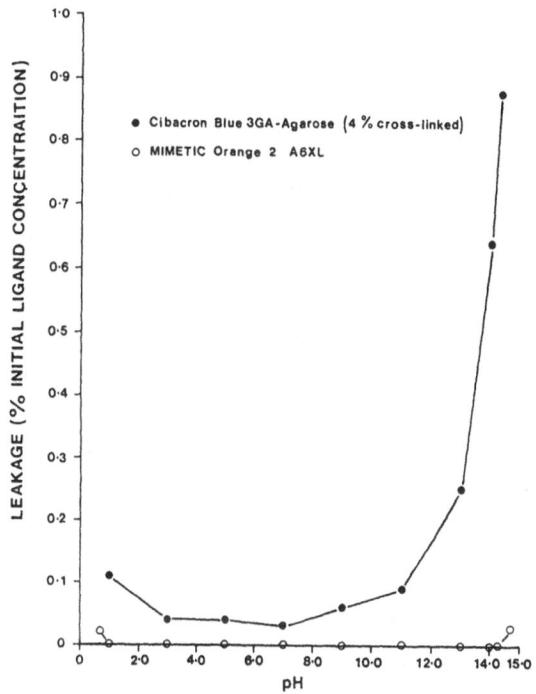

Figure 2. Analysis of ligand leakage from MIMETIC ligand A6XL and conventional textile dye-agarose adsorbents.

Affinity adsorbents incorporating biomimetic dye ligands are now available commercially. MIMETIC^{TM} ligand adsorbents encompass a diverse range of ligands which are synthesised purely for use in protein separations. This has the advantage that defined adsorbents are now readily available which considerably eases the burden of dye screening. The adsorbents incorporate a spacer arm linkage which not only improves ligand specificity, but also eliminates the unstable triazine-matrix linkage which is largely responsible for the leakage of conventional reactive textile dyes from chromatographic media (figure 2). MIMETIC ligand adsorbents provide high degrees of purification. For example, human serum albumin can be purified from whole plasma to near homogeneity in a single step by affinity chromatography on MIMETIC Blue 1 A6XL (figure 3). Similarly, <u>Pseudomonas putida</u> morphine dehydrogenase has been purified over 1,200-fold by a two-step procedure involving chromatography on MIMETIC Orange 3 A6XL and MIMETIC Red 2 A6XL to yield a homogeneous enzyme [16].

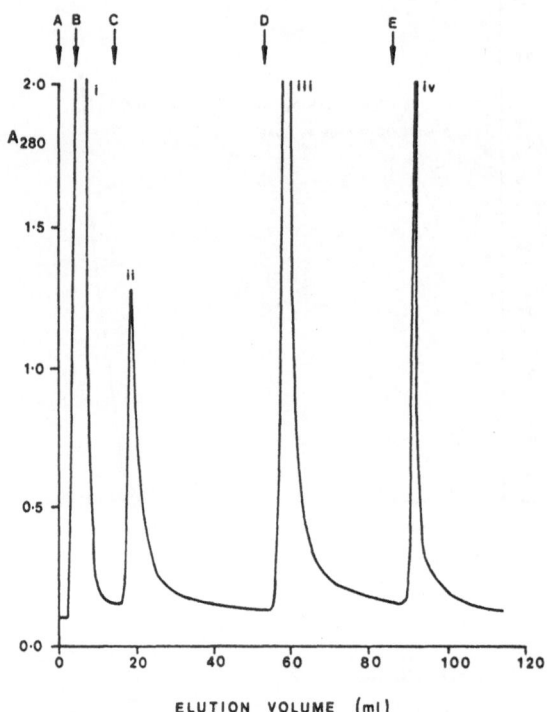

Figure 3. Purification of human serum albumin by affinity chromatography on MIMETIC Blue 1 A6XL. Column: 1 cm dia. x 5 cm; flow rate: 0.33 ml/min; applied solutions: (A) 2 ml pooled plasma buffered to pH 6.0 (65 mg protein); (B) 25 mM MOPS buffer pH 6.0; (C) 50 mM sodium phosphate buffer, pH 8.0; (D) 50 mM sodium phosphate buffer, pH 8.0 containing 2 M NaCl; (E) 1 M NaOH. Eluted peaks: (i) non-bound proteins; (ii) eluted contaminants; (iii) human serum albumin (>96% pure); (iv) residual contaminants.

CONCLUSION

Custom designed biomimetic dye ligands are extremely useful for the purification of proteins by affinity chromatography. The adsorbents provide high degrees of purification, are inexpensive, and are biologically and chemically inert. For example, MIMETIC affinity adsorbents can be cleaned, sterilised and depyrogenated by treatment with 1 M sodium hydroxide. Consequently, these adsorbents are ideally suited to use in industrial protein fractionation processes and are a logical replacement for the traditional low-specificity techniques such as ion exchange, hydrophobic or gel permeation chromatography.

REFERENCES

1. Burton, S. J., Affinity Chromatography. In Techniques in Molecular Biology, eds J. M. Walker and W. Gaastra, Croom Helm, London, 1987, pp. 55-81.

2. Clonis, Y. D., Large scale affinity chromatography. Biotechnology, 1987, 5, 1290-1293.

3. Lowe, C. R., Burton, S. J., Pearson, J. P. Clonis, Y. D. and Stead, C. V., Design and application of bio-mimetic dyes in biotechnology. J. Chromatogr., 1986, 376, 121-130.

4. Burton, S. J., Stead, C. V. and Lowe, C. R., Design and applications of biomimetic anthraquinone dyes. II: The interaction of C.I. Reactive Blue 2 analogues bearing terminal ring modifications with horse liver alcohol dehydrogenase. J. Chromatogr., 1988, 455, 201-206.

5. Lindner, N. M., Jeffcoat, R. and Lowe, C. R., Design and applications of biomimetic anthraquinone dyes. Purification of calf intestinal alkaline phosphatase with immobilised terminal ring analogues of C.I. Reactive Blue 2. J. Chromatogr., 1989, 473, 227-240.

6. Clonis, Y. D., Stead, C. V. and Lowe, C. R., Novel cationic triazine dyes for protein purification. Biotechnol. Bioeng., 1987, 30, 621-627.

7. Lowe, C. R., Applications of reactive dyes in biotechnology. In Topics in Enzyme and Fermentation Biotechnology, 5, ed. A. Wiseman, Ellis Horwood, Chichester, 1984, pp. 78-161.

8. Bonnerjea, J., Oh, S., Hoare, M. and Dunnill, P., Protein purification: the right step at the right time. Biotechnology, 1986, 4, 954-958.

9. Biellmann, J. F., Samama, J. P., Branden, C. I. and Eklund, H., X-ray studies on the binding of Cibacron Blue F3GA to liver alcohol dehydrogenase. Eur. J. Biochem., 1979, **102**, 107-110.

10. Burton, S. J., Stead, C. V. and Lowe, C. R, Design and applications of biomimetic anthraquinone dyes. I: synthesis and characterisation of terminal ring isomers of C.I. Reactive Blue 2. J. Chromatogr., 1988, **435**, 127-137.

11. Scopes, R., Stratgies for enzyme isolation using dye ligand and related adsorbents. J. Chromatogr., 1986, **376**, 131-140.

12. Burton, S. J., Stead, C. V. and Lowe, C. R., Design and applications of biomimetic anthraquinone dyes. III: Anthraquinone immobilised C.I. Reactive Blue 2 analogues and their interaction with horse liver alcohol dehydrogenase and other adenine nucleotide binding proteins. J. Chromatogr., 1990, in press.

13. Burton, S. J., Biomimetic anthraquinone dyes. PhD thesis, University of Cambridge, 1986.

14. Burton, N. PhD thesis, University of Cambridge, 1990.

15. Small, D.A.P., Lowe, C. R., Atkinson, T. and Bruton, C. J., Affinity labelling of enzymes with triazine dyes. Eur. J. Biochem., 1982, **128**, 119-123.

16. Bruce, N. C., Wilmot, C. J., Jordan, K. N. Gray-Stephens, L. and Lowe, C. R., Purification and characterisation of morphine dehydrogenase from Pseudomonas putida. Biochem. J., 1990, submitted for publication.

THE UTILITY OF POLYMERIC REVERSED PHASE PACKINGS FOR THE PURIFICATION OF PEPTIDES, PROTEINS AND ANTIBIOTICS

P.G. Cartier, K.C. Deissler, J.J. Maikner
Rohm and Haas Research Lab., 727 Norristown, Rd., Spring House PA 19477
M. Kraus
TosoHaas GmbH, Zettachring 6, D–7000 Stuttgart 80, West Germany

ABSTRACT

The utility of relatively large diameter (35 micrometers and up) porous polymeric packings for the purification of peptides, proteins and antibiotics is discussed.

Chromatographic separations of a number of model peptide, protein and antibiotic mixtures were developed to demonstrate the chromatographic selectivity of polymeric packings. High separation selectivity is a very valuable property in a packing. Frontal analysis (column capacity) with a series of probe molecules is used to determine the maximum adsorption capacity. Porous polymeric packings are shown to combine high selectivity and high capacity with excellent stability. This means that large diameter particles can be used in a selective desorption (gradient or step gradient) process to obtain a high degree of product purification.

The effect of pore size and particle diameter on column efficiency and adsorption capacity is shown. Evaluated are Amberchrom™ CG–161 (styrenic), Amberchrom™ CG–162 (wide pore styrenic) and Amberchrom™ CG–71 (methacrylic ester) polymeric reversed phase packings. Three particle diameter ranges are studied: 20–50, 50–100 and 80–160 micrometers.

An understanding of the performance and stability of polymeric packings aids the user in the selection of the optimum packing for a given production purification.

INTRODUCTION

Porous polymeric adsorbents have been used by the pharmaceutical industry for many years at the front end of downstream purification for product recovery and rudimentary product purification. These adsorbents are typically highly crosslinked, high surface area styrenic or methacrylic ester, spherical resins prepared by the method of Meitzner and Oline (1). A typical particle size range is 50–20 mesh or 297 to 840 micrometers. These adsorbents are typically exposed to "dirty" process streams and are cleaned frequently with acid, base and solvents.

Further along in the purification process, silica based particles at the other end of the particle size spectrum (5–100 micrometers) dominate the reversed phase purification of drugs, peptides and proteins. Based on the success of C18 silica reversed phase media in analytical applications, many process scientists turn to silica based media when high performance purification is needed in downstream processing. The paramount example is the use of a 10 micrometer C8 reversed phase packing for the final chromatographic purification of insulin (2).

Polymeric based media have a number of intrinsic features which should make them useful at all stages of purification. Those features give rise to the following benefits:

KEY FEATURES AND BENEFITS OF POROUS POLYMERIC MEDIA

FEATURE	BENEFIT
Chemical Stability	Cleanup in Acid or Alkaline Environment
Physical Stability	Good Hydraulic Performance
Thermal Stability	Steam Sterilization Potential
Inert Polymeric Material	Resistant to Degradation by Microbial Growth
Same Composition Throughout	No Coating to Degrade Away

The Amberchrom™ family of polymeric packings was conceived to bring the above benefits to the process chromatographer in the form of high performance, smaller particle size, spherical versions of Rohm and Haas Amberlite adsorbents. Available are small particle size analogs of Amberlite® XAD–16 and Amberlite® XAD–7 polymeric adsorbents. The analogs are called Amberchrom CG–161 and Amberchrom CG–71.

The small size of the Amberchrom packings allows the user to go beyond the adsorption/desorption mode typical of Amberlite XAD resin operation and run true chromatographic purifications. The physical properties of the Amberchrom reversed phase media are summarized in Table 1.

The "wide pore" Amberchrom CG–162 packing is designed for purification of proteins and peptides and does not have an analog in the Amberlite XAD family. The surface area of CG–162 is quite a bit lower than that of CG–161, and total capacity for small molecules has been sacrificed in favor of porosity in the 20–60 nanometer diameter range. The increase in average pore size and the resultant elimination of microporosity leads to improved mass transfer characteristics in CG–162 compared to CG–161. Amberchrom CG–162 offers superior chromatographic performance, as demonstrated by the sharper peaks and better resolution seen in the gradient separation comparisons with CG–161.

TABLE 1

PHYSICAL PROPERTIES OF AMBERCHROM RESINS

	Surfece Area M^2/Gram	Porosity Vol %	Pore Size Diameter Nanometers Average	Range	Exclusion Limit Daltons	Grams Dry Polymer/ Wet ML
Styrenic Adsorbents						
CG–161	800–950	65–69	11–17	1–30	70,000	0.22
CG–162	200–300	60–70	30–40	1–300	800,000	0.22
Acrylic Ester Adsorbent						
CG–71	450–550	58–69	20–30	1–50	188,000	0.23

Available in three size grades:
S 20–50 Micrometers, M 50–100 Micrometers, C 80–160 Micrometers

CHROMATOGRAPHIC SELECTIVITY OF AMBERCHROM PACKINGS

Amberchrom packings have unique surface chemistries which give rise to selectivities useful for chromatographic purification. Demonstrations of this selectivity are shown in Table 2 and Figure 1. Table 2 shows a retention comparison of Amberchrom CG–161, Amberchrom CG–162 and Amberchrom CG–71 media, Toyopearl® HW–40 (a hydrophilic acrylic reversed phase/SEC packing) and an ODS (C18) silica packing. Not unexpectedly, the styrenic CG–161 and CG–162 are very retentive of the aromatic test molecules. The aliphatic CG–71 is much less retentive of the aromatic test probes and similar to the C18 silica phase.

TABLE 2

SELECTIVITY COMPARISON RETENTION OF AROMATICS

Product	Retention Time Relative to Phenol Phenol	Toluene	Naphthalene	Anthracene
Amberchrom™ CG–161	1.0	2.0	3.7	13.1
Amberchrom™ CG–162	1.0	1.3	2.2	—
Amberchrom™ CG–71	1.0	1.1	1.4	1.9
Toyopearl® HW–40	1.0	1.2	1.0	1.2
TSK–GEL® ODS–120T	1.0	1.2	1.2	1.5

A systematic comparison of the retention difference between styrenic and alky-bonded silica phases shows that unique selectivity is obtained with styrenic phases for compounds containing polarizable groups (3). The study finds that the retention difference between styrenic and C18 phases is greatest with methanol and decreases if mobile phases that bind strongly to the packing (i.e. tetrahydrofuran) are used.

The long retention times (high selectivity) in the isocratic test reported in Table 2 frequently translate into excellent resolution in a gradient separation. Figure 1 shows the gradient separation of para-hydroxybenzoic acid and a series of its esters (parabens) on CG–161 and CG–162. Resolution is good between the similar ester compounds for the separation using CG–161 and excellent for the CG–162. This excellent resolution means that the benefits of low pressure operation with large particles can be realized without sacrificing purification performance.

TARGET APPLICATIONS

Gradient chromatography (selective desorption) is an ideal way to use the relatively large diameter Amberchrom packings.

Figures 2 thru 5 show model peptide, protein and antibiotic separations using Amberchrom media. The first example is the separation of a mixture of peptides called the Alberta peptide mixture, purchased from Synchrom, Inc. The composition of the mixture provided by the manufacturer is shown at the top of Figure 2. The mixture is used for retention time calibration in peptide analysis. Figure 2A shows that complete resolution of the mixture is achieved with an 8 micrometer C8 column. Figure 2B shows that a much larger diameter polymeric packing (50–100 micrometers) can resolve a decapeptide with a free N-alpha amino group vs. an N-alpha acetylated group (S1 from S3), and it can resolve on the basis of H– (Gly), CH3– (Ala) and (CH3)2CH– (Val) residues.

Figure 3 compares Amberchrom CG–161, CG–71 and CG–162 media for the separation of a mixture of peptides. This example demonstrates the peak sharpness and retention differences between the three packings.

Figure 4 demonstrates the use of Amberchrom CG–161 for penicillin purification. The penicillin examples show a model separation run with purchased compounds and the recovery and purification of penicillin V from a clarified fermentation broth.

Figure 5 shows the separation of a protein test mixture on CG–162 as a function of particle size. Greater than 95% mass recovery of proteins is obtained when mixtures of acetonitrile and water are used to regenerate CG–162. The effect of regeneration on protein activity has not been studied.

FIGURE 1

SEPARATION OF ESTERS
OF HYDROXYBENZOIC ACID

- COLUMN: 6.3 X 1.0 CM DIAMETER
- FLOW RATE: 0.5 ML / MIN (36CM / HR)
- SAMPLE: 100µl CONTAINING 100µg EACH
  PARABEN
  500µg PHENYL SALICYLATE

- SOLVENT A: 0.1% TFA IN WATER
- SOLVENT B: ACETONITRILE
- GRADIENT: 4 MINUTES ISOCRATIC, 30% B
  64 MINUTES TO 100% B

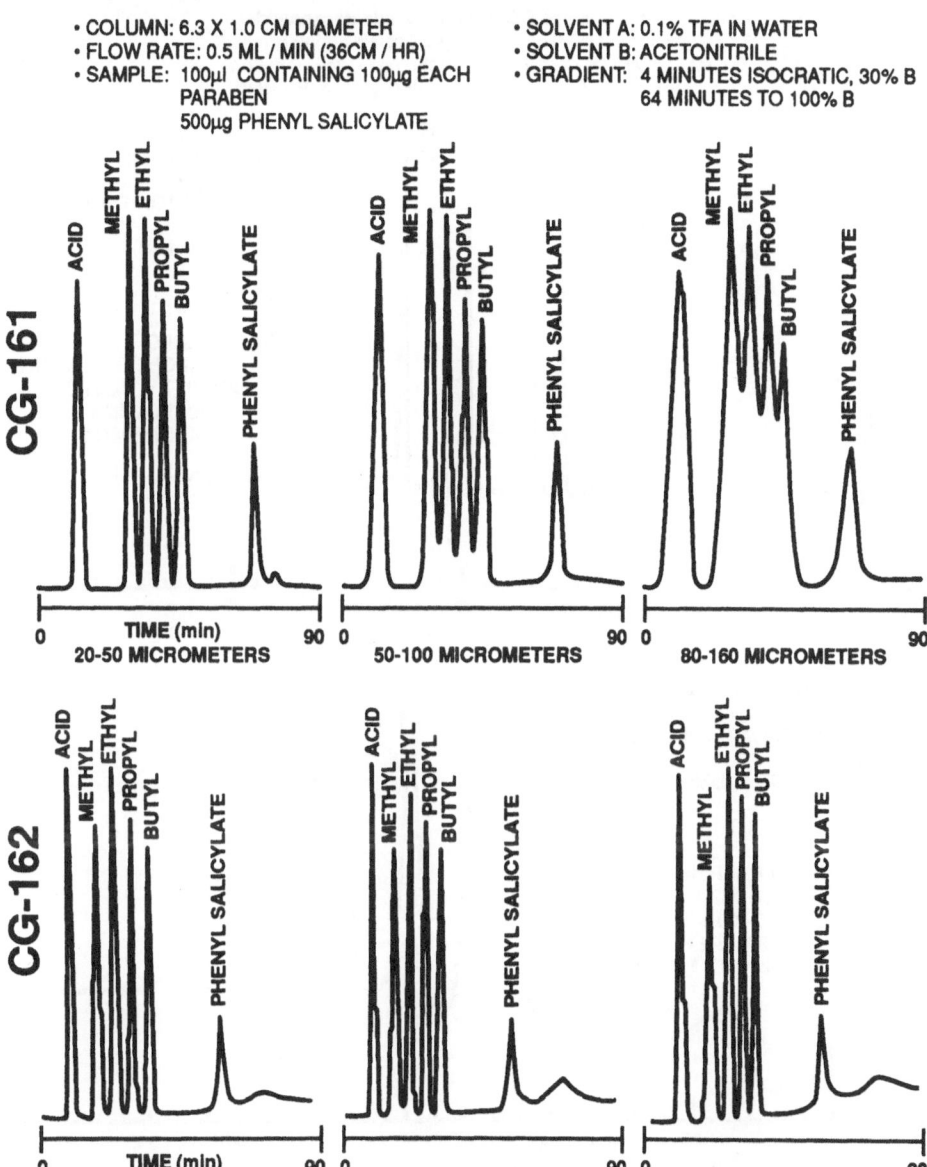

FIGURE 2

COMPARISON WITH SILICA BASED RP

ALBERTA PEPTIDE MIXTURE — Contains 1 mg each of 5 C-terminal amide decapeptides. The S1 peptide, Ala³-Gly⁴, contains a free Nᵅ-amino group. The S2 - S5 are Nᵅ-acetylated with the sequence variation as follows: Gly³-Gly⁴, Ala³-Gly⁴, Val³-Gly⁴ and Val³-Val⁴.

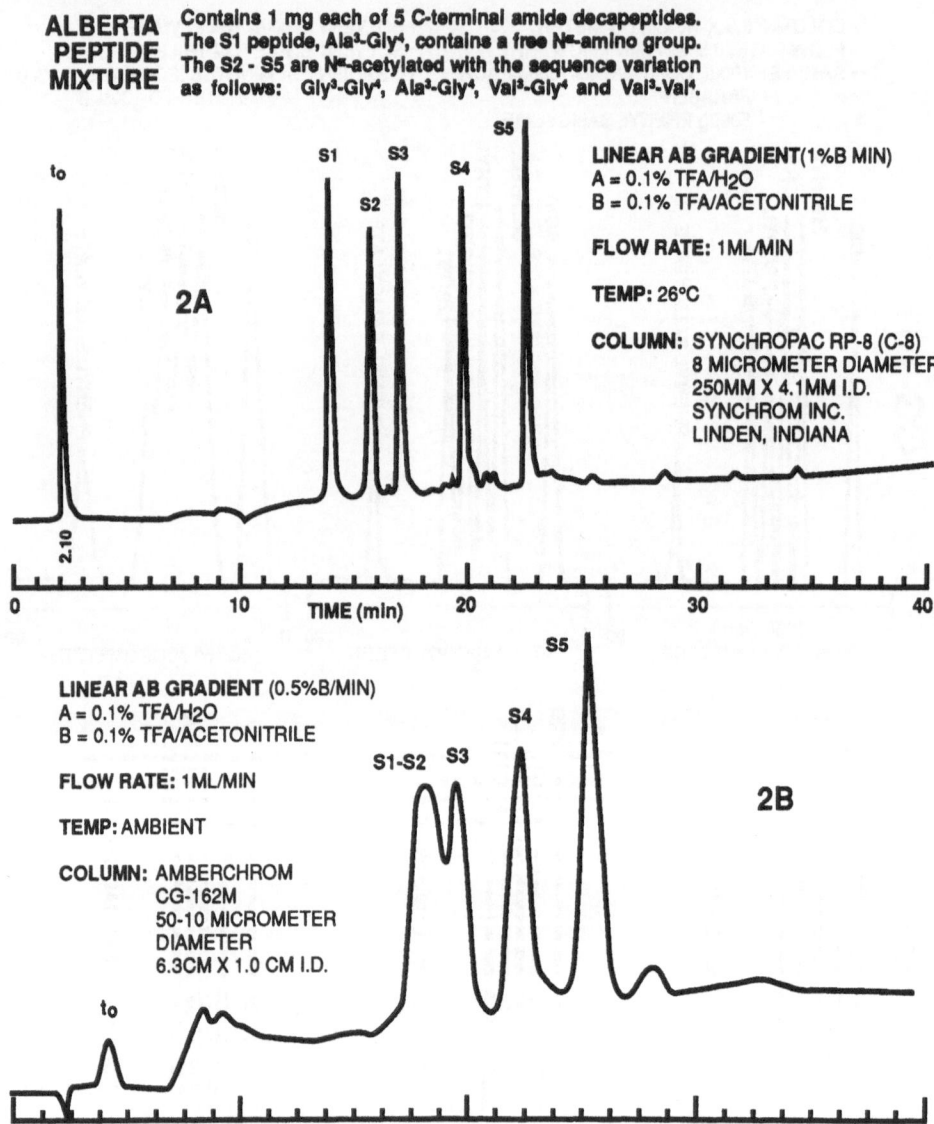

2A

LINEAR AB GRADIENT(1%B MIN)
A = 0.1% TFA/H2O
B = 0.1% TFA/ACETONITRILE

FLOW RATE: 1ML/MIN

TEMP: 26°C

COLUMN: SYNCHROPAC RP-8 (C-8)
8 MICROMETER DIAMETER
250MM X 4.1MM I.D.
SYNCHROM INC.
LINDEN, INDIANA

LINEAR AB GRADIENT (0.5%B/MIN)
A = 0.1% TFA/H2O
B = 0.1% TFA/ACETONITRILE

FLOW RATE: 1ML/MIN

TEMP: AMBIENT

COLUMN: AMBERCHROM
CG-162M
50-10 MICROMETER
DIAMETER
6.3CM X 1.0 CM I.D.

2B

FIGURE 3

GRADIENT SEPARATION OF A PEPTIDE MIXTURE

- COLUMN: 6.3 X 1.0 CM DIAMETER
- SOLVENT A: 0.1% TFA IN WATER
  B: 0.1% TFA IN ACN
- GRADIENT: 15% B TO 50% B
  IN 90 MIN
- FLOW RATE: 2 ML / MIN
- DETECTION: UV 220 NM

RETENTION TIMES (MIN) FOR	CG-71	CG-161	CG-162
1. NEUROMEDIN C	13.7	22.7	14.6
2. BOMBESIN	22.3	29.6	22.0
3. SOMATOSTATIN	35.1	41.8	33.9

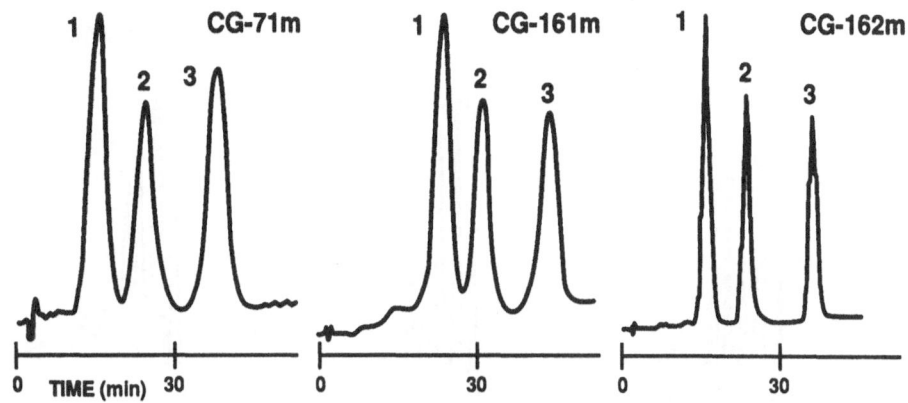

FIGURE 4

PENICILLIN BROTH AND DERIVATIVES ON AMBERCHROM™ CG-161m

- COLUMN: 6.3 X 1.0 CM DIAMETER

- SOLVENT A: 50mM PHOSPHATE, pH 8.0
  B: METHANOL

- FLOW RATE: 1.0 ML / MIN

1. 6-AMINOPENICILLANIC ACID

2. PENICILLIN G

3. PENICILLIN V

GRADIENT:
  10 MIN ISOCRATIC IN A;
  0-100% B IN 25 MINUTES

GRADIENT:
  4 MIN ISOCRATIC 60% B
  25 MIN TO 100% B

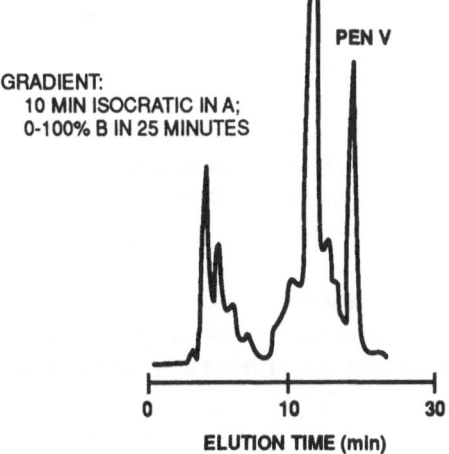

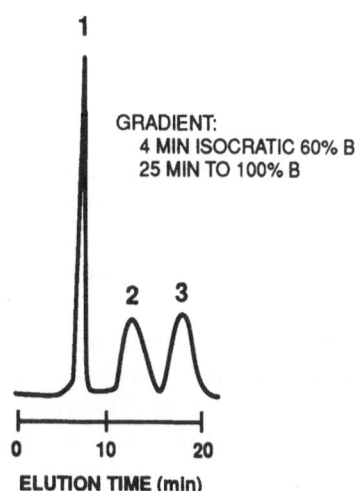

FIGURE 5

SEPARATION OF PROTEINS USING AMBERCHROM™ CG-162 WIDE PORE PACKING

PROTEIN	M.W.
1. RIBONUCLEASE	12,500
2. INSULIN	6,000
3. CYTOCHROME - C	12,400
4. LYSOZYME	14,000
5. BOVINE SERUM ALBUMIN	67,000
6. MYOGLOBIN	17,200
7. CHICKEN EGG ALBUMIN	~60,000

• COLUMN: 6.3 X 1.0 CM DIAMETER
• SOLVENT A: 0.1% TFA IN WATER
        B: 0.1% TFA IN ACN
• GRADIENT: 20% B TO 60% B IN 70 MINUTES
• FLOW RATE: 1 ML / MIN
• DETECTION: UV 280 NM
• INJECTION: 20μl, 1MG / ML EACH PROTEIN

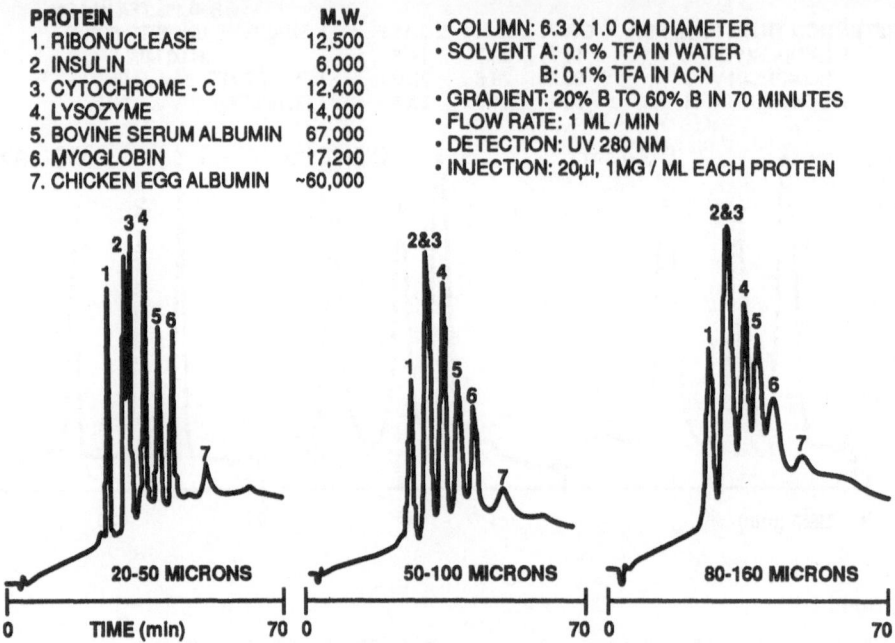

STABILITY OF POLYMERIC PACKINGS

Two important aspects of packing performance are stability with respect to chemical cleaning, and column stability with respect to repeated cycling between solvent and water. The total acid and base stability of polymeric media often means that little pretreatment or polishing of the influent stream is required. Media clean-up can be accomplished by either concentrated acid or base. Experience with the Amberlite XAD adsorbents indicates that Amberchrom media will be totally stable in nonoxidizing acids or bases.

Figure 6 shows the NaOH clean-up of a CG–162 fouled with bovine serum. Panel 2 shows the performance of the fouled column. A solvent wash (panel 3) recovered some performance and an 18 hour soak in 0.5 M NaOH followed by a solvent wash restored the column to its original condition.

The methyacrylic ester packing CG–71 was subjected to a 60 hour soak in 0.5 M NaOH at 60 C and found to be completely stable. Packing appearance was unchanged and no hydrolysis was detected. The hydrolysis test was a titration to detect the presence of COOH groups; the test has a sensitivity of 0.1 meg/g of dry polymer.

FIGURE 6

NaOH CLEANING OF FOULED AMBERCHROM™ CG-162

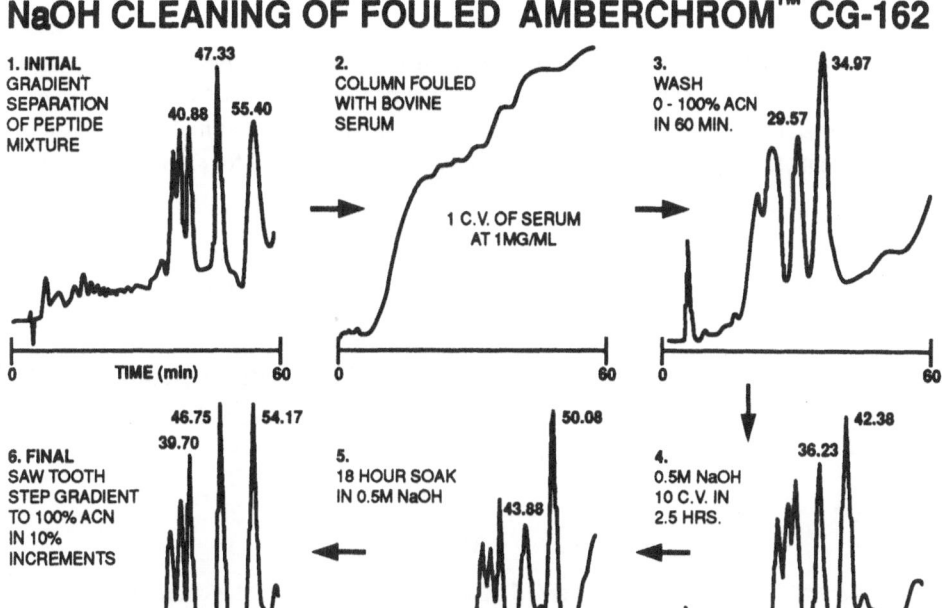

Amberchrom CG–162 does experience a small change (<10%) in volume when cycled between common solvents and water. This volume change does not degrade its column performance. A 1.0 cm diameter x 6.3 cm long column of CG–162 was cycled between water and acetonitrile for 40 cycles. Column performance was monitored by following plate count. Plate count actually increased from 1300 to 2200 plates per meter during the course of the cycling.

ADSORPTION CAPACITY/FRONTAL ANALYSIS

The chromatograms shown above demonstrate the separation selectivity of Amberchrom media in a variety of applications. Most of the examples were run at low (analytical) loadings. It is difficult to predict how a separation will change as sample load is increased. For the separation of methyl paraben from butyl paraben (see Figure 1), it was found that base line separation could be maintained at a total loading greater than 13 mg/ml or 25% of saturation capacity (4).

The saturation adsorption capacity of the molecule to be purified is the upper limit to the chromatographic loadability of a column. Batch capacity or column capacity/frontal analysis can be used to determine the maximum capacity. Table 3 lists the saturation column capacities of Amberchrom CG–161, CG–162 and CG–71 media for a series of probe molecules. The probes were chosen to cover a range of molecular sizes.

The saturation capacity was determined by running the column until the effluent concentration equaled the influent concentration. Then the column was regenerated with the indicated solvent. The saturation capacity was determined from the amount of test molecule present in the regenerant stream. Greater than 95% mass recovery was obtained for each test molecule.

TABLE 3

SATURATION COLUMN CAPACITY OF AMBERCHROM RESINS

Resin	Total Capacity, MG/ML		
	Phenol	Insulin	Bovine Serum Albumin
CG–71M	55	110	40
CG–161M	62	63	27
CG–162S	17	39	33
CG–162M	18	33	27
CG–162C	18	33	33

Experimental Conditions	Influent, 5 MG/ML	Regenerant
Phenol	Water	Methanol
Insulin	Water, 0.1% TFA	Methanol, 0.1% TFA
Bovine Serum	0.05 M TrisHCL, pH=8	50:50 Water: ACN, 0.1% TFA
TFA = Trifluoroacetic Acid	Column: 1.0 X 6.3 CM	Flow: 150 CM/HR

REFERENCES

1. E. F. Meitzner and J. A. Oline, U.S. Patent 4,297,220 Oct. 27, 1981.

2. E. P. Kroeff, R. C. Owens, E. L. Campbell, R. D. Johnson and H. I. Marks, Journal of Chromatography, 462 (1989) 45-61.

3. S. Pedigo and L. D. Bowers, Journal of Chromatography, 499 (1990) 279-290.

4. P. G. Cartier, K. C. Deissler, J. J. Maikner and W. G. Schwartz, Poster 106, 9th International Symposium of Proteins, Peptides and Polynucleotides, Philadelphia, PA, November 6, 1989.

ACKNOWLEDGEMENT

We thank the TOSOH Corporation for running the retention screen, Table 1, and supplying size exclusion chromatography characterization of the Amberchrom packings.

BATCH STIRRED TANK ADSORPTION OF IMPURITY OF CEPHALOSPORIN C, DESACETYLCEPHALOSPORIN C, USING MODIFIED AMBERLITE XAD-2 RESIN

J L Casillas[1], J L Garrido[1], J Aracil[1], M Martinez[2],
F Addo-Yobo[3], C N Kenney[3],
Departamento de Ingenieria Quimica, Universidad Complutense (Spain)[1]
Departmento de Inginieria Quimica (ETSIM), Universidad Politecnica (Spain)[2]
Department of Chemical Engineering, University of Cambridge (UK)[3]

ABSTRACT

The adsorption of desacetylcephalosporin C from an aqueous solution onto particles of a modified macroporous Amberlite XAD-2 resin, XAD-2-CH2-CH2-Br, using a batch stirred tank is described. The pH studies used 2.8, at 293.15 K and an ionic strength 0.1 M. Equilibrium adsorption experiments showed that a linear adsorption isotherm is appropriate to describe the adsorption on the modified resin. A theoretical model has been developed to simulate adsorption in a batch stirred tank and includes the effects of mass transfer from the bulk liquid to the surface of the particle and diffusion within the pores of the resin, with a linear isotherm. Basic parameters such as the isotherm relationships, effective diffusivities and tortuosity factor have been determined.

NOTATION

C   Concentration of desacetylcephalosporin C in the bulk liquid (g/1)

C_i adsorbate concentration at time t (g/l)

C_{in} true inlet concentration g

C_o initial concentration of desacetylcephalosporin C (g/1)

d_p particle diameter (cm)

D   molecular diffusivity (cm^2/s)

D_e effective diffusivity in porous particle (cm^2/s)

K_f mass transfer coefficient between bulk liquid and other surface of particle, cm/s

q   concentration of adsorbed desacetylcephalosporin C g/g on adsorbent

r   radial coordinate in particle (assumed spherical, cm)

t   time (s)

V   liquid phase volume

W   mass of resin (g)

Greek letters

ϵ_p porosity of particles

τ tortuosity factor, defined by Equation (5)

INTRODUCTION

Advances in biochemical engineering have resulted in the development of processes capable of producing biocompounds which are of interest as pharmaceuticals or medical diagnostic reagents and which would be too complex or too costly to produce by chemical synthesis.

The efficient isolation and purification of these products when prepared by fermentation is essential for commercial success. Downstream recovery often represents a large portion of the product price and this expense may be a major production cost. The competitive edge of a given manufacturer may depend on the optimisation of the downstream processing.

Commercial antibiotics are produced by microbial fermentation: the process involves three stages, fermentation, purification and chemical modification and the process can either be batch of continuous. The most important of the cephalosporins, produced in fermentation process, is cephalosporin C. Byproducts are generally compounds of similar structure and size and have similar physical properties to the desired antibiotics such as desacetylcephalosporin C, desacetoxycephalosporin C and penicillin N.

The purification of the antibiotic cephalosporin C from its fermentation broths involves a number of complex unit operations. In fermentation broths, the product of interest is often produced at very low concentration, for example, in the case of cephalosporin C, at about 15 g/1. High degrees of product purity are required for safe medical applications.

The techniques used for cephalosporin C purification are: solvent extraction, ion-exchange and adsorption. The most commonly used industrially is adsorption. The adsorption technique utilises Van der Waals and hydrophobic forces, which bind the sorbate to the surface. The adsorbents commonly used industrially are non-polar macroporous resins, such as XAD-2 and an account of the batch and fixed bed adsorption of cephalosporin on this resin has been described (1).

In this research, as part of a collaborative programme to develop more effective separation processes, the use of a modified Amberlite XAD-2 resin, XAD-2-CH2-CH2-Br, for the adsorption of the most important impurity Cephalosporin C, desacetylcephalosporin C is described. Equilibrium adsorption experiments showed that a linear adsorption isotherm is appropriate in the range of concentrations employed to describe adsorption on the modified resin.

A theoretical model has been developed to simulate adsorption in a batch stirred tank and includes the effect of mass transfer from the bulk liquid to the surface of the particle, diffusion within the pores of the resin, with a linear isotherm adsorption.

MATHEMATICAL MODELLING IN A STIRRED TANK

The model describing the kinetics of adsorption of desacetylcephalosporin C on the resin adsorbent in a stirred tank assumes the following mass transfer steps:

(i)   film mass transfer
(ii)   diffusion within the adsorbent particles
(iii)   local equilibrium at the pore surface

As a basis for the construction of the model it is also assumed that:

(i)   the adsorption process is isothermal
(ii)   the adsorbent is made of a porous material into which the solute must diffuse in a manner described by an effective diffusivity, De
(iii)   the effective diffusivity is independent of concentration
(iv)   the adsorbent particles are spherical, of uniform size and density
(v)   the adsorption process is represented by an isotherm relationship $q = f(c)$
(vi)   mass transfer to the surface of the adsorbent is governed by a film model characterised by a mass transfer coefficient K_f

For diffusion within the adsorbent particle, the point concentration is given by:

$$\epsilon_p \frac{\partial C_i}{\partial t} + (1 - \epsilon_p)\frac{dq}{dt} = \epsilon_p D\left(\frac{\partial^2 C_i}{\partial r^2} + \frac{r}{2}\frac{\partial C_i}{\partial r}\right) \tag{1}$$

The rate of mass transfer through the external film relates the bulk liquid concentration to the concentration in the pore liquid at the surface of the particle:

$$-D\epsilon_p \left(\frac{\partial C_i}{\partial r}\Big/_{r=R_p}\right) = K_f \epsilon_p (C - C_i/_{r=R}) \tag{2}$$

At the centre of the particle

$$\frac{\partial C_i}{\partial r} = 0 \text{ for } r = 0 \tag{3}$$

with the initial conditions

$C_i = q = 0$	at $t = 0$	(4)
$C_{in} = C_o$	at $t \geq 0$	(5)

The above equations were solved numerically by discretising the spatial derivatives using a second order difference approximation to give predicted concentration versus time curves in terms of De. Then comparison with the experimental curves will give the most appropriate value of De.

EXPERIMENTAL APPARATUS & PROCEDURES

APPARATUS

A diagram of the apparatus used to carry out the batch stirred tank experiments is shown in Figure 1.

The stirred tank consisted of a Pyrex vessel of volume 500 cm^3 with a hemispherical base of 5 cm diameter and 7 cm deep. The stirred tank was equipped with stationary baffles attached along the circumference. The method of agitation chosen involved using a marine-type mixing propeller. This method proved very effective in keeping the fluid well mixed and the particles in suspension. The impeller speed was set at 1000 rpm. Ancillary instruments were a temperature recorder and controller, a pH recorder and controller and a stirrer speed controller.

The stirred tank was immersed in a constant-temperature bath capable of maintaining the temperature to within 0.1°C of that desired for the experiment.

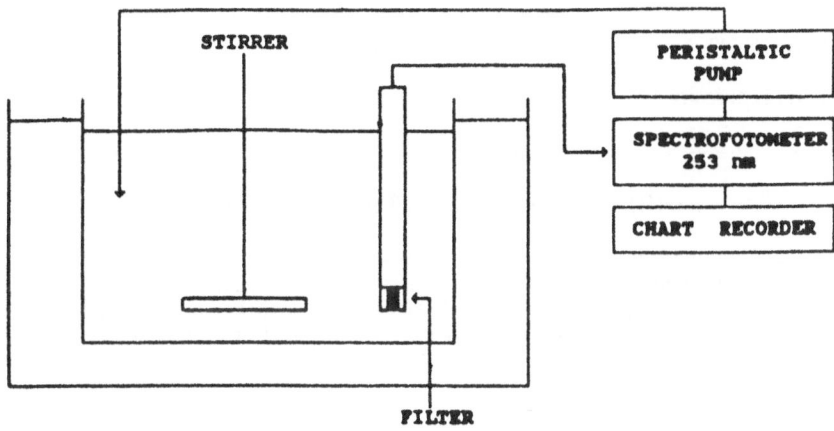

Figure 1:   Apparatus for batch stirred tank experiments

The sampling device consisted of a filter through which liquid from the vessel was continuously extracted then passed through a spectrometer to detect the absorbence at 253 nm and returned to the vessel. The average residence time of the fluid in the recycle loop was four seconds. The mass of resin added to the tank was determined by weighing, after the system had attained equilibrium.

Adsorption runs typically lasted at least four hours and during an experiment, temperature, impeller speed and pH remained constant.

REAGENTS & BUFFER

The method of synthesis of the modified Amberlite XAD-2-CH2-CH2-Br has been fully described previously (2). The monomer trapped in the resin pores was removed by Soxhlet extraction using methanol as solvent. The properties of the XAD-2-CH2-CH2-Br resin are given in Table 1.

The phosphate-buffered aqueous solutions of controlled pH and ionic strength were prepared by dissolving known quantities of pure Na_2HPO_4, KH_2PO_4 and NaCl in distilled deionised water in proportions specified by Perrin (3).

The desacetylcephaolsporin C was supplied by Glaxo Group Research Ltd.

TABLE 1
Properties of the XAD-2-CH2-CH2-Br resin

Solid phase density	1.10	g/cm^3
Wet phase density	1.06	g/cm^3
Surface area	3.40	m^2/g
Particle porosity	0.44	

EQUILIBRIUM ADSORPTION EXPERIMENTS

A set of equilibrium adsorption experiments was conducted to determine the isotherm type and parameters. The adsorption isotherm was measured by introducing known samples of a solution of desacetylcephalosporin C in a 0.01 M buffer of pH 2.8 into a flask containing the slurry typically 1 g of resin (dry weight) and operating until equilibrium was achieved. The initial and final desacetylcephalosporin C concentrations in solution were recorded and the amount of retained desacetylcephalosporin C was determined by difference.

The results were fitted to a linear isotherm, in Figure 2.

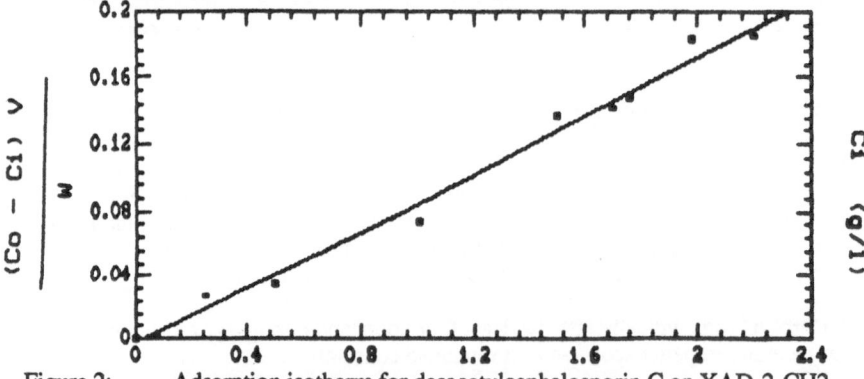

Figure 2:   Adsorption isotherm for desacetylcephalosporin C on XAD-2-CH2-CH2-Br resin.

The linear expression for the isotherm is q = 0.087 C.

A few tests were made in which the liquid concentration was measured versus time when pure water was added to particles containing desacetylcephalosporin C. The results showed that the uptake process was reversible.

RESULTS & DISCUSSION

Figure 3 (solid line) shows typical experimental plots of dimensionless concentration, C/Co, versus time. The dotted line in Figure 3 was predicted using the mathematical model described above. Figure 3 illustrates the agreement of the experimental and predicted dimensionless concentration, C/Co, versus time using the effective diffusivity, De, values from Table 2. Agreement for other runs was similar.

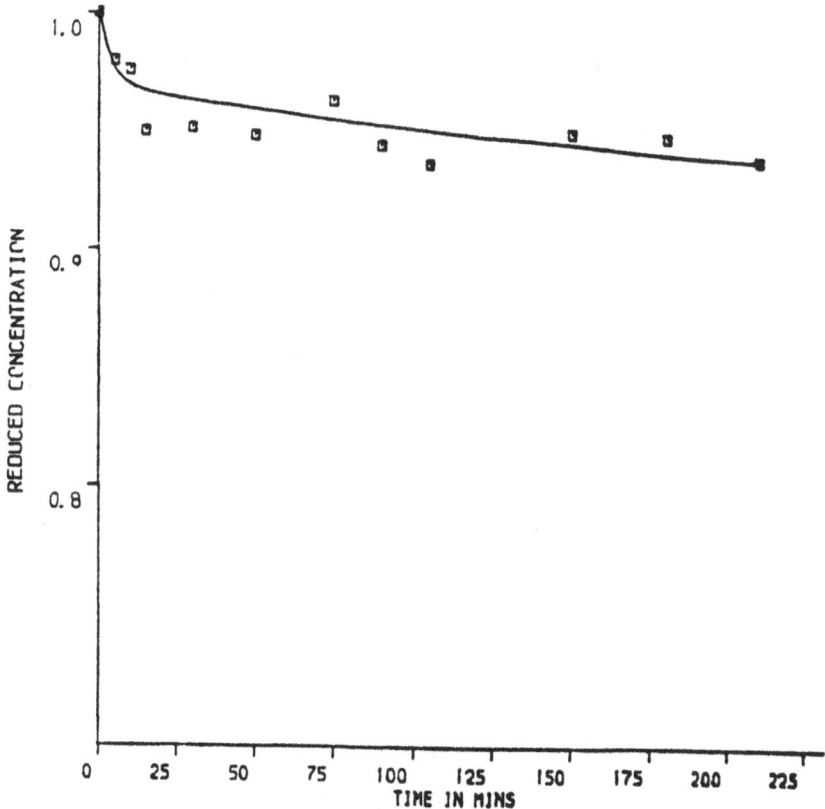

Figure 3:   Typical variation of concentration of desacetylcephalosporin C with time

An effective diffusivity, D_e, was obtained by comparing numerical solutions of the equations 2 - 6 with the experimental dimensionless concentration versus time curves. The derived diffusivities are given in Table 2 for all the data. There appears to be no regular variation in effective diffusivity with a particle size and initial concentration. The calculations showed that the initial slope of the breakthrough curve near the origin increased as the Sherwood number was increased from low values, it became relatively insensitive at high values, as shown in Figure 4.

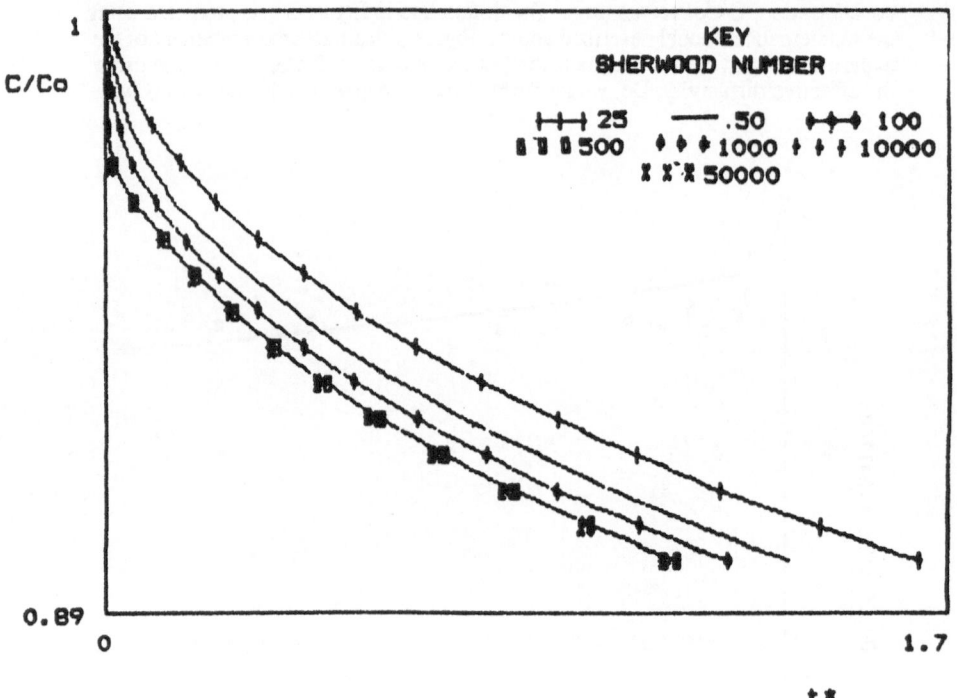

Figure 4

In this work large values of Sherwood number were required to fit the data suggesting that agitation was effective. This enabled the determination of D_e values which were independent of the mass transfer coefficients.

The molecular diffusivity of desacetylcephalosporin C in water at 20°C was calculated to be 4.1 10^{-5} cm^2/s using the Othmer-Thakar equation recommended by Reid and Sherwood (4). Tortuosity factors calculated from the expression

$$\tau = \varepsilon_p \frac{D}{D_e}$$

(6)

are given in the last column of Table 2.

TABLE 2
Experimental conditions and Results

Sums of residuals 10^2	Experimental number	Particle radius (cm) 10^2	Concentration (g/1)	Effective diffusivity (cm^2/s) 10^6	Tortuosity factor
2.63	1	1.425	1.0	1.30	1.40
3.75	2	2.025	1.0	1.28	1.41
1.90	3	2.500	1.0	1.31	1.38
1.01	4	2.025	1.5	1.33	1.36
1.25	5	2.025	2.0	1.34	1.35

The optimisation problem was to find isotherm and effective diffusivity coefficient parameters which minimized the sum of squares of errors between the theoretical and experimental decay profiles.

SUMMARY & CONCLUSIONS

Over the range of adsorbate concentration employed in this investigation, the isotherm of desacetylcephalosporin C on XAD-2-CH2-CH2-Br resin is linear.

The kinetics of uptake on a modified XAD-2 resin of desacetylcephalosporin C, the major impurity in the manufacture of cephalosporin C, can be modelled using a single internal diffusion resistance and an external mass transfer resistance. By using a CSTR it was possible to achieve high mass transfer coefficient and hence high Sherwood numbers and leading to a method sensitive to internal diffusion resistances.

The value of the effective diffusivity found here is less than the molecular diffusivity of both desacetylcephalosporin C and also cephalosporin C which will be discussed in the near future (5).

ACKNOWLEDGEMENTS

The authors wish to thank British-Spanish Integrated Action Project QT-288 (1989-90) for supporting this collaborative investigation.

REFERENCES

1)   K Weisenberger. Downstream Processing of Cephalosporin C. PhD dissertation, University of Cambridge, 1987.

2)   M Martinez, J Aracil, F Addo-Yobo, C N Kenney. Dechema Biotechnology Conference, 83-90 (1988).

3)   Perrin, D D. Anit J Chem. 16. 572 (1963).

4)   Reid, R G, Sherwood, T K. The Properties of Gases and Liquids. 2nd edition, McGraw-Hill, New York (1966).

5)   M Martinez, J Aracil, F Adoo-Yobo, C N Kenney. Proceedings of the 5th Mediterranean Congress on Chemical Engineering, Barcelona (1990).

MODIFICATION OF MICROFILTRATION MEMBRANES AS DYE-LIGAND ADSORBENTS FOR THE ISOLATION OF ENZYMES FROM CRUDE EXTRACTS

B. Champluvier and M.-R. Kula
Institut für Enzymtechnologie,
Heinrich-Heine-Universität Düsseldorf,
Postfach 2050, D-5170 Jülich, FRG

ABSTRACT

Triazine dyes, such as Cibacron blue F3G-A or the Procion series (ICI), could be coupled to various membranes based on nylon (Ultipor, Immunodyne, Loprodyne; Pall). The pore size of the material showed only a weak influence on the amount of dye coupled or on the capacity for pig heart MDH taken as a test enzyme. The use of polyethylenimine as a spacer gave a tenfold higher ligand concentration and increased drastically the MDH capacity. Adsorption values in the range 200-400 μg/cm² were obtained with the best membranes. A proprietary dye ligand membrane (Sartorius) gave similar values. On a wet weight basis the membranes had the same capacity as Sepharose CL-4B dyed with Cibacron blue.

The dyed membranes have been tested as stationary phase in a FPLC system. Proteins from a Baker's yeast crude extract could be repeatedly loaded and fully recovered in a KCl gradient. The elution profile was strongly affected by the particular specimen of modified membrane. Dye concentration and spacer type appeared to control the desorption process.

INTRODUCTION

Integration of unit operations for an enzyme purification is highly desirable with regard to speed of operation, overall recovery, and economic reasons. Integrated recovery should use at best a highly selective separation principle and should also allow to process feed streams containing solids. The usefullness of affinity interactions taking place in solution has been demonstrated for the first step of downstream processing [1].

Triazine dyes have been used successfully as group specific ligands in affinity partitioning for the recovery of formate dehydrogenase from *Candida boidinii* [2], and numerous other enzymes from crude extracts [3]. On the other hand, dye ligand chromatography has been proven as a powerful technique for enzyme purifications at laboratory scale [4].

Microfiltration membranes carrying dye ligands can provide a means to integrate purification operations. They offer high mechanical strength, large interfacial area and large degree of freedom to choose conditions and configurations for processing. Also the pore dimensions and the convective mass transfer towards the ligands should allow higher flow rates than conventional liquid chromatography media. The resulting process would thus be either an affinity filtration which mimics affinity chromatography or affinity cross-flow filtration. Results are presented here on the design and evaluation of dye ligand microfiltration membranes for such purposes.

MATERIALS AND METHODS

Reagents and buffer
Trichloro-S-triazine (Cyanuric chloride, C-5393) and polyethylenimine (PEI, P-3143) were purchased from Sigma, glutaraldehyde (49629) and 1,6-diaminohexane (33000) from Fluka. Cibacron blue F3G-A was a gift from Ciba Geigy, the Procion dyes from ICI Germany. All other reagents were of analytical grade. Unless otherwise stated, the buffer was potassium phosphate, 25 mM, pH 7.5, 1 mM EDTA.

Membrane material
Nylon-based membranes (Ultipor, Loprodyne, Immunodyne) were generously supplied by Pall (Dreieich, FRG). According to technical information, these are anisotropic supported membranes, respectively made of (i) pure nylon 66 with a high surface concentration of amine and carboxylate groups in 1:1 ratio, (ii) coated with a hydrophilic polymer or (iii) pre-activated for protein coupling. Two samples from a proprietary membrane (345 and JN138), carrying the blue dye, were gifts from Drs. Nußbaumer and Weiss (Sartorius, Göttingen, FRG).

Ligand coupling
The dye-ligand was coupled to the nylon-based membranes either directly or using 1,6-diaminohexane (C6) or polyethylenimine (PEI) as spacer. In the latter cases, the membrane was first activated, when required, with triazine according to a published procedure [5] or by incubating with 5 % glutaraldehyde for 2 hours at room temperature in 25 mM potassium phosphate buffer at pH 8. C6 or PEI were coupled to the activated membranes for two hours at pH 8 at room temperature. Large excess of reagents were used throughout and the membranes were repeatedly rinsed after every reaction step. The dyeing was made in two steps; first one hour dye adsorption at room temperature from a 6% NaCl solution, then 1 hour reaction at 50°C after raising the pH to 10.8 with a carbonate buffer. The last rinsing involved several steps with water and one with methanol. The membranes were stored dry at

room temperature. Blue dextran coupling was performed as described above using triazine-activated membranes.

The amount of dye coupled to the membrane was determined colorimetrically after dissolution of known amounts of dry membrane in phenol, HCl 6N or formic acid. Concentrations of crude dyes were calculated from apparent absorptivity determined in similar conditions. For Cibacron blue, molar concentrations could also be calculated using the published extinction coefficient in water of 13600 $M^{-1}cm^{-1}$ [3].

Enzyme samples

Pig heart mitochondrial malate dehydrogenase was from Boehringer (nr. 855 693). Before the experiments, samples were desalted on Sephadex columns (PD10, Pharmacia) and recovered in the standard buffer.

A crude enzyme extract was prepared from commercial Baker's yeast (Saccharomyces cerevisiae). A 40% w/v suspension in 100 mM potassium phosphate buffer, pH 6.0, 1 mM EDTA, 5 mM mercapto-ethanol was disintegrated by two passes in a Manton Gaulin homogenizer (MC4-10TBSX, APV Schröder GmbH), centrifuged and filtered through a 0.2 μm cellulose acetate membrane (Sartorius). The extract was stored at -20°C. Before use, samples were again clarified by centrifugation and diluted four times with 1 mM EDTA.

Batch adsorption experiments

The experiments were performed in micro test tubes, Eppendorf type, using a 0.5 cm^2 piece of membrane and 1 ml of liquid. The tubes were shaken at 1000 rpm, which produced a sharp vortex. Blanks without membranes were incubated in parallel. Adsorption values were obtained through a mass balance. About 1 mg MDH were introduced per cm^2 membrane for adsorption studies, unless otherwise stated.

For the desorption, the membrane was drained and put in 1 ml of buffer containing 0.5 M KCl.

Chromatography with membranes

Two 25 mm disks of dyed-membrane were inserted in a housing, connected to a FPLC system (Pharmacia) and equilibrated with buffer. The housing was constructed with a minimal internal volume, using two plates of sintered stainless steel as flow equalizers. The accessible membrane area was about 3.3 cm^2. Enzyme samples of 500 μl were injected. The membrane was washed with 10 ml buffer. Elution was carried out with a 10 ml linear gradient of 0 to 0.25 M KCl; a 5 ml step with 2 M KCl in buffer and a 5 ml step with start buffer. A short re-equilibration with start buffer was allowed before the next run. The flow rate was 1 ml/min throughout. Fractions of 2.5 ml were collected.

Assays

Proteins were assayed with the Coomassie or with the micro-BCA test using kits purchased from Pierce. Malate dehydrogenase (MDH) activity was measured according to Bergmeyer [6]. The samples were diluted in potassium phosphate buffer 100 mM, pH 7.5, 0.1% BSA.

RESULTS AND DISCUSSION

Amount of dye coupled

The effect of membrane nominal pore size and the coupling method were systematically investigated using the Ultipor membranes. The results are summarized in Table 1. An effect of pore size could not be detected, only PEI loaded-membranes showed the expected trend to a limited extent. The coupling method produced however decisive differences: ten times as much dye was bound using PEI as spacer compared to the other procedures. A similar trend was observed with Loprodyne membranes having pore ratings in the range 0.2 to 5.0 μm (data not shown). Whether the amine groups available for reaction at the nylon surface were too few or the surface was sterically saturated with dye molecules is difficult to distinguish. Indeed, the higher dye substitution with PEI-treated membranes could have resulted either from the introduction of a higher number of amine groups or from their better accessibility. The dye concentration per unit volume of membrane could be calculated using the manufacturer value of porosity and assuming a density of one for the wet material. Values in the range 0.5 $\mu mol.ml^{-1}$ (no spacer or C6) to 7 $\mu mol.ml^{-1}$ (PEI) were found. Chambers [7] found 0.87 and 0.64 $\mu mol.ml^{-1}$ in packed beds of Blue Sephadex G-150 and Blue dextran-Sepharose while Lowe and Pearson gave 1 to 10 $\mu mol.ml^{-1}$ as typical figures for agarose gels [8]. Obviously high concentrations could be attained as well on the membranes.

Various dyes of the Procion series were tested and gave similar concentrations. Blue dye concentration was the lowest for coupling Blue dextran on Ultipor (6 $nmol.cm^{-2}$), it attained about 100 and 600 $nmol.cm^{-2}$ on C6- and PEI-Immunodyne respectively. For the Sartorius JN138, 90 $nmol.cm^{-2}$ were found.

TABLE 1

Amount of Cibacron blue F3G-A ligand coupled to Ultipor membranes of different nominal pore sizes, in nmol. per cm^2 external membrane area, average of triplicate measurements and standard error

Pore size	Spacer */ activation method			
(μm)	without	C6/triazine	C6/glutar-aldehyde	PEI/glutar-aldehyde
0.1	15 ± 1	17 ± 5	8 ± 2	243 ± 18
0.2	16 ±0.2	18 ± 1	8 ± 1	301 ± 73
0.45	14 ±0.0	13 ± 1	7 ±0.5	159 ± 6
1.2	15 ± 2	16 ± 3	10 ± 1	151 ± 25

* C6, 1,6-diaminohexane; PEI, polyethylene-imine.

Batch adsorption of MDH

The adsorption test performed used standardized conditions aimed to approach saturation of the support. The solution was never depleted in MDH as at least 40% of the amount introduced remained in the solution.

The non specific adsorption on membranes in the absence of dye was always very low, typically under 20 $\mu g/cm^2$. The highest values, about 50 $\mu g/cm^2$, were obtained with PEI-Ultipor and PEI-Immunodyne. This was however much lower than the adsorption on the corresponding dyed membranes. These values do not represent true blanks or background adsorption of the final membranes, since the original binding sites may be masked upon coupling of the dye ligands.

Table 2 gives the amount of MDH as protein or activity adsorbed and desorbed from membranes with high capacities. Both measurements were in good agreement.

The capacity and recovery decreased in the order Satorius, C6-Immunodyne, PEI-Ultipor. In the latter case a large irreversible binding was observed. It should be noted also that a comparison based on surface, though practical in membrane applications, is biased by the fact that not all the membranes have the same thickness. A comparison made on dry weight basis gives close values for the tested membranes also, but a reversed order.

TABLE 2

Amount of pig heart MDH activity or protein adsorbed and desorbed from membranes carrying Cibacron blue F3G-A ligands, average from three experiments (± standard deviation).

Membrane	Adsorption		Desorption		Recovery	
	Activity U/cm^2	Proteins $\mu g/cm^2$	Activity U/cm^2	Proteins $\mu g/cm^2$	U /Protein (%)	
Sartorius 345	363±34	427±36	297±17	309±8	82±10	72±6
C6-Immunodyne 0.45 μm	364±44	240±16	226±8	173±15	62±8	72±8
PEI-Ultipor 0.1 μm	252±45	186±70	57±26	14±4	22±11	8±3

The adsorption for dyed Ultipor membranes without or with C6 spacers were too low to be measured by difference of concentration in solution under the standard conditions. Performing the assay with five times lower MDH concentrations gave values in the range 40-50 $\mu g/cm^2$ with C6 spacer and glutaraldehyde activation, and about 10-20 $\mu g/cm^2$ with C6/triazine or with no spacers. Using membranes with pore ratings from 0.1 to 1.2 μm yielded no detectable differences in adsorption of MDH.

Table 3 shows the results observed with PEI-Ultipor membranes of different pore size. Though nominal pore size

differed by a factor of ten, only a twofold difference in the
MDH capacity was found. The trend in MDH adsorption capacity
was parallel to that in the amount of dye immobilized
(Table 1). The efficiency of the ligand, estimated by the
ratio moles MDH / moles dye, was relatively stable around ca.
0.015 for the membranes with a high capacity for which it
could only be determined. From such a low value, one can
suppose that the interface area should be the limiting factor.
Further information is expected from determinations of the
surface area or observations on thin cross-sections.

TABLE 3.
Effect of pore size of the dye-ligand membrane (PEI-Ultipor)
on capacity and reversibility of adsorption for pig heart MDH.
Errors were calculated from assay's precision.

Pore size	U/cm^2		Protein, µg/cm^2		Recovery %	
(µm)	adsorbed	desorbed	adsorbed	desorbed	U/Protein	
0.1	363±35	82±5	290±23	85±0.2	23	29
0.2	333±4	71±5	274±16	70±2	21	26
0.45	197±4	67±0	164±11	65±2	34	40
1.2	193±18	40±0.4	139±11	43±2	21	31

Membrane capacity compared to Sepharose
Membranes were compared to beads as ligand carriers. Therefore
Cibacron blue F3G-A was coupled to Sepharose CL-4B using a
classical procedure. Drained gel instead of membrane was used
in the standard adsorption experiment. Table 4 shows the
amount MDH adsorbed and desorbed, per unit wet weight of
carrier. The actual concentration of adsorbed protein was
almost the same whatever the carrier. On the other hand, the
reversibility was nearly 100% with Sepharose.

TABLE 4.
Comparison of membranes and Sepharose CL-4B as ligand carrier
(Cibacron blue F3G-A). Amounts of pig heart MDH units or pro-
tein adsorbed or desorbed per unit wet weight of carrier[*].

Carrier	U/mg WW		µg proteins/mg WW	
	adsorbed	desorbed	adsorbed	desorbed
Sartorius 345	18±2	14.8±0.8	21 ±2	15.5±0.4
PEI-Ultipor(0.1µm)	13±1	2.6±0.1	9 ±1	3 ±0.0
C6-Immuno.(0.45µm)	14±2	9 ±1	9 ±1	7 ± 1
Sepharose CL-4B	14±0.5	11.9±0.4	12.2±0.2	12.0±0.4

[*] data used for the estimation: PEI-Ultipor and C6-Immunodyne
75% porosity; Sartorius 345, 1 cm^3 for 50 cm^2, (information of
manufacturers); density was taken as one.

Modified microfiltration membranes as chromatographic support
The binding of proteins in flow conditions was studied using
one to ten membranes stacked as stationary phase and the crude
extract from Baker's yeast. Note first that, although the
module was constructed with a minute internal volume, a severe
band spreading was observed when the system was tested with a
cellulose acetate membrane and non retained solutes such as
blue-dextran or BSA. A 500 µl injection could be almost
quantitatively recovered in 1.5 ml eluate.

In order to compare different membrane materials with
regard to their capacity and reusability, several succesive
loading and elution cycles were made using two membranes.
Elution was performed with a KCl gradient taken as the
simplest and cheapest method for early experiments. The total
amount of protein recovered was calculated, as well as the
amount recovered after starting the gradient. The latter data
are plotted in Figure 1.

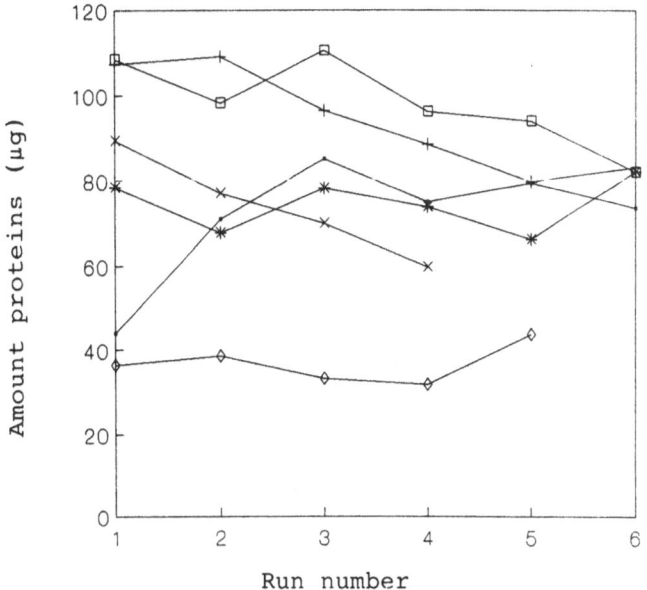

Figure 1. Amount of protein recovered after starting the KCl
gradient in successive runs made with modified microfiltration
membranes carrying Cibacron blue F3G-A. 500 µl of yeast crude
extract were injected in each run. Key to membranes:
•, Sartorius JN138; +, C6-Immunodyne 0.45 µm; *, PEI-Ultipor
1.2 µm; □, PEI-Immunodyne 0.45 µm; ×, PEI-Ultipor 0.1 µm;
◇, Blue dextran Ultipor 1.2 µm.

The recovery was around 100 % ±20, the spread of data was
mainly due to experimental imprecision. One can not conclude
from the data that a severe irreversible binding occured in
the first run. On the other hand, the amount of protein
recovered in the gradient was typical for the support;

Immunodyne membranes gave the highest values followed by PEI-Ultipor and Sartorius JN138. Blue dextran Ultipor gave about half the capacity of the former probably because of the much lower dye concentration. The Sartorius membrane showed a very low capacity in the first run. This run yielded also a totally different elution profile, most of the protein having been recovered in fractions 10 and 11. The second run also showed this profile, although less pronounced. Figure 2 gives the average elution pattern over six successive runs except for Blue dextran Ultipor (five runs) and for PEI-Ultipor 1.2 μm and Sartorius JN138 (four runs).

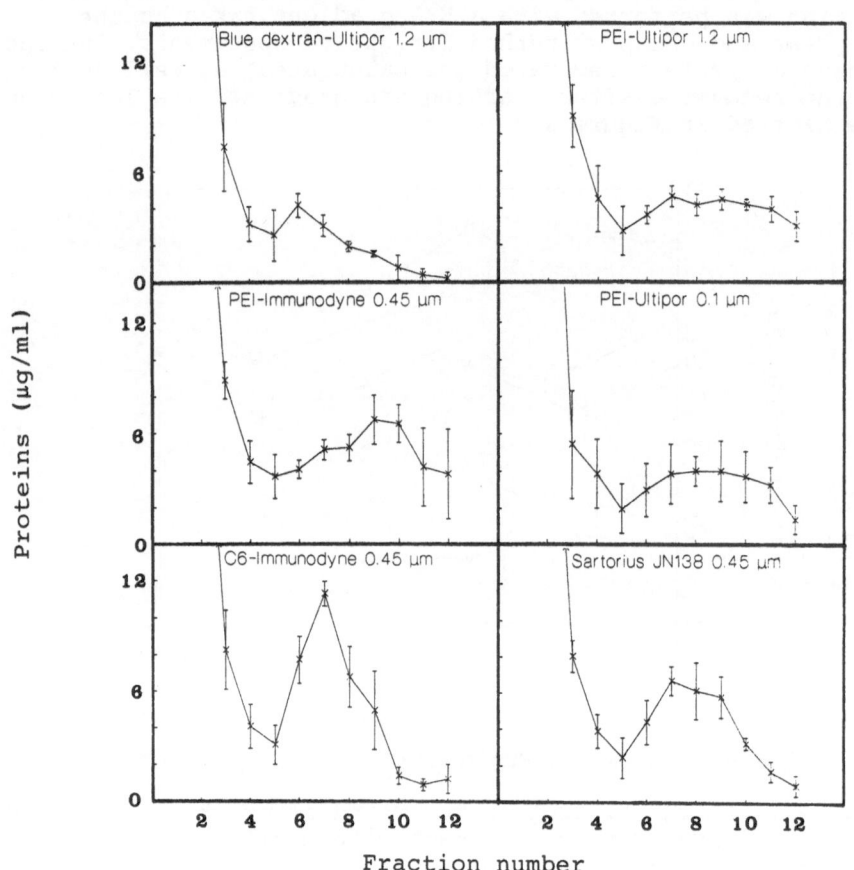

Figure 2. Elution profiles, averaged over up to six successive chromatograms, using various microfiltration membranes modified with Cibacron blue F3G-A. 500 μl of crude extract from yeast were injected. Elution was made with a linear gradient starting after fraction 4 and reaching 0.25 mM at the end of fraction 8, followed by a 5 ml step with 2 M KCL and another 5 ml step with start buffer. Fraction size was 2.5 ml. The error bars represent twice the standard deviation.

The elution pattern was characteristic of the support material considering the spreading of eluted protein, as well as the median elution time. Blue dextran Ultipor produced an early elution, PEI-Immunodyne a late one and the other exhibited intermediate values. The sharpness of elution decreased in order Blue dextran-Ultipor, C6-Immunodyne, Sartorius. Proteins were spread out most when PEI was used as spacer. A possible controlling factor could be the dye concentration, which at the highest values may give rise to multipoint attachement of part of the proteins, therefore retarding their elution and spreading the profile. Indeed, it was observed with gels that multivalence of the binding readily occurs at high ligand concentration [9]. One could therefore optimize the dye concentration using PEI-treated membrane or limit the amount of polymer coated in the preceeding step.

Additional experiments, using ten disks of the Sartorius membranes reproduced accurately the elution profile shown in Figure 2. The amount of protein recovered in the gradient was proportionally higher.

It can be expected that the spreading upon elution resulted also partially from a selectivity of the support. Column developments using simple KCl gradient allowed indeed to separate yeasts enzymes on blue gels [10]. Preliminary measurements of MDH and adenylate kinase activities in the fractions showed no pronounced enrichement for either one. Further work will aim to elute enzymes specifically by competitive ligands and to analyse the fractions using electrophoresis.

CONCLUSION

Through modification with spacers and dye ligands, micro-filtration membranes could be designed as affinity adsorbents that could be substituted for traditional supports. The dye concentration can be chosen by selecting the membrane material itself or creating a sufficient number of reactive groups on the surface by means of grafted polyethylenimine. The dye concentration could thus be optimized. Several membranes offered a similar capacity compared to dyed Sepharose for malate dehydrogenase adsorption. Improvements are desirable concerning the reversibility of adsorption and the desorption characteristics in a flow system. Irreversible fouling was not observed, however, in the latter. Also other elution methods than increasing the ionic strenght with KCl can be reasonably expected to produce significant improvements. A definitive advantage of membranes is the possibility to cope with unclarified broth, applying a concept of affinity cross-flow filtration. In this regard, the relative insensitivity of binding capacity to the pore size will be a great advantage to maximize permeate flow and protein transmission from broths or cell homogenates.

304

AKNOWLEDGEMENTS

B. Champluvier was supported by a grant from the European Commission, DG XII. We thank the companies Pall and Sartorius for the gift of membranes.

REFERENCES

1. Mattiasson B., Olson U., Senstad C., Kaul R. and Ling T., Affinity interactions in free solution as an initial step in Downstream Processing. Proc. 4th Eur. Congr. Biotechnol., eds. O.M. Neijssel et al., Elsevier Science Publ. B.V., Amsterdam, 1987, pp. 611-614.
2. Cordes A. and Kula M.-R., Process design for large-scale purification of formate dehydrogenase from Candida boidinii by affinity partition. J. Chromatogr., 1986, 376, 375-384.
3. Johansson G., Dye-ligand aqueous two-phase systems, In Reactive Dyes in Protein and Enzyme Technology, eds Y.D. Clonis et al., Stockton Press, New York, 1987, pp. 101-124.
4. Kopperschläger G., Böhme, Hofmann E., Cibacron blue F3G-A and related dyes as ligands in affinity chromatography. In Adv. Biochem. Eng., ed. A. Fiechter, Springer-Verlag, Berlin, 1982, vol. 25, pp. 102-138.
5. Dean P.D.G., Johnson W.S. and Middle F.A., Affinity chromatography, a practical approach, IRL Press Ltd., Oxford, 1985, pp. 31-59.
6. Bergmeyer H.U., Methods of Enzymatic Analysis, Verlag Chemie, Weinheim, 1983, vol. II, pp. 246-247
7. Chambers G.K., Determination of Cibacron blue F3GA substitution in blue Sephadex and blue dextran-Sepharose. Anal. Biochem., 1977, 83, 551-556.
8. Lowe C.R. and Pearson J.C., Affinity chromatography on Immobilized dyes. In Methods in Enzymology, ed. W.B. Jacoby, Academic Press, New York, 1984, vol. 104, pp. 97-113.
9. Stellwagen E. and Liu Y-C., Quantitative considerations of chromatography using immobilized biomimetic dyes. In Analytical Affinity Chromatography, ed. I.M. Chaiken, CRC Press, Boca Raton, Florida, 1987, pp. 157-184.
10. Scawen M.T. and Atkinson T., Large scale dye-ligand chromatography, In Reactive Dyes in Protein and Enzyme Technology, eds Y.D. Clonis et al., Stockton Press, New York, 1987, pp. 51-86.

ADSORPTION-DESORPTION OF PROTEINS IN RP-HPLC RNASE ON C_4 SUPPORTS

EZIO D'ADDARIO, MICHELE PATRISSI
Eniricerche S.p.A.
Via E. Ramarini, 32 - 00015 Monterotondo (ROMA) ITALY

MAURIZIO MASI, GIUSEPPE STORTI, SERGIO CARRA'
Dipartimento di Chimica Fisica Applicata, Politecnico di
Milano, P.za Leonardo da Vinci, 32 - 20133 MILANO ITALY

ABSTRACT

The adsorption-desorption of RNase on a C_4 RP-HPLC support
has been investigated in order to elucidate the effect of
the solvent (isopropyl alcohol) and of the ion pairing
(trifluoroacetic acid) on the equilibrium data. A fundamental role of the ion pairing has been evidenced by the experimental data. The adsorption equilibrium data have been
described using a multicomponent Langmuir-Freundlich type
isotherm.

NOTATION LIST

k	kinetic constants (see Theoretical Aspects).
K	equilibrium constants, $(ml/mg)^n$.
C	concentration in the liquid phase, (mg/ml).
m	number of active sites interacting with RNase.
n	number of solvent molecules solvating RNase.
n1	n+1
r	step rate, (mg/ml/s).
q	concentration in the adsorbed phase, (mg/g).
q^∞	loading capacity of the support, (mg/g).
ϵ	relative error, $\Sigma_i = \|(q_{ic} - q_i)/q_i\| \cdot 100$.
σ	active site on the adsorption surface.
θ	coverage degree, q/q^∞.

Subscripts

a	adsorption.	P	Protein (Ribonuclease).
c	calculated.	S	Solvent (isopropyl alcohol).
d	desorption step.	v	vacant active sites.

INTRODUCTION

The RP-HPLC technique offers numerous potential advantages for the recovery and the purification of proteins, but its use is mainly limited by protein denaturation during the elution and by the lack of knowledge of the basic phenomena involved in the process. Up to now, an experimental analysis, tailored to select media and elution conditions which can be tolerated by the molecules under examination, is the only pragmatic approach to the problem. On the other hand, fundamental studies are required in order to develop descriptive models useful for designing the optimal large scale apparatus.

The thermodynamic model of the reversed phase retention mechanism is one of the most rigorous reported in the literature [1]. Unfortunately, it requires the knowledge of a number of physico-chemical constants available only for low molecular weight non polar solutes [2]. In the case of proteins, simpler models, generally based on linear reversible equilibrium data, are available for conventional chromatographic separation processes [3]. For instance, informations suitable to industrial applications for the separation of proteins by ion exchange chromatography (retention volumes, peak widths, etc.) can be determined by the model of Furusaki et al. [4], in the case of linear equilibrium. On the other hand, chromatographic processes based on hydrophobic bindings show complicated mechanism of interaction between the protein and the stationary phase. For example, the binding of phosphorilase B on immobilized low pressure butyl-agarose residues, exhibits hysteresis phenomena due to irreversible adsorption [5]. This context indicates that problems of interpretation and correlation of the equilibrium and kinetic adsorption-desorption data, cannot be generalized. Furthermore, these problems have to be properly solved in order to develop mathematical models for the description of protein separation processes based on hydrophobic chromatographic interactions.

In the present work the above mentioned studies have been extended to RP-HPLC technique. In particular, the effects on the protein adsorption-desorption process of the ion-pairing content and of the eluting solvent concentration in the liquid phase have been investigated. A behavior with hysteresis has been evidenced in the examined system.

EXPERIMENTAL

Equipment

Experiments have been performed using a Varian analytical HPLC system equipped with a double pump apparatus, a 2020 gradient programmer, a 2050 UV detector and a 4290 integrator. The solvent has been analyzed via GC using a 8500 Perkin-Elmer chromatograph.

Materials

RNase (P) (Boheringer 109134) has been used as a model protein. Before the experiments a water solution of this enzyme has been eluted through the chromatographic column in order to eliminate traces of impurities. Analytical grade isopropyl alcohol (S) (Rudi-Pont) and purified water from a Milli-Q system have been used as solvents. Analytical grade trifluoroacetic acid (TFA) (0.1% v) (Merck 8262) has been used as ion-pairing. A 300 Å; 6.5 μm macroporous spherical silica covalently bonded with a polyamide coating and derivatized with a ligand which terminates with a butyl (C$_4$) moiety (Synchron, USA), has been utilized as packing in a 4x100 mm column.

Procedure

The equilibrium adsorption-desorption data have been obtained using the dynamic method [6] operating in a conditioned room (20°C). According to this procedure the column has been continuously fed (0.8-1 ml/min) with solutions at constant concentration. The feeding went on till plateau values of the output concentration were reached (RNase was detected at 280 nm and isopropyl alcohol was discontinuously analyzed via GC). The amount of the adsorbed solute has been calculated on the base of a material balance. The RNase desorption test has been carried out by eluting the column with a liquid phase containing the enzyme at concentration of 0.5 mg/ml lower than the one used in the corresponding adsorption test.

THEORETICAL ASPECTS OF THE ADSORPTION

The adsorption of RNase on the C_4 support can be treated as a competitive process between the protein and the solvent, and it is generally accepted that the protein in the solid phase is adsorbed on several active sites [2]. Furthermore it can be assumed that solvation phenomena, as well as the presence of the ion pairing, can alter both the protein structure and the available active sites. Here, these proc-esses have been simplified into the following main steps: (i) adsorption of protein, (ii) adsorption of solvent, (iii) desorption of protein.

Introducing the coverage degree for the protein (θ_P) and for the solvent (θ_S), the following equations, describ-ing the kinetic of each step, have been derived [7,8]:

i) $\quad P + m \sigma \rightleftharpoons P_a$; $\qquad r_1 = k_{1a} C_P \theta_v^m - k_{1d} \theta_P$

ii) $\quad S + \sigma \rightleftharpoons S_a$; $\qquad r_2 = k_{2a} C_S \theta_v - k_{2d} \theta_S$

iii) $\quad P_a + n S \longrightarrow P$; $\qquad r_3 = k_3 \theta_P C_S^n$

where $\theta_v + \theta_S + \theta_P = 1$. Assuming the equilibrium conditions $(r_1 - r_3 = 0$ and $r_2 = 0)$, the following relationship for the coverage degree of the RNase has been obtained:

$$\frac{\theta_P}{(1 - \theta_P)^m} = \frac{K_1 C_P}{(1 + K_3 C_S^n) \cdot (1 + K_2 C_S)^m} \tag{1}$$

where: $K_1 = k_{1a}/k_{1d}$, $K_2 = k_{2a}/k_{2d}$, and $K_3 = k_3/k_{1d}$. If $m=1$, defining $K_4 = K_3 \cdot K_2$, eq. (1) can be written as follows:

$$\theta_P = \frac{K_1\, c_P}{1 + K_1\, c_P + K_2\, c_S + K_3\, c_S^n + K_4\, c_S^{n+1}} \tag{2}$$

Eq. (2), in the case of $K_3 = K_4 = 0$ and $K_2 = K_4 = 0$, reduces to the Langmuir (**L**) and to the Langmuir-Freundlich (**LF**) isotherm, respectively [9].

RESULTS

Adsorption

Eq. (2), Langmuir and Langmuir-Freundlich isotherms, have been used to fit the experimental results. The parameter values have been obtained by non linear regression [10], and their values are summarized in Tables 1 and 2.

TABLE 1
Adsorption equilibrium parameter values for RNase
in the absence of TFA.

Isotherm	q^∞ mg/g	K_1 ml/mg	K_2 ml/mg	K_3 $(ml/mg)^n$	K_4 $(ml/mg)^{n1}$	n	$\epsilon\%$
L	54.45	2.944	0.417	-----	----	----	26.7
LF	47.98	4.897	----	0.00171	----	2.13	16.9
Eq.(2)	47.93	4.935	0.001	0.00001	0.00174	1.13	16.9

TABLE 2
Adsorption equilibrium parameter values for RNase
in the presence of TFA.

Isotherm	q^∞ mg/g	K_1 ml/mg	K_2 ml/mg	K_3 $(ml/mg)^n$	K_4 $(ml/mg)^{n1}$	n	$\epsilon\%$
L	24.51	7.559	0.240	-----	----	----	19.6
LF	26.85	4.566	----	4.307	----	.333	5.3
Eq. (2)	26.89	4.527	0.021	7.598	0.0152	.177	5.3

The Langmuir-Freundlich isotherm (LF), as well as the more complicated eq. (2), fits the RNase adsorption experimental data. A better agreement between the calculated and the experimental data is obtained in the case of the presence of TFA, as it is shown in Figs. 1 and 2.

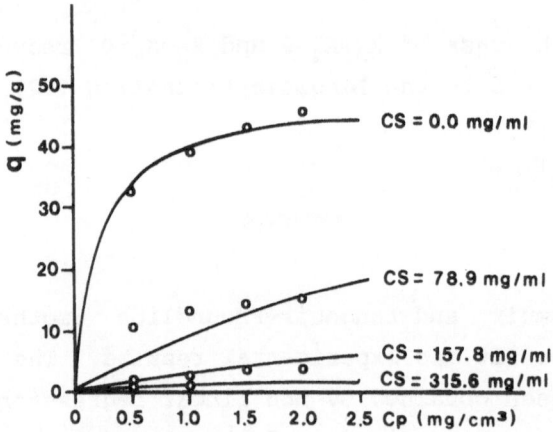

FIGURE 1. Comparison of calculated (Eq.(2) and LF model) and experimental data of RNase adsorption isotherm in the absence of TFA for different values of solvent concentration.

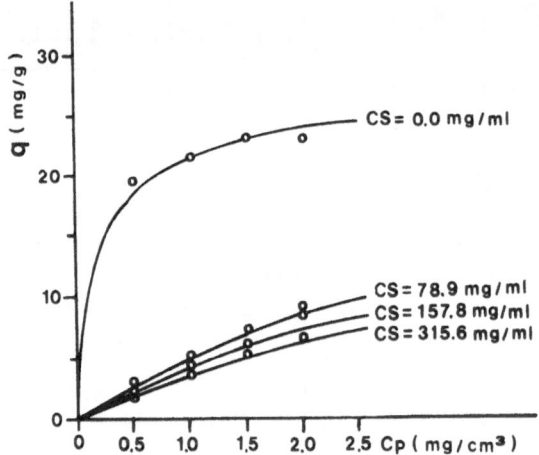

FIGURE 2. Comparison of calculated (Eq.(2) and LF model) and experimental data of RNase adsorption isotherm in the presence of TFA for different values of solvent concentration.

Finally, experimental equilibrium data for the pure solvent, in the presence and in the absence of TFA, have been described through a simple Langmuir isotherm, according to step ii) of the kinetic scheme reported above. The calculated parameter values are reported in Table 3 and the corresponding comparison between the calculated and the experimental data is shown in Figure 3. Note that the estimated K_2 values are in satisfactorily agreement with the same values reported in Table 1 and 2 with reference to Equation (2). Thus, despite the similar behavior with respect to the description of the equilibrium data shown in Figure 1 and 2, equation (2) only is consistent with the equilibrium data of the pure solvent, substantiating the proposed adsorption-desorption scheme.

TABLE 3
Adsorption equilibrium parameter values for isopropyl
alcohol in the absence of RNase.

Isotherm	TFA	q^{∞} mg/g	K_1	K_2 ml/mg	K_3	K_4	n	$\epsilon\%$
Langmuir	NO	3538.1	--	0.00298	-	-	-	4.5
Langmuir	YES	4663.2	--	0.00195	-	-	-	10.9

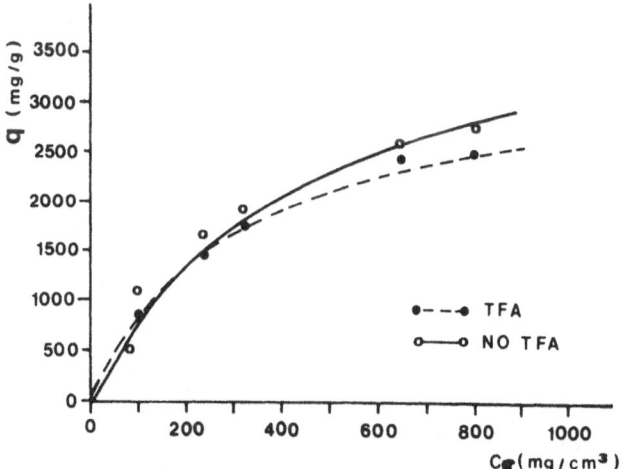

FIGURE 3. Comparison of calculated (L model) and experimental data of isopropyl alcohol adsorption isotherm in the absence of RNase (in the absence and in the presence of TFA)

Desorption

Typical experimental adsorption-desorption isotherms obtained both in the absence and in the presence of TFA are
shown in Figs. 4 and 5, respectively. These figures indicate
significant hysteresis phenomena which tend to disappear at
high solvent concentration and in presence of ion pairing.
As the experimental desorption equilibrium data were stable
through the time (the feeding of the eluting media went on
for at least 80 column volumes after the output plateau
values were reached), the hysteresis phenomena may be due to
irreversible adsorption of RNase. A theoretical study of
this aspect is still in progress.

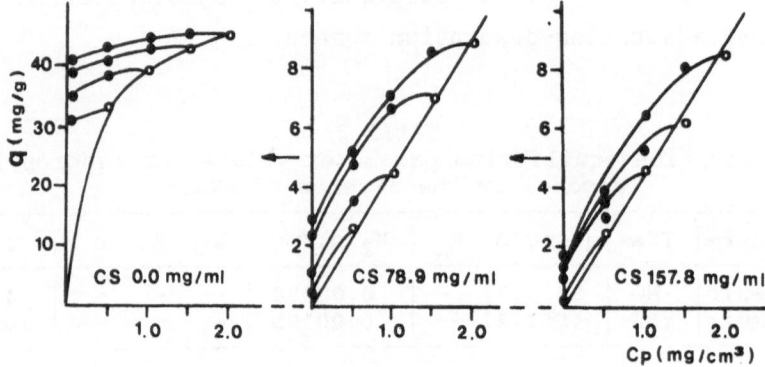

FIGURE 4. RNase adsorption-desorption experimental isotherms
in the absence of TFA for different values of solvent
concentration.

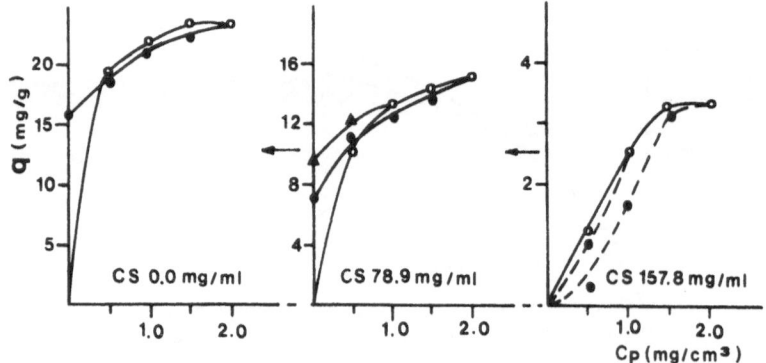

FIGURE 5. RNase adsorption-desorption experimental isotherms
in the presence of TFA for different values of solvent
concentration.

DISCUSSION AND CONCLUSIONS

The experimental results show that the ion pairing plays a fundamental role on the adsorption-desorption of RNase. This molecule strongly reduces: (i) the maximum support loading capacity and (ii) the irreversible RNase binding capacity and significantly increases (iii) the protein solvent competitive adsorption.

Theoretical multicomponent isotherms (Langmuir, Langmuir-Freundlich and eq. (2)) satisfactorily describe the adsorption of: (i) RNase, (ii) isopropyl alcohol and, (iii) their mixtures in presence of a typical RP-HPLC concentration (0.1% v) of TFA.

Unfortunately, the above mentioned isotherms do not appear valid for the description of the RNase - isopropyl alcohol adsorption in the absence of TFA at intermediate concentration of the solvent in the liquid phase. In this case the solvent adsorption is strongly favorite and, at high solvent concentration, the RNase experimental isotherm assumes a linear profile.

In conclusion, the typical RP-HPLC applicative conditions (TFA 0.1% v) can be satisfactorily described by the theoretical model considered here, even if they do not account for the ion pairing contribution. However, considering the decisive role played by the ion pairing on the RP-HPLC adsorption-desorption processes and its denaturing properties, different TFA concentration levels are under examination following the approach here reported.

REFERENCES

1. Horvarth Cs., Melander W., Molnar I., J. Chromatogr., 1976, **125**, 129.

2. Geng X., Regnier F.E., J. Chromatogr., 1984, **296**, 15.

3. Standlius M.A., Gold H.S., Snyder L.R., J. Chromatogr., 1985, **327**, 27.

4. Furusaky S., Haraguchi E., Nozawa T., Bioprocess Engineering, 1987, **2**, 49.

5. Jennisen H.P., J. of Colloid and Interface Science, 1986, **2**, 49.

6. Jacobson J.M., Frenz J.H. and Horvath C., Ind. Eng. Chem. Res., 1987, **26**, 43.

7. Carberry J.J., "Chemical and Catalytic Reaction Engineering", McGraw-Hill, New York, 1970.

8. Butt J.B., "Reactor Kinetics and Reactor Design", Prentice-Hall, Englewood Cliffs N.J., 1980.

9. Ruthven D.M., "Principles of Adsorption and Adsorption Processes", J. Wiley, New York, 1984.

10. Buzzi Ferraris G., Fortran Routine BURENL, 1985, private communication.

AFFINITY SEPARATIONS USING MICROCAPSULES

Andrew J. Daugulis*, Charlene P. Wight, Mattheus F.A. Goosen, and
Kelly E. Chesney

Department of Chemical Engineering, Queen's University, Kingston, Ont.,
Canada, K7L 3N6.

and

Q-Life Systems Inc., 1180 Clyde Court, Kingston, Ont., Canada, K7P 2E4

ABSTRACT

In order to overcome some of the problems associated with column affinity
chromatography, two different types of liquid core, semi-permeable
microcapsules (poly-L-lysine and chitosan) have been used for the
encapsulation of blue dextran, containing the dye Cibacron blue F3GA.
These capsules have been used to isolate bovine serum albumin (BSA) from a
solution of BSA alone, and from whole bovine serum. Conventional blue
columns were also used, and the results from the two different methods
compared. The performance of these affinity microcapsules and their
possible method of use are discussed.

INTRODUCTION

Affinity purification of biological materials usually involves the use of a
column containing the affinity molecule bound to a support matrix.
Problems with the use of such columns, especially for large scale
applications, include plugging, channeling, high pressure drops, and bed
compaction, as well as the need for expensive (and expansive) equipment.
Problems also occur because of the way in which the affinity molecule is
attached to the matrix material. These include steric hindrance,
attachment of the affinity molecule in the area of its active site, and
uneven distribution of the affinity molecules over the matrix surface.

Others have tried to overcome these problems by using techniques such
as ultrafiltration [1]. In this technique, the affinity molecule is

coupled to a particle which is larger than the molecular weight cut off of the ultrafiltration membrane. However, this method still requires the removal of any particulate material before the affinity separation can be done [1].

Affinity ligands have also been microencapsulated. Sakoda and Wang have encapsulated IgG agarose in a calcium alginate gel membrane [2]. The presence of particulates does not interfere with this method. However, all of the techniques mentioned still require the binding of the affinity molecule to the surface of a relatively large carrier particle. Thus, problems with steric hindrance, etc., remain. These problems could be overcome if microcapsules were used which could retain the affinity molecule without requiring it to be bound to a large carrier particle, or if solid gel capsules were not required. Using encapsulated affinity molecules could also be very efficient as the volume of a sphere has greater capacity to retain molecules than does its surface, and more of the affinity molecule could be contained in a given volume of material.

We have encapsulated an affinity molecule within a microcapsule bounded by a semipermeable membrane of controlled molecular weight cut off. The capsule core, while initially a solid gel, can be liquefied, thus leaving the affinity molecule freely suspended within the capsule. As our model system, we have encapsulated blue dextran, which has bound to it the dye Cibacron blue F3GA. Liquid core poly-L-lysine (PLL) or chitosan capsules were used. These microcapsules have been used to isolate bovine serum albumin (BSA) from a saline solution and from whole serum. The results have also been compared to those obtained using a conventional blue sepharose column.

MATERIALS AND METHODS

Materials

Poly-L-lysine hydrobromide (PLL-21,500 average molecular weight), dextran, blue dextran, Reactive blue 2-sepharose CL6B, Cibacron blue F3GA, bovine serum albumin (fraction V powder), and whole bovine serum (lyophilized powder) were from Sigma Chemical Co., St. Louis, MO, USA. Sea Cure-F (medium viscosity) chitosan was from Protan Laboratories Inc., Redmond, WA, USA, and sodium alginate powder (Kelco Gel LV) was from Kelco Specialty Colloids Ltd., Chicago, IL, USA.

Methods

Chitosan preparation and treatment: A 0.1% (w/v) solution of
chitosan was made by dissolving 0.1 g of chitosan in 100 ml of 1.0% acetic
acid. Aliquots of 0.1% $NaNO_2$ were added to the chitosan such that the
mole ratio of $NaNO_2$ to chitosan monomers ranged from 0.01 to 0.10. This
treatment randomly breaks the chitosan chain in a stoichiometric fashion,
thus reducing its average molecular weight [3]. After continued stirring
overnight, the pH of each chitosan solution was adjusted to 6.5. The
viscosity average molecular weight of each treated chitosan solution was
measured using a Canon Ubbelohde viscometer, according to the procedure
outlined in ASTM Standards D 445 [4].

Capsule formation: The methods used here for capsule formation are
modified from the techniques described by McKnight, et al. [3] and King, et
al. [5]. Alginate alone or containing 2.5% (w/v) blue dextran (4 mL) was
extruded through a droplet generator [5] into either 35 ml of one of the
treated chitosan solutions containing 1.5% (w/v) $CaCl_2$, or 35 ml of 1.5%
$CaCl_2$ containing 15 mg PLL. The capsules were incubated in the polymer
solution for 14 min from the beginning of the extrusion. After allowing
1 min for the capsules to settle, the supernatant was removed using a
vacuum pump, and the capsules washed once in 1.1% $CaCl_2$, then twice in 0.9%
NaCl (saline). To reduce stickiness, they were reacted for 4 min with
0.03% sodium alginate in saline. After another saline wash, the capsules
were reacted for 10 min with 0.05M sodium citrate in 0.45% NaCl. This
converts the solid calcium alginate cores back into liquid sodium alginate.
After one more saline wash, the capsules were incubated in saline for at
least half an hour before use. Capsule size ranged from 1.0-1.2 mm.

Use of capsules: For diffusion studies, dextran was dyed with Cibacron
blue F3GA according to Ashton and Polya [6]. Each experiment was done in
triplicate. One ml of settled capsules was added to 2.0 ml of saline
containing dyed dextran of a certain molecular weight. The contents of
each tube were mixed quickly, the capsules allowed to settle, and 0.5 ml
zero time samples collected. The tubes were then incubated at room
temperature with shaking for 4 h. Samples were again collected, and the
differences in absorbance at a wavelength of 595 nm determined.

Each affinity binding experiment was also done in triplicate. Two ml
of saline, 0.5% (w/v) BSA in saline, or 0.2% (w/v) whole bovine serum in

saline was added to each set of capsules. The tubes were incubated for 4 h, and samples collected as described above. The capsules were then washed twice with 10 ml of saline, and incubated overnight in another 10 ml of saline. Samples from the beginning and end of the overnight wash were collected. After the wash solution was removed, 2.0 ml of 0.2M NaSCN was added to each sample and zero time samples collected. This eluent has been shown to be effective in eluting bound serum albumin from Cibacron blue F3GA [7,8]. After 4 h incubation, the final samples were collected.

All of the samples were assayed for protein content using the Bradford assay [9] (BioRad Protein Assay Kit) with BSA as a standard. As they became stirred up during transfer, the actual volume of settled capsules in each tube had to be determined. These volumes were used to calculate the amount of protein available during uptake and released during washing and elution. The composition of the protein in each sample was determined using polyacrylamide gel electrophoresis [10].

Capsule reuse: Used capsules were combined and washed thoroughly with saline. They were then reused in experiments as described above.

Blue sepharose column: A 3.0 ml blue sepharose column was prepared in saline. It was loaded with 6.0 ml of either 0.5% (w/v) BSA in saline, or 0.5% (w/v) whole bovine serum in saline. The column was washed with 18 ml of saline, and then any bound protein eluted with 6.0 ml of 0.2M NaSCN. Fractions of 0.8 ml were collected and analyzed for protein content using the Bradford assay. The composition of the protein in these samples was also determined using polyacylamide gel electrophoresis. The column was thoroughly washed with saline before being used again.

RESULTS AND DISCUSSION

Capsule formation

PLL is available commercially in a wide range of average molecular weights. The length of the polymer used and the time of reaction during capsule formation determines the molecular weight cut off of the membrane formed. Previously, calcium alginate beads were formed before the addition of polymer [5]. Because the molecules being encapsulated could easily diffuse out of these beads, the polymer here was included at the beginning of the encapsulation procedure, such that the membrane would begin to form as soon as the alginate drops had hardened. To keep the variation in capsule permeability low, only a small volume of capsules could be extruded from a single droplet generator at any one time.

It was necessary to reduce the average molecular weight of the chitosan molecules using NaNO$_2$, which randomly cleaves the chitosan chain [3]. Different mole ratios of 0.1% NaNO$_2$ to chitosan monomers gave different average molecular weights of chitosan, as determined using viscosity measurements (Table 1). Of these different preparations, only those indicated formed stable capsules.

TABLE 1
Viscosity Average Molecular Weights of Treated Chitosan

NaNO$_2$: Chitosan	Viscosity Average MW
0.0100	609,000
0.0250	289,000
0.0375	243,000
0.0500	170,000*
0.0750	123,000*
0.1000	101,000*

*these preparations of chitosan could be used to form stable capsules.

Using blue dextran of different average molecular weights, the permeabilities of the chitosan capsules were determined (Table 2). These data indicate that capsules prepared using chitosan prepared with a 0.1 to 1.0 mole ratio of NaNO$_2$ to chitosan monomers (average molecular weight = 101,000) have a molecular weight cut off between 70.8 and 150 Kd. It was decided that these capsules would be the most useful when encapsulating smaller affinity molecules and excluding undesirable substances of high molecular weight.

TABLE 2
Chitosan Capsule Permeabilities to Dextran

Dextran MW	Chitosan MW	% Uptake
35,000	170,000	14.1
	123,000	12.3
	101,000	13.6
70,800	170,000	14.8
	123,000	14.2
	101,000	13.1
150,000	170,000	7.8
	123,000	4.7
	101,000	0.4

Isolation of BSA from solution

Table 3 presents the results of a typical experiment using PLL or chitosan
microcapsules containing blue dextran to isolate BSA from a saline
solution. Because a certain amount of the protein will diffuse into the
capsules until an equilibrium is reached, plain alginate (blank) capsules
were used as a control. Controls of blue or blank capsules in saline
alone were also included, as were controls to check for any nonspecific
binding of protein to the plastic tubes. Zero to 2.4% of the protein did
bind initially, but was removed during the subsequent wash steps.

From Table 3 it can be seen that, with both PLL and chitosan blue
capsules, more protein was taken up from solution than would have been
expected because of the migration of BSA into the capsule volumes.

TABLE 3
Uptake, binding, washing, and release of BSA from blue or blank PLL or
chitosan capsules.

Uptake

Capsule type	Available protein[a]	% Taken up	% Expected[b]
Blue PLL	6.35 mg	50.4	30.0
Blank PLL	7.02	9.5	30.8
Blue chitosan	7.31	46.4	24.0
Blank chitosan	6.89	6.8	28.0

Wash

Capsule type	% Released	% Expected	Amount bound
Blue PLL	22.2	92.7	2.49 mg
Blank PLL	84.3	92.5	0.11
Blue chitosan	19.8	93.3	2.72
Blank chitosan	93.6	92.2	0.03

Elution

Capsule type	% Eluted	% Expected	% Recovery
Blue PLL	57.4	66.7	22.5
Blank PLL	0.0	0.0	0.0
Blue chitosan	51.1	71.4	19.0
Blank chitosan	0.0	0.0	0.0

[a]Determined using the Bradford assay with BSA as standard, and corrected
for differences in volume.

[b]Based on the assumption of equilibrium after diffusion of protein from
a given volume into empty (no alginate) capsules, and assuming that no
binding is occuring.

This indicates that binding of BSA to the blue dye was occuring. This
binding can be seen to be even greater when it is realized that much less
protein diffused into the blank capsules than was expected. Less protein

may have diffused in because the alginate in the core occupies a large proportion of the capsule volume. Another possibility is that, even though the core has been liquefied, the alginate forms a matrix which is difficult for the protein to penetrate. In any case, five to seven times more protein was taken up into the capsules containing blue dextran than into the blank capsules.

During washing, it can be seen that almost all of the protein taken up by the blank capsules was released. Only about 20% of the protein taken up by the blue capsules was washed out, indicating that the remaining protein was bound within the capsules. This bound protein was eluted using 0.2M NaSCN. Because only a small volume of eluent was added to the capsules, only about 65-70% of the bound protein could be eluted in this case. Slightly more protein was eluted from the PLL than the chitosan capsules, but in both cases this represented almost all of the release expected. The final recoveries, therefore, were 19.0% from the blue chitosan capsules, and 22.5% from the blue PLL capsules. No protein was eluted from the blank capsules, indicating that the protein recovered from the blue capsules had been bound specifically by the blue dye.

When a standard blue column using the same loadings and reagents was used to isolate BSA from solution, a recovery of 23% was obtained. The amounts of blue dye in the volumes of blue capsules and blue column material used were calculated to be equal, therefore, the recovery of BSA using blue capsules is at least as good as that from a blue column, and should be better once the system is further refined.

Isolation of BSA from whole bovine serum

Blue capsules were also used to isolate BSA from whole bovine serum. In this case, the chitosan capsules dissolved when they were mixed with the reconstituted serum. Thus, the stability of the two kinds of capsules in different solutions must be considered. This is also true of binding buffers and eluents. High pH, for example, causes both types of capsule to dissolve, and divalent cations such as $CaCl_2$ cause the gel inside the capsule to resolidify.

Using the blue PLL capsules, protein was taken up from the whole serum and eluted as for the experiments with pure BSA (Table 4).

TABLE 4

Uptake and release of BSA from whole bovine serum using blue or blank
PLL capsules

Capsule type	Available protein	Amount bound, then eluted	% Recovery
Blue PLL	2.69 mg	0.60 mg	22.3
Blank PLL	3.38	0.05	1.5

The recovery of protein was 22.3% of the total protein available, 60% of which is BSA. Using a blue column the recovery was 21.7%.

Polyacrylamide gel electrophoresis was done to determine which proteins from whole serum were being bound and eluted in each case (Figure 1).

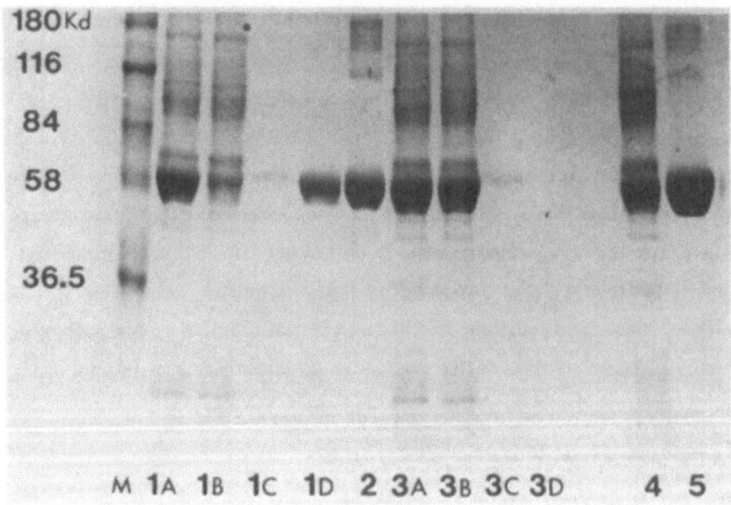

FIGURE 1:   Identification of proteins present in samples taken during the
isolation of BSA from whole bovine serum.

Lanes 1A-D contain samples from the experiment performed using blue PLL
capsules, and lanes 3A-D samples from the control experiment using blank
PLL capsules.   Lane A samples are from time zero; B, after 4 h incubation;
C, after washing; and D, after 4 h of elution.   Lane 2 contains BSA as a
marker.   Lane 4 contains a sample of the flowthrough from the blue column,
and lane 5 contains a sample of the protein eluted from the column.   Lane
M contains molecular weight markers.

Not all of the BSA available was bound, but when the intensities of the
different bands in lanes 1A, 1B, 3A, and 3B are compared, it can be seen
that more albumin was removed when blue capsules were used.   Contaminating

bands, such as those seen in a commercial preparation of BSA (lane 2), are present in the material eluted from the column (lane 5). These do not appear in the lane containing protein eluted from blue capsules (lane 1D), although this protein is less concentrated than the protein eluted from the column.

Protein is present in the samples obtained after washing both blue and blank capsules (lanes 1C and 3C). However, no protein was eluted from the blank capsules after washing (lane 3D), indicating that any protein bound nonspecifically does not interfere with the elution of the desired protein. Thus, blue capsules can be used to isolate a specific protein from solution, and that protein may be purer than that isolated using a conventional blue column.

Capsule reuse

After washing the used capsules extensively with saline, blue and blank PLL and chitosan capsules were reused for the isolation of BSA from a pure solution. A recovery of 21.7% was obtained using PLL capsules, compared to 22.5% originally. Using chitosan capsules, the recovery was 12.9%, compared to 19.0% originally. The column was also reused, from which a recovery of 25.9% was obtained, compared to the previous recovery of 23.1%. This suggests that microcapsules containing blue dextran can be reused effectively, as is the case for conventional blue sepharose columns.

Conclusions

We conclude, therefore, that liquid core semipermeable microcapsules containing an affinity molecule are useful in the the isolation of specific molecules from solution. Because capsules can potentially be used in a number of contacting configurations (suspended bed, fluidized bed, etc.), the use of capsules circumvents the need for packed columns and the accompanying problems such as plugging, channeling, high pressure drops, bed compaction and expensive equipment. While the dye Cibacron blue F3GA is a small molecule, and therefore needs to be bound to some kind of matrix to prevent its diffusing from the capsules, the use of dextran as a carrier instead of a gel or silica bead makes more effective use of the volume available. Carriers may not be needed for larger molecules such as antibodies, whose active sites would then always be freely available. The liquid core capsule also allows for more even distribution of the affinity molecule.

Future work

Blue capsules are now being used to isolate enzymes from bacterial and
plant material. The maximum amount of blue dextran in the capsules, as
well as the optimum times for uptake, washing, and elution, must still be
determined. Future work also includes modifying the polymers used in
capsule formation such that more stable capsules can be made. The
molecular weight cut offs of our capsules are also being refined, so that
smaller affinity molecules such as antibodies (150 Kd) can be
encapsulated.

REFERENCES

1. Ling, T.G.I. and Mattiasson, B., Membrane filtration affinity
   purification (MFAP) of dehydrogenases using Cibacron blue. Biotech.
   Bioeng., 1989, 34, 1321-1325.

2. Sakoda, A. and Wang, H.Y., A new isolation and purification method for
   staphlococcal protein A using membrane encapsulated rabbit IgG-agarose.
   Biotech. Bioeng., 1989, 34, 1098-1103.

3. McKnight, C.A., Ku, A. and Goosen, M.F.A., Synthesis of
   chitosan-alginate microcapsule membranes. J. Bioact. Biocomp. Pol.,
   1988, 3, 334-355.

4. American National Standards Institute (ASTM), Annual Book of ASTM
   Standards. Designation D 445, 1971.

5. King, G.A., Daugulis, A.J., Faulkner, P. and Goosen, M.F.A.,
   Alginate-polylysine microcapsules of controlled membrane molecular
   weight cutoff for mammalian cell culture engineering. Biotech.
   Progress, 1987, 3, 231-240.

6. Ashton, A.R., and Polya, G.M., The specific interaction of Cibacron and
   related dyes with cyclic nucleotide phosphodiesterase and lactate
   dehydrogenase. Biochem. J., 1978, 175, 501-506.

7. Hill, E.A. and Hirtenstein, M.D., Affinity chromatography: Its
   application to industrial scale processes. Adv. Biotech. Proc., 1983,
   1, 31-66.

8. Travis, J., Bowen, J., Tewksbury, D.J. and Pannell, R., Isolation of
   albumin from whole human plasma and fractionation of albumin-depleted
   plasma. Biochem. J., 1976, 157, 301-306.

9. Bradford, M., A rapid and sensitive method for the quantitation of
   microgram quantities of protein utilizing the principle of protein-dye
   binding. Anal. Biochem., 1976, 72, 248-254.

10. Laemmli, U.K., Cleavage of structural proteins during the assembly
    of the head of the bacteriophage T4. Nature, 1970, 227, 680-687.

MODELLING OF PROTEIN ADSORPTION IN LIQUID FLUIDIZED BEDS

NICHOLAS M. DRAEGER and HOWARD A. CHASE
Department of Chemical Engineering
University of Cambridge
Pembroke Street, Cambridge, CB2 3RA, UK

ABSTRACT

The adsorption, in a liquid fluidized bed, of Bovine Serum Albumin, (BSA), onto an ion exchange adsorbent, consisting of quarternary (Q) amino groups, bound to an agarose matrix, (Sepharose®, Pharmacia LKB Biotechnology, Uppsala, Sweden), has been studied. The solid hydrodynamics have been examined with bed expansion tests. The equilibrium adsorption characteristics of the system have been assessed by the measurement of equilibrium adsorption isotherms which fitted well to the Langmuir equation. The rate of adsorption of BSA onto Q-Sepharose has been studied in a batch stirred tank, a fixed bed and a fluidized bed system. The data from the fluidized bed system has been used to assess the validity of a theoretical model adapted from one that predicts the performance of the adsorption phase in fixed bed systems.

INTRODUCTION

In most cases, adsorption and separation of proteins is carried out using fixed bed techniques. However, although the isolation of proteins from the process liquid is relatively efficient, the liquid has to be free of particulate matter before being passed through the bed, otherwise the particulates will be filtered out by the bed and the filter cake formed will cause a rise in the pressure drop across the bed. Removal of particulates necessitates the inclusion of an additional separation step upstream of the adsorption column, which is generally expensive and may result in the loss of a considerable amount of protein. A technique that would enable the protein to be adsorbed from the bulk liquid, without prior removal of particulates, is of major interest. The use of fluidized bed adsorbers may be the solution to this problem. The essential concept is that, as the bed expands under the force of the liquid passing through it, the spaces between the adsorbent particles become larger allowing particulates to pass through the bed unhindered.

The optimization and scale up of fluidized bed adsorbers requires that the liquid and particle hydrodynamics in the bed, and the equilibrium and mass transfer characteristics of the adsorbent are fully understood.

In fixed or fluidized bed adsorption, the variation of adsorbate concentration, at the column outlet, with time, is known as the breakthrough curve. Prediction of the shape of this curve is one of the keys to modelling the adsorption performance of the bed (Ruthven,1984). Parameters which affect the shape of the curve include liquid flowrate, bed height (or voidage), and characteristics of the adsorbent particle which affect the diffusivity of adsorbate in the pores of the particle. Prediction of the shape of the breakthrough curve also requires knowledge of the equilibrium behaviour of the system as characterized by an adsorption isotherm and characteristics of mass transfer from the bulk liquid phase to the adsorption sites.

Adsorption of a protein from the bulk solution onto the adsorbent involves a number of discrete mass transfer resistance steps, i.e. (a) the transfer of protein from the bulk liquid to the outer surface of the adsorbent particle - liquid film mass transfer, (b) diffusion of protein into the adsorbent pores - pore diffusion, followed on the adsorbent by, (c) the chemical interaction of protein with the binding site - surface reaction, (Ruthven, 1984, Arve and Liapis, 1987).

The aim of this work was to examine the adsorption of BSA by the ion exchange adsorbent Q-Sepharose Fast Flow, ®(Pharmacia, Uppsala, Sweden) in fixed and fluidized beds. Although ion-exchangers are unlikely to be used in the direct recovery of proteins from fermentation broths due to the high ionic strength of such liquids resulting from very low adsorbent capacities, this is a cheap and convenient adsorbent for studying protein adsorption.

EXPERIMENTAL ADSORPTION SYSTEM

Adsorbent Properties and Size Analysis
Samples of the adsorbent particles were analysed on a count and volume basis using an OPTOMAX particle size analyzer, to measure the particle spherical diameters. The adsorbent was dyed so that they could be viewed more easily on the analyzer. 546 particles were measured in a size range of 44 μm - 180 μm.

TABLE 1
Physical properties of Q-Sepharose Fast Flow

Mean spherical diameter	93.5 ± 31^1	(μm)
Particle porosity	0.96^2	
Particle density	1131^3	(kgm^{-3})

1 Standard deviation   2 Estimate used in the simulations, based on agarose content
3 Measured to an accuracy of ± 5%

Adsorbate and Buffers
Bovine Serum Albumin, (BSA) was used as the adsorbate throughout these studies. It has a molecular mass of 66000 daltons, (Peters and Reed, 1978) and is 99% pure. The adsorption properties of ion-exchangers are sensitive to pH. Therefore, all experiments were carried out in the presence of 0.01M Tris(hydroxymethyl)aminomethane, (Tris), buffer at pH 7.0. The viscosity of this buffer was measured as 0.93×10^{-3} Pas at 25°C in a capillary viscometer. The molecular diffusivity of BSA in free solution, (D_{AB}), was estimated to be 7.4×10^{-11} m^{2}s^{-1}, using the equation of Polson, (1950), which compares well with experimental values in the literature for proteins of similar molecular mass. The correlation given by Geankoplis, (1983), was used to estimate the fluid film mass transfer coefficient, (k_f), for BSA to the surface of the Q-Sepharose Fast Flow adsorbent particles in stirred tank batch experiments.

BED EXPANSION CHARACTERISTICS

Method
Liquid was pumped through adsorbent contained in a 2.2x45 cm glass column. Each bed was 4 cm high, when settled, giving a bulk volume of adsorbent of 15.2 cm^3. Tubing of internal diameter 1 mm was used for plumbing the system. The liquid flowrate was initially increased in steps, until the surface of the expanded bed just became unstable. Then the flowrate was

decreased in steps back to zero. 20 minutes were allowed after each alteration in flowrate for the bed to stabilize. Bed height was measured as a function of the liquid flowrate in the bed, measured at the column outlet. Q-Sepharose FF was fluidized with 0.01M Tris buffer, pH 7.0.

Results and Analysis

The expansion of the fluidized bed of Q-Sepharose FF with increasing velocity of flow through the bed is shown in fig. 1.

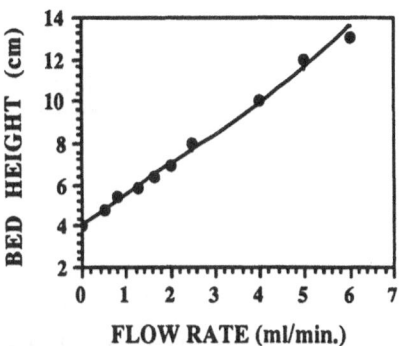

Figure 1. Bed expansion characteristics for Q-Sepharose Fast Flow adsorbent particles fluidized by:
- 0.01M Tris buffer pH 7.0 - (i) viscosity = 0.93×10^{-3} Pas   (ii) density = 1000 kgm^{-3}

Accurate experimental values of the minimum fluidization velocity, u_{mf}, could not be obtained as the associated pressure drop was found to be small. The u_{mf}, calculated for Q-Sepharose FF, (Ch.3, eq.21, Kunii and Levenspiel, 1969), was 7.3×10^{-3} mm/s (with a Re_{mf} of 7.34×10^{-4}, (Ch. 3, eq.20, Kunii and Levenspiel, 1969)) which is much lower than the superficial liquid velocities used in the bed expansion tests so that it can be assumed that the beds were fully fluidized. In order to test the validity of using the Richardson-Zaki equation ie:

$$u = u_t \, \varepsilon^n \tag{1}$$

to describe the expansion of the adsorbent bed as a function of superficial liquid velocity, the experimentally measured bed height data had to be transformed into values of bed voidages. A value for e_0 of 0.4 was assumed, based on voidage data detailed in the work of Kunii and Levenspiel, (1969), and Zenz and Othmer, (1960). A plot of ln(u) versus ln(ε) fitted well to a straight line and the Richardson-Zaki parameter, (n), was found from the slope of the best fit line. The terminal velocity was found from the y-intercept of the line.

The value normally used for the Richardson-Zaki parameter, in the laminar flow regime, is 4.8, (Chong et al, 1979, Richardson and Meikle, 1961, and Gibilaro et al, 1985), so that the experimentally determined value of 4.9 for Q-Sepharose FF, fluidized with 0.01M Tris buffer, is close to this value. A maximum particle Reynolds number, Re_p, of .019 was calculated for the liquid flowrates used in the tests, so that laminar liquid flow around each particle in the bed can be assumed, since laminar flow occurs when the particle Reynold's number is less than 0.2, (Kunii and Levenspiel, 1969). Therefore Stokes' law can be used for the calculation of the particle terminal velocities. Hence, using Stokes' law, a theoretical value for the terminal velocity for Q-Sepharose FF in Tris buffer was calculated to be 0.67 mm/s and the experimental value of 0.68 mm/s for the adsorbent fluidized with Tris buffer agrees well with this value. Although the adsorbent has a range of particle diameters, the mean spherical diameter of the particle was used in the Stokes' equation and appears to be appropriate.

EQUILIBRIUM ADSORPTION TESTS

Method
A range of concentrations of BSA were made up in 20ml of 0.01M Tris buffer pH 7.0 in separate flasks. The adsorbent, in a 1:1 buffer slurry (2 ml in each test), was added to each flask. Controls were also performed with systems without added adsorbent. The flasks were placed in a shaking water bath, at 25°C, for 50 to 70 hours. A 1.5 ml sample was removed from each flask, and centrifuged to pellet any suspended adsorbent. The optical adsorbance of each supernatant was measured in a spectrophotometer at 280 nm and used to obtain the protein concentration in the liquid phase.

Results and Analysis
The equilibrium isotherm, for the adsorption of BSA onto Q-Sepharose Fast Flow, fitted excellently to a Langmuir type expression, (Langmuir, 1916), and a typical isotherm is shown in fig. 2. Approximately 2-3 days were required before adsorption appeared to reach an equilibrium value. The maximum adsorption capacity (q_m) of 80 mg/ml, and the dissociation constant, (K_d), of 8.3×10^{-2} mg/ml for BSA were determined by a least squares fit to a linearized form of the Langmuir equation, (Horstmann and Chase, 1989). The maximum capacity is somewhat greater than that found by Chase and Skidmore, (1989) for the same adsorption systems although their experiments were carried out in 0.05M Tris buffer pH 7.2, i.e. at a higher ionic strength and a higher pH than were used in the present studies.

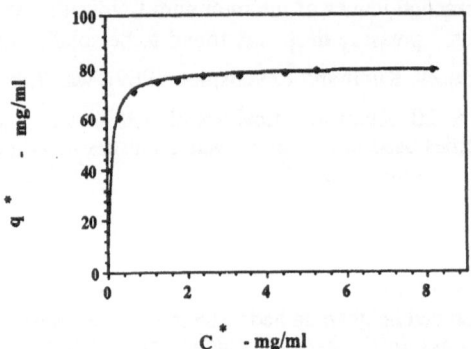

Figure 2. Equilibrium adsorption isotherm for the adsorption of BSA in 0.01M Tris buffer pH 7.0 to anion exchanger Q-Sepharose Fast Flow at 25°C.:
♦ experimental data — best fit theoretical simulation using Langmuir equation

The lower values for the dissociation constant also found in the present work suggests that buffer conditions alter the strength of the interaction of the protein with the adsorbent.

MASS TRANSFER STUDIED IN STIRRED TANK SYSTEMS

Method
The rates of protein uptake were measured in a batch stirred tank with a volume of 100 ml. The

adsorbate solution was continuously circulated through the system using the same apparatus as used by Horstmann and Chase, (1989). 40ml of a 0.01M Tris buffer solution, (pH=7.0), were agitated in a beaker in a shaking water bath at 25°C, and pumped through a flow spectrophotometer, and the optical adsorbance at 280nm was adjusted to zero. A concentrated solution of adsorbate was injected into the beaker and the rise in adsorbance was recorded on a chart recorder whilst the solution was pumped continuously through the spectrophotometer.

When the adsorbance had risen to a steady maximum value, 1 ml of a 1:1 $^v/_v$ slurry of adsorbent in Tris buffer was added to the beaker, and the subsequent fall in adsorbance was recorded on the chart recorder. The fall in optical adsorbance was monitored until it had leveled off. The beaker was shaken throughout the process to keep the adsorbent in suspension.

Results and Analysis
A knowledge of the rate of protein adsorption is important in modelling adsorption processes, as equilibrium is unlikely to be achieved in most fixed or fluidized bed operations, and these kinetic tests indicate how quickly the equilibrium is approached. The data from these batch uptake experiments were plotted as $\frac{C}{C_0}$ versus time. A mathematical model, discussed elsewhere, (Horstmann and Chase, 1989, and Skidmore, Horstmann and Chase, 1989), was used to fit a curve to these data assuming that film mass transfer and intraparticle diffusion are the major resistances to mass transfer. The model is based on the following assumptions:

(1) The adsorbent is porous, into which the solute must diffuse, in a manner described by an effective diffusivity, D_p. D_p is assumed to be independent of concentration .

(2) Mass transfer to the surface of the adsorbent is governed by a film model characterized by a mass transfer coefficient, k_f.

(3) Surface reaction between the adsorbate and the adsorbent site is described by a reversible second order reaction. Adsorption is isothermal, and its equilibrium behaviour can be represented by the Langmuir equation.

(4) The adsorbent particles are spherical, with uniform size and density. The functional groups of the ion-exchanger are distributed evenly throughout the interior of the particle.

(5) Axial dispersion, D_x, is negligible in packed bed simulations.

Such an approach allows values of the effective diffusivity of the protein within the porous adsorbent to be determined.

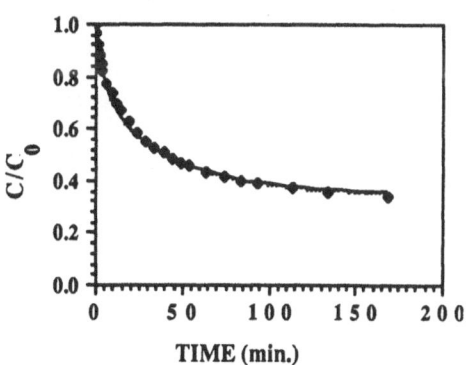

Figure 3: Analysis of batch stirred tank adsorption of BSA to Q-Sepharose FF in 0.01M Tris buffer pH 7.0.
$C_0 = 2.5$ mg/ml $\quad\quad D_{AB} = 7.4\times10^{-11}$ m^{2}s^{-1} $\quad\quad k_f = 7.65\times10^{-6}$ m^{2}s^{-1} $\quad$ ♦ experimental data
— best fit theoretical simulation assuming film diffusion and pore diffusion govern mass transfer.

As values of the film mass transfer coefficient can be estimated from correlations,

(Geankoplis, 1983), the model is essentially a one parameter model with only the effective pore diffusivity being unknown. Hence the value for D_e of 1.0×10^{-11} m^2s^{-1} is that which achieved the best fit curve to the experimental data. A typical plot is shown in fig. 3 for the adsorption of BSA onto Q-Sepharose FF. As in previous work with similar adsorption systems, (Horstmann and Chase, 1989), the surface reaction rate was assumed to be fast compared to the other two rates. It was possible to obtain an excellent fit to the experimental data using values of the maximum capacity, (80.0 mg/ml), and K_d, (0.02 mg/ml), obtained in the isotherm tests. The effective diffusivity for BSA in Q-Sepharose Fast Flow was found to be about 7 times lower than the diffusivity in free solution, ($D_{AB} = 7.4 \times 10^{-11}$ m^2s^{-1}). This reduction would be expected since the diffusion of BSA within the particle is restricted by the porous matrix, and is similar to the 6-7 fold reduction in diffusivity found for the diffusivity of BSA in QMA Spherocil, (van der Wiel, 1989).

FRONTAL ANALYSES IN FIXED AND FLUIDIZED BED SYSTEMS

Method

The same experimental protocol was used throughout these tests as was used in the previous work of Draeger and Chase, (1990). During the fluidized bed runs, the bed was expanded to a height of 12 cm, and it was observed that the height of the expanded bed dropped slightly, as a result of protein becoming bound to the adsorbent and increasing the effective density of the particles. Therefore, the flowrate had to be increased gradually, from 5.5 to 6.5 ml/min., to maintain a constant extent of bed expansion.

Results and Analysis

The experimental data are plotted in the form of $\frac{C}{C_0}$ versus volume of adsorbate solution applied to the bed. This was considered to be more appropriate than the use of time as the ordinate, since the flowrate was altered slightly during the fluidized bed runs as detailed above. The washing and elution profiles for each run have not been included in the plots since the primary aim of these tests was to investigate the shape of the protein breakthrough curve. These experiments were all carried out with the same amount of adsorbent at the same liquid flowrate. A qualitative analysis of the curves for both the fluidized and fixed bed adsorption of protein onto the adsorbent indicates that an ideal step breakthrough of protein is not being achieved, although the curves are relatively close to such a step change. The shapes of the two curves are very similar, fig. 4, indicating that the characteristics of the fixed and fluidized bed adsorption systems are almost the same, even though the height of the bed, during each fluidized bed run, was 3 times the height of the bed during each fixed bed run, and the voidage of the fluidized bed system was approximately twice that of the fixed bed system. The apparent maximum adsorption capacity calculated from both runs was 80 mg/ml which is consistent with the values from the batch isotherm experiments.

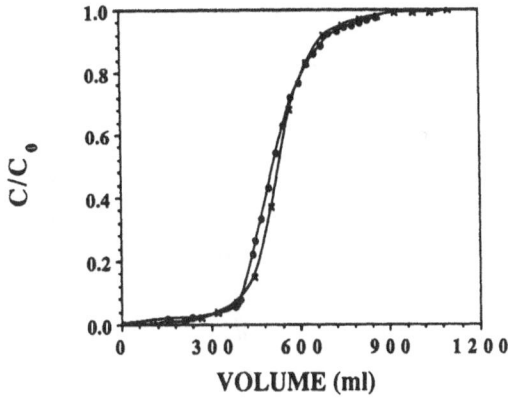

Figure 4: Fixed, and fluidized bed adsorption of BSA to Q-Sepharose Fast Flow in 0.01M Tris buffer pH 7.0
• Fixed bed   × Fluidized bed

A theoretical computer model for predicting adsorption performance in a fixed bed adsorption system, detailed in the work of Horstmann and Chase, (1989), was used to fit theoretical curves to the data from both the fluidized and fixed bed runs. The model was adapted for predicting protein adsorption in fluidized bed systems by using the greater voidage and increased height of the fluidized bed. In every other respect, the model was the same, when used to predict protein breakthrough in the fixed and fluidized bed systems. The data used in each fit are detailed in table 2. The film mass transfer coefficient, k_f, was estimated to be 5.7×10^{-6} ms^{-1} from the correlation of Foo and Rice, (1975), for packed bed systems:

$$Sh = 2 + 1.45 \, Re_p^{1/2} \, Sc^{1/3} \tag{2}$$

TABLE 2
Data used in simulations of protein breakthrough curves

Maximum adsorbent capacity [1]	80.0	(mgml^{-1})
Dissociation constant	0.083	(mgml^{-1})
Film mass transfer coefficient	5.7×10^{-6}	(ms^{-1})
Initial protein concentration	2.5	(mgml^{-1})
Superficial liquid velocity	2.47×10^{-4}	(ms^{-1})
Bed height [2]	0.12/0.04	(m)
Bed porosity	0.8/0.4	

1 Determined by least squares fit to linearized form of Langmuir equation   2 Fluidized / fixed heights

The theoretical fit to the fluidized bed breakthrough curve is very close except in the region of initial breakthrough of protein, ($0 < C/C_0 < 0.05$), and as protein breakthrough approaches equilibrium, ($0.95 < C/C_0 < 1.0$). Breakthrough occurs earlier than predicted theoretically, which may be due to either a small degree of liquid dispersion in the bed, or film mass transfer being slower than predicted by equation 2, or it may be a combination of both factors. Correlations for k_f in fluidized beds, (Udadhyay, 1975, Gamson, 1951, and Chu et al,

1953), are unsuitable for use in this work as they are all based on systems involving much larger particles, higher fluid flowrates and larger bed volumes. The correlation of Gamson gives a value for k_f which is a couple of orders of magnitude larger than that given by the correlation of Foo and Rice.

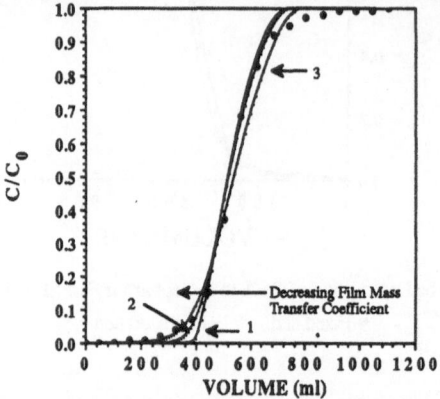

Figure 5: Fluidized bed adsorption of BSA to Q-Sepharose Fast Flow in 0.01M Tris buffer pH 7.0
• experimental data   — theoretical simulations using data in table 3
Simulations with (1) $k_f = 5 \times 10^{-5}$ ms^{-1} (2) $k_f = 5.7 \times 10^{-6}$ ms^{-1} (3) $k_f = 2 \times 10^{-6}$ ms^{-1}

The effect of k_f on the fit of the theoretical curve to the experimental data from the fluidized bed runs, was analysed by varying k_f between 5×10^{-5} ms^{-1} and 2.0×10^{-6} ms^{-1}. It is evident from fig. 5 that, if k_f is decreased from 5.7×10^{-6} ms^{-1} to 2.0×10^{-6} ms^{-1}, the theoretical fit is much closer in the region of initial breakthrough of protein. However the discrepancy must also be examined using a theoretical model incorporating a fluid dispersion term.

There also is a difference between theory and experiment as protein breakthrough approaches equilibrium. A possible explanation is that as protein molecules adsorb onto the adsorbent, the effective radius of the pores in the adsorbent decreases. Therefore, diffusion of protein molecules to the adsorbent sites will be slower, and adsorption will take longer to reach equilibrium. Such effects are not accounted for by the pore diffusion model in use in this work.

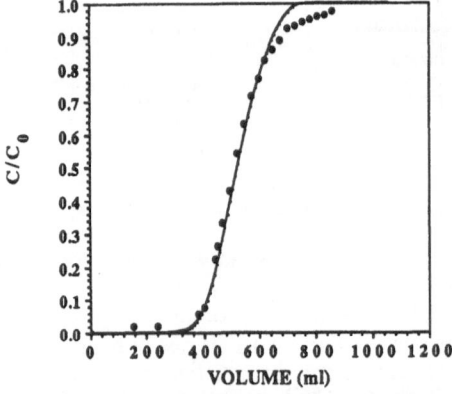

Figure 6: Fixed bed adsorption of BSA to Q-Sepharose Fast Flow in 0.01M Tris buffer pH 7.0
• experimental data   — theoretical simulations using data in table 3

The theoretical fit to experimental data from the fixed bed run is also very close - fig. 6. In the fixed bed simulations, there is not as large a difference between theory and experimental data in the region of initial protein breakthrough, as in the fluidized bed simulations, as expected, since there will be less liquid dispersion in a fixed bed system than in a fluidized bed system and k_f would be expected to be predicted accurately for the fixed system by equation 2.

CONCLUSIONS

Expansion of fluidized beds of Q-Sepharose Fast Flow appears to obey the theory of Richardson and Zaki relatively closely and the liquid terminal velocities can be accurately predicted from Stokes' law.

The batch stirred tank experiments indicate that the rate of adsorption of protein onto the adsorbent is governed by pore diffusion and film mass transfer. The rate of surface reaction between adsorbent and protein can be assumed to be fast, since there was no need to include a term describing this process in order to obtain a fit to the experimental data.

It would appear that the breakthrough of BSA, in both fixed and fluidized beds of Q-Sepharose FF, is very similar suggesting that the fluidized bed system is behaving in much the same way as the fixed bed process. There is very close agreement between the theoretical prediction of the performance of the performance of protein adsorption and data from frontal analyses. Although the model was designed for fixed bed systems, it predicts the performance of protein adsorption in fluidized bed systems very accurately, indicating again the similarity between the fluidized and fixed bed systems. It is not yet clear whether either film mass transfer or liquid dispersion cause discrepancies between theory and experiment. These parameters will have to examined further using a model which takes account of liquid dispersion in the bed. Further tests will have to be carried out using liquids which contain biological cells, to examine the effect of particles in the process liquid on protein adsorption in fluidized bed systems.

NOMENCLATURE

C	- bulk protein concentration	- mgml^{-1}
C^*	- equilibrium concentration of adsorbate in solution	- mgml^{-1}
C_0	- protein concentration in solution at t=0	- mgml^{-1}
d_p	- particle diameter	- m
D_{AB}	- diffusivity of protein in free solution	- m^{2}s^{-1}
D_p	- effective diffusivity	- m^{2}s^{-1}
k_f	- film mass transfer coefficient	- ms^{-1}
K_d	- dissociation constant	- mgml^{-1}
n	- Richardson - Zaki exponent	-
q	- amount of adsorbate adsorbed per unit volume of adsorbent	- mgml^{-1}
q^*	- equilibrium adsorbed-phase concentration	- mgml^{-1}
q_m	- max. amount of adsorbate adsorbed per unit vol. of adsorbent	- mgml^{-1}
Re_p	- particle Reynolds number at superficial liquid velocity - $\dfrac{\rho u d_p}{\mu}$	-
Re_{mf}	- particle Reynolds number at min fluidization velocity - $\dfrac{\rho u_{mf} d_p}{\mu}$	-

Sc	- Schmidt number - $\dfrac{\mu}{\rho D_{AB}}$	-
Sh	- Sherwood number - $\dfrac{k_f d_p}{D_{AB}}$	-
t	- time	- s or min.
u_t	- particle terminal unhindered settling velocity	- ms^{-1}
u_{mf}	- velocity at the point of minimum fluidization	- ms^{-1}
u	- superficial liquid velocity	- ms^{-1}
ε	- voidage in bed	-
μ	- liquid viscosity	- Pas
ρ	- liquid density	- kgm^{-3}

ACKNOWLEDGEMENTS

The authors would like to thank Miss B. J. Horstmann and Mrs S. Harrison for all their help and advice, and Pharmacia LKB Biotechnology for the generous provision of experimental materials and apparatus.

REFERENCES

1. Arve B.H. and Liapis A.I., J. AICHE, 33, (2), pg 179, (1987)
2. Chase H.A. and Skidmore G.L., Streat M. - Ion Exchange in Industry, pgs 520-532, Ellis Horwood, (1988)
3. Chong Y.S., Ratkowsky D.A. and Epstein N., Powder Tech., 23, pgs 55-66, (1979)
4. Chu J.C., Kalil J. and Wetteroth W.A., Chem. Eng. Prog., 49, (3), pgs 141-149, (1953)
5. Draeger N.M. and Chase H.A., I. Chem. E. Symp. Ser., No. 118, pgs 12.1-12.12, (1990)
6. Foo S.C. and Rice R.G., J. AICHE, 21, pg 1149, (1975)
7. Gamson B.W., Chem. Eng. Prog., 47, (1), pgs 19-28, (1951)
8. Geankoplis C.J., Transport Processes and Unit Operations, 2nd Ed., Allyn and Bacon, (1983)
9. Gibilaro L.G., Di Felice R., Waldram S.P. and Foscolo P.U., Chem. Eng. Sci., 40, pgs 1817-1823, (1985)
10. Horstmann B.J. and Chase H. A., Chem. Eng. Res. Des., 67, (5), (1989)
11. Kunii D. & Levenspiel O., Fluidization Engineering, pgs 72-79, Wiley, (1969)
12. Langmuir I., J., Amer. Chem. Soc., 38, pg 2221, (1916)
13. Peters T. and Reed R., in Albumin: Structure, Biosynthesis, Function: Colloq. B9, FEBS 11th Meeting, Copenhagen, eds T. Peters and I. Sjoholm, Pergamon Press, Oxford, (1978)
14. Polson A., J., Phys. Colloid. Chem., 54, pg 649, (1950)
15. Richardson J.F. and Meikle R.A., Trans. Instn Chem. Engrs, 39, pgs 348-356, (1961)
16. Ruthven D.M., Principles of Adsorption and Adsorption Processes, Wiley & Sons, (1984)
17. Skidmore G.L., Horstmann B.J. and Chase H.A., J. Chromatogr., 498, pgs 113-128, (1990)
18. Udadhyay S.N. and Tripathi G., J. Chem. Eng. Data, 20, pgs 20-26, (1975)
19. van der Wiel J.P., PhD Thesis - Continuous recovery of bioproducts by adsorption, Academisch Boeken Centrum, (1989)
20. Zenz F.A. and Othmer D.F., Fluidization and Fluid-Particle Systems, Reinhold Publishing Corp., New York, (1960)

AFFINITY SEPARATION OF PROTEINS IN AQUEOUS TWO-PHASE SYSTEMS

**T.Franco, B.A.Andrews, O.Cascone, C.Hodgson, A.T.Andrews[*]
and J.A.Asenjo**

Biochemical Engineering Laboratory
University of Reading
P.O. Box 226, Reading RG6 2AP, England

[*]Institute of Food Research, Shinfield, England

ABSTRACT

This paper investigates means of obtaining selective affinity for specific target proteins in aqueous two-phase systems. This can be achieved by exploiting specific properties of the protein such as biological affinity. By using specific salts (eg. NaCl) the charge and hydrophobic properties of the proteins can be exploited thus changing the partition between the two phases quite dramatically.

Finally our present work on chemical protein modification to study the effect of individual protein properties is described. This will allow the development of appropriate correlations that can be used for prediction of protein behaviour in aqueous two-phase systems.

INTRODUCTION

Aqueous two-phase systems are an extremely attractive procedure to separate and purify proteins on a large scale. Two crucial issues that have to be resolved for industrial applications are: the question of selectivity of protein partitioning and the question of maximum recycling of the phase forming materials (i.e. polymers and salts) [1, 2, 3]. The ideal system is one with an extremely high selectivity for the target protein: that is a very high partition coefficient for the target protein and an extremely low partition coefficient for all the contaminating proteins.

Assuming that the main factors that contribute to the overall partition coefficient (K) of a protein, between the lighter and heavier phases are independant from each other, then equation (1) can be considered as valid

$$\ln K = \ln K_{hphob} + \ln K_{el} + \ln K_{size} + \ln K_{conf} + \ln K_{lig} \qquad (1)$$

This equation includes a hydrophobic term, a term that includes electostatic effects mainly determined by the net charge of the protein and hence pH, the charge of the polymer and ion distribution. Protein size and conformation are also important but it appears more difficult to independently manipulate these parameters, particularly conformation. Finally the presence of ligands in one of the phases can have an important effect by exploiting biological affinity for increasing the partition coefficient of a particular protein.

Three ways of increasing selectivity will be discussed in this paper; exploitation of biological affinity and manipulation of charge and/or hydrophobicity.

MATERIALS AND METHODS

Polyethylene glycol (PEG) with MWs of 600 and 4000 were obtained from Fluka Chemicals Ltd., MWs of 1450 and 8000 from Sigma and MWs of 6000 and 10,000 from BDH. Dextran T500 was supplied by Pharmacia, Sweden. Epoxy-oxirane (1,4-butanediol diglycidyl ether) was supplied by Sigma as was glutathione, BSA, ethanolamine, trypsin and trypsin inhibitor. Thaumatin was kindly donated by Tate and Lyle, Reading, UK. All other chemicals were analytical grade.

Methods of PEG activation and ligand binding have been described previously [4, 5]. Preparation of phase components and aqueous two-phase systems has also been recently described [1].

Thaumatin was acetylated according to the method of Riordan and Valee [6]. The acetylation reaction consists of reacting acetic anhydride with the free amino groups of the protein to decrease its isoelectric point. The family of modified proteins obtained was separated by isoelectric focusing [7] in a preparative bed with Sephadex IEF and ampholytes with a pH range from 3 to 10.

Hydrophobic Interaction chromatography was carried out using an FPLC system from Pharmacia with a phenyl superose column HR5/5. Gradients of $(NH_4)_2SO_4$ from 1.5 M to 0 M and from 0.5 M to 0 M were used.

Protein concentrations were measured by the Bradford protein assay [8] using standard curves prepared with each protein.

RESULTS AND DISCUSSION

Biological Affinity

There are many procedures available for immobilization of ligands to insoluble polymers. Considerable interest in methods of activation and ligand binding to soluble polymers now exists for biotechnological and biomedical applications. Two recent studies from our laboratory [4],[5] describe methods of activation of PEG and ligand binding for use in affinity partitioning in aqueous two-phase systems. PEG was activated using three methods: epoxy-oxirane, epichlorohydrin and periodate. All have moderate or low toxicity and excellent stability.

Glutathione is a well known ligand for the protein flavour enhancer and sweetener of plant origin, thaumatin (M.W. 28,000). Thaumatin has recently been cloned in several different hosts, hence proper purification processes have to be developed and aqueous two-phase systems are ideal candidates. Table 1 shows the partitioning of thaumatin in a PEG/dextran system using non-modified PEG and glutathione bound to PEG as an affinity ligand. The PEG was activated using epoxy-oxirane.

The partition coefficient (K) of thaumatin is increased 20 fold in the system with glutathione bound to the PEG. In a PEG/phosphate system with glutathione-modified and non-modified PEG the increase in K was much smaller (3 fold). A PEG/dextran system could thus be used for thaumatin separation into the PEG phase and a PEG/phosphate system for elution of the protein from the PEG into the phosphate phase.

TABLE 1
Affinity Partitioning of Thaumatin using Glutathione as a Ligand
in PEG 8000/Dextran Systems (pH 7)

	PEG	Glutathione-PEG
K	0.27	4.6
[Thaumatin] in PEG, g/l	1.45	7.86
[PEG] %	11	11

In table 2 the partitioning of trypsin (M.W. 23,300) is shown in a
PEG/dextran system with non-modified PEG and with trypsin inhibitor bound
to PEG (activated with epoxy-oxirane) as an affinity ligand.

TABLE 2
Affinity Partitioning of Trypsin using Trypsin Inhibitor as a Ligand
in PEG 8000/Dextran Systems (pH 7)

	PEG	Trypsin Inhibitor-PEG
K	0.5	16
[Trypsin] in PEG, g/l	5	16
[Ligand] g/l	0	8
[PEG] %	10	10

The concentration of ligand bound to the PEG was high (8 g/l). The
partition coefficient increased 32 fold, from 0.5 in a non-modified PEG
system to 16 in the affinity system.

Other proteins have been bound to PEG (e.g. protein A and BSA) and
successfully used as affinity ligands in aqueous two-phase systems (for Ig
G and monoclonal anti-BSA antibodies respectively) [5]. The partition
coefficient of IgG was increased 12 fold using protein A as an affinity
ligand.

Charge and Hydrophobicity

Use of affinity ligands such as peptides and proteins that exploit
bioaffinity interactions should not be carried out in isolation from other
important interactions in aqueous two-phase systems.

As equation (1) shows hydrophobic and electrostatic interactions
which are usually very important in aqueous two-phase systems should also
be investigated. This is similar to the use of ion-exchange and hydrophobic
interaction chromatography in addition to affinity chromatography, which
are all important on a large scale.

The effect of adding small concentrations of NaCl to PEG/Dextran systems, which affects the distribution of ions and hence of charges in the system is well known [9]. Similarly high concentrations of NaCl will affect the hydrophobicity between the PEG and dextran rich phases. The effect of NaCl in PEG/salt systems has also been documented.

We have investigated the effect of NaCl in both PEG/dextran and PEG/phosphate systems on the partition of thaumatin. Thaumatin has a very high isoelectric point, pI, so it will be positively charged at virtually all pH values.

The partition behaviour of pure thaumatin in aqueous two-phase systems has been studied to investigate the effects of changes in phase components on K. Figure 1 shows the effect of PEG molecular weight on the partition coefficient of thaumatin in PEG/phosphate systems at pH 7 and pH 9. A dramatic decrease in the value of K is seen in the range of about 1,000 to 3 – 4,000 M.W. of PEG, from 12.4 (pH 7) and 18.2 (pH 9) in systems with PEG M.W. 600 to 0.76 (pH 7) and 0.85 (pH 9) with PEG M.W. 4,000 and to 0.30 (pH 7) and 0.25 (pH 9) with PEG M.W. 10,000.

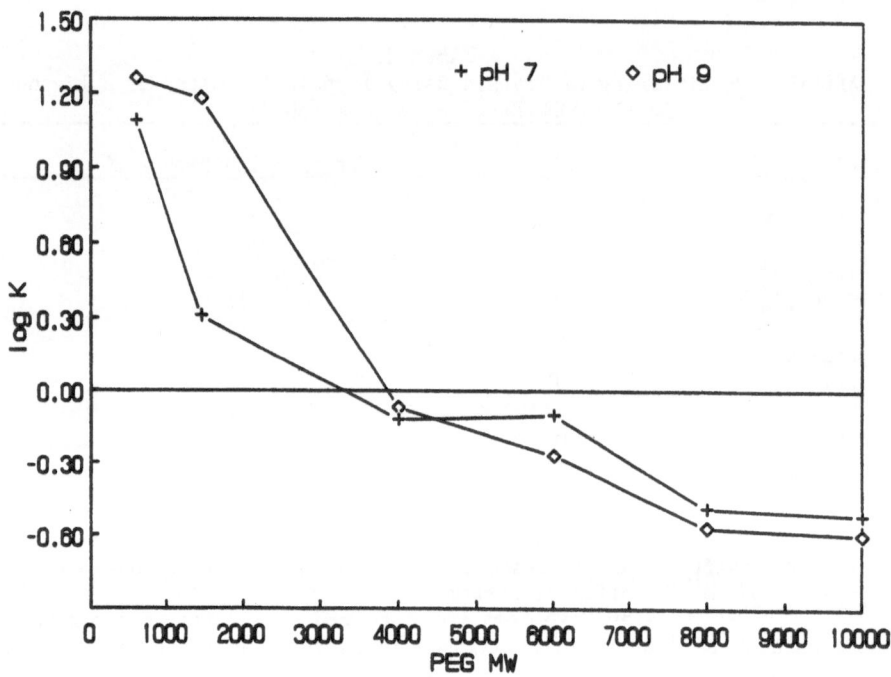

FIGURE 1. Effect of PEG Molecular Weight on the Partition Coefficient of Thaumatin in PEG/Phosphate Systems at pH 7 and pH 9.
(20% PEG 600, 10% PEG 1,450, 8% PEG 4,000, 6% PEG 8,000 and 5% PEG 10,000, 1 M phosphate)

Figure 2 shows the partition of BSA and thaumatin in PEG 6000/dextran two-phase systems as a function of concentration of NaCl. Thaumatin partitions preferentially to the top phase at all concentrations of NaCl used (0 to 1.5 M). The partition coefficient rises from 0.92 at 0 M NaCl to 16 at 1.5 M NaCl at pH 3, a 17 fold increase. At the other pH values used (5, 7 and 9) the K values are similar and range from around 1 with no NaCl to 5.2 with 1.5 M NaCl. BSA exhibits partition coefficients well below 1 up to an NaCl concentration of 1 M. At salt concentrations greater than 1 M BSA is salted out from the bottom phase and hence its partition coefficient is greatly increased. This is observed at both pH 3 and pH 7.

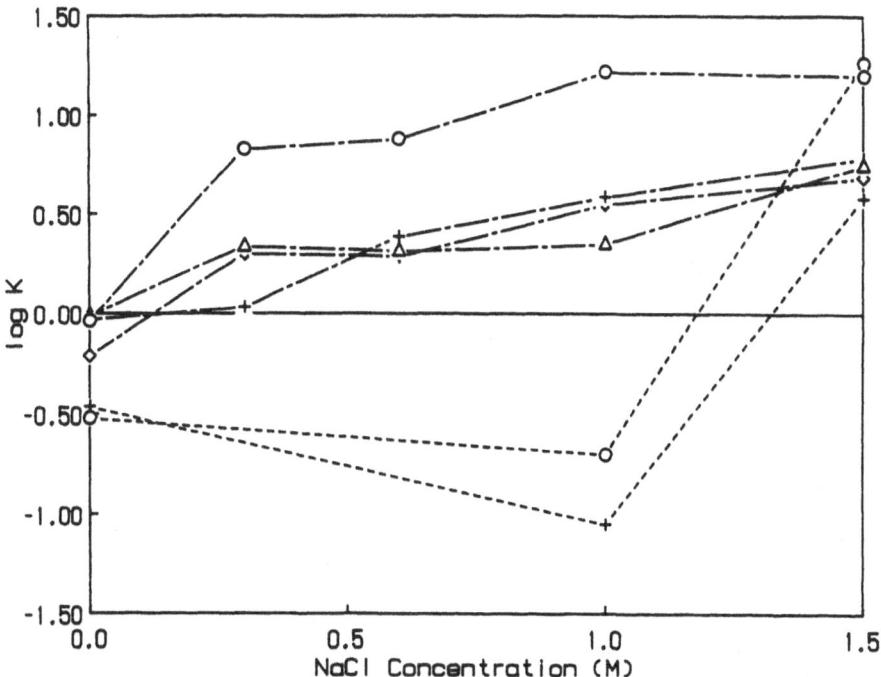

FIGURE 2. Effect of NaCl Concentration on the Partition Coefficient of Thaumatin (— - —) and BSA (- - - - -) in PEG 6000/Dextran Systems at pH 3 (o), pH 5 (▲), pH 7 (+) and pH 9 (◇) (6.5% PEG, 7% dextran)

In Figure 3 the partition of E. coli proteins, BSA and thaumatin in PEG 6000/phosphate systems is shown as a function of NaCl concentration at pH 7 and pH 9. Both E. coli proteins and BSA have partition coefficients well below 1. The partition coefficient of E. coli proteins at pH 9 remains constant with concentrations of NaCl up to 1.5 M, at pH 7 the partition coefficient rises from 0.16 with no NaCl to 0.26 with 1.5 M NaCl. BSA shows lower partition coefficients than E. coli proteins and K does not vary significantly over the range of NaCl concentrations from 0 to 2 M. The partition coefficients of thaumatin rise with increasing concentration of NaCl at both pH 7 and pH 9, from 0.79 with no NaCl to 31 with 1.5 M NaCl (pH 7), a 39 fold increase and at pH 9 from 0.53 with no NaCl to 33 with 2.0 M NaCl, a 62 fold increase.

Most experiments on partition of thaumatin were carried out at a relatively low protein concentration (1-2 g/l). We have used concentrations up to 40 g/l in PEG 6000/phosphate systems with 1.5 M NaCl and found that the partition coefficient is virtually constant.

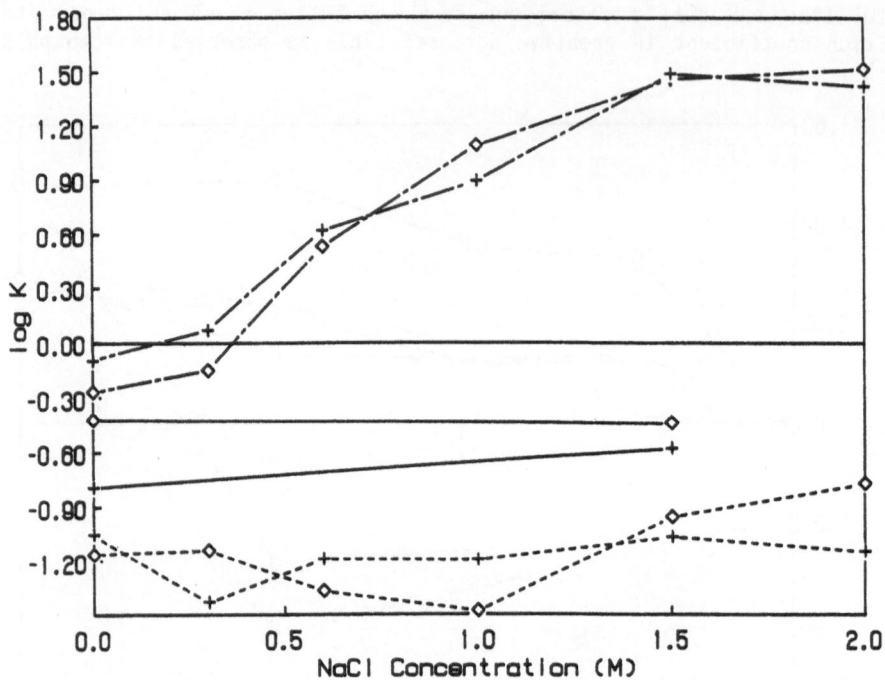

FIGURE 3. Effect of NaCl Concentration on the Partition Coefficient of
Thaumatin (— - —), BSA (- - - -)and E. coli (———)
Proteins in PEG 6000/Phosphate Systems at pH 7 (+)
and pH 9 (◇), (8% PEG, 0.75 M phosphate)

Chemical Modification of Proteins
In order to study the effect of protein charge on its partition behaviour
in an aqueous two-phase system in the absence of changes in other molecular
and physico-chemical properties we have chemically modified the charges on
the protein thaumatin by acetylation to lower its isoelectric point. A
similar approach is being used to study the effect of protein
hydrophobicity. We have synthesized a number of thaumatin molecules with
increased negative or decreased positive charges. Families of modified
thaumatin molecules with isoelectric point values (pIs) of 8.5, 5.7 and 4.3
were separated by preparative isoelectric focusing. The original thaumatin
molecule has an isoelectric point of 10.5. Figure 4 shows the behaviour of
the modified thaumatin molecules in a PEG/dextran system in the presence of
low concentrations of NaCl. In such a system Cl^- ions partition
preferentially to the top phase thus increasing the tendency of more
positively charged molecules to partition to this phase.

At pH 9 the original thaumatin molecules are positively charged and
thus show an increase in partition coefficient with a 0.1 M NaCl
concentration. The modified thaumatin with a pI of 5.7 shows a slight
decrease in the value of K with NaCl whereas the more negative thaumatin
(pI 4.3) shows a more dramatic decrease from a K of 6 with no NaCl to a K
of 2.4 with 0.1 M NaCl.

Figure 5 shows the behaviour of the modified thaumatin molecules in the presence of low concentrations of Na_2SO_4.

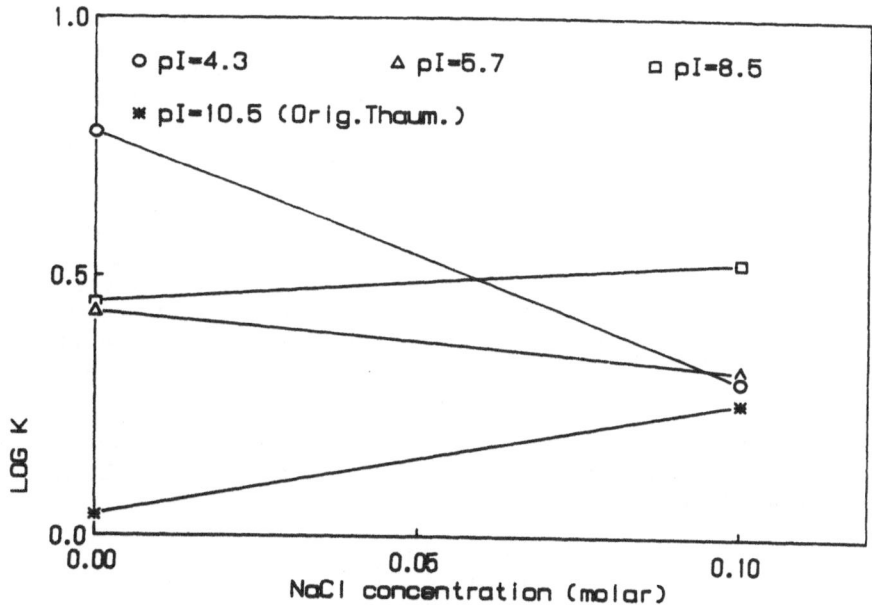

FIGURE 4. Effect of NaCl (0.1 M) on the Partition Coefficient of
Modified Thaumatins (with different pI) in a PEG 4000/Dextran
System at pH 9 (6.5% PEG, 7% dextran)

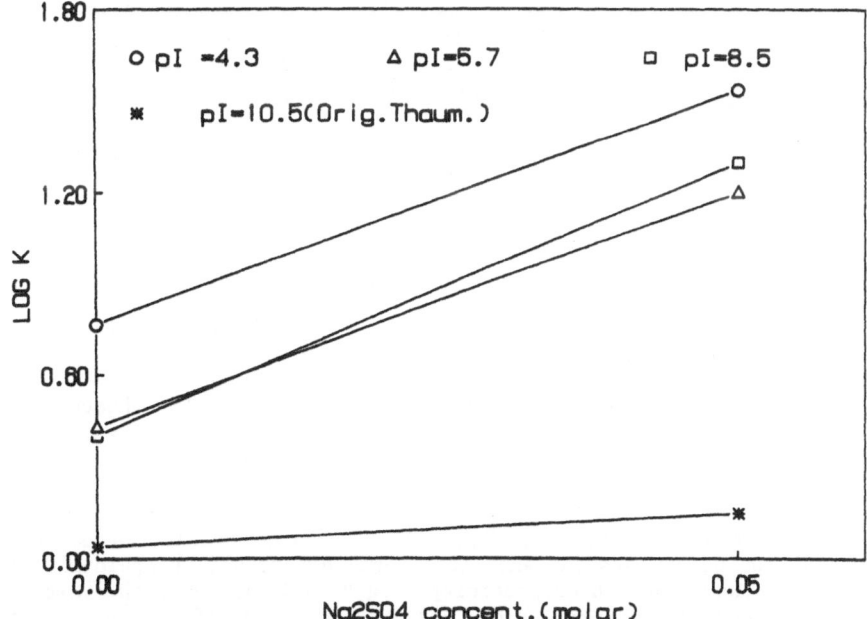

FIGURE 5. Effect of Na_2SO_4 (0.05 M) on the Partition Coefficient
of Modified Thaumatins (with different pI) in a PEG 4000/Dextran
System at pH 9 (6.5% PEG, 7% dextran)

As SO_4 ions are divalent the concentrations are half of those used in figure 4 (0.05 M). Here we can see almost the opposite behaviour from that shown in Fig. 4. As SO_4 ions partition preferentially to the bottom phase [9] this phase has an affinity for the positively charged molecules and the top phase for the more negative ones. We can thus see an increase in the value of K in the presence of Na_2SO_4. This was found for the thaumatin families with pIs of 8.5, 5.7 and 4.3 for which the K was increased from 2.8 to 20, from 2.7 to 13 and from 6 to 35 respectively.

The maintenance of hydrophobicity between chemically modified thaumatins was checked using hydrophobic interaction chromatography.

Models for Charge and Hydrophobicity

The prediction of partition coefficients of individual proteins is a crucial aim for the design of separation and purification processes based on aqueous two-phase systems. The most important properties of such a system that can be manipulated in the phases as shown in equation (1), without chemically modifying the polymers as is usually done for biological affinity, are hydrophobicity and charge. This requires the development of appropriate models and correlations to predict their behaviour. Chemical modification of an individual protein to alter a specific property allows the study of the effect of one property in the absence of other effects. The effect of charge on the partition of a protein can be described by equation (2) [10].

$$\ln K_{el} = \frac{Z_p}{(Z^+ + Z^-)} \ln \left(\frac{K^-}{K^+}\right) \tag{2}$$

where: Z_p = net charge of protein

Z^+, Z^- = net charge on cation and anion

K^+, K^- = partition coefficient of cation and anion

This equation takes into account the effect of electrostatic factors only (K_{el}). The equation predicts no effect on the partition coefficient if the ions of the added salt (e.g. NaCl, Na_2SO_4) partition equally between the top and bottom phases, $\ln K^-/K^+ = 0$, or if the protein is at its isoelectric point $Z_p = 0$.

Similarly the effect of hydrophobicity differences between the phases can be described by equation (3) [11].

$$\ln K_{hphob} = D \Delta W_2 \log \left(\frac{P}{P_0}\right) \tag{3}$$

where: ΔW_2 = concentration difference of one component between the top and bottom phases (e.g. PEG)

$\log P/P_0$ = hydrophobicity of the solute relative to the hydrophobic difference between the two phases

D = discrimination factor which represents the linear variation of a solute's (protein) hydrophobicity with the value of K at constant ΔW_2.

Intrinsic hydrophobicity ($\log P_0$) and D are characteristics of a given aqueous two-phase system and they are determined graphically [11]. Increasing the solutes hydrophobicity, $\log P$, will increase the value of $\ln$ K. For proteins an expression has to be found for the effect on hydrophobicity as a function of the changing phase compositions [11]. The hydrophobicity of a solute is inversely proportional to the log of its solubility ($\log P = 1/\log S$); the use of the empirical Cohn equation valid for salting out proteins has been proposed [11]

$$\log P = \frac{1}{B - k \, \Delta W_2} \tag{4}$$

where B and k are the salting-out constants.

Mammalian and Recombinant Proteins

Our present work is directed at establishing a rational basis for manipulating partition coefficients based on physico-chemical properties of the proteins and thus allowing the selection of the most appropriate two-phase systems.

This is particularly important for the development of large scale processes for the isolation of novel proteins which are being developed by the biotechnology industry today.

Three proteins on which we are carrying out work in aqueous two-phase systems at present to increase the phase affinity and selectivity and develop large scale purification procedures are

IgG
tPA
VLPs

Monoclonal antibodies (IgGs) are in very wide use today both for analytical and also now for therapeutic use. Tissue plasminogen activator, tPA, a recombinant thrombolytic agent, was the drug or pharmaceutical with the largest first year sales ever and VLP's, Virus Like Particles, cloned in yeast are presently being used for the development of an AIDS vaccine.

Aqueous two-phase systems show many advantages for the development of large scale separation processes for the separation and purification of novel therapeutic, diagnostic and reagent proteins including bulk enzymes as shown in Table 3.

TABLE 3.
Advantages of Aqueous Two-Phase Systems

- Easy to Scale-up
- Continuous Processing
- Mild Processing Conditions (polymers stabilize proteins)
- Can Handle Cell Debris and Particulate Material
- Inexpensive
- High Yields
- Large Scale Equipment Available
  (speed of processing, short contact times)
- Polymers and salts can be recycled

In Conclusion, we have seen that affinity ligands, hydrophobicity and charge can be used to manipulate the partition coefficient of target proteins in aqueous two-phase systems quite dramatically.

Our present work is directed at establishing a rational basis for manipulating such partition coefficients based on physico-chemical properties of the proteins and thus choosing the most appropriate two-phase systems to exploit these properties. Our work on chemical protein modification to study the effect of individual protein properties has been described. This will allow the development of appropriate correlations that can be used for prediction of behaviour. Recovery from the PEG phase can be efficiently obtained in a second stage where conditions for elution and partition of the target protein into a non-polymer bottom phase are mantained.

REFERENCES

1. Andrews, B.A. and Asenjo, J.A., Aqueous Two-Phase Partitioning. In Protein Purification Methods, a Practical Approach, eds. E.L.V. Harris and S. Angal, IRL Press, Oxford, 1989, pp. 161 - 174

2. Kula, M.R., Extraction Processes. Proc. of 8th Int. Biotechnology Symposium, 1, eds. G. Durand, L. Bobichon and J. Florent, Societe Francaise de Microbiologie, 1988, pp. 612 - 622

3. Asenjo, J.A., Andrews, B.A., Cascone, O., Franco, T. and Hodgson, C., Affinity Separation of Proteins in Two-Phase Aqueous Systems. Paper presented at 6th Int. Conference on Partitioning in Aqueous Two-Phase Systems, Assmannshausen, Germany, 1989

4. Head, D., Andrews, B.A.and Asenjo, J.A., Biotech. Techniques, 1989, 3, 27 - 32

5. Andrews, B.A., Head, D., Dunthorne, P. and Asenjo, J.A., Biotech. Techniques, 1990, 4, 49 - 54

6. Riordan, J.F. and Valee, B.L., Acetylation. In Methods in Enzymology, Enzyme Structure, 25, part B. Eds. C.H. Hirs and S.N. Timasheff. Academic Press, New York, 1972

7. Andrews, A.T., Electrophoresis: Theory, Techniques and Biochemical and Chemical Applications, Oxford University Press, 2nd Edition, 1986

8. Bradford, M., Anal. Biochem., 1976, 72, 248 - 254

9. Johansson, G., Partition of Proteins. In Partitioning in Aqueous Two-Phase Systems. Ed. H. Walter and D.E. Brooks. Academic Press, 1985, pp. 161 - 225

10. Albertson, P.A., Johansson, G. and Tjerneld, F. Aqueous Two-Phase Separations. In Separation Processes in Biotechnology. Ed. J.A. Asenjo, Marcel Dekker, 1990, pp. 287 - 327

11. Eitman, M.A. and Gainer, J.L., A Model for the Prediction of Partition Coefficients in Aqueous Two-Phase Systems. Paper presented at 6th Int. Conference on Partitioning in Aqueous Two-Phase Systems, Assmannshausen, Germany, 1989

EFFECT OF pH ON POLYGALACTURONASE ADSORPTION IN CM-SEPHADEX GELS.

S. Harsa, D.L. Pyle, C.A. Zaror
Biotechnology Group
Department of Food Science and technology
University of Reading, Reading, UK

ABSTRACT

This paper reports the effect of pH on the adsorption
characteristics of fungal polygalacturonase in CM-Sephadex ion
exchangers, in dilute solutions and in batch operations. The
partition coefficient increases sharply at pH below 4.3,
reaching a peak at about pH 3.2. Equilibrium isotherms follow
the Langmuir law and are strongly affected by pH; the maximum
adsorption capacity of adsorbent increases with decreasing pH.
These findings reflect the effect of pH on the relative
degrees of ionisation of adsorbate and adsorbent, and
highlight the need to determine the optimum pH for every
specific enzyme system.

INTRODUCTION

Polygalacturonase production from K. marxianus fermentations
has attracted considerable research in recent years, since it
may offer a feasible route for whey utilisation. About 90%
of the proteins secreted by this organism under anaerobic
conditions are polygalacturonases (PG). Enzyme preparations
with higher pectinolytic specificity are in demand in the food
industry: currently, commercial polygalacturonase is produced
using Aspergillus niger fermentation; however, a wide range of
enzymes and other metabolites are also excreted which gives
poor specificity to the final product.

Preliminary assessments show that downstream processing is a
key to process economic feasibility. Adsorption chromatography
has been identified as an important stage in the purification
process, in particular, using low cost ion exchangers.

Polygalacturonases have been purified using adsorption chromatography, on matrices based on cross-linked pectates [1], Separon [2] and alginate beads [3]; however, enzymatic biodegradation and lack of operational stability may not favour the industrial use of these adsorbents. Laboratory scale experiments [4], [5] have shown that CM-Sephadex is a suitable low cost ion exchanger for polygalacturonase purification; unfortunately, there is little experimental information on the effect of key process variables (pH, ionic strength, temperature) on the phase equilibria and adsorption/desorption kinetics. This paper focusses on the effect of pH on the adsorption characteristics of polygalacturonase in CM-Sephadex ion exchangers, in dilute PG solutions concentrations (less than 15 mg/ml) and in batch operations.

Firstly, some theoretical considerations on adsorption chromatography are reviewed. The effect of pH on partition coefficients and on adsorption isotherms are presented and discussed, together with some preliminary findings on adsorption kinetics. Finally, the implications of these results for process design are outlined.

ADSORPTION OF ENZYMES BY ION EXCHANGE CHROMATOGRAPHY

Sorption processes have been widely used in the chemical industry for a long time [6]. Industrial applications of adsorption techniques for enzyme purification have attracted considerable attention in the last two decades since they provide a high selectivity/cost ratio. In particular, ion exchange chromatography and affinity chromatography are key steps in protein purification schemes in biotechnology [7], [8].

Adsorption involves the transfer of one or more solutes from a fluid phase to an adsorbent (solid) matrix. The nature of any attachment is dictated by the physical and chemical properties of solutes and adsorbent, and binding may be due to Van der Waals forces (eg. physical adsorption), electrostatic attraction (eg. ion exchange) or covalent bonds (eg. affinity chromatography).

In ion exchange chromatography, the adsorbent - usually a cross linked polymer, eg. dextran, containing ionised functional groups - and the (ionic) solute are bound by electrostatic forces.

In the case of enzymes, we note that they may interact with the adsorbent at more than one binding site; moreover, the adsorbent may have a heterogeneous distribution of binding sites, some of which may be inaccessible due to exclusion and steric hindrance.

Cross-linked dextran gels (eg. Sephadex) are particularly
suitable as ion exchanger matrices since they are hydrophilic
and show very low non-specific adsorption. In the case of CM-
Sephadex, carboxymethyl (CM-) groups are introduced and
attached to the glucose polymer by stable ether linkages.
This gives rise to a weakly acidic cation exchanger, ie. at a
suitable pH range (say, above pH 5) it will be negatively
charged and, therefore, interact electrostatically with
positively charged molecules. CM-Sephadex has been used for
polygalacturonase purification since it does not affect the
enzyme activity.

Typically, gels are highly porous, filled with solvent, and
enzyme molecules can bind both to sites on the external and
internal surfaces. Mass transfer is important since the
solute has to diffuse from the bulk to the gel binding sites.
At any time during the process, the enzyme may be in two
different states: as a free solute, either in the liquid
bulk (solvent) or within the gel porous structure (but not
adsorbed to its surface); or as an adsorbed molecule, ie.
interacting electrostatically with the adsorbent.

The interaction between adsorbate and adsorbent is usually
described on the basis of a simplified equilibrium reaction,
ie:

(free binding sites) + (free protein) <——> (adsorbed
protein)

with an associated equilibrium constant, k:

$$k = \frac{q}{m \; p} \qquad (1)$$

where q, p and m are the equilibrium concentrations of
adsorbed protein, free protein and free binding sites,
respectively. Alternatively, the relationship between these
concentrations is expressed as a partition coefficient [9], C:

$$C = \frac{q}{p} \qquad (2)$$

which indicates the fraction of adsorbed protein at
equilibrium. When comparing literature values of k and C it
is important to note the basis on which these concentrations
are expressed. Some confusion may arise from the fact that
the gel is composed of both the matrix and solvent. The
original state of commercial adsorbents is usually as a dry
powder; after hydration, part of the water finds its way into
the molecular structure forming the gel, and part will be
trapped, as free water, within the pores. It is reasonable
to assume that at equilibrium, the concentration of free
enzyme in the bulk is the same as the concentration of unbound

enzyme in the pores. However, in some cases the pore size may be smaller than the enzyme and, therefore, no enzymes will be present there. These effects are difficult to quantify and authors use either the total volume (ie. bulk solvent plus gel) or the bulk volume only, as the basis of the unbound enzyme concentration, p. On the other hand, the concentration of adsorbed protein, q, is taken on the basis of the gel bulk volume (or weight), either before or after hydration.

The conventional description of phase equilibrium in protein adsorption processes makes use of the Langmuir isotherm [10]:

$$q = \frac{q_M \, p}{K_p + p} \qquad\qquad (3)$$

where q_M is the binding capacity of the gel and K_p is an appropriate constant. If the term $K_p \gg p$ (eg. in the case of dilute solutions) a linear isotherm is obtained:

$$q = C \, p \qquad\qquad (4)$$

where

$$C = \frac{q_M}{K_p}$$

C is equal to the partition coefficient defined above.

The design and control of adsorption systems relies, amongst other factors, on these equilibrium relationships. For a given liquid-solid system, they are strongly dependent on pH, ionic strength and temperature.

The binding capacity of the gel and the ionisation state of the enzyme are directly affected by the operating pH, and new experimental data on these effects are reported below for the case of polygalacturonase/ CM-Sephadex system.

MATERIALS AND METHODS

Adsorption experiments were conducted in batch mode in stirred solutions (total volume about 30 ml) with a typical volumetric ratio of gel/bulk liquid of 0.1. All experiments took place at 25°C in shaking water bath over 24 to 48 hours. Samples (1 ml) were taken, centrifuged and analysed. Sodium acetate 0.1M buffers, pH in the range 6.5 to 3.2 were used in this study. The initial and equilibrium concentration of enzyme/total protein was determined in each run.

Preparation of ion exchange media:
The adsorbent was CM-Sephadex C50 (Pharmacia), supplied as a dehydrated dry powder (40-120 microns diameter).
CM-Sephadex (0.05 g per run) was repeatedly hydrated with

buffer until the pH of the spent solution was equal to the fresh buffer pH; the excess liquid was then removed by filtration. The bulk volume of the hydrated gel was around 30 ml/g dry powder. The hydrated gel is highly porous and can adsorb molecules with molecular weight up to 150000 Daltons.

Preparation of polygalacturonase samples:
Commercial PG (supplied by Sigma Chemicals, from Aspergillus niger, 3-9 Units/mg protein) were diluted in buffer solution (pH 3.2-6.5), to concentrations in the range 0-15 mg enzyme/ml. 30 ml of enzyme solution were used in each run.

Analysis:
PG activities were estimated using the viscosity method [11] and the reducing sugars method [12]. 1 Unit of activity was defined as the amount of enzyme required to produce one μmole of reducing groups in 1 minute. PG adsorbed in the gel was estimated by difference between the initial and final concentrations of free enzyme. Total protein was measured by the Lowry method [13]. PG samples were characterised by isoelectric focussing using a LKB flat bed gel electrophoresis kit, in the range pH 4-10. PG was analysed on polyacrylamide gels, in a Bio-Rad Mini Protean II dual slab cell, using the method described by Laemmli [14]. The isoelectric point of the enzyme preparation used in this study is in the range 5.2-5.4 (which is within the range reported for A.niger enzyme: 3.2-5.9 [15] and 5.2 [16]; however, this is somewhat lower than values reported in the literature for K.marxianus [4].

The binding capacity of CM-Sephadex as a function of pH was obtained by pH titration with HCl 0.1M. Similar pH titration curves were obtained for a typical polygalacturonase preparation in the pH range 2-9.

RESULTS AND DISCUSSION

The main results of this study are shown in Figures 1 to 3 and Table 1. Figure 1 shows the extent of ionisation in the enzyme and the CM-Sephadex gel at different pH levels. The effect of pH on the partition coefficient of the PG/CM-Sephadex adsorption system (initial PG concentration 2.4 mg/ml; ie. 4.5 U/ml enzyme) is presented in Figure 2. Adsorption isotherms in the pH range 3.5-5.0, and PG concentrations below 15 mg/ml (30 Units/ml), are shown in Figure 3. The isotherm parameters are reported in Table 1.

Ionisation of adsorbent and adsorbate:
As can be seen in Figure 1, the gel is fully (negatively) charged at pH above 6.5; below that level, its net charge decreases sharply, and at pH 3.4 only half of its potential binding capacity is active. On the other hand, at the isoelectric point (here 5.2-5.4), the net charge of the enzyme is zero. At lower pH, the molecule becomes (positively) charged: at pH 5 about 5% of its total ionisation capacity is

present, whereas at pH 4.0 this increases to 50%. The PG is
fully charged at very low pH. At pH above the isoelectric
point, the net charge of the enzyme is negative (ie. the same
as the adsorbent) and no electrostatic binding will occur.

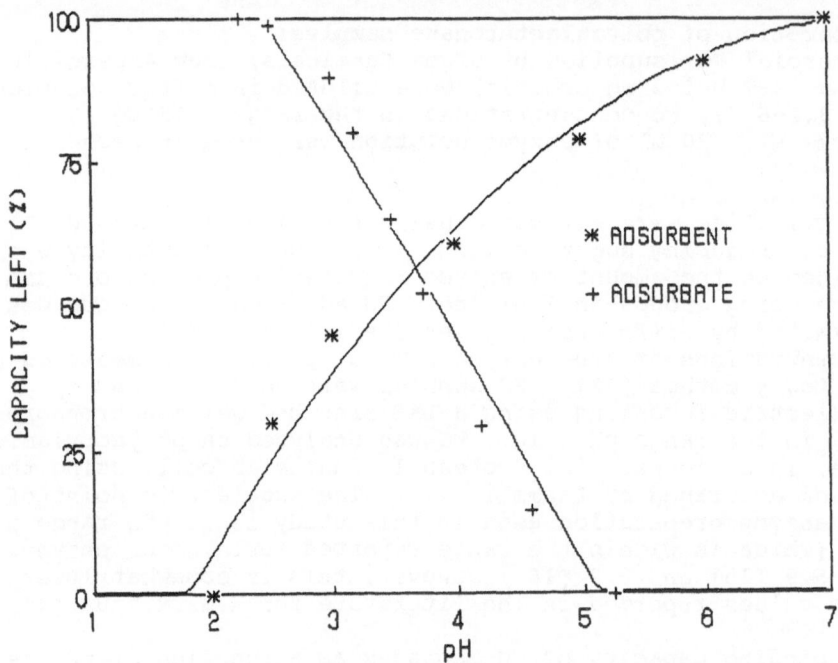

Figure 1: Ionisation of Adsorbate and Adsorbent

These results indicate that under the conditions used here,
there is no value of pH where both the enzyme and the gel are
fully charged (with opposite charges).

Partition Coefficients:
Figure 2 shows that the partition coefficient of PG adsorption
in CM-Sephadex gels is strongly affected by pH.
The minimum value is obtained at pH 7 where, as shown above,
both the enzyme molecules and the gel will have the same
(negative) charge. Under these conditions, a small partition
coefficient is expected, since adsorption should be negligible
(Van der Waals forces will be countered by electrostatic
repulsion). Since calculations do not allow for a distinction
between adsorbed molecules and those which are free within the
pores, the partition coefficient at pH 7.0 can be taken as a
measure of the fraction of (free) enzyme molecules which are
trapped within the porous structure.
As pH decreases below the isoelectric point, the partition
coefficient remains fairly constant; however, at pH below
4.3, the partition coefficient increases sharply and reaches

its peak at about pH 3.2. Although at this pH level less than
55% of the gel carboxymethyl groups are ionised, more than 80%
of enzymes are ionised and can be adsorbed.

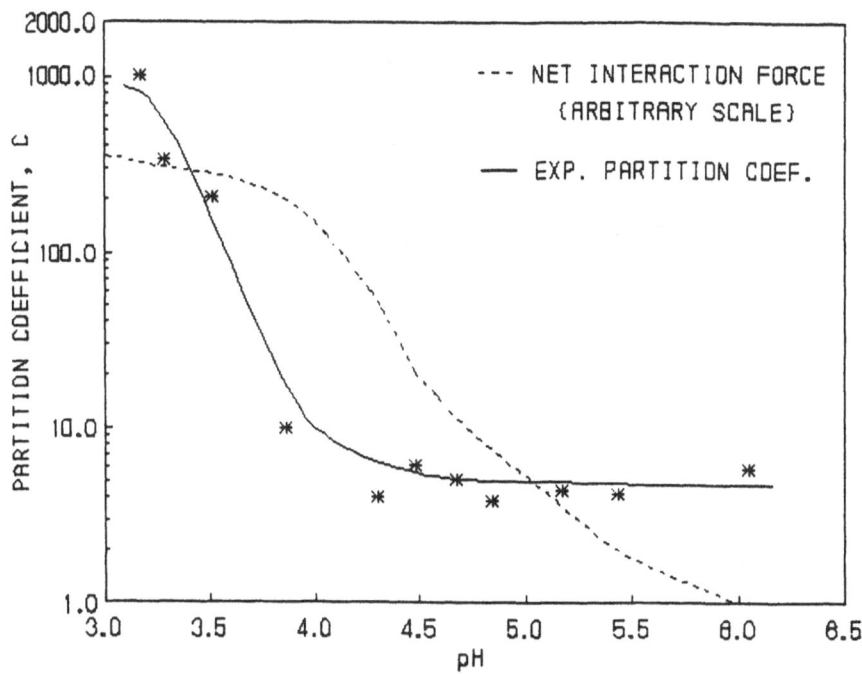

**Figure 2: Partition Coefficient of PG Adsorption in
CM-Sephadex Gels**

From Coulomb's law the interaction force between unlike
charges e_1, e_2 is proportional to e_1e_2. Thus, the net
interaction force between the protein and matrix should be
proportional to the product of the fractional degree of
ionisation of each component, given in Figure 1. This product
is shown on Figure 2 (arbitrary units). The region of maximum
partition coefficient corresponds, as predicted, to the region
of maximum ionic interaction.

Adsorption isotherms:
Figure 3 and Table 1 show equilibrium isotherms at pH 3.75,
4.0, and 4.75. It can be seen that all isotherms follow a
Langmuir relationship. A linear approximation can be obtained
for free enzyme concentrations below 10 (U/ml). The ratio
between equilibrium concentrations in the gel and the liquid
bulk increases with pH, particularly at pH 3.75, which is
consistent with the trend shown in Figure 2.

TABLE 1
ADSORPTION ISOTHERMS FOR POLYGALACTURONASES/CM-SEPHADEX
Free Polygalacturonase Concentration (< 30 U/ml)

pH	q_M	K_p
3.75	178.0	6.2
4.00	125.0	9.6
4.70	24.0	1.6

q_M (U adsorbed enzyme / ml hydrated gel)
K_p (U free enzyme / ml solvent)

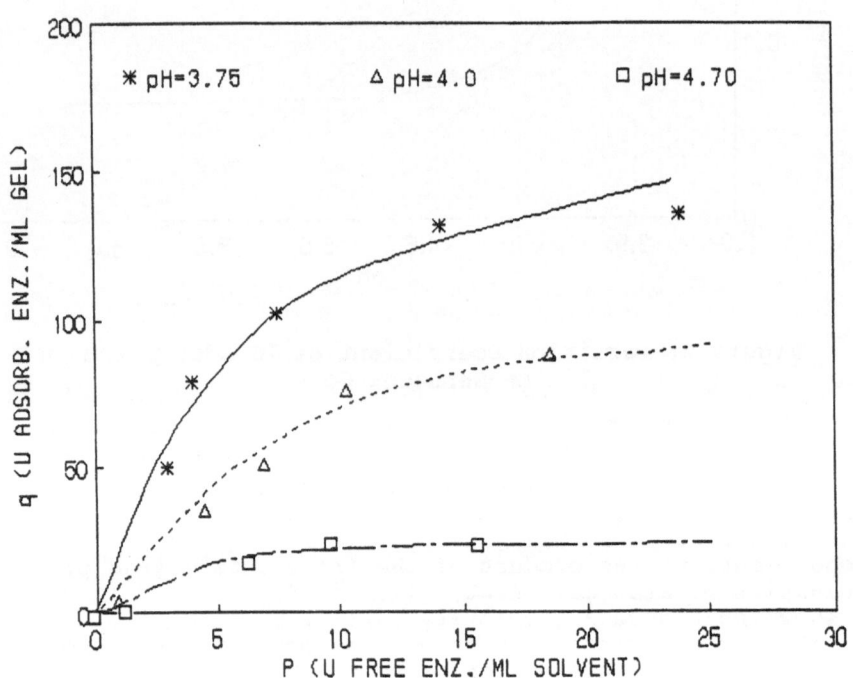

Figure 3: Equilibrium Isotherms

It is interesting to note that laboratory practices report
operating pH above 5 which, as shown in Figure 2, do not
correspond to the point of maximum partition coefficient

obtained in this study. This may reflect differences in the isoelectric points of their preparations, reflecting the different isoenzyme systems from different sources, or a preference for conservative pH levels to prevent enzyme denaturation. Our findings stress the need for routine determination of optimum pH levels for every specific system.

CONCLUSIONS

The results reported demonstrate the importance of pH for the performance of adsorption chromatography. In this case, the pH range for maximum adsorption is somewhat narrow and low. It implies that operating pH should be kept within less than 0.5 units around the operating value.
The pH in the immediate vicinity of the solid surface may be up to 1 pH unit lower that in the bulk, due to Donnan effects [7]. This effect may be minimised in the case of well stirred tank reactors. However, in the case of column chromatography, the volumetric ratio between liquid bulk and gel is very small and, therefore, the bulk pH may vary, particularly at low values of ionic strength. This affect the stability of the enzyme and produce loss of activity due to denaturation. In this study, maximum adsorption occurs at very low pH values; if the Donnan effect was significant, then there would be a significant loss of activity.

Usually, the Langmuir expression is one of the main sources of nonlinearities in the unsteady mass balance (differential) equation, which makes analytical solution very involved [17]. On the other hand, linear isotherms can be easily incorporated in models of ion exchange chromatography. Moreover, further simplifications may be possible since preliminary kinetic experiments show that the rate of PG adsorption/desorption is very fast and most of it can be approximated by a zero order process. Experiments on adsorption/desorption kinetics are currently being undertaken, and will be the subject of a future publication.

REFERENCES

1. Rexova-Benkova, L. and Tibensky, V., Biochim. Biophys. Acta, 1972, 268, 173-193.

2. Rexova-Benkova, L., Omelkova, J., Filka, K., and Kocourek, J., Carbohydr. Res.,1983, 122, 269-281.

3. Rozie, H., Somers, W., Bonte, A., Visser, J., Van't Riet, K., and Rombouts, F.M., Biotechnology and Applied Biochemistry, 1988, 10, 346-358.

4. Barnby, F.M., PhD Dissertation University of Reading,

Reading, UK., 1987.

5. Barnby, F.M., Morpeth, F.F. and Pyle, D.L., Enzyme Microb. Tech., (accepted for publication), 1990.

6. Perry, R. and Green, D., Perry's Chemical Engineer's Handbook, 6th ed., McGraw Hill Book Co., Singapore., 1984.

7. Scopes, R.K. , Protein Purification: Principles and Practice, 2nd. ed., Springer-Verlag New York Inc., 1987.

8. Liberti, L. and Helfferich, F.C., ed., Mass transfer and kinetics of Ion Exchange, Martinus Niijhoff Pub., 1983.

9. Brown, P.R. and Krstulovic, A.M., Ion-Exchange Chromatography. In Separation and Purification, ed. Perry and Weissberger, John Wiley and Sons. Vol.XII., 1978, pp. 197-255.

10. Langmuir L., J. Am. Chem. Soc., 1916, 38, 2221-2295.

11. Wimborne, M.F. and Rickard, A.D., Biotechnology and Bioengineering, 1978, 20, 231-242.

12. Gross, K.C., Hort. Sci., 1982, 17, 933-934.

13. Lowry, O.H., Rosebrough, N.J., Farr, A.L. and Randall, R.J., J. Biol. Chem., 1951, 193, 265-275.

14. Laemmli, U.K., Nature, 1970, 227, 680-685.

15. Kester, C.M. and Visser, J., Biotech. and App. Biochem., 1990, 12, 150-160.

16. Bailey, M.J. and Pessa, E., Enzyme Microb. Tech., 1990, 12, 266-271.

17. Cowan, G.H., Gosling, I.S., Laws, J.F., and Sweetenham, W.P., Journal of Chromatography, 1986, 263, 37-56.

NOMENCLATURE

C	Partition coefficient (U ml^{-1} gel / U ml^{-1} solvent)
k	Associated equilibrium constant
K_p	Dissociation constant for adsorbent-adsorbate complex, (U / ml solvent)
m	Concentration of free binding sites
p	Equilibrium concentrations of free enzyme, (U free /ml solvent)
q	Equilibrium concentration of adsorbed enzyme, (U adsorbed /ml gel)
q_M	Maximum adsorption capacity, (U adsorbed / ml gel)

PURIFICATION AND CHARACTERISATION OF SARCOSINE DEHYDROGENASE

R.J. HINTON*, T. ATKINSON*, C.P. PRICE** AND M.D. SCAWEN*

*Division of Biotechnology
Centre for Applied Microbiology and Research
Salisbury, Wiltshire, SP4 0JG

**Department of Clinical Biochemistry
London Hospital Medical College
London, E1 2AD

ABSTRACT

Sarcosine dehydrogenase was purified to homogeneity from *Pseudomonas fluorescens* induced in TSB medium by sarcosine. The enzyme in a cell-free extract was initially bound to the hydrophobic matrix phenyl-Sepharose in low ionic strength buffer and eluted with Triton X-100. FPLC anion-exchange and gel-filtration was performed using Mono-Q and Superose-12 matrices respectively. Final chromatofocusing with Mono-P gave homogeneous enzyme, as shown by a single protein band silver-stained on a PAGE gel.

The purified enzyme catalysed the dehydrogenation of the N-methyl amino-acids, methylhydantoin, N,N-dimethylglycine, N-methyl-DL-alanine, N-methyl-DL-valine and N-methyl-L-leucine. The K_m and V_{max} values for sarcosine were 1.9mmolL^{-1} and $6.60 \mu \text{mol min}^{-1}$ respectively. Maximum initial activity was obtained at pH 9.0 and 40°C. The isoelectric point was determined as pH 4.8. Native molecular weight was estimated at about 180,000 consisting of two sub-units of 95,000. Enzyme activity was inhibited by a variety of sulphydryl reagents including p-hydroxymercuribenzoate and phenylmercuric nitrate. Activity was not stimulated by the presence of metal ions nor was their presence required for activity.

INTRODUCTION

Sarcosine dehydrogenase (SDH) belongs to the group of demethylating enzymes and is the first link in the sarcosine-oxidoreductase system (E.C.1.5.99.1) catalysing the reaction:

$$\text{sarcosine} + H_2O + FAD^+ \longrightarrow \text{glycine} + HCHO + FADH + H^+$$

The enzyme has previously been partially purified and characterised from Rhesus monkey (1) and rat liver mitochondria (2). The enzyme is reported to be a flavoprotein (3) and contains a covalently bound non-acid extractable FAD (4,5). The most studied SDH has been the mitochondrial enzyme, and there have been few reports of bacterial sarcosine dehydrogenase. Frisell (6) isolated an unidentified Gram-negative rod utilising sarcosine as its sole source of carbon and nitrogen. The organism was later identified as a *Pseudomomas sp.* and the peptide-bound flavin from the enzyme was purified and characterised (7). Similarly Oka *et al* (8) isolated a *Pseudomonas putida* grown in a medium containing creatinine or betaine as the carbon and nitrogen source. They carried out a preliminary characterisation of enzyme purified by affinity chromatography.

This report describes the purification to homogeneity of SDH, its subsequent characterisation and a proposed use in a linked assay for the estimation of serum creatinine.

MATERIALS AND METHODS

Culture conditions
Pseudomonas fluorescens (RJ126) was grown aerobically for 14 hours at $25^{\circ}C$ in a medium containing 2% Tryptone Soya broth (Oxoid Ltd., UK) and 1.2%(w/v) sarcosine, pH7.6. Cells were harvested by centrifugation (4000 x g 30 minutes, $4^{\circ}C$) and the resulting paste stored frozen at $-20^{\circ}C$ until required.

Enzyme assay
SDH activity was determined from a kinetic assay by measuring the formation of a coloured formazan (E_{590nm} 12.78 x 10^3 mol. cm^{-1}) produced by the reduction of a thiazolyl blue tetrazolium (MTT), coupled to the intermediate electron acceptor, phenazine methosulphate (PMS). To a semi-micro cuvette was added, 0.8ml 500mM sarcosine in 50mM Tris-HCl, 0.1%(v/v) Triton X-100, pH8.0 and 100µl MTT/PMS (2.4 and 0.33mM in H_2O respectively). The reaction was initiated by the addition of 100µl enzyme and the increase in absorbance monitored at 570nm. One unit of activity is defined as the quantity of enzyme required to reduce one µmole of substrate per minute at $37^{\circ}C$.

Enzyme extraction
All extraction buffers contained 100µM PMSF. Frozen cell-paste was disrupted by rapid agitation with 150µm glass beads using a bead-beater (Biospec Products, USA) in 50mM Tris-HCl pH8.0 containing 0.3%(v/v) Triton X-100 at $4^{\circ}C$. Cell debris was removed by centrifugation (5000 x g, $4^{\circ}C$ for 20 minutes. Protein concentrations were determined using the BCA assay system (Pierce Chemical Co., USA).

Enzyme purification

Phenyl-Sepharose, Mono-Q HR 10/10, Superose-12 HR 16/50 and Mono-P HR 5/20 were obtained from Pharmacia/LKB. The prepacked columns were equilibrated and loaded using the FPLC (Pharmacia) according to the manufacturer's recommendations. All buffers contained 100μM PMSF and 1mM DTT.

Polyacrylamide gel electrophoresis

All PAGE was performed with the Phast System (Pharmacia). Precast gels of polyacrylamide gradients, 8-25% and 10-15% were used according to manufacturer's recommendations with both SDS and native buffer strips.

Isoelectric focusing was performed with homogeneous polyacrylamide gels containing carrier ampholytes for the generation of pH gradient between 4.0-6.5. All gels were electrophoresed at 15°C and stained by Coomassie Blue R250 automatically by the Phast system or manually using silver stain (Quicksilver, Amersham).

Estimation of molecular weight by gel filtration

A Superose-12 HR 10/30 FPLC column was equilibrated with 50mM Tris-HCl pH7.5/100mM NaCl. Molecular weight marker proteins (Sigma), dissolved in 100μl equilibration buffer containing 5% (v/v) glycerol were loaded and eluted with a linear flow rate of 14cm hr^{-1}. Protein in the eluate was monitored at 280nm and SDH by specific assay. The column void volume was determined using blue dextran. The logarithm of molecular weight vs Ve/Vo for each marker protein was plotted and the native molecular weight of SDH determined.

Inhibition studies

Compounds were prepared in 50mM Tris-HCl pH8.0 at varying concentrations, 20μl enzyme was added to 0.4ml and stored on ice for 30 minutes. Samples were then taken, diluted 1 in 10 in buffer and assayed for residual activity.

RESULTS AND DISCUSSION

Purification of Sarcosine Dehydrogenase

The cell-free lysate from 20g cell-paste was loaded onto a 45 x 75mm, 100ml column of phenyl-Sepharose equilibrated with 50mM Tris-HCl pH8.0 at a linear flow rate of 5.3cm hr^{-1}. The matrix was washed with 4 column volumes of loading buffer and eluted with 300ml 2%(v/v) Triton X-100/50mM-Tris-HCl pH8.0.

Fractions containing enzyme were loaded directly onto a preparative Mono-Q HR

10/10 column equilibrated with 50mM Tris-HCl containing 0.6%(v/v) Triton X-100 pH8.0. It was eluted with a 160ml 0-350mM NaCl gradient in 50mM Tris-HCl containing 0.6% Triton X-100 pH8.0 at a linear flow rate of 300cm hr^{-1} (Figure 1).

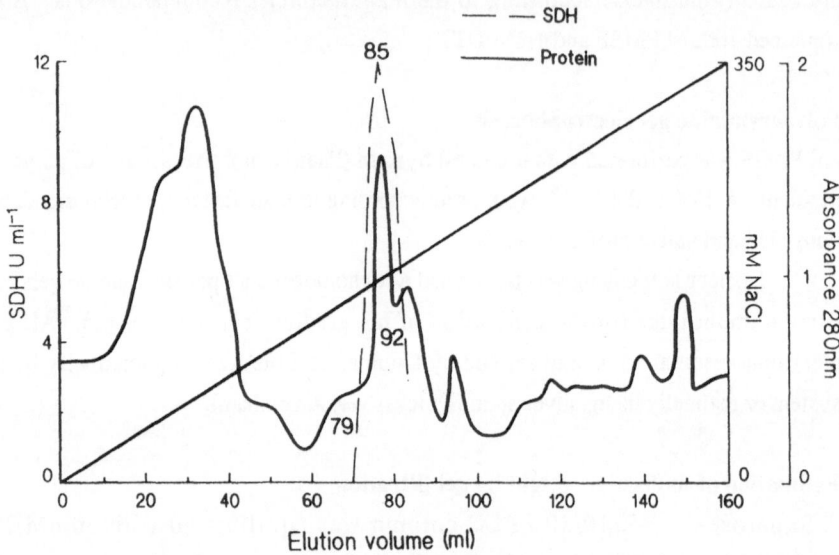

Figure 1. Anion-exchange on Mono-Q HR 10/10.

Fractions were concentrated by ultrafiltration to 1.0ml and loaded onto a preparative Superose-12 HR 16/50 gel filtration column (15,500 plates M^{-1}) equilibrated with 50mM Tris-HCl containing 100mM NaCl/0.6%(v/v) Triton pH8.0 at a linear flow rate of 15cm hr^{-1} (Figure 2).

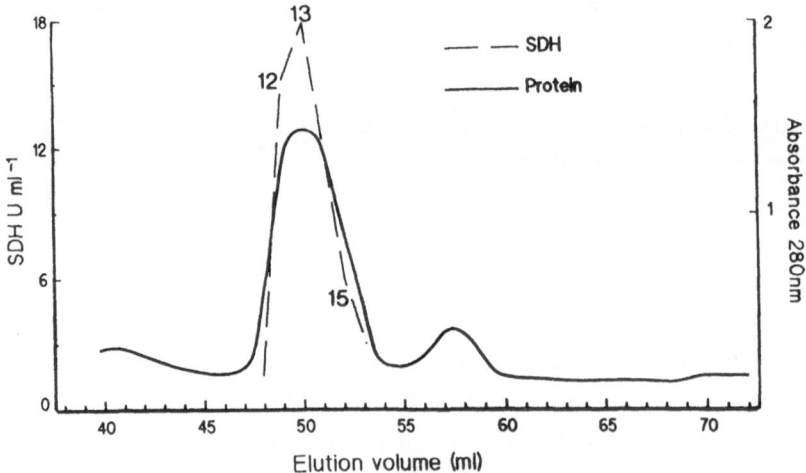

Figure 2. Gel filtration on Superose-12 HR 16/50.

Fractions 12-15 were pooled, and after buffer exchange (PD-10, G25 Pharmacia), were loaded onto a Mono-P HR 5/20 chromatofocusing column equilibrated with 25mM piperazine containing 9%(w/v) betaine titrated to pH6.3 with iminodiacetic acid. The column was eluted with Polybuffer 74 pH4.5 at a linear flow rate of 150cm hr^{-1} (Figure 3)

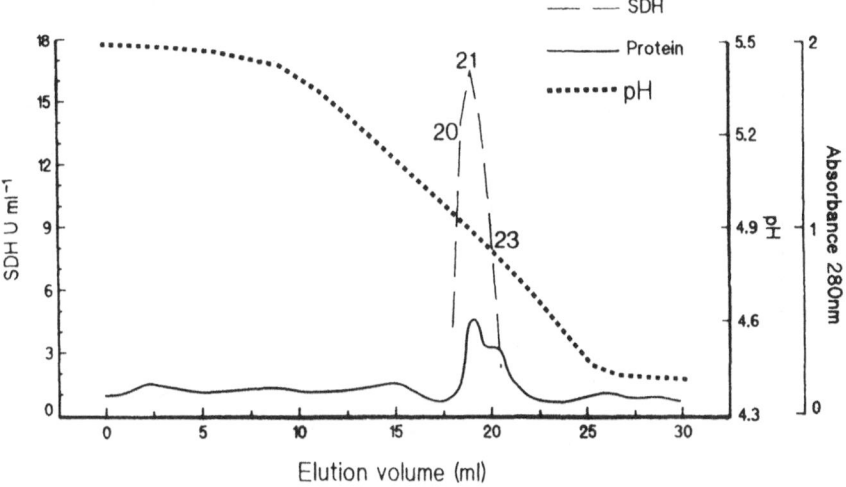

Figure 3. Chromatofocusing on Mono-P HR 5/20.

Peak enzyme activity eluted at pH4.9 with a specific activity of 7.1Umg^{-1} protein. It was homogeneous as indicated by a single protein band silver stained on a SDS-PAGE gel. Table 1 summarises the results obtained.

Table 1
Summary of purification results

	Enzyme (Units)	Protein (mg)	Specific activity (U mg^{-1})	Purification factor	Recovery %
Cell extract	135	1,160	0.12	0	100
Phenyl-Sepherose	104	148	0.70	6	77
Mono-Q	88	22	4.0	34	65
Superose-12	51	7.5	6.8	59	37
Mono-P	19.4	2.7	7.1	61	14

Enzyme Properties

The native molecular weight of SDH was determined by FPLC gel filtration to be 180,000 (Figure 4). This was confirmed by native PAGE.

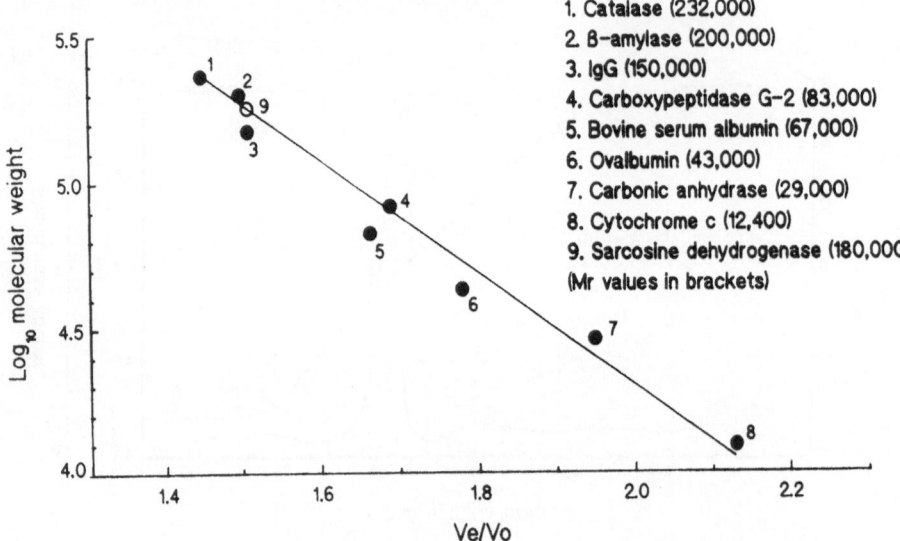

1. Catalase (232,000)
2. ß-amylase (200,000)
3. IgG (150,000)
4. Carboxypeptidase G-2 (83,000)
5. Bovine serum albumin (67,000)
6. Ovalbumin (43,000)
7. Carbonic anhydrase (29,000)
8. Cytochrome c (12,400)
9. Sarcosine dehydrogenase (180,000)
(Mr values in brackets)

Figure 4. Determination of native molecular weight by gel filtration

The subunit molecular weight was determined by SDS-PAGE as 95,000 (Figure 5) showing that the native enzyme exists as a dimer.

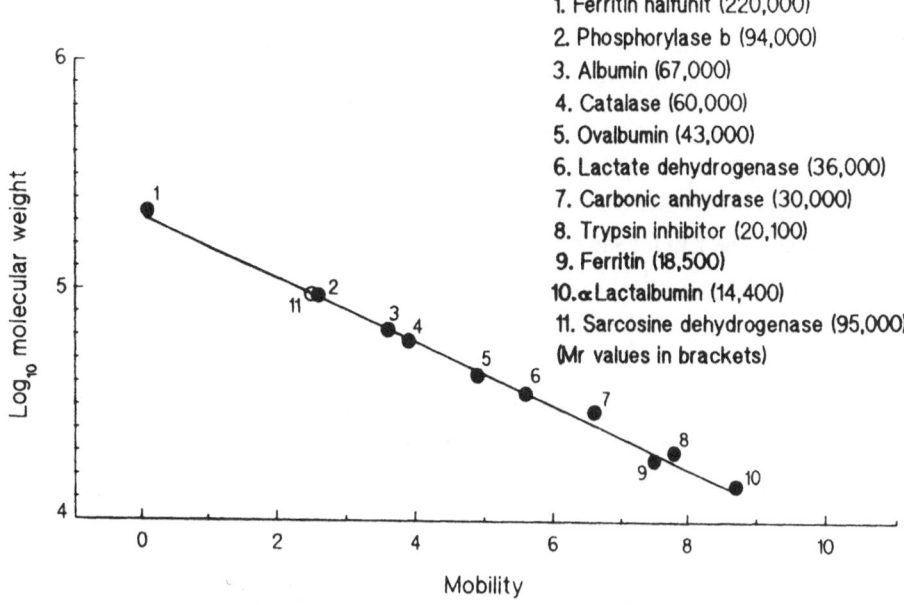

1. Ferritin halfunit (220,000)
2. Phosphorylase b (94,000)
3. Albumin (67,000)
4. Catalase (60,000)
5. Ovalbumin (43,000)
6. Lactate dehydrogenase (36,000)
7. Carbonic anhydrase (30,000)
8. Trypsin inhibitor (20,100)
9. Ferritin (18,500)
10. αLactalbumin (14,400)
11. Sarcosine dehydrogenase (95,000)
(Mr values in brackets)

Figure 5. Determination of subunit molecular weight.

The isoelectric point of SDH was estimated as 4.8 after isoelectric focusing, using a Phast IEF 4.0-6.5 gel (Figure 6).

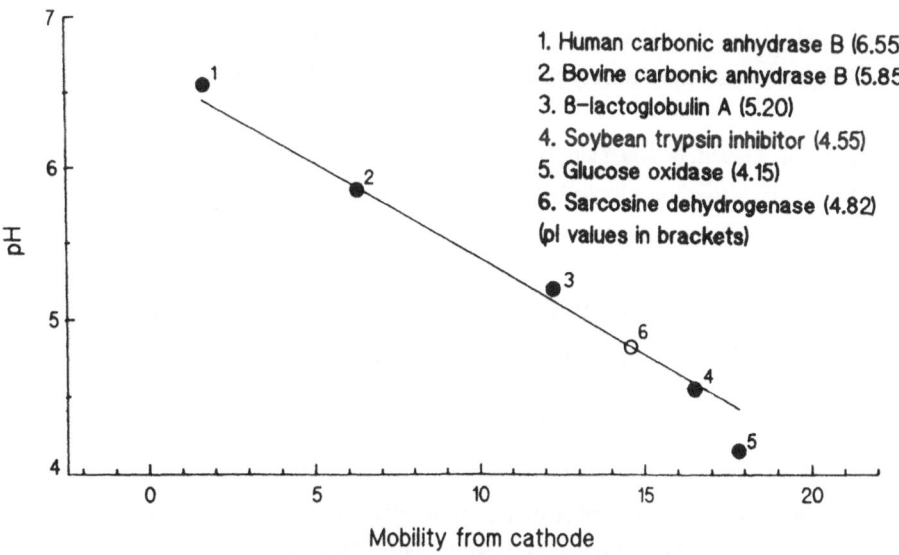

1. Human carbonic anhydrase B (6.55)
2. Bovine carbonic anhydrase B (5.85)
3. β-lactoglobulin A (5.20)
4. Soybean trypsin inhibitor (4.55)
5. Glucose oxidase (4.15)
6. Sarcosine dehydrogenase (4.82)
(pI values in brackets)

Figure 6. Determination of isoelectric point.

The enzyme showed maximal activity between pH8.5 and 9.0, above that pH, activity was lost rapidly. Similarly maximal initial activity occurred at $40^{\circ}C$ with rapid loss above this temperature and no activity at $55^{\circ}C$.

Sarcosine dehydrogenase was active against a variety of N-methyl amino acids including N-methyl-DL-alanine, N-methyl-DL valine, N-methyl-L-leucine, methylhydantoin and N,N-dimethyl glycine. The Km for sarcosine, using the Eisenthal and Cornish-Bowden (9) plot was estimated as 1.9mmol L^{-1} and a Vm of 6.60μmol min^{-1}.

Enzyme activity was inhibited by a variety of metal ions which react with sulphydryl groups, such as Ag^+, Zn^{2+}, Cd^{2+}, Co^{2+} and Pb^{2+}. Similarly phenylmercuric nitrate, p-hydroxymercuribenzoate and 5,5'dithiobis(2-nitrobenzoic acid) were also inhibitors. However, iodacetate and iodoacetamide up to 100mM did not inhibit activity.

The determination of serum creatinine can give a valuable indication of renal function (10), the sarcosine dehydrogenase purified and characterised in this report can be conveniently used in conjunction with creatinine amidohydrolase and creatine amidinohydrolase for the enzymatic determination of creatinine in body fluids over the clinical important ranges (11).

REFERENCES

1.   Hoskins, D.D. and Bjur, R.A., The oxidation of N-Methylglycines by Primate Liver Mitochondria. J. Biol. Chem., 1964, **239**, 1856-63.

2.   Frisell, W.R. and Mackenzie, C.G., Separation and Purification of Sarcosine Dehydrogenase and Dimethylglycine Dehydrogenase. J. Biol. Chem., 1962, **237**, 94-8.

3.   Sato, M., Ohishi, N. and Yagi, K., Identification of a Covalently Bound Flavoprotein in Rat Liver Mitochondria with Sarcosine Dehydrogenase. Biochem. Biophys. Res. Comm., 1979, **87**, 706-11.

4.   Patek, D.R. and Frisell, W.R., Purification and Characterisation of the Flavin Prosthetic Group of Sarcosine Dehydrogenase. Arch. Biochem. Biophys., 1972, **150**, 347-54.

5.   Hoskins, D.D. and Mackenzie, C.G., Solubilisation and Electron Transfer Flavoprotein Requirement of Mitochondrial Sarcosine Dehydrogenase and Dimethylglycine Dehydrogenase. J. Biol. Chem., 1961, **236**, 177-83.

6. Frisell, W.R., One-Carbon Metabolism in Micro-organisms. <u>Arch. Biochem. Biophys.</u>, 1971, **142**, 213-22.

7. Pinto, J.R. and Frisell, W.R., Characterisation of the Peptide-Bound Flavin of a Bacterial Sarcosine Dehydrogenase. <u>Arch. Biochem. Biophys.</u>, 1975, **169**, 483-91.

8. Oka, I., Yoshimoto, T., Rikitake, K., Ogushi, S. and Tsuru, D., Sarcosine Dehydrogenase from *Pseudomonas putida* Purification and Some Properties. <u>Agric. Biol. Chem.</u>, 1979, **43**, 1197-203.

9. Eisenthal, R. and Cornish-Bowden, A., <u>Biochem. J.</u>, 1974, **139**, 715-20.

10. Doolan, P.D., Alpen, E.L. and Theil, G.B., <u>Am. J. Med.</u>, 1962, **32**, 65-79.

11. Hinton, R.J., Enzymes in the determination of creatinine-sarcosine dehydrogenase. Ph.D. thesis, CNAA, 1989.

SEPARATION OF SOME SULPHATED OLIVANIC ACIDS USING IMMOBILISED CYCLODEXTRINS

MICHAEL J. HUDSON and SIMON HEMMINGS
University of Reading, Chemistry Department,
Whiteknights, P.O. Box 224,
Reading, Berks RG6 2AD

M.S. VERRALL
Extraction Section, Beecham Pharmaceuticals,
Betchworth, Surrey RH3 7AJ

ABSTRACT

Samples of three sulphated olivanic acids were prepared from a culture of *Streptomyces olivaceous*. The side chains at C3 were $-S-CH=CH.NHCOCH_3$ (MM13902); $-S(O)CH=CH.NHCOCH_3$ (MM4550) and $-S.CH_2CH_2NHCOCH_3$ (MM17880). A mixture of the three metabolites was partially resolved by HPLC using a reversed-phase C18 silica column. The separation was improved by incorporating in the mobile phase tetra-alkylammonium compounds as ion-pairing reagents. By these means MM13902 was separated from the others. Improved separations of MM4550 and MM17880 were achieved using β-cyclodextrins covalently bound onto silica columns. The more flexible side chain of MM17880 was more able than that of MM4550 to fit into the truncated β-cyclodextrin cone.

INTRODUCTION

The sulphated olivanic acids studied are shown in figure 1. These antibiotics are classified as carbapenams and are characterised by the 7-oxo-1-azabicyclo [3,2,0] hept-2-ene nucleus. Their relationship to penicillins (having a penam nucleus) is indicated in the trivial name carbapenam, the heterocyclic sulphur atom having been replaced by carbon. The three microbial metabolites depicted in figure 1 were prepared from cultures of *Streptomyces olivaceous*. The separation of two of the metabolites, MM4550 and MM17880 is usually difficult. A fairly effective analytical HPLC system was available [1] but resolution was such that preparative scale operation could be performed only at low column loadings. Moreover, the analytical system operated at pH 4.7 where the metabolites of interest were somewhat unstable.

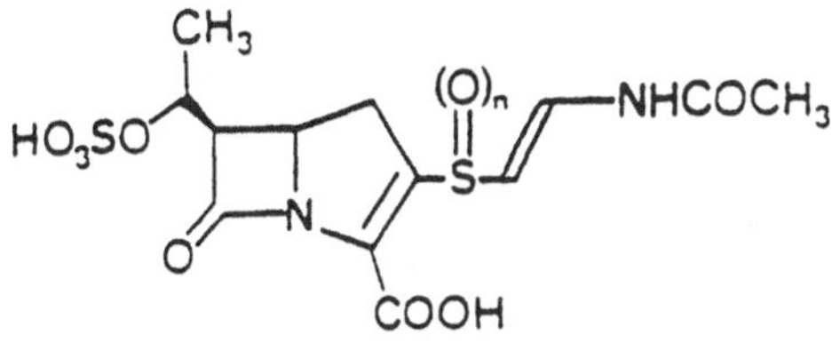

(n - 0: MM 13902)
(n = 1: MM 4550)

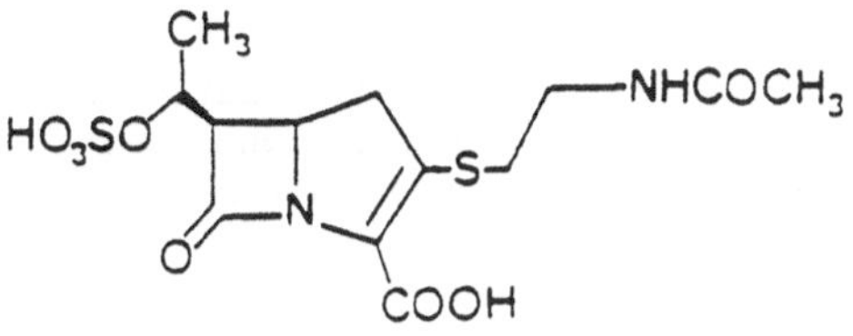

MM 17880

Figure 1. The three sulphated olivanic acids.

EXPERIMENTAL

Standard aqueous solutions of the individual olivanic acids or mixtures of the three were prepared and pH adjusted with 0.1 M NaOH. These pH values were MM4550 6.5; MM13902 7.5 and MM17880 7.5; mixture 6.8. The wavelengths used for detection were selected from the U.V. spectra MM4550 (285nm); MM13902 (308nm); MM17880 (285nm); mixtures (297nm). The injection size was 10 µL and the flow rate was 1ml/min.

Analyses were performed on a Waters µ Bondapak C_{18} column (10µm) with an Altex 110A pump/421 controller. Column effluents were monitored by U.V. absorbance as above (Perkin Elmer LC75) and recorded on a Hewlett-Packard HP 3390A integration/plotter. For investigation of the effects of temperature a Kipp Analytica 9222 HPLC column oven was used.

RESULTS AND DISCUSSION

Reversed-phase Chromatography

The parameters used for the conventional reversed-phase silica chromatography were revaluated, retaining the same basic system which consisted of 0.05M sodium propionate, pH 4.7 containing up to 5% CH_3CN.

Higher pH values gave progressively poorer resolution;
at pH 5.5 MM4550 and MM17880 and were coeluted. Lower
pH values were not investigated due to instability of
the analyte.

The concentration of organic modifier had a small effect
on the resolution of MM4550 and MM17880. In addition
the ionic strength of the mobile phase has little effect
on the resolution of MM4550 and MM17880 as indicated in
figure 2. Increasing the temperature also impaired
resolution. The best resolution obtained was 1.62 on
the analytical column.

Ion-Pair Chromatography

Ion-pair chromatography [2] involves the elution of the
analyte as an ion-pair using a suitable counter-ion in
the mobile phase. A typical ion-pair is formed thus:

$$RCOO^- + Bu_4N^+ \rightleftharpoons [RCOO^-N^+Bu_4]$$

in which the ion-pairs are partitioned between the mobile
phase and stationary phases. Table 1 shows the influence
of the hydrophobicity of the quaternary ammonium counter
ion on t_R, the retention time. It can be seen that
retention time increases with the length of the alkyl chain
in the eluting reagent. Resolution was improved from 1.62,
the best achieved with simple reversed-phase chromatography,
to a maximum of 2.75.

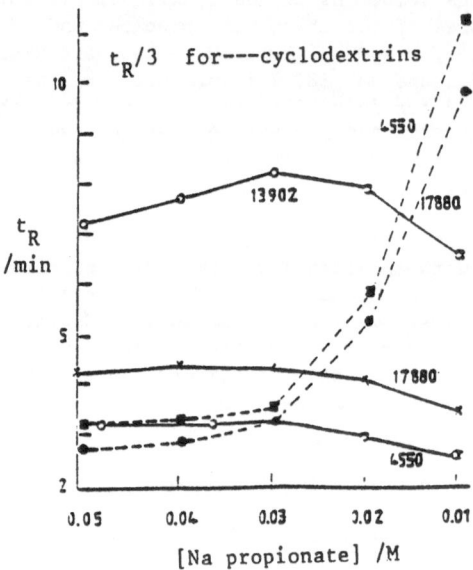

Figure 2. Reversed-phase HPLC; influence of sodium
propionate on retention time. Compare with
β-cyclodextrins shown with continuous lines for which
tR/3 is plotted.

TABLE 1
Influence of the size of the counter-ion on the
separation of the olivanic acids.

Ion-pair Reagent Concentration	t_R (Minutes)			t_R(MM17880) $-t_R$(MM4550)
	MM4550	MM17880	MM13902	
-	3.58	4.42	7.60	0.84
1mM-Me$_4$NCl	4.06	5.05	8.98	0.99
0.5mM-Et$_4$NI	4.54	5.62	9.17	1.08
0.3mM-Bu$_4$NH$_2$PO$_4$	5.21	6.42	10.51	1.21
0.3mM-(C$_5$H$_9$)$_4$NBr	5.31	6.86	10.83	1.55

Separation Using Immobilised Cyclodextrins(CD)

Cyclodextrins are natural macrocyclic polymers of glucose
containing from 6 to 12 D-(+)-glucopyranose units bonded
via α-(1,4) linkages, with all glucose units in a chain
configuration. The overall appearance of the cyclodextrin
(CD) is that of a truncated cone the dimensions of which
are governed by the number of glucose moieties present.
The β-CD structure is shown in Figure 3. The interior of
the cavity thus defined has an internal diameter of 7.8A.

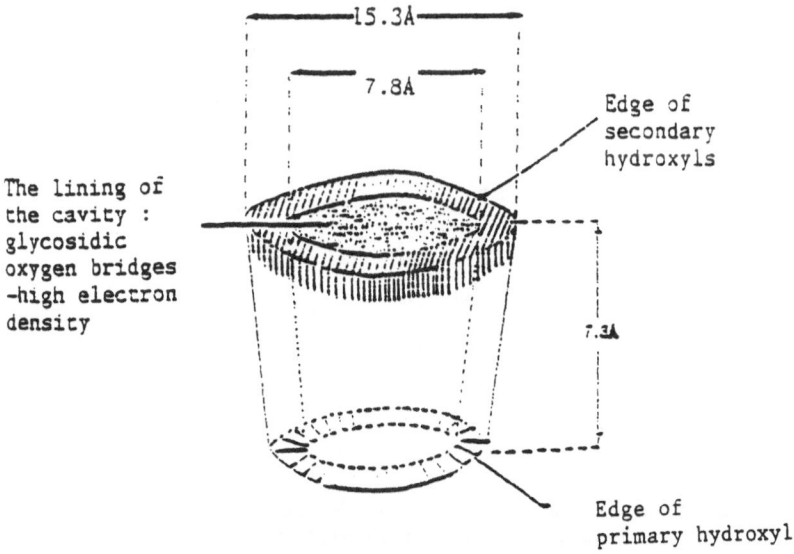

Figure 3. Schematic illustration of the structure of
β-cyclodextrins.

Figure 3 contains two rings of CH-groups with a
ring of glycosidic oxygens in between. As a result,
[3,4,5] the cavity is relatively hydrophobic whereas the
external faces are hydrophilic. The truncated, hydrophobic
cone has a well defined geometry which has not hitherto
been used to separate MM17880 and MM4550. Figure 4 shows
the simplified geometries of the two side chains.

Clearly, the radius of the side chain of MM4550, in which
there is restricted rotation about the carbon double bond,
is greater than that of MM17880. Quantitative calculations
using van der Waal radii have confirmed that the chain on
MM17880 could fit into the cavity. Consequently, it was
predicted that MM17880 would be retained longer than
MM4550 using a silica column with immobilised
β-cyclodextrins. This was indeed found to be the case. It
was found that varying certain parameters has a more-
pronounced effect on retention time a β-cyclodextrin column

1. MM 17880

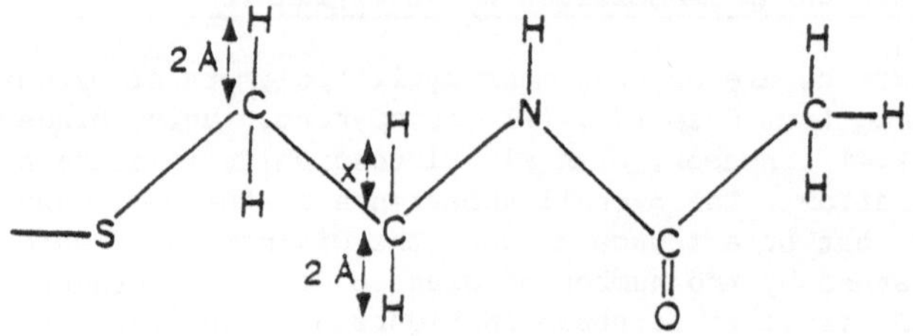

2. MM 4550

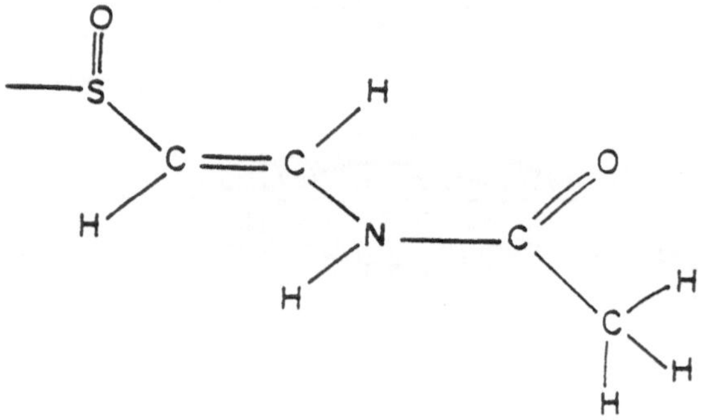

Figure 4. C3 side chains

('Cyclobond I', Advanced Separation Technologies Inc.) than
on octadecyl silica. Large effects were noted for pH,
ionic strength and temperature, The latter two effects
are illustrated and compared with C18-silica in
figures 2 and 5. In the case of temperature, the

resolution of MM4550 and MM17880 is less affected by
increasing temperature in β-cyclodextrin chromatography
than in C18-silica chromatography. The effect of ionic
strength on retention time is the reverse of that seen
on C18-silica. Thus, since the C18-silica is presumably
operating in a largely hydrophobic mode, the
β-cyclodextrin mediated separation may have a substantial
H-bonding component. Hydrophobic interactions occur
within the truncated cone whereas H-bonding may involve
the secondary hydroxyls. In the current studies the best
resolution of MM4550 and MM17880 obtained in the
β-cyclodextrin column was 3.32.

The separation was achieved by differential insertion
of the side chains into the cyclodextrin cavity.

CONCLUSIONS

Separation of the three sulphated olivanic acids can be
achieved using ion-pair chromatography and immobilised
cyclodextrins. The immobilised cyclodextrins were much more
effective at producing a separation. The
hydrophilic/hydrophobic character and the geometries of the
side chains were important factors.

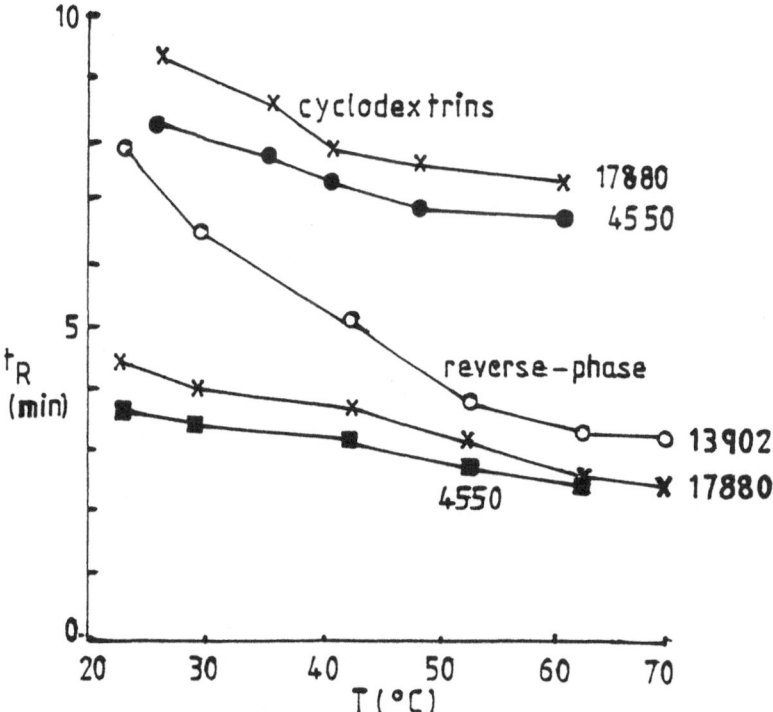

Figure 5. Separation of two sulphated olivanic acids
using an immobilised β-cyclodextrin and three using a
reversed phase column.

Acknowledgements

Beecham Pharmaceuticals and the SERC for a CASE studentship to SH. Mrs. G. Boffey for typing.

REFERENCES

1. Fox S.J., in "Discovery and Isolation of Microbial Metabolites", Verrall M., (Editor) 1985, Ellis Horwood.

2. Frausson B., Wahlund, K.G., Johansson I.M. and Schill G., J. Chromatography, 1976, 125, 327.

3. Bender M.L. and Komiyama T., "Cyclodextrin Chemistry", Springer-Verlag, New York 1978.

4. Griffiths D.W. and Bender M.L., Adv. Cat., 1973, 23. 209.

5. Saenger W., Angew. Chem., Int. Edition (English), 1980, 19, 34.

SEPARATION WITH POROUS GLASS MEMBRANES AND CONTROLLED PORE GLASS (CPG) CHROMATOGRAPHY

P. Langer, R. Schnabel

Schott Glaswerke

6500 Mainz, Fed. Rep. of Germany

ABSTRACT

Membrane processes as well as chromatography are important unit operations in biotechnological production. Porous glass is a material which can be employed successfully for both techniques. In this paper, a characterization of glass membranes and CPG stationary phases will be given, and some applications will be discussed.

INTRODUCTION

In the form of a stationary phase for chromatography, so-called controlled pore glass (CPG) has been used for analytical purposes for many years. It is increasingly being used on a preparative scale or in batch extraction processes where the main advantages of porous glass - high mechanical stability, homogeneous pore size, chemical surface variability - have a significant impact on process performance. This development has been furthered by the recent availability of large, homogeneous lots of BIORAN-CPG.

BIORAN porous glass membranes are a more recent development. They are manufactured in the form of thin-walled capillaries with an inner diameter of 0.3 and and outer diameter of 0.4 mm. Such capillaries are assembled into modules of a length of 25 cm and a total membrane area of between 0.05 and 1.0 m^2, permitting convenient scaling-up of separation processes.

With the high pore volume and the correspondingly large specific surface area of porous glass, both products are ideal carriers for the immobilization of enzymes and other proteins. Typical processes are affinity chromatography or membrane based enzyme reactors.

MATERIAL PROPERTIES

It is important to distinguish the manufacturing process of porous glasses, which will be dealt with in this paper, from a sintering process which is used to make ceramics and sinter glasses. While the latter results in materials with pores larger than approximately 1 μm, porous glasses have pores on the order of 10 to 100 nm in diameter. They are produced from a suitable, initially homogeneous, non-porous glass by a temperature treatment which causes a phase separation. The subsequent leaching of the soluble phase results in interconnected pores between 10 and 100 nm in diameter, depending on process conditions. The pores thus generated are very homogenous in size. Fig. 1 shows a typical pore size distribution of porous glass in comparison with a commercial silica gel, a widely used material in chromatography. The pore size distribution is characterized by a very steep slope towards larger pores, ensuring a precise cut-off.

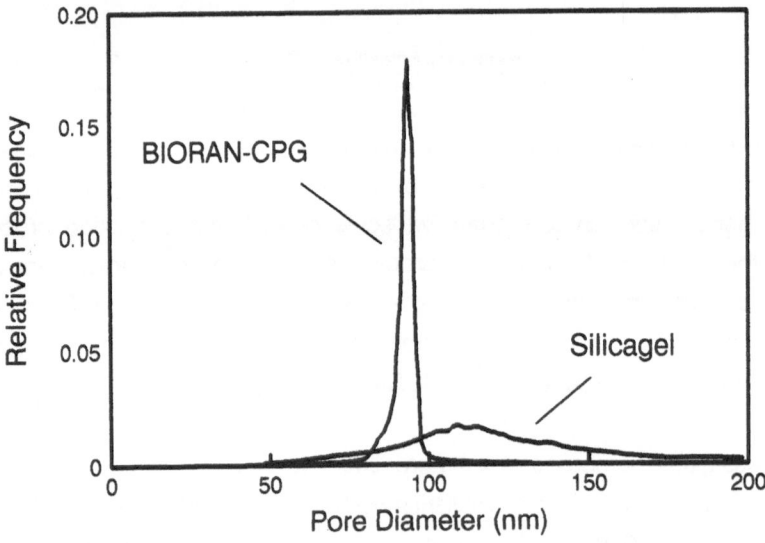

Figure 1. Pore size distribution of BIORAN porous glass in comparison with a commercial silicagel, measured by mercury porosimetry.

For membrane applications in biotechnology, the adsorption of proteins to the membrane has to be avoided in order to minimize fouling problems. This is accomplished by a chemical modification of the glass surface. Extensive adsorption studies [1, 2] have shown the Diol modification to be the best choice for this purpose. A comparison with other membrane materials (Table 1) demonstrates the superiority of BIORAN porous glass in this respect: Diol-modified porous glass shows the lowest level of adsorption.

TABLE 1

Adsorption of BSA (bovine serum albumin, 10 g/l, equilibrium experiment) to different membrane materials, referred to the BET specific surface area of the materials (part of the data was taken from the literature)

Membrane material	Amount adsorbed (mg/m^2)
Carbon	12.9
Aluminium Oxide	4.0
Nylon	3.5
PES	1.0
Glass (unmodified)	0.4
Regen. Cellulose	0.15
Polysulfon	0.11
PVDF	0.064
PES modified	0.044
BIORAN-Diol	0.002

Other surface modifications, such as the hydrophobic C_8 and C_{18} groups, are more relevant to chromatography. Table 2 lists the various chemical surface modifications currently available. The typical coverage is between 2 and 3 $\mu mol/m^2$.

In biotechnology it is often necessary to work under sterile conditions or with sterile equipment in order to prevent contamination. BIORAN membranes as well as the granular material may be steam sterilized at 121°C repeatedly without changing properties. Even the chemical surface modification is unchanged after such treatment. In steam sterilization runs up to a total time of

400 h the specific surface coverage as measured by elemental analysis has been shown to remain constant between 2 and 3 $\mu mol/m^2$. In the case of Chloropropyl, the coverage decreased after the first sterilization cycle from 5 to 4 $\mu mol/m^2$ and then remained constant.

TABLE 2

Chemical surface modifications on BIORAN porous glass membranes and CPG

$\equiv$ Si-OH	Silanol	native glass surface
$-(CH_2)_3-O-CH_2-CH-CH_2$ $\quad\quad\quad\quad\quad\;$ OH OH	Diol	recommended for membranes, low protein adsorption
$-(CH_2)_3-NH_2$	Amino	for covalent binding of proteins
$-(CH_2)_3-Cl$	Chloropropyl	hydrophobic
$-(CH_2)_7-CH_3$	Octyl	hydrophobic
$-(CH_2)_{17}-CH_3$	Octadecyl	hydrophobic

The pure water permeability of a membrane is an important criterion in comparing membranes. It is often overlooked, however, that different membrane materials with identical water fluxes may show very different permeabilities in protein solutions. As BIORAN glass membranes are symmetric membranes, their water permeabilities (as shown in Table 3) may not appear to be very high. As will be shown below, however, they retain high flow rates even in difficult solutions and at long operation times.

TABLE 3

Properties of BIORAN porous glass membranes

Pore Diameter (nm)	Water Permeability (l/m^2h bar)	Approx. Cut-Off (D)
10	2.5	10 000 - 50 000
13	8	20 000 - 70 000
19	10	50 000 - 150 000
27	40	150 000 - 1 000 000
44	120	> 1 000 000
90	500	Microfiltration

BIORAN CPG is a stationary phase for chromatography. It is available in different size fractions between 32 and 250 μm and with pore diameters between 30 and 100 nm (up to 300 nm upon request). The use of controlled pore glass (CPG) for chromatography has been described extensively in the literature. As has been shown in Fig. 1, the sharp pore size distribution is a remarkable feature of porous glass which distinguishes it from other materials. Especially in process chromatography the high mechanical stability of CPG will be appreciated. It is practically incompressible and therefore the ideal packing material for large, preparative chromatography columns. The different chemical surface modifications permit the binding of affinity ligands and the use of the material in affinity chromatography.

ULTRAFILTRATION

Capillary membrane modules are preferably operated in the cross-flow mode. A typical setup for an ultrafiltration process consist of two pumps at the inlet and the outlet of the module and one or two pressure transducers. A recycling loop may either be operative from retentate to feed or from permeate to feed, depending on where the desired components accumulate. For an efficient cross-flow process it is important that the flow yield, i. e. the ratio of permeate to feed flux, is kept low; this results in a high axial flow velocity in the capillaries which helps to reduce a boundary layer on the membrane surface. Nevertheless, due to their excellent surface properties, even at flow yields up to 0.9 BIORAN membranes show a very constant retention behaviour [3].

Many membrane manufacturers characterize their UF-membranes by a nominal molecular weight cut-off (MWCO), indicating the molecular weight which is retained by the membrane to a degree of 90 %. While these figures are useful approximations, the only meaningful characterization of a membrane has to come out of the user's own lab with the solutions which he has to process. For this reason, and because they can be characterized reliably by mercury porosimetry, porous glass membranes are specified by their pore diameters. Actual pore diameters of any membrane batch are guaranteed to be within 10 % of the rated mean values, keeping lot-to-lot variations much smaller than is common with other membranes.

As testing a membrane in the user's lab is so important, small scale experiments ought to scale up reliably. In order to facilitate the development of large-scale processes from laboratory experiments, porous glass membranes are offered in three different sizes of 0.05, 0.2 and 1.0 m^2 membrane area. These modules are identical in length and contain different numbers of capillaries. Therefore the optimum process parameters like pressure, flow yield and flow velocity as determined in the laboratory experiment can be taken over to the production scale. - In scaling up chromatographic separations, laboratory experiments may be conducted with small columns and small granular size fractions, while larger particle sizes may be used in process chromatography.

In the ultrafiltration range, porous glass membranes may be used for the fractionation of protein solutions. In order to obtain realistic results, a mixture of proteins at a total concentration of 5 g/l has been used instead of single substances. This mixture was fractionated in an ultrafiltration setup by four BIORAN membranes of different pore sizes and a Diol surface modification. The retention coefficient of a substance is defined as

$$R = 1 - c_p/c_r$$

where c_p is the permeate and c_r the retentate concentration of that substance. Fig. 2 shows retention coefficients of the four proteins. The data are mean values of 10 - 30 runs with each membrane at pressures between 0.2 and 0.8 bar and flow yields between 0.2 and 0.9.

As already mentioned, these results as well as the molecular weight cut-off indicated in Table 3 should be taken as guidelines for the selection of an appropriate membrane for a certain process. However, depending on the shape of a particular molecule, the retention may vary. Interactions between molecules have to be considered, too, and may affect the separation of two different species [3].

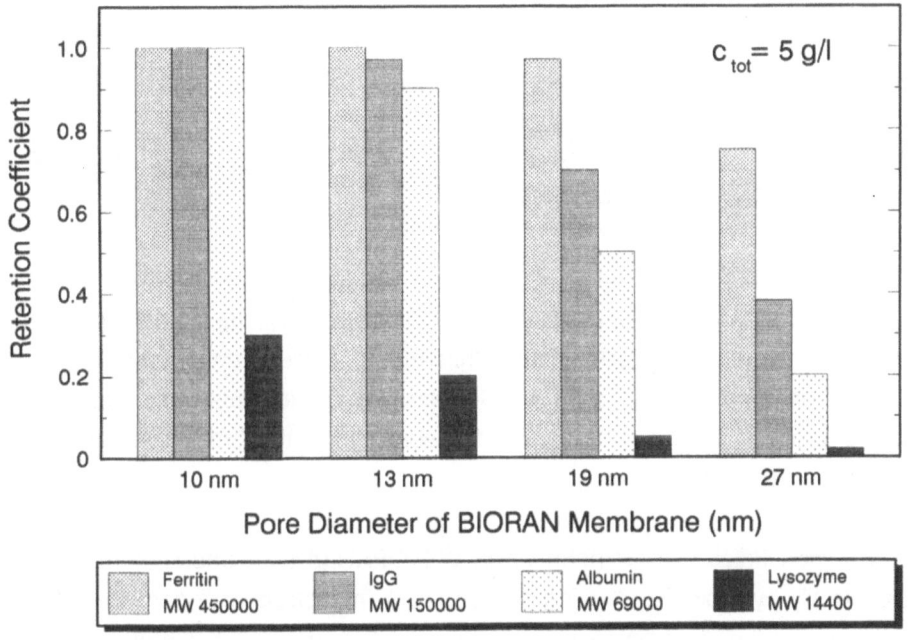

Figure 2. Retention coefficients of four different proteins used in a mixture. Each bar is the mean of 10-30 experiments at pressures between 0.2 and 0.8 bar, a flow yield between 0.1 and 0.9 and at filtration periods up to 6 h.

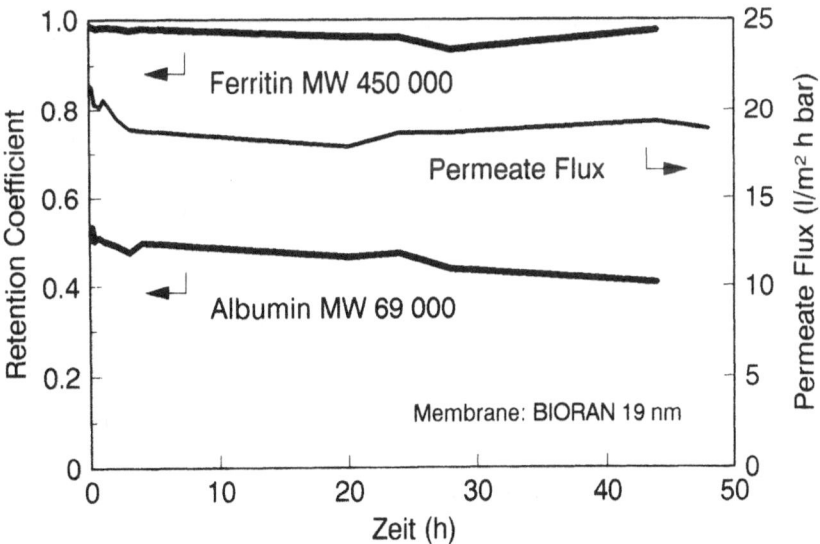

Figure 3. Long term stability of a protein separation by a BIORAN membrane.

The long term stability of a separation is important for process control. While the data shown in Fig. 2 have been accumulated during filtration runs lasting up to 6 hours, Fig. 3 shows the long term performance of a glass membrane over 48 hours. The retention coefficients remain practically constant over this period of time, and the permeate flux shows only a small decrease at the beginning before it remains constant.

Processes in the pharmaceutical industry have to qualify under Good Manufacturing Practice regulations. It is an asset of BIORAN glass membranes that they may be cleaned reliably and steam sterilized in place. Furthermore, they are delivered dry and are therefore not loaded with surfactants which can leach into the process stream. If not used for some time, they may also be stored dry for any length of time.

Fig. 4 demonstrates that the permeability of BIORAN glass membranes may be restored easily by a simple cleaning procedure. A membrane of a pore diameter of 16 nm was used periodically in a protein solution (5 - 15 g/l protein) for several hours and then cleaned by a 30 % urea solution. After each cycle, the membrane permeability reaches its initial value.

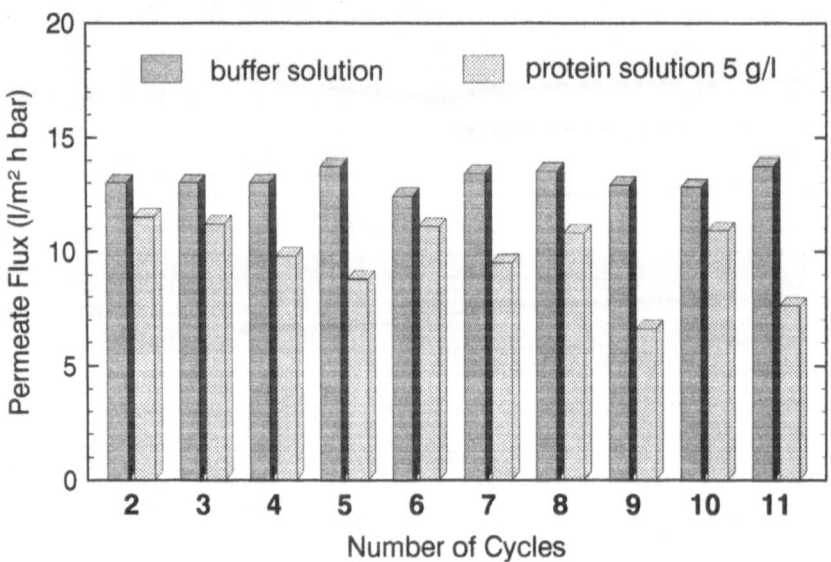

Figure 4. Regeneration of a 16 nm-BIORAN membrane after repeated use in a protein solution (5-15 g/l) by rinsing with a 30 % urea solution.

MICROFILTRATION

It is common to characterize microfiltration membranes by bubble point measurements which allow to specify a maximum pore diameter. For sterile filtration a maximum pore size of 0.2 μm is required, and such membranes are
commonly used for cell harvesting in biotechnology. Most microfiltration membranes are prone to severe fouling which often leads to a decrease in flux by
a factor 10 or more within a short time. This affects, of course, the recovery
of proteins from the fermentation broth.

BIORAN porous glass membranes with a pore size of 0.09 μm can be used for such
separations. Due to the low-fouling surface modification, cells do not form a
solid layer on the surface and may be removed by periodic reversal of flow, by
using pulsating flow or by backflush from the permeate side. In the experiments described below, a 2.5 % suspension of baker's yeast in 0.3 % NaCl solution with a content of 5 g/l BSA (bovine serum albumin) was used to examine
the separation of cells from the protein. At a constant pressure of 0.4 bar
and a flow yield of 0.1, the flow velocity in the capillary was 25 cm/s. The
pore diameter of the membrane was 60 nm. The retentate was continuously recirculated into the feed. The permeate was collected in fractions to estimate the
amount of BSA recovered as a function of time.

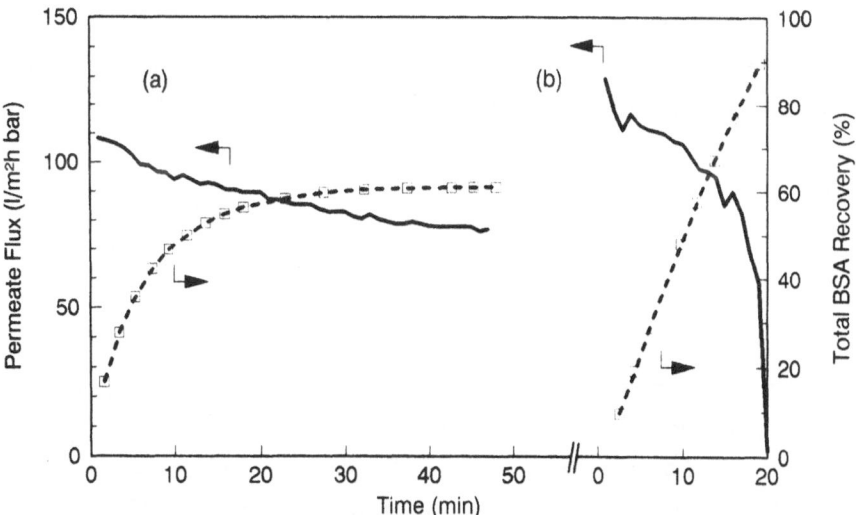

Figure 5. Separation of baker's yeast (2.5 %) from BSA (5 g/l) by a 60 nm BIO-
RAN-membrane; (a) cell conc. in feed kept constant by maintaining constant
feed vol., (b) batch filtration, concentrating feed volume from 500 to 35 ml.

In the case of Fig. 5a, the cell concentration in the feed solution was kept constant by adding NaCl solution to the feed at the same rate as permeate was collected. The permeate flux decreased slightly, and the cumulative recovery of BSA in the permeate reached about 60 % after approximately 30 min. By rinsing the membrane module with water, the initial flux was restored within a few minutes.

In the experiment of Fig. 5b, a limited volume of 500 ml was processed without adding NaCl solution to the feed. This meant a more than tenfold increase in cell concentration in the feed during the course of the filtration, which was tolerated perfectly well by the membrane. The total BSA recovery in this case was as high as 90 %. The sharp decrease in permeate flux at the end of the experiment was due to the breakdown of filtration pressure when the feed reservoir was being emptied, not to blocking of the membrane. Rinsing with water restored the initial membrane properties.

CONCLUSIONS

BIORAN porous glass membranes and CPG are versatile materials for separation and purification in biotechnology. An asset is the high level of reproducibility with which these products are made. The special surface properties of membranes ensure biocompatibility, long term operation and convenient cleanability. The chemical surface modifications, especially NH_2-groups, constitute an interface to specialized, individual applications which require the binding of affinity ligands to a membrane or a stationary phase to be used in chromatography.

REFERENCES

1. P. Langer, R. Schnabel, Porous Glass Membranes for Downstream Processing, Chem. Biochem. Eng. Q. 2, 242-44 (1988)
2. P. Langer, R. Schnabel, Adsorption of Proteins on Surface Modified Porous Glass, Third Int. Conf. on Fundamentals of Adsorption, Sonthofen, 1989
3. P. Langer, S. Breitenbach, R. Schnabel, Ultrafiltration with Porous Glass Membranes, Int. Techn. Conf. on Membrane Separation Processes, Brighton, 1989

PERFORMANCE OF ANION-EXCHANGE CELLULOSE AT PROCESS-SCALE

PETER R. LEVISON, STEPHEN E. BADGER, DAVID W. TOOME, DIANE CARCARY and EDWARD T. BUTTS
Whatman Specialty Products Division,
Springfield Mill, Maidstone, Kent, ME14 2LE, U.K. and
9 Bridewell Place, Clifton, NJ, 07014, U.S.A.

ABSTRACT

Whatman DE52 (tertiary amine substituted anion-exchange cellulose) has been shown to exhibit fast kinetics of adsorption and high capacity for proteins of molecular mass up to 670 kDa. These features lend themselves to process-scale use and we have investigated the operational parameters of process-scale chromatography of egg-white proteins in batch and column modes using DE52. We report that in a single pass adsorption step the efficiency of feedstock binding was 70% for batch adsorption compared with 78% for column adsorption. Desorption is more efficient in the column mode.

The binding capacity of DE52 was unaffected after 5 successive process-scale loadings although chromatographic performance was slightly reduced. This was restored using a clean-in-place procedure which simultaneously effects sterilisation and depyrogenation of the column and media.

INTRODUCTION

Ion-exchange chromatography is a major tool in the downstream processing of commercially important biopolymers. While other techniques including affinity, hydrophobic interaction and gel permeation chromatography each have their place in protein purification, perhaps the most versatile and cost effective technique is that of ion-exchange.

By selection of a suitable ion-exchange medium and appropriate operating parameters a very high degree of purification can be achieved in a single process-step. In general there are two approaches to process-scale ion-exchange chromatography, these involving batch or column operation. Where the chromatographic step is to be integrated into a process system for repetitive use then the latter technique is more widely employed, but in instances where, for example, large volumes of a dilute feedstock are used then batch stages may be preferred.

We have previously investigated a number of parameters which affect the economics of process-scale column chromatography using Whatman QA52 (quaternary-amine substituted anion-exchange cellulose). These include studies on capacity and kinetics [1], feedstock composition [2], process optimisation [3] and scale-up [4, 5]. In the present study we have extended these investigations to DE52 and examine the efficiency of batch versus column techniques for the process-scale chromatography of hen egg-white proteins. The re-use of the ion-exchanger has been investigated in a column mode and a procedure for clean-in-place (CIP) has been validated.

MATERIALS AND METHODS

1) Capacity and kinetics of DE52

DE52 (Whatman Specialty Products Division, Maidstone, England) was equilibrated with 0.01M-sodium phosphate buffer, pH 7.5. The adsorption of 2.0 mg/ml soybean trypsin inhibitor (20.1 kDa), 1.0 mg/ml bovine serum albumin (67 kDa) and 0.3 mg/ml porcine thyroglobulin (670 kDa) (Sigma Chemical Co. Ltd., Poole, England) was carried out in a batch mode (50 ml volume) using 50 mg, 50 mg and 200 mg DE52 respectively over a 60 minute period.

2) Egg-white feedstock preparation

The whites of 600 size 2 eggs (Barradale Farms, Headcorn, England) were separated and diluted with 0.025M-Tris/HCl buffer, pH 7.5 (BDH Chemicals Ltd., Dagenham, England) to give a 14% (^{V}/v) suspension. the suspension was treated with a total of 22 kg of CDR (Whatman Specialty Products Division, Maidstone, England) in a batch mode, to give a clear solution (200 l) containing ~ 1.6 kg of protein.

3)   Process-Scale Chromatography

DE52 (25 kg) was equilibrated with 0.025M-Tris/HCl buffer, pH 7.5.
The ion-exchanger was used with the egg-white feedstock (200 l)
accordingly:-

(a)   Batch adsorption/Batch desorption.
(b)   Batch adsorption/Column desorption.
(c)   Column adsorption/Column desorption.

All procedures were carried out at room temperature (15-20°C).

(a) Batch adsorption/Batch desorption

DE52 (25 kg) was stirred with the CDR-treated feedstock (200 l).
The DE52 was collected by centrifugation and washed with
0.025M-Tris/HCl buffer, pH 7.5 (200 l). Bound material was
eluted using 0.025M-Tris/HCl buffer, pH 7.5 containing 0.5M-NaCl
(200 l). In this study the elution was carried out on the
centrifuge wall for half of the DE52 and in a stirred tank for
the remainder of the DE52.

(b) Batch adsorption/Column desorption

DE52 (25 kg) was stirred with the CDR-treated feedstock (200 l).
After standing, supernatant was removed by decantation to give a
30% (^{W}/v) slurry. The slurry was pump-packed into a PREP-25
column (45 cm i.d. x 16 cm, 25.4 l capacity; Whatman Specialty
Products Division, Maidstone, England) at a pressure of $\sim$ 10psi.

The bed was washed with 0.025M-Tris/HCl buffer, pH 7.5 (100 l)
and bound material eluted using a linear gradient of 0-0.5M-NaCl
in 0.025M-Tris/HCl buffer, pH 7.5 (200 l) at a flow rate of
0.8 l/min.

(c) Column adsorption/Column desorption

DE52 (25 kg) was slurried to 30% (^{W}/v) in 0.025M-Tris/HCl buffer, pH 7.5 and pump-packed into the PREP-25 column as described above. The egg-white feedstock (200 l) was loaded onto the column and non-bound material removed by washing with 0.025M-Tris/HCl buffer, pH 7.5 (100 l). Bound material was eluted using a linear gradient of 0-0.5M-NaCl (200 l). A flow rate of 0.8 l/min was maintained throughout.

4) DE52 Re-Use

In a column mode, five consecutive process-scale runs were carried out as described in 3(c) above.

The bed was equilibrated with 0.5M-NaOH (50 l) and stood for 12 hours. The bed was washed successively with H$_2$0 (50 l), 0.1M-Tris/HCl buffer, pH 7.5 (50 l) and 0.025M-Tris/HCl buffer, pH 7.5 (200 l) maintaining a flow rate of 0.8 l/min.

A further post-CIP process-scale run was then carried out as described above.

Analytical loadings (100 g, 12 l feedstock) were carried out in order to assess chromatographic performance of the bed. Runs were carried out as in 3(c) above interspersed throughout the 6 process-scale runs.

5) Assays

Pooled fractions obtained at various stages of chromatography were assayed for protein content using A$_{280}$ measurement against standard solution of ovalbumin (Sigma Chemical Co. Ltd., Poole, England).

Throughout the column procedures the effluent was monitored for absorbance at 280 nm and by conductivity.

RESULTS AND DISCUSSION

The influence of molecular mass on the kinetics of protein adsorption to DE52 at pH 7.5 are summarised in Figure 1:

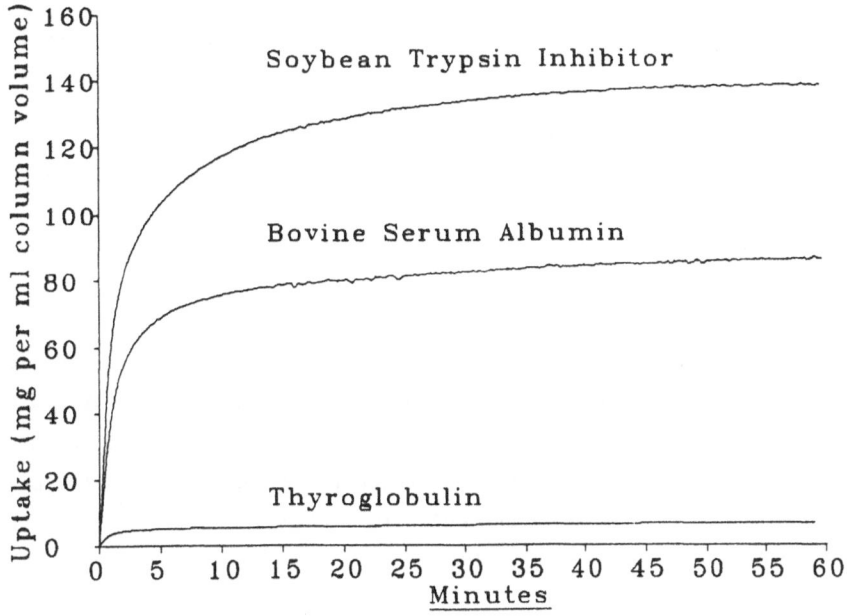

Figure 1. Kinetic uptake of soybean trypsin inhibitor, bovine serum albumin and porcine thyroglobulin by DE52 in 0.01M-sodium phosphate buffer, pH 7.5. (50 ml) in a stirred batch mode.

The binding capacity of DE52 for soybean trypsin inhibitor, bovine serum albumin and porcine thyroglobulin was 6.6, 86 and 138 mg/ml swollen medium respectively. Under these experimental conditions, where protein was in excess, the data indicate that fast adsorption kinetics are maintained for proteins of molecular mass up to 670 kDa, although binding capacity reduces as molecular mass increases. This reduction in capacity is presumably due to size exclusion factors. In terms of rates of adsorption/desorption in each case the time required for 50% of the total protein to be adsorbed or desorbed ranged from 45-90 seconds. This data demonstrates the suitability of DE52 as a bioprocessing medium, and while we have not defined the specific rate limiting processes in protein binding, the ion-exchanger is nevertheless likely to exhibit good binding efficiency for the target within the contact-time restraints of the adsorptive process.

In this study we have compared batch and column techniques under similar process-scale chromatography conditions. The loading capacity data for these studies are summarised in Table 1. The batch loading data indicate that ~ 70% of applied protein is adsorbed to the DE52, and this compares to a binding efficiency of ~78% for column loading. This is to be expected since the batch step is in fact a simple equilibrium process and can be regarded as having a theoretical plate count of one. On the other hand a column has a higher plate count and consequently will be more efficient during the loading procedure.

Following elution of bound material we observed ~ 100% recovery of protein in both column steps.

The batch elution when carried out on the centrifuge wall gave 100% recovery of adsorbed protein whereas the stirred tank system provided only 86% recovery. This can be related to partitioning equilibria since the stirred tank system contains only one theoretical plate whereas the DE52 pad on the centrifuge wall is analogous to a radial flow column and would therefore have a higher plate count.

TABLE 1
Protein capacities for batch vs column operations of
process-scale chromatography of egg-white proteins

Adsorption Mode	Desorption Mode	Protein loaded (kg)	Protein bound (kg)	Protein desorbed (kg)	Binding efficiency (%)
Batch	Batch Centrifuge Wall	1.479	0.520	0.531	70.3
Batch	Batch Stirred Tank	Split into 2 lots after loading	0.516	0.444	69.8
Batch	Column	1.574	1.086	1.192	69.0
Column	Column	1.703	1.321	1.360	77.6

In terms of the chromatography we did not observe any obvious differences in the separation of egg-white proteins in either batch or column use. Under process-scale conditions the strongly bound ovalbumin component (M.W. 45 kDa) of egg-white displaces the weakly bound conalbumin (M.W. 77 kDa) during loading such that the eluted product is ovalbumin at a relatively high degree of purity. These observations were consistent throughout the investigation.

The effect of five successive loadings of egg-white feedstock on the binding capacity of DE52 is summarised in Table 2:

TABLE 2
Protein capacities during successive process-scale
column chromatography of egg-white proteins

Run	Protein loaded (kg)	Protein bound (kg)	Protein desorbed (kg)	Binding efficiency (%)
1	1.757	1.315	1.351	74.8
2	1.742	1.331	1.360	76.4
3	1.700	1.378	1.384	81.0
4	1.905	1.486	1.482	78.0
5	1.770	1.383	1.401	78.1
6	2.059	1.591	1.575	77.3

During runs 1-5 ~ 1.38 kg of protein was adsorbed to the DE52 which reflects ~ 78% binding efficiency in a single pass through the column. In terms of binding capacity and efficiency of loading the CIP after run 5 had little benefit (Table 2). As mentioned above, under process-scale conditions, displacement chromatography occurs such that the ovalbumin component of egg-white displaces the less acidic conalbumin. It is therefore difficult to establish chromatographic performance of the bed under these conditions. In order to identify if chromatographic performance was affected by successive process-scale loadings, analytical loadings of egg-white (100 g) were carried out before run 1, after run 5 and after run 6. Under these conditions all acidic components of the egg-white bind to the DE52 and consequently peak shape gives a visual indication of column performance and product purity. These runs demonstrated that chromatographic performance of the bed was slightly reduced during the 5 process-scale loadings, but the CIP had the effect of restoring chromatographic performance to that observed at run 1. The conditions used for CIP i.e. storage in 0.5M-NaOH for 12 hours have been demonstrated to render highly pyrogenic columns of DE52, which had been contaminated with ~ 10^9 E.coli/ml bed, sterile, pyrogen-free and endotoxin-free [6,7].

In any chromatographic process there are two key factors which affect the success and economics of the separation. Firstly the binding efficiency of the medium during a single adsorption cycle and secondly the purity of the product. We have demonstrated that DE52 has physical properties which confer a high protein capacity and fast adsorption kinetics on the medium (Figure 1). These properties together with choice of a suitable mobile-phase for the feedstock [2] enable the anion-exchange cellulose studied to be used at process-scale with a high binding efficiency during the adsorption stage. In this study it is clear that column loading is more efficient than the batch technique although the latter is potentially more time efficient particularly where large volumes of dilute feedstock are used since the lengthy column-loading stage would be eliminated.

In terms of product elution a simple stirred-tank desorption is relatively inefficient compared with continuous flow desorption either on the centrifuge wall or in a column. There do not appear to be any marked differences in product purity in either of the techniques described here.

During ion-exchange processes it is important to pretreat the feedstock to protect the chromatography medium from fouling. In this study no reduction in protein binding was seen after 5 successive process-scale loadings. This is a reflection of the effectiveness of CDR in removing materials from the diluted egg-white which, if not removed, would foul the DE52. Although some reduction in chromatographic performance of the DE52 was seen after 5 loadings the performance could be restored using a CIP procedure which simultaneously effects sterilisation and depyrogenation of the column and media.

It is concluded that Whatman DE52 is suitable for process-scale chromatography in either batch or column operation. Only binding and desorption efficiency are significantly affected by the choice of separation mode with column adsorption/column desorption being the most efficient.

REFERENCES

1. Levison, P.R., Toome, D.W., Badger, S.E., Carcary, D. and Butts, E.T., Capacity and kinetics of protein adsorption by anion-exchange celluloses. Influence of molecular mass and pH. In <u>Pittsburgh Conference Abstract Book</u>, Pittsburgh Conference, Pittsburgh, 1990, p. 1079.

2. Levison, P.R., Toome, D.W., Badger, S.E., Brook, B.N. and Carcary, D., Influence of mobile-phase composition on the adsorption of hen egg-white proteins to anion-exchange cellulose. <u>Chromatographia</u>, 1989, **28**, 170-178.

3. Schwartz, W.E., Clark, F.M. and Sabran, I.B., Process-scale isolation and purification of immunoglobulin G. <u>LC-GC Magazine</u>, 1986, **4**, 442-448.

4. Schwartz, W.E., Clark, F.M., Koscielny, M.L., Powell, S.J., Levison, P.R. and Ralston, C., Factors of importance in chromatographic scale-up. In <u>Proteins Purification Technologies Vol. 3</u>, eds. C. Doinel, A. Faure and O. Robert, G.R.B.P., Villebon/Yvettes, 1988, pp. 448-449.

5. Levison, P.R., Koscielny, M.L. and Butts, E.T., A simplified process for large-scale isolation of IgG from goat serum. <u>Bioseparation</u>, 1990, in press.

6. Levison, P.R. and Clark, F.M., Validation of a technique for in situ sterilisation and depyrogenation of ion-exchange cellulose. In <u>Pittsburgh Conference</u> Abstract Book, Pittsburgh Conference, Pittsburgh, 1989, p. 754.

7. Levison, P.R., Badger, S.E., Toome, D.W., Carcary, D. and Butts, E.T., Studies on the re-use of anion-exchange cellulose at process-scale. In <u>Downstream Processing in Biotechnology II</u>, eds. R. de Bruyne and A. Huyghebaert, The Royal Flemish Society of Engineers (k.viv.), Antwerp, 1989, pp. 2.11-2.16.

HIGH SPEED ANALYTICAL AND PREPARATIVE SEPARATIONS OF BIOLOGICAL MACROMOLECULES

Linda L. Lloyd   Frank P. Warner

Polymer Laboratories Ltd., Essex Road, Church Stretton, Shropshire, SY6 6AX, U K

ABSTRACT

A wide pore, 4000A, macroporous resin has been developed which is rigid and able to operate under HPLC conditions of flow rate and pressure. The poly(styrene/divinylbenzene) matrix can be used for reversed-phase separations and for both anion and cation exchange chromatography after surface coating/derivatization.

The open pore structure of this resin makes it suitable for the HPLC analysis of biological macromolecules due to the significant improvement in mass transfer. The rigidity and rapid equilibration enables small columns, 50 x 4.6mm, to be operated with steep gradients at high flow rates for both high resolution analytical separations and high load preparative fractionations.

INTRODUCTION

The increased commercial availability of HPLC adsorbents specifically designed for the analysis/purification of biological macromolecules has stimulated interest in the technique with a proliferation of HPLC methodologies, column packings and instrumentation. However, even with the wide pore 300A silica based materials or wide pore microporous resins which are used for the analysis of biological macromolecules, kinetic and thermodynamic factors severely limit the speed of separation of large molecules compared with small molecule HPLC. The diffusion coefficient of a solute decreases as its molecular size in solution increases, for large molecules values of 10^{-6} and 10^{-11} m/s have been measured with the effective diffusion coefficient being lower due to tortuosity effects. The magnitude of the reduction being dependent upon the actual pore shape and size, with small pores adsorbed molecules can significantly reduce the effective pore size (1). When the ratio of solute diameter to pore diameter exceeds 0.2, the pore diffusion is restricted (2). This results in increased band spreading/reduced efficiency at flow rates normally used for small molecule HPLC.

In order to maintain efficiency/reduce band spreading at high flow rates non-porous materials have been evaluated where the intraparticle stagnant mobile phase mass transfer is eliminated. K. K. Unger et al have evaluated 1.5um non-porous silica matrices for reversed-phase (3) and hydrophobic interaction chromatography (4) and D. J. Burke et al used non-porous polymers for ion exchange (5). However, even when small particle size materials are used the capacity is low when compared to porous matrices due to the loss of internal pore surface area. It has been shown that it is possible to produce a highly porous polymer matrix which even for large biomolecules does not exhibit a significant decrease in efficiency/increased band speading as the flow rate is increased (6). As this material is porous and the intraparticle surface area is accessible to large molecules capacities of 10-30 mg/ml are achieved for reversed phase and ion exchange matrices.

This paper evalutes the suitability of this highly porous material for reversed-phase and, in a derivatised form, ion exchange high speed analytical separations and preparative fractionations.

EXPERIMENTAL

HPLC System : A modular binary Knauer HPLC system (Knauer, Berlin, F RG) was used with a Rheodyne 7125 injection valve fitted with a 200ul loop (supplied by HPLC Technology Ltd., Macclesfield, UK). To improve mixing at high flow rates whilst minimising hold-up volume a Lee Visco-Jet (R) micro-mixer with a 10ul internal volume was inserted between the pumps and the static mixing chamber (Lee Products Ltd., Gerrards Cross, UK).

Columns : All HPLC materials were evaluated in 50 x 4.6mm ID stainless steel columns with 2um porosity frits (Polymer Laboratories Ltd, Church Stretton, UK). The chromatographic characteristics of the matrices are summarized in table 1.

TABLE 1
Chromatographic characteristics of the HPLC adsorbents

Adsorbent	Pore Size A	Particle Size um	Functionality	Surface Area m2/g	Counter Ion
PLRP-S	4000	8	PS/DVB	139	-
PLRP-S	300	5	PS/DVB	384	-
PL-SCX	4000	8	-SO3-	-	Na+
PL-SCX	1000	8	-SO3-	-	Na+
PL-SAX	4000	8	-N+(CH3)3	-	Cl-
PL-SAX	1000	8	-N+(CH3)3	-	Cl-

Mobile Phases and Test Solutes   :   Water used for sample and eluent preparation was purified using an Elgastat UHP system (Elga Ltd., High Wycombe, UK) and the buffer salts were of analytical or HPLC grade (FSA Laboratory Supplies Ltd., Loughborough, UK). Proteins of varying molecular weight and isoeletric point were used as chromatographic test probes (Sigma Chemical Company Ltd., Poole, UK). Properties are summarized in table 2.

TABLE 2
Properties of globular protein test probes

Protein	molecular weight	Isoelectric point pI
ribonuclease A	14,000	7.8
cytochrome c	13,000	10.6
lysozyme	14,300	11.0
bovine serum albumin	67,000	5.0
myoglobin	17,000	7.0
ovalbumin	40,000	4.6
⍺-chymotrypsinogen A	25,000	9.1
soybean trypsin inhibitor	22,460	4.6

RESULTS AND DISCUSSION

It is possible to use the hydrophobic surface of a macroporous poly(styrene/divinylbenzene) matrix for reversed-phase separations (7). Therefore, in order to evaluate the chromatographic performance of the highly porous, poly(styrene/divinylbenzene) resin, PLRP-S 4000A for high speed macromolecule analysis a reversed phase separation of a representative selection of globular proteins was performed. The resolution of the protein mixture on a conventional reversed-phase material, PLRP-S 300A operated at a flow rate of 1.0 ml/min with a 20 minute linear gradient and the PLRP-S 4000A operated at 4.0 ml/min with a 1 minute linear gradient Figure 1B and 1A respectively is compared. No deterioration in separation efficiency was observed for the highly porous material PLRP-S 4000A when operated repeatedly with steep gradients at high flow rate. Rapid equilibration was achieved, less than 30 seconds between the end of the first run and the start of a second.

To evaluate the highly porous material in non-denaturing conditions for its suitability as a rapid purification matrix for biologically active solutes a strong cation exchanger, PL-SCX 4000A and a strong anion exchanger, PL-SAX 4000A was produced. The corresponding 1000A materials were used to obtain comparative data. Figure 2 compares the resolution achieved for a 4 protein mixture, myoglobin, ⍺-chymotrypsinogen A, cytochrome c and lysozyme using the strong cation exchangers PL-SCX 4000A and PL-SCX 1000A and figure 3 the resolution of the 2 protein mixture ovalbumin and soybean trypsin inhibitor with the strong anion exchange materials PL-SAX 4000A and PL-SAX 1000A.

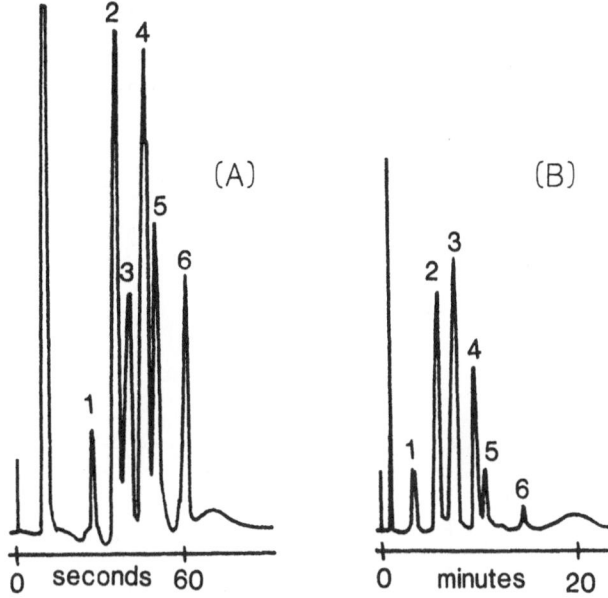

Figure 1. Reversed phase separation of 6 globular proteins.
Proteins, 1 : ribonuclease A, 2 : cytochrome c, 3 : lysozyme, 4 :
bovine serum albumin, 5 : myoglobin, 6 : ovalbumin; detector, UV at
280 nm; eluent C, 0.1% TFA in 5% acetonitrile : 95% water v/v;
eluent B, 0.1% TFA in 95% acetonitrile : 5% water v/v. (A) column,
PLRP-S 4000A 8um 50 x 4.6mm I.D.; gradient, linear 18 - 60% B in 1
minute at a flow rate of 4.0 ml/min. (B) column, PLRP-S 300A 5um 50 x
4.6mm I.D.; gradient, linear 15 - 55% B in 20 minutes at a flow rate
of 1.0 ml/min.

It is apparent from figures 1, 2 and 3 that the highly porous
4000A poly(styrene/divinylbenzene) matrix either in the unmodified form
for reversed-phase or after derivatization to produce either anion or
cation exchange materials is capable of operating at high flow rates
with short gradient development times. High resolution analytical
separations of standard proteins covering a range of molecular weights
and isoelectric points, table 2, can be obtained in a fraction of the
time. To further evaluate the effect of gradient development time and
flow rate on resolution the separation of ovalbumin and soybean trypsin
inhibitor was performed using the PL-SAX 4000A and 1000A materials
under various conditions. The resolution factor Rs for the separation
was calculated from:

$$Rs = \frac{tr(1) - tr(2)}{1/2 \, (\, w(1) + w(2) \,)} \tag{8}$$

where $tr(1)$ and $tr(2)$ are the retention volumes of soybean trypsin
inhibitor and ovalbumin respectively and $w(1)$ and $w(2)$ are the
respective peak widths at the base. As would be expected, figure 4,
with porous HPLC materials a decrease in gradient development time at

constant flow rate results in a decrease in resolution factor. However at increased flow rate the resolution factor is significantly improved with the PL-SAX 4000A material but not the PL-SAX 1000A. Similar resolution can be obtained with the PL-SAX 4000A in 2 minutes and the PL-SAX 1000A at 20 minutes.

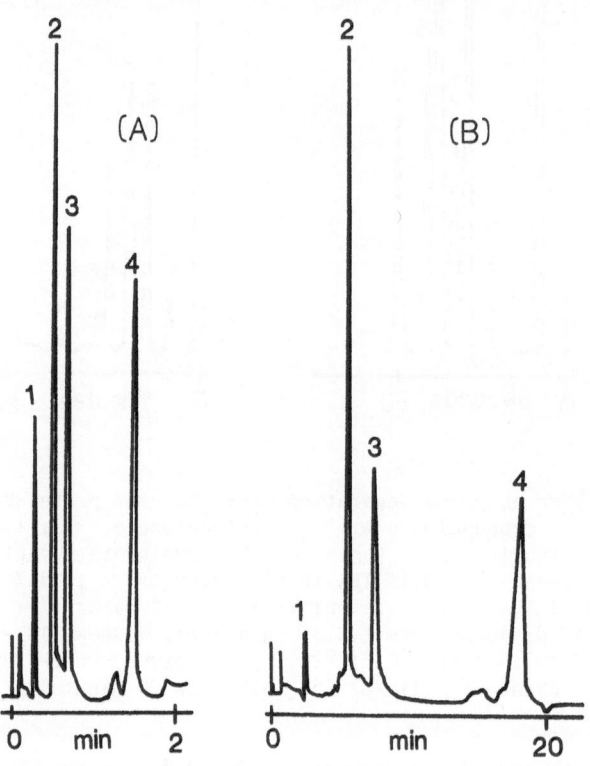

Figure 2. Separation of 4 standard proteins using the strong cation exchanger, PL-SCX

Proteins, 1 : myoglobin 2 : α-chymotrypsinogen A, 3 : cytochrome c, 4 : lysozyme; detector, UV at 280 nm; eluent C, 0.02M KH2 PO4, pH 6.0; eluent B, 0.02M KH2 PO4, 0.5M NaCl, pH 6.0. (A) column, PL-SCX 4000A 8um 50 x 4.6mm I.D.; gradient, linear 0 - 100% B in 2 minutes at a flow rate of 4.0 ml/min. (B) column, PL-SCX 1000A 8um 50 x 4.6mm I.D.; gradient, linear 0 - 100% B in 20 minutes at a flow rate of 1.0 ml/min.

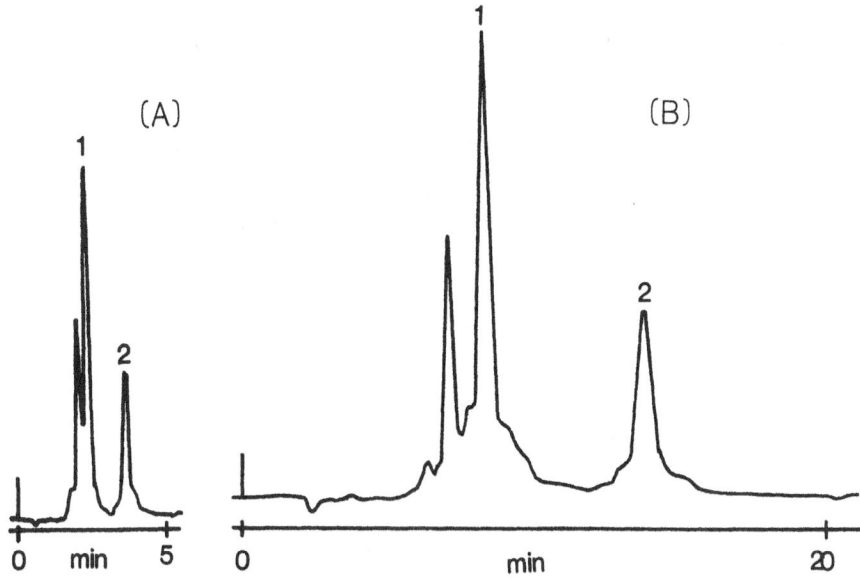

Figure 3. Anion exchange separation of ovalbumin (1) and soybean trypsin inhibitor (2). Detector UV at 280nm; eluent C, 0.01M Tris HCl, pH 8.0; eluent B, 0.01M Tris HCl, 0.35M NaCl, pH 8.0. (A) column, PL-SAX 4000A 8um, 50 x 4.6mm I.D.; gradient, linear 0 - 100% B in 5 minutes at a flow rate of 4.0ml/min. (B) column, PL-SAX 1000A 8um 50 x 4.6mm I.D.; gradient, linear 0 - 100% B in 20 minutes at a flow rate of 1.0 ml/min.

The improvement in the resolution factor at increased flow rate with the PL-SAX 4000A material suggests that there is a considerable difference in the mass transfer, intraparticle diffusion, of this matrix and a conventional wide pore material. The highly porous 4000A matrix is rigid and does not compress when operated under high pressure/flow rate and therefore would be expected to be suitable for high speed, high resolution analytical separations and preparative fractionations of large biomolecules.

The purification of biomolecules from complex matrices, media, fluids, extracts etc. often requires several stages including one or more chromatographic steps. The product purity/homogeneity may need to be determined at one or several stages in the purification protocol. An HPLC packing able to perform high resolution separations in seconds would enable such measurements to be carried out during the purification and hence enable optimization of product yield to be achieved. Examples of such applications include the purification of antibodies and the isolation of enzymes from cell culture filtrates.

Affinity chromatography using immobilized Protein A can be used to selectively isolate the group of antibodies in serum from non-antibody proteins. The antibodies are retained by the immobilized Protein A and after equilbration eluted by decreasing the pH. It is necessary to determine the purity of the bound fraction, i.e. albumin, transferrin contamination and also if any valuable antibodies are present in the unbound fraction. This can be accomplished using the PL-SAX 4000A 8um 50 x 4.6mm column in under 5 minutes. The 2 fractions, unbound and bound were collected from a Protein A separation and after adjusting the pH an aliquot injected. Figure 5 shows the resolution achieved for diluted serum and the 2 fractions.

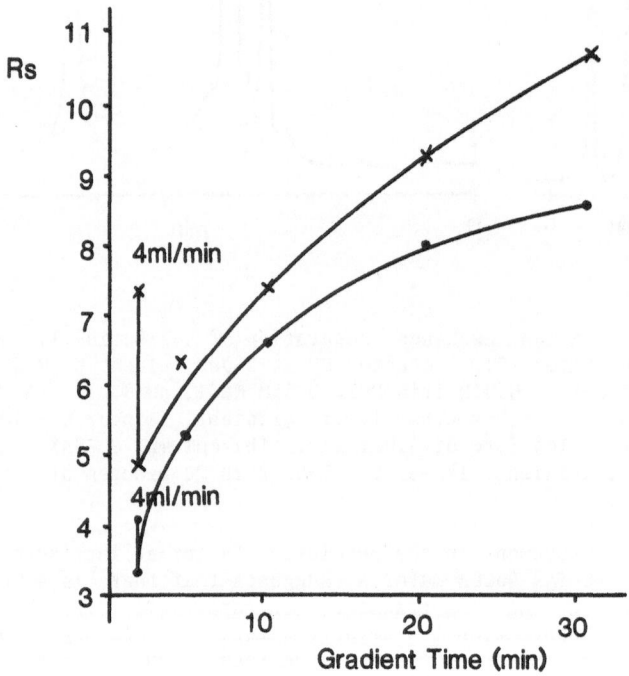

Figure 4. Plot of resolution factor for the anion exchange separation of ovalbumin and soybean trypsin inhibitor against gradient development time. Detector, UV at 280 nm; eluent C, 0.01M TrisHCl, pH 8.0; eluent B, 0.01M Tris HCl, 0.35M NaCl, pH 8.0; gradient, linear 0 - 100% B; at a flow rate, 1.0 ml/min unless specified. (x) PL-SAX 4000A 8um 50 x 4.6mm I.D.; (.) PL-SAX 1000A 8um 50 x 4.6mm I.D..

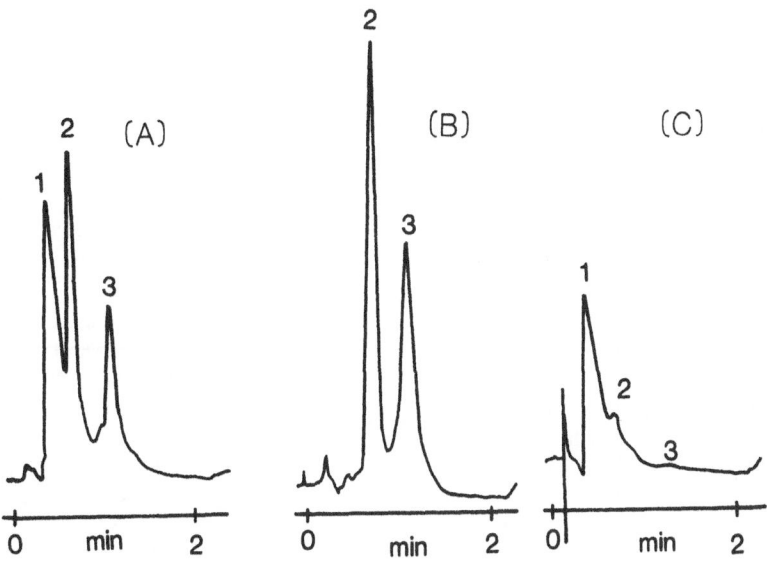

Figure 5.   Anion exchange QC of a Protein A affinity separation of antibodies from serum.   Detector,   UV at 280nm;   eluent C,   0.01M Tris HCl,   pH 8.0;   eluent B,   0.01M Tris HCl,   0.5M NaCl, pH 8.0; gradient, linear 0 -100% B in 2 minutes at a flow rate of 4.0ml/min;   column, PL-SAX 4000A 8um 50 x 4.6mm I.D.   (A) diluted serum;   (B) unbound fraction (C)   bound   fraction.   Peak   identification,   (1)   antibodies;   (2) transferrin; (3) albumin.

The   enzyme   amyloglucosidase produced   by   fermentation   of Aspergillus niger was fractionated from the cell culture filtrate using preparative HPLC.   The 2 fractions I and II were collected and enzyme activity established in both fractions.   The purity was determined using the PL-SAX 4000A high speed matrix, figure 6, in under 2 minutes. The enzyme is known to occur in 2 forms with molecular weights of 99KD and 112KD   (peaks 1 and 2 respectively) of the   same amino acid composition but with different carbohydrate content.   Resolution of the 2 isoenzymes is achieved with the high speed matrix.

It   is also possible to use the high speed matrix for preparative fractions of large biomolecules.   An example is the isolation of amyloglucosidase from the Aspergillus niger cell culture filtrate with a protein concentration determined by UV at 280nm of 395   mg/ml, figure 7.

With the high speed matrix it is possible to purify 1.4mg of the 2 isoenzymes or 20mg of total enzyme in under 2 minutes.

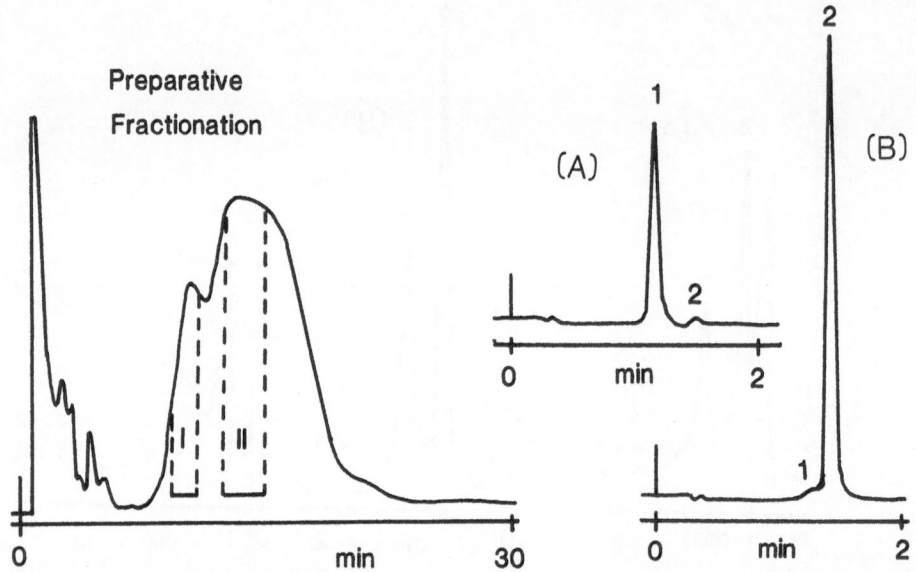

Figure 6. Amyloglucosidase isolation and purity determination. Detector UV at 280nm; eluent C, 0.01M Tris HCl, pH 8.0; eluent B, 0.01M Tris HCl, 0.5M NaCl, pH 8.0; gradient, linear 0 - 100% B in 2 minutes at a flow rate of 4.0 ml/min; column, PL-SAX 4000A 8um 50 x 4,6mm I.D. (A), Fraction I; B, Fraction II. Peak identification, 1, isoenzyme 99KD; 2, isoenzyme 112KD.

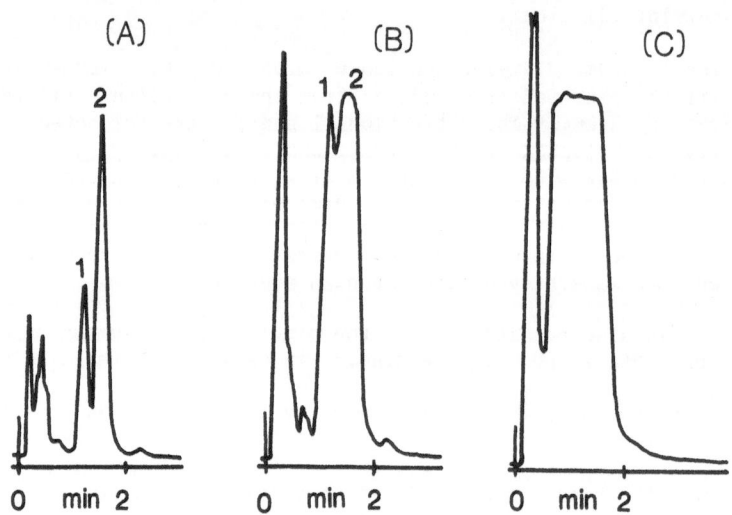

Figure 7. High speed fractionation of cell culture filtrate. Detector, UV at 280nm; eluent C, 0.01M Tris HCl, pH 8.0; eluent B, 0.01M Tris HCl, 0.5M NaCl, pH 8.0; gradient, linear 0 - 100% B in 2 minutes at a flow rate of 4.0 ml/min; column, PL-SAX 4000A 8um 50 x 4.6mm I.D. (A) 3.6mg protein load; (B) 10mg protein load; (C) 20mg protein load.

CONCLUSION

A highly porous, rigid polymer matrix has been produced which is capable of performing high speed HPLC separations without a reduction in resolution. It is anticipated that this material will be particularly useful in the following area:

i)   High speed purity determinations of fractions from a large LC system.

ii)   Optimization of product yields from fermentation broth and process separation systems.

iii) Rapid preparative fractionation of labile biomolecules where minimal exposure to denaturing conditions is required to maximise yield of biologically active material.

REFERENCES

1. Guiochan G., Martin M., J. Chromatogr., 326 (1985) 3

2. Unger K.K., Janzen R., Jilge G., Chromatographia; , 24 (1987) 144

3. Unger K.K., Jilge G., Kinkel J.N., Hearn M.T.W., J. Chromatogr., 397 (1987) 91

4. Anspach B., Unger K.K., Giesche H., Hearn M.T.W., 11th International Symposium on HPLC of Proteins, Peptides and Polynucleotides, Baltimore, December 1984.

5. Burke D.J., Duncan J.K., Siebart C., Ott G.S., J. Chromatogr., 359 (1986) 533

6. Lloyd L.L., Warner F.P., J. Chromatogr; in press

7. Dawkins J.V., Lloyd L.L., Warner F.P., J. Chromatogr., 352 (1986) 157

8. Bristow P.A., LC in Practise, hetp, Cheshire, 1976, p. 41.

PURIFICATION AND CHARACTERISATION OF GENTAMICIN ACETYL TRANSFERASE

JULIE MILLER*, HELEN C. MORRIS**, CHRISTOPHER P. PRICE**,
T. ATKINSON* AND PETER M. HAMMOND*

* Division of Biotechnology
PHLS Centre for Applied Microbiology and Research
Porton Down, Salisbury, Wiltshire, SP4 0JG

** Department of Clinical Biochemistry
London Hospital Medical College
Turner Street, London, E1 2AD

ABSTRACT

Homogeneous gentamicin acetyl transferase from *Escherichia coli* JR225 was isolated using four major steps: ion-exchange chromatography on DEAE-Sepharose, dye ligand chromatography on Procion Scarlet H-2G Sepharose, ion exchange by FPLC MonoQ HR 10/10 and gel-filtration by FPLC Superose 12 10/30. The purified enzyme had a specific activity of approximately 34 units/mg of protein and was shown to exist as a dimer of subunit molecular weight 30,000. The isoelectric point of the enzyme was calculated to be 5.34. The enzyme was maximally active at pH8.5 and at a temperature of 50°C. The enzyme shows Michaelis-Menten kinetics with Km values of: gentamicin (20µM) apramycin (83µM), kanamycin (202µM), neomycin (31µM), netilmicin (2µM), ribostamycin (190µM), soframycin (21µM). The enzyme shows some activity towards 2-deoxystreptamine.

INTRODUCTION

Gentamicin is a widely used aminoglycoside antibiotic used in the treatment of serious infections due to Gram-negative bacilli. Because of its narrow therapeutic index, therapy should be controlled by carefully calculated dosage regimens and accurate monitoring of serum peak and trough levels [2,3]. Peak levels of 5-10mg/l are necessary for maximum therapeutic efficiency [4]. Trough levels of more than 2mg/l may be associated with nephrotoxicity [5] and ototoxicity [6].

Many methods are available for measuring aminoglycoside levels, including, bioassay, radioimmunoassay, high-pressure liquid chromatography, enzyme immunoassay and fluorescent immunoassay [7-11].

Enzymatic methods of detection employ aminoglycoside - modifying enzymes. The aminoglycoside acetylating enzymes catalyse the transfer of an acetate group from acetyl-CoA to an amino group on the antibiotic:

aminoglycoside + acetyl-CoA $\longrightarrow$ N-acetyl-aminoglycoside + CoA

Although enzymatic assays are highly specific and accurate, they require highly purified enzyme [11]. This paper describes the extraction, purification to homogeneity and characterisation of gentamicin acetyl transferase.

MATERIALS AND METHODS

Materials
Chromatography matrices (MonoQ, Superose, DEAE-Sepharose), PhastGel gradient media 10-15, 8-25, IEF 4-6.5 and calibration proteins were purchased from Pharmacia. Procion scarlet H-2G was a gift from ICI, Blackley. All other chemicals were purchased from Sigma.

Bacteria
The bacteria used as a source of enzyme was *Escherichia coli* JR225.

Gentamicin Acetyl Transferase Assay
Gentamicin acetyl transferase activity was determined by a spectrophotometric assay utilising the reduction of DTNB by CoA to produce the thionitrobenzoate (E_{412nm} = 13,600 litre^{-1}mol^{-1}cm^{-1} at pH7.8) [12]. The reaction mixture contained 8.5ml 60mM Tris, pH7.8, 1ml 100mM MgCl$_2$, 10mM DTNB (dissolved in reaction buffer), 200μl 5mM acetyl CoA, 20μl 5mM gentamicin sulphate. To a cuvette was added 975μl of reaction mix. This was incubated for 5 minutes and the reaction was initiated by addition of 1-20μl enzyme. The absorbance was measured at 412nm with a Perkin Elmer 5525 UV/VIS spectrophotometer.

One enzyme unit is defined as that amount of enzyme catalysing the production of one μmol of product per minute at 30°C.

Protein Assay
Protein concentrations of samples were determined using the BCA protein assay method [13]. The protein contents of column eluates were monitored at 280nm.

Purification of Gentamicin Acetyl Transferase
All purification steps were carried out at 4°C unless otherwise stated. Column dimensions are expressed as internal diameter x bed height. All buffers contained 0.1mM phenylmethyl-sulphonyl fluoride.

<u>Disruption of cells</u>: Bacterial cells (200g wet weight) were resuspended in 25mM Tris/HCl, pH8.0, containing 0.5µg of DNase and 0.5µg of RNase per ml. The cell suspension was disrupted using a 15M-8BA Manton Gaulin homogeniser at 55MPa, followed by centrifugation at 9000g for 50 minutes at 4°C.

<u>Ion exchange chromatography</u>: The extract was applied to a 1 litre DEAE-Sepharose column (9cm x 45cm) equilibrated with 25mM Tris/HCl pH8.0. The enzyme was eluted with a 5 litre gradient from 25-600mM Tris/HCl, pH8.0 containing 10mM KSCN. Fractions of 10ml were collected.

<u>Pseudo-affinity chromatography</u>: The main active fractions from the ion exchange column were pooled and dialysed against 50mM Tris/HCl, pH6.5 and applied to a 500ml Procion Scarlet H-2G Sepharose 4B column [14] (4.4cm x 50cm) equilibrated with 50mM Tris/HCl, pH6.5. Enzyme was eluted with 2 litres of 50mM Tris/HCl, pH8.5. Fractions of 10ml were collected.

<u>FPLC MonoQ HR 10/10</u>: The active fractions from the Procion dye-Sepharose column were pooled and applied to the MonoQ HR 10/10 column, previously equilibrated with 50mM Tris/HCl, pH8.5. Enzyme was eluted with a 45ml gradient from 140-315mM NaCl, pH8.5 in loading buffer. Fractions of 0.5ml were collected.

<u>FPLC Superose 12 10/30</u>: Active fractions from repeat runs on the MonoQ column were concentrated using an Amicon 8010 stirred cell fitted with a PM10 membrane (nominal cut-off MW = 10000). The Superose 12 column was equilibrated with 50mM Tris/HCl, pH7.5 containing 200mM NaCl. An aliquot of sample containing 10mg protein was loaded and eluted in the same buffer. Fractions of 0.5ml were collected. Active fractions from repeat runs were pooled and concentrated by ultrafiltration as above, followed by dialysis against 50mM Tris, pH8.0. This material was used for characterisation studies.

Polyacrylamide Gel Electrophoresis

The molecular weight of gentamicin acetyl transferase was determined using the PhastSystem (Pharmacia). Subunit molecular weight was determined using Phastgel 10-15 gradient. The native molecular weight was determined using the Phastgel 8-25 gradient, in 25mM Tris/HCl, pH8.0. All gels were stained with Coomassie Brilliant Blue R-250.

Determination of Isoelectric Point

The isoelectric point of the enzyme was determined using IEF 4-6.5 Phastgels. Gels were stained with Coomassie Brilliant Blue R-250.

Effect of pH on Enzyme Activity

The effect of pH on enzyme activity was determined using the following buffers: Na_2HPO_4/Citric acid (pH3.7-7.5), Tris/HCl (pH7.1-8.9), $Na_2B_4O_7$/NaOH (pH9.5-10.0), Na_2HPO_4/NaOH (pH10.4-11.2), at concentrations of 50mM. Tris/HCl, pH7.8, used in the assay of gentamicin acetyl transferase, was replaced by each of the above buffers and a pH profile for the enzyme was produced.

A summary of gentamicin acetyl transferase purification is shown in Table 1.

TABLE 1
Purification of gentamicin acetyl transferase to homogeneity from 200g bacterial cell paste

Sample	Enzyme (units)	Protein (mg)	Specific Activity Umg^{-1} Protein	Recovery %	Purification Fold
Cell extract	6,421	13,662	0.47	100	0
DEAE-Sepharose	3,917	3,377	1.20	61	2.5
Procion Scarlet H-2G-Sepharose 4B	3,339	1,433	2.30	52	5.0
MonoQ	2,889	114	25.30	45	54.0
Superose 12	2,569	75	34.30	40	73.0

The enzyme was shown to be homogeneous, giving a single protein band on SDS/polyacrylamide gel electrophoresis.

Molecular Weight of Native Enzyme and Subunits

The relative molecular mass of the native enzyme was shown to be 60,000 with a subunit relative molecular mass of 29,800.

Isoelectric Point Determination

The pI of gentamicin acetyl transferase was shown to be 5.34.

Temperature Stability

The effect of temperature on enzyme activity was determined over the range 10-100°C. The maximal enzyme activity was shown to be at 50°C (Figure 5).

Thermal inactivation of the enzyme was monitored over a period of 450 minutes. There was no loss of activity at 4°C and 20°C. At temperatures of 37°C, 45°C, 56°C and 60°C enzyme activity decreased with time. The $t_{1/2}$ at 45°C was 16 minutes (Figure 6).

RESULTS AND DISCUSSION

Enzyme Purification

DEAE-Sepharose: Gentamicin acetyl transferase was eluted at 292mM Tris/HCl (Figure 1), fractions 240 to 290 were pooled and dialysed against 50mM Tris/HCl, pH6.5.

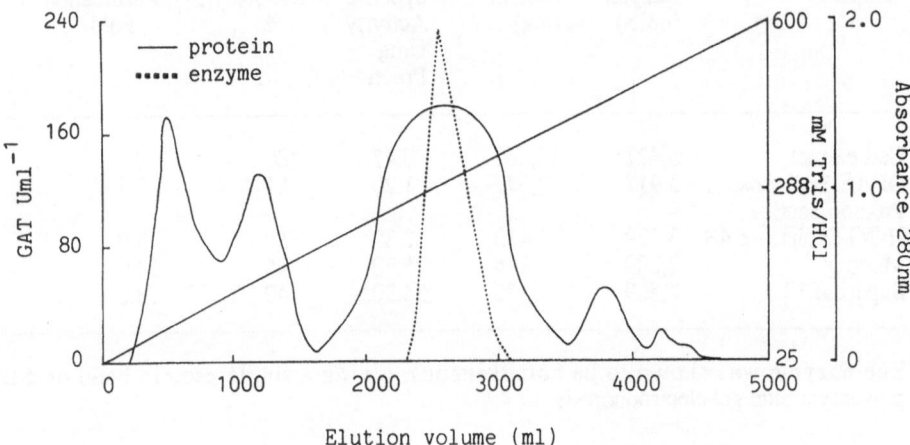

Figure 1. Anion exchange chromatography on DEAE-Sepharose.

Procion Scarlet H-2G: The enzyme was loaded in 50mM Tris/HCl, pH6.5 and eluted in 50mM Tris/HCl pH8.5 (Figure 2) Fractions 60-95 were pooled.

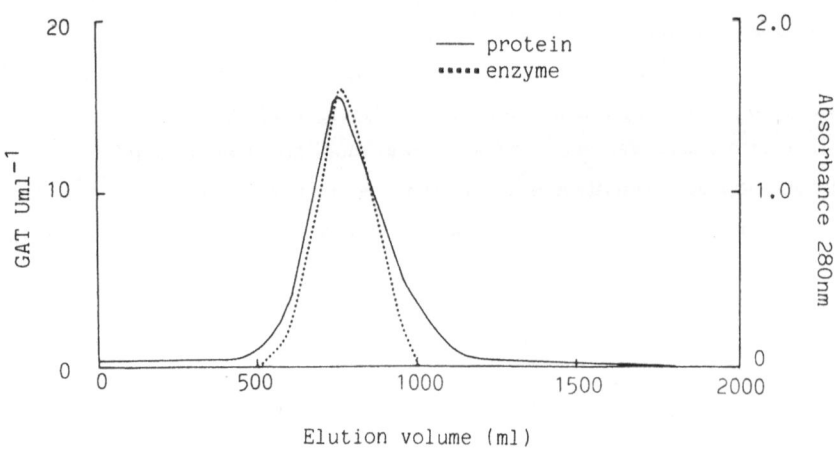

Figure 2. Pseudo-affinity chromatography on Procion scarlet H-2G.

<u>MonoQ HR 10/10</u>: The enzyme was eluted at 210mM NaCl (Figure 3), fractions 55 to 63 were pooled.

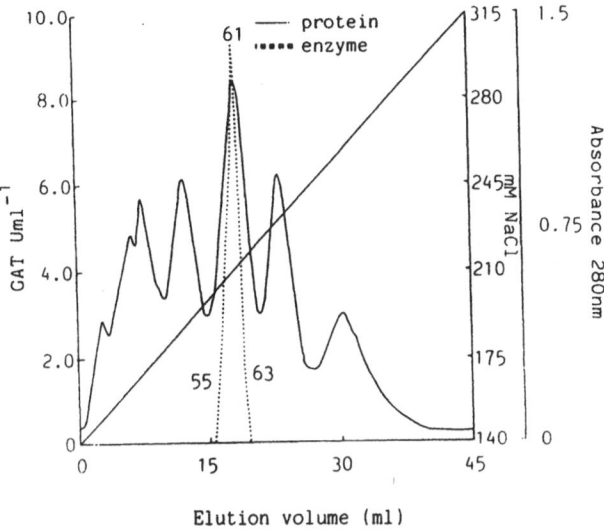

Figure 3. Anion-exchange chromatography on MonoQ HR 10/10.

<u>Superose 12 10/30</u>: The enzyme was loaded and eluted in 50mM Tris/HCl, pH7.5 containing 200mM NaCl (Figure 4) fractions 15 to 17 were pooled.

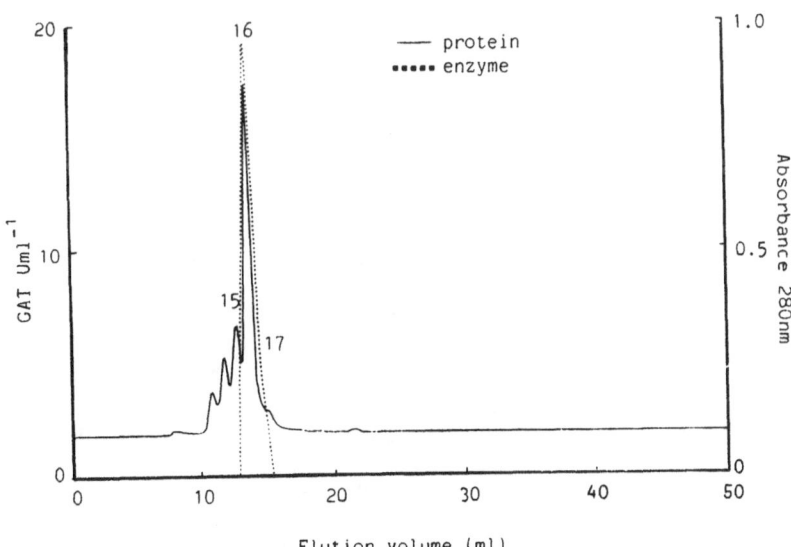

Figure 4. Gel filtration chromatography on Superose 12 prep 10/30.

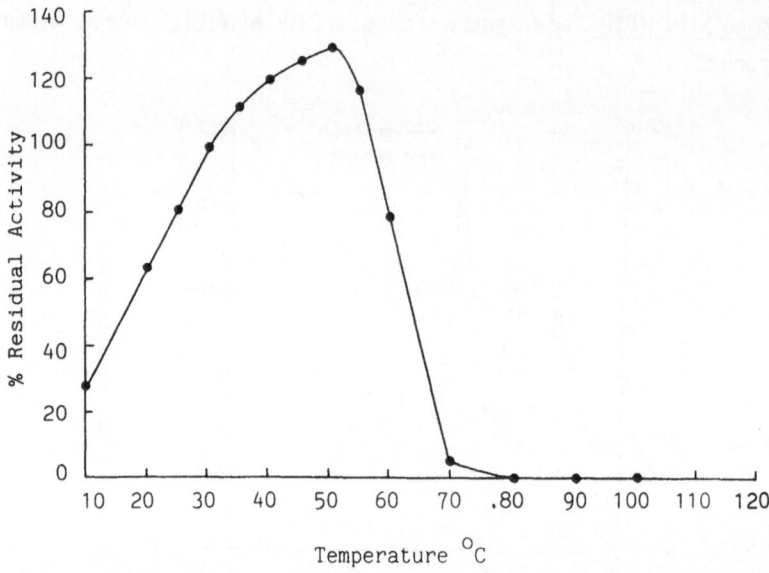

Figure 5. The effect of temperature on gentamicin acetyl transferase activity.

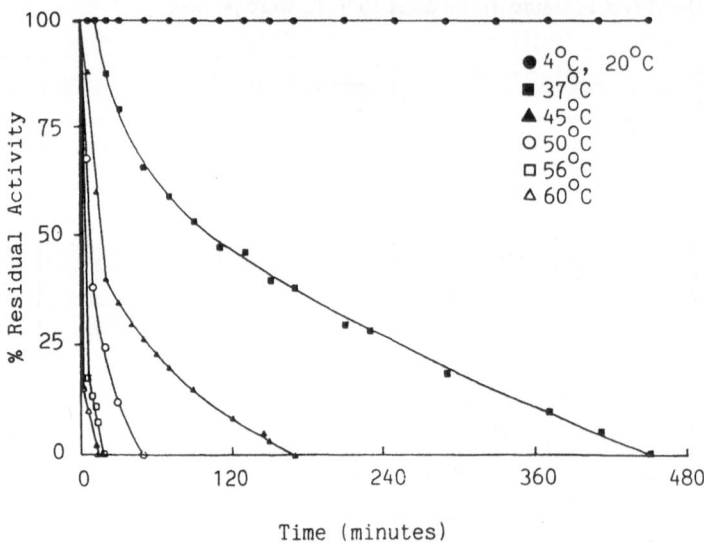

Figure 6. Thermal inactivation of gentamicin acetyl transferase.

Effect of pH

Gentamicin acetyl transferase showed activity over a broad pH range and was optimally active at pH8.5 (Figure 7).

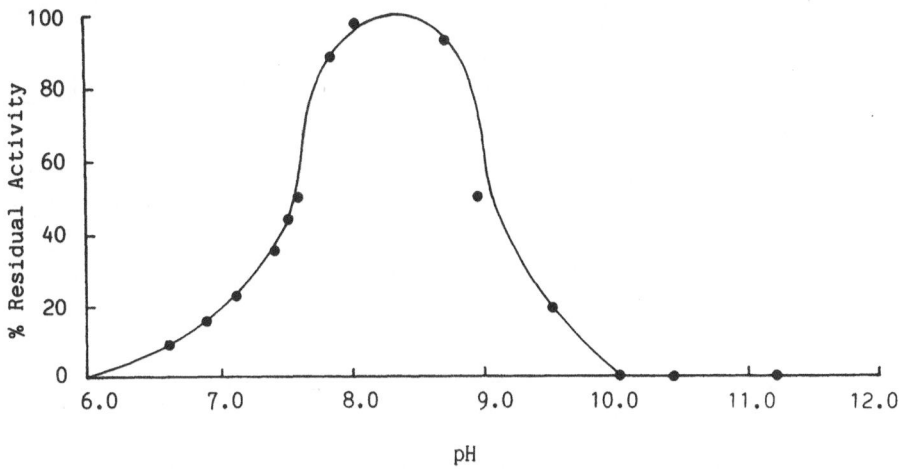

Figure 7. The effect of pH on gentamicin acetyl transferase activity.

Kinetic Studies

Kinetic parameters were evaluated by the double-reciprocal-plot method. The Km for gentamicin was determined as 20μM at 30°C. The Km for other aminoglycosides at 30°C were determined as: apramycin 83μM, kanamycin 202μM, neomycin 31μM, netilmicin 2μM, ribostamycin 190μM, soframycin 21μM. The enzyme showed some activity to 2-deoxystreptamine. This was calculated as 10%, taking gentamicin activity as 100%.

CONCLUSION

Gentamicin acetyl transferase may be purified to homogeneity using the techniques described in this paper. Due to its high affinity for several aminoglycosides it may be useful in the determination of these antibiotics in clinical samples, or in the sterility testing of pharamceutical preparations [15].

ACKNOWLEDGEMENTS

Many thanks to Professor Julian Davies (Pasteur Institute, Paris) for the *E.coli* JR225 strain, and to Dr. C.V. Stead for supplying the Procion Dyes.

REFERENCES

1. Reeves, D.S., Gentamicin therapy. Brit. J. Hosp. Med., 1974, **12**, 837-850.

2. Reeves, D.S. and Wise, R., Antibiotic assays in clinical microbiology. In Laboratory Methods in Antimicrobial Chemotherapy, ed. D.S. Reeves, Churchill Livingstone, Edinburgh, 1978, pp.137-143.

3. Barza, M. and Lauermann, M., Why monitor serum levels of gentamicin? Clin. Pharmacokinet., 1978, **3**, (3), 202-215.

4. Schentag, J.J., Cumbo, T.J., Jusko, W.J. and Plaut, M.E., Gentamicin tissue accumulation and nephrotoxicity reactions. J.A.M.A.., 1978, **240**, (19), 2067-2069.

5. Jackson, C.G. and Arcieri, G., Ototoxicity of gentamicin in man. A survey and controlled analysis of clinical experience in the United States. J. Infect. Dis., 1971, **124**, Suppl. S130-137.

6. Buchanan, A.G., Witwicki, E. and Albritton, W.L., Serum aminoglycoside monitoring by enzyme immunoassay, biological and fluorescence immunoassay procedures. Am. J. Med. Technol., 1983, **49**, (6), 437-441.

7. Delaney, C.J., Opheim, K.E., Smith, A.L. and Plorde, J.J., Performance characteristics of bioassay, radioenzymatic assay, homogeneous enzyme immunoassay and high-performance liquid chromatographic determination of serum gentamicin. Antimicrob. Agents Chemother., 1982, **21**, (1), 19-25.

8. Ratcliff, R.M., Mirelli, C., Moran, E., O'Leary, D. and White, R., Comparison of five methods for the assay of serum gentamicin. Antimicrob. Agents Chemother., 1981, **19**, (4), 508-512.

9. White, L.O., Scammell, L.M. and Reeves, D.S., Serum aminoglycoside assay by enzyme-mediated immunoassay (EMIT): correlation with radioimmunoassay, fluoroimmunoassay and acetyltransferase and microbiological assays. Antimicrob. Agents Chemother., 1981, **19**, (6), 1064-1066.

10. Witebsky, F.G., Sliva, C.A., Selepak, S.T., Ruddel, M.E., MacLowry, J.D., Johnson, E.E. and Elin, R.J., Evaluation of four gentamicin and tobramycin assay procedures for clinical laboratories. J. Clin. Microbiol., 1983, **18**, (4), 890-894.

11. Williams, J.W., Langer, J.S. and Northrop, B.D., A spectrophotometric assay for gentamicin. J. Antibiotics, 1975, **28**, 982-987.

12. Shaw, W.V., Chloramphenicol acetyltransferase from chloramphenicol-resistant bacteria. Methods in Enzymology, 1975, **43**, 737-755.

13. Smith, P.K., Krohn, R.I., Hermanson, G.T., Mallia, A.K., Gartner, F.H., Provenzano, M.D., Fujimoto, E.K., Goeke, N.M., Olson, B.J. and Klenk, D.C., Measurement of protein using bicinchoninic acid. Anal. Biochem., 1985, **150**, 76-85.

14. Atkinson, T., Hammond, P.M., Hartwell, R.D., Hughes, P., Scawen, M.D., Sherwood, R.F., Small, A.P., Bruton, C.J., Harvey, M.J. and Lowe, C.R., Triazine-dye affinity chromatography. Biochem. Soc. Trans., 1981, **9**, 290-293.

15. Simpson, A.M. and Breeze, A.S., The sterility testing of aminoglycosides: A new approach using enzyme inactivation. J. Appl. Bacteriol., 1981, **50**, 469-474.

AFFINITY PURIFICATION USING RADIAL STREAMING AFFINITY CHROMATOGRAPHY

Y. PLANQUES and H. PORA
Cuno Europe/Life Sciences Division
1, Boulevard de l'Oise, 95030 Cergy-Pontoise, FRANCE

ABSTRACT

A new support for affinity chromatography based on a cellulose-acrylic network was developed.

Using differently substituted supports, we have been able to purify trypsin, antibodies, and trypsinogen with their corresponding ligand. In a single pass, we were able to obtain good purity (> 85%) with good yields. Using coupled p-amino-benzamidine, we have shown that this purification method can be scaled-up very easily on radial-flow cartridges.

INTRODUCTION

With the advances in techniques based on recombinant DNA, hybridomas, etc. the biomedical and pharmaceutical industries are demanding purification methods which can achieve high product purity in a few simple steps. So affinity chromatography has become a method of choice, since high purity can be achieved in a single step [1,2]. But for large scale purification, there are two major drawbacks in using this technique : it is difficult to predict the chromatographic behaviour after scaling up; the gel support commonly used has disadvantages in hydrodynamic strength and rigidity. We have solved the two impediments by introducing radial flow chromatography rigid supports [3].

A modified cellulose and special techniques in fabricating cartridges for radial flow chromatography are described here.

Cellulose, which is hydrophilic in nature, is grafted with
gel type polymers. The degree of polymerization and the
reaction conditions control the porosity and the other
physical properties of the composite matrix. The matrix
thus has the physical rigidity of cellulose and
chemical functions resembling gel polymers.
The modified cellulose possesses large numbers of
modified groups for preparing highly substituted
derivatives.

To reduce steric hindrance, spacer arms have been
introduced. We describe herein the use of various
substituted matrices for the purification of trypsin,
antibodies and trypsinogen. This method presents great
advantages as we have obtained good purity and it gives
the capability of easy scale-up.

MATERIALS AND METHODS

Coupling conditions :
Two different kinds of cartridges containing modified cellulose have been tested for ligand coupling.

Amino derivatives with C_{18} spacer arm were used for
coupling of protein A. Activation of amino derivatives is
accomplished by 0.25% glutaraldehyde in 0.1M borate buffer
at pH 8.0 for 2 hours. The ligand of interest is dissolved
in 0.05M sodium phosphate+ 0.25M NaCl solution and recycled through the cartridges overnight. The cartridge is
drained and the unbound reactive sites are blocked by 2%
glycine ethylester hydrochloride solution. The amount of
ligand attached to the matrix is determined by the changes
of ligand concentration before and after the coupling reaction.

Hydroxyl derivatives with C_8 spacer arms were used
for coupling of p-aminobenzamidine and Lysine. The activation of hydroxyl derivatives is accomplished by 1.5%
sodium metaporiodate solution. Other steps remained
unchanged.

Chromatographic procedure for trypsin purification :
Crude trypsin from porcine pancreas (Sigma) is dissolved in
equilibration buffer = 0.05M Tris-HCl pH 7.6 containing
0.25 M NaCl, and 2mM $CaCl_2$ and purified on p-aminobenzamidine Zetaffinity. After washing with equilibration
buffer, elution was performed by the use of 0.2M Glycine
-HCl at pH 2.3.
Enzyme activity was measured by using the substrate
Bz-lle-Glu-Gly-Arg-b-nitroaniline from Kabi.

Chromatographic procedure for plasminogen purification

Lysine Zetaffinity was equilibrated with 0.1M sodium
phosphate pH 7.4 containing 0.05 NaCl and 0.01 % Tween
80. Plasma was applied to the cartridge after dilution
with 0.5 v of equilibration buffer. After washing with
equilibration buffer, elution was performed by means
of 0.2 M 6-amino caproic acid in equilibration buffer.
Purity was assessed by SDS-PAGE ; plasminogen activity
was measured with S-2251 Kabi substrate.

Chromatographic procedure for monoclonal antibodies
purification from hybridomas.

Protein A Zetaffinity was equilibrated with 150 mM glycine
NaOH pH 8.9 containing 300 mM NaCl. Cell culture supernatant containing 5% (v/v) IgG free FCS and 25 to 50 pg of
murine IgG2b was diluted with 0.1 v of 10X equilibration
buffer. After washing with equilibration buffer, elution
was performed by the use of 300 mM glycine - HCl pH 2.7.
Purity was assessed by PAGE-SDS, and activity by specific
ELISA.

RESULTS AND DISCUSSION

Trypsin purification by p-aminobenzamidine

The performance of a radial flow system on a membrane
based material was investigated. P-aminobenzamidine
whose molecular weight is about 100, was used for trypsin
purification although not a very specific ligand [4]. In
fact, it has been previously used in affinity chromatography to separate urokinase [5], bovine thrombin [6] and
kallikrein [7] besides trypsin.

As the crude preparation of trypsin used contained
chymotrypsin activity, the chromatogram showed that the
eluted peak contained trypsin of much higher specific activity.

Figure 1 = Effluent history in trypsin purification
on a 800 Zetaffinity cartridge.

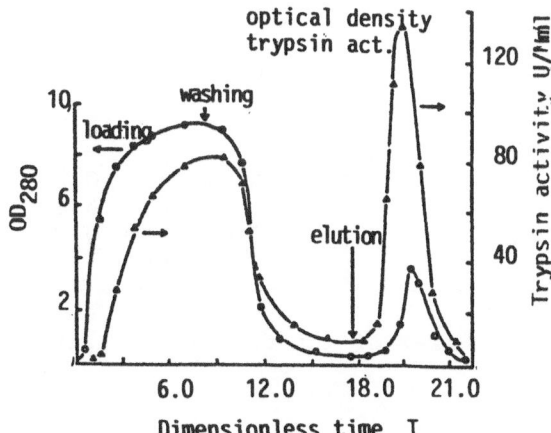

As the system has been designed to sustain high flow-rates, we performed two experiments using the cartridge at 84 ml/min. and at 295 ml/min. The results are summarized in table 1.

	EXPERIMENT I	EXPERIMENT II
Flow-rate:(ml/min)	295	83
Ligand	1 260 mg PAB	1 260 mg PAB
Crude Solution :		
V (Liters)	4.00	4.00
Conc.(mg/ml)	7.68	7.75
Act. (U/ml)	90.6	84.3
Spec.Act.(U/mg)	11.8	10.9
Purified Solution:		
V (Liters)	2.76	2.70
Conc.(mg/ml)	0.68	0.75
Act. (U/ml)	50.9	56.6
Spec.Act.(U/mg)	74.9	75.5
Purification factor	6.3	6.9

Table 1 : Trypsin purification by p-aminobenzamidine

The results showed that the differences between the two experiments were extremely small. Thus, using such a method extremely high flow-rates compared to those usually achieved with soft gels can be used, and up to 50 liters of sample can be processed within a few hours.

The use of this radial flow cartridge system [8] in ion exchange has shown linear scale-up capability,we investigated the scaling-up on p-aminobenzamidine. Radial flow cartridges of different sizes (Table 2)were used for this study.Parameters such as flow-rate, and ligand capacity (Trypsin) were evaluated. Results are summarized in Figure 2 and 3;

Nominal size	250 ml	800	3 200
Outer Diameter	7.0 cm	12.7	12.7
Inner Diameter	0.6 cm	0.9	0.9
Height	6.4 cm	6.4	23.8
Actual Size	210 ml	810	3 020

Table 2 : Sizes of the different cartridges.

Figure 2 = Linear Scaling-up
of radial flow cartridges;

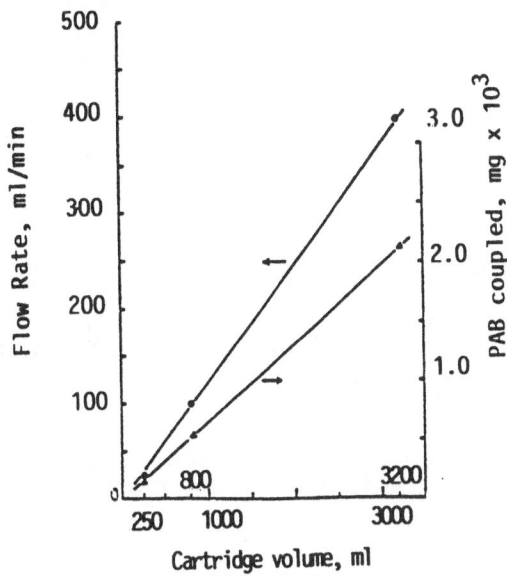

Figure 3 = Linear Scaling-up of cartridge capacity.

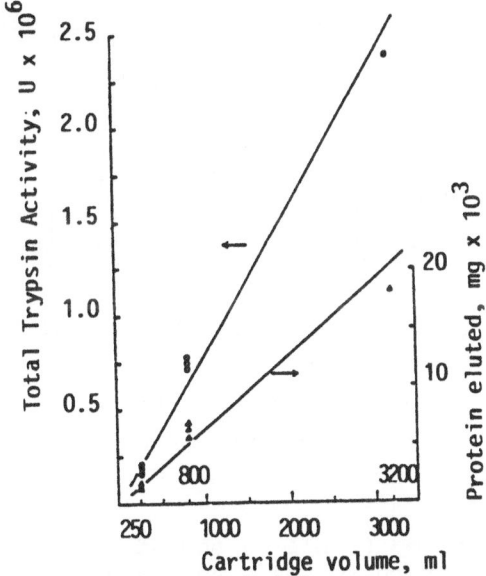

We have shown that a linear relation between cartridge
size and flow-rate; p-aminobenzamidine loading,
protein capacity and trypsin activity existed within the
limits of experimental error. Results of replicate experiments
are shown by multiple points.

The advantages of having such a linear scale-up
capability reside in the fact that by applying
simple scaling-up factors one can increase very
easily the size of its purification scale from lab
scale to more industrial scale. To really investigate
all the possibilities of such a cartridge chromatographic system, we had to evaluate its capability system
on more complex samples where the component of interest
is present at very low concentration in a biological
sample.

Plasminogen purification on a cartridge containing bound lysine

A 250 lysine Zetaffinity cartridge was used for the purification of plasminogen from human plasma (400ml). The
entire volume was applied in a single pass at 20 ml/min.
(Figure 4).

Figure 4 = Plasminogen purification from human
plasma with lysine Zetaffinity 250 cartridge.

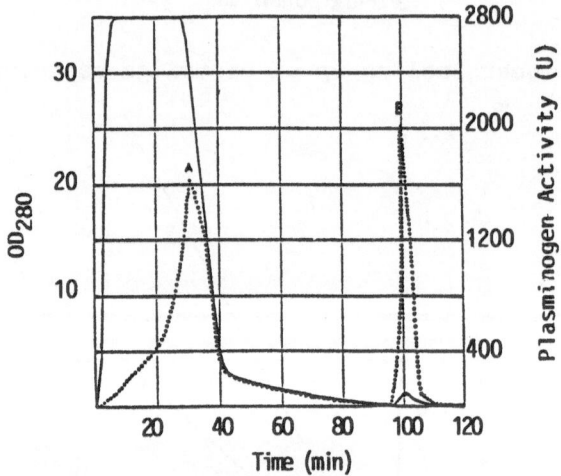

Bound plasminogen was eluted (Peak B)
with 0.2M 6-aminocaproic acid in
equilibration buffer.

The results obtained (Table 3) showed that the recoveries
were good and that there was a very significant increase
in purity. Analysis by SDS-PAGE showed that in the
elution peak, there was a single band whose molecular
weight is consistent with that of plasminogen.

Device Size	Plasma Volume	Plasminogen Activity Eluted	Specific Activity Eluted	Purif. Fold	Total Rcvry
250Capsule (1)	400mL	8948units	358	144X	57%
250Capsule (2)	250mL	6045units	410	124X	74%

(1) 400mL plasma diluted to 600mL in equilibration buffer.
(2) 250mL plasma diluted to 500mL in equilibration buffer.
* Entire fractionation done at 4°C.

Table 3 : Plasminogen purification using Lysine Zetaffinity.

So using this method, with cartridges of a larger size,
litres of plasma can be easily treated within a few
hours.

Purification of monoclonal antibodies on a cartridge containing bound protein A

Using a 100 Protein A Zetaffinity cartridge, we purified
murine IgG 62b from cell culture supernatant. Due to the
very large volumes we had to process (6 to 12 liters), we
studied the influence of the loading flow-rate on the
cartridge capacity. The cartridge loadings were
performed at room temperature at flow-rates ranging from
5 to 100 ml/min. To provide maximum interactions, the
samples were recirculated once at the same flow-rate.

Figure 5 : Influence of the loading flow-rate
on 100 Protein A Zetaffinity capacity. Bound
material was quantified by specific ELISA and
Coomassie G-250.

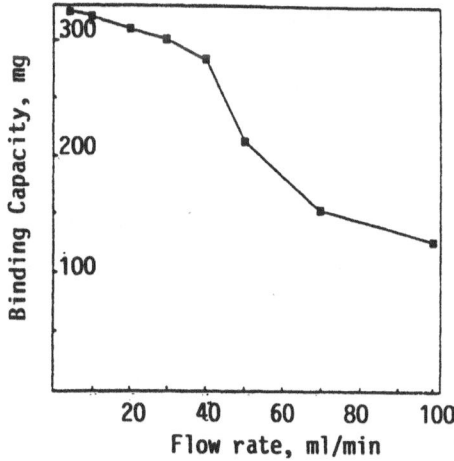

Up to 30ml/min, the cartridge capacity remains high
(300 mg of mAb). For higher flow-rates, the binding
capacity showed a severe reduction. PAGE-SDS analysis
of the eluted fraction showed that the purified immunoglobulins were free of transferin and albumin.
The reactivity of the antibodies was also maintained.

We have shown that this cartridge method is a very
attractive solution for affinity purification since
due to the very high flow rates, we can still purify
quantities of product even if it stands in a very
dilute sample.

CONCLUSION

We have described a new chromatographic technique which
uses a membrane based support built up into a cartridge
system. The advantages of this technique reside in the
very high flow rates which can be applied to the system
due to the open structure of the matrix. So large volumes
can be chromatographied within a few hours. It has then
the same capabilities as the radial flow ion exchange
chromatography already designed by us such as : high
dynamic capacity, high flow-rates [9,10]. We have demonstrated using radial flow cartridges of various sizes that
this system shows a linear scale-up in the same way as the
ion exchange [8].

This means that once the experimental conditions have
been set at lab scale they can be easily scaled-up to
a production scale. In the future, this technique will
certainly be a good choice for large scale production
of high added value products and other more specific
media will be developed to fulfill the requirements
of end users.

BIBLIOGRAPHY

1. Cuatrecases P., Adv. Enzymol. $\underline{36}$, 29, (1972).

2. Dean P.D.G, Johnson W.S and Middle F.A. Eds, Affinity Chromatography - a practical approach - IRL Press Limited, Oxford, England, (1985).

3. Hou K.C., Mandaro R.M., Biotechniques $\underline{4}$, 358 (1986).

4. Jany K.D., Keil W., Meyer H., Kiltz H.H, Biochem. Biophys. Acta, $\underline{453}$, 62 (1976).

5. Holmberg L., Bladh B., Astedt B., Biochem. Biophys. Acta, $\underline{445}$, 215, (1976).

6. Schma G. Physiol. Chem., 353, $\underline{810}$ (1972).

7. Heimark R.L., Dane E.W., Biochemistry, $\underline{18}$, 5743 (1979).

8. Jungbauer A., Unterluggauer F., Uhl K., Buchacher A., Steindl F, Pettauer D. and Wenish E., Biotech. Bioengin., $\underline{32}$, 326, (1988).

9. Menozzi F.D., Vanderpoorten P., Dejaiffe C. and Miller A.O., J. Immunol. Meth., $\underline{99}$, 229,(1987).

10. Benanchi P.L ., Gazzei G. and Giannozzi A., J. Chromatography, $\underline{450}$, 133,(1988).

MULTI-COMPONENT ADSORPTION OF PROTEINS TO ION EXCHANGERS

GRAHAM L. SKIDMORE & HOWARD A. CHASE
Department of Chemical Engineering, University of Cambridge
Pembroke Street, Cambridge, CB2 3RA, U.K.

ABSTRACT

A model system, consisting of bovine serum albumin (BSA), lysozyme and the cation-exchanger S Sepharose FF, was used to investigate multicomponent protein adsorption to ion-exchangers. The equilibrium characteristics of the adsorption of the pure proteins to the strong cation-exchanger have been characterised and shown to obey a Langmuir type equation. Experiments have also been carried out with both proteins present simultaneously. Two models, one based on a complete absence of competition between adsorbing molecules and the other a competitive model, based on the assumption that all adsorption sites are available to both proteins have been compared to experimental results. Evidence for competitive adsorption was seen in experiments in which breakthrough curves and the profiles of adsorbed proteins in packed beds were determined. However, some discrepancies were found suggesting that when considering multicomponent protein adsorption to ion-exchangers it may also be necessary to take account of factors such as the molecular size of the adsorbing proteins and inter-protein interactions.

INTRODUCTION

Ion exchange adsorbents have found widespread use in the purification of proteins [1], both in the laboratory and in the production plant, since the introduction in 1956 of the first ion-exchanger specifically designed for proteins. We have been studying the adsorption of proteins to ion-exchangers to investigate how different properties of the ion-exchangers affect their adsorption performance, and with the aim of producing simple models of the purification process to assist in process design. Previous studies examined how the properties of different anion-exchangers, such as functional groups and the particle matrix, affected performance [2]. Whilst there are many published studies of the adsorption of single proteins to different adsorbents, few studies have been reported using more realistic multicomponent systems which involved more than one adsorbing protein. A theoretical discussion of an approach to modelling multicomponent protein adsorption has been published [3] although there was no associated experimental data. Recent studies of the adsorption of a mixture of albumin and β-lactamase to encapsulated ion-exchangers showed that competition between the two adsorbing proteins occurred, with albumin displacing the more weakly binding β–lactamase from the ion-exchanger [4]. Methods have also been described for the measurement of competitive adsorption isotherms by frontal chromatography for small solutes adsorbing to HPLC columns [5].

As a simple start to the study of multicomponent adsorption, we have examined a model system consisting of two adsorbing species, BSA and lysozyme, and the strong cation-exchanger S Sepharose FF. We have studied the characteristics of the adsorption of each pure protein and show that the Langmuir adsorption isotherm could be used to describe the equilibrium adsorption characteristics of both proteins. We also present the results of studies of multicomponent adsorption of BSA and lysozyme to S Sepharose FF and compare these to two models, one based on competitive adsorption to the ion-exchanger and the other based on non-competitive adsorption.

The choice of proteins for these studies was determined on two grounds. Firstly the high capacity of ion-exchangers for proteins meant that gram quantities of pure proteins were required at reasonable cost. Secondly there had to be a method of quantifying the amounts of each protein present in a mixture. The difference in molecular size between BSA and lysozyme enabled the proteins to be separated by molecular exclusion chromatography and the amounts of the two proteins could be determined by analysis of the resulting chromatogram.

THEORY

Proteins adsorb to ion-exchangers as a result of ionic interactions between charged groups on the surface of the protein and oppositely charged groups on the ion-exchanger. As the three-dimensional distribution of ionic groups on the surface of the adsorbent is random, the actual protein adsorption site is not a unique entity. Hence the adsorption site on a protein ion-exchanger cannot strictly be treated in the same manner as that postulated for affinity adsorption where molecules of the immobilised affinity ligand constitute adsorption sites with identical properties. However, experimental results from systems in which a single protein is adsorbed to an ion-exchanger yield equilibrium adsorption isotherms which can be described by a Langmuir equation of the form shown below [6,7]:

$$q^* = \frac{c^* q_m}{c^* + K_d} \tag{1}$$

The observation of a Langmuir type shape can probably be explained by protein adsorption to the ion-exchanger continuing until there is no longer room on the surface of the adsorbent for further molecules of adsorbate to bind. Hence further adsorption ceases once monolayer coverage has occurred.

The situation is further complicated when the adsorption of two or more proteins is being considered. As a result of the different sizes and distribution of charges on the surfaces of different proteins, the number of ionic groups that will participate in the adsorption interaction and the amount of adsorbent surface which interacts with the different proteins will vary. Whilst recognising the complexities of multicomponent adsorption of proteins to ion-exchangers, we have adopted two extreme views to analyse two-component protein adsorption, namely a non-competitive model and a totally-competitive model.

Non-competitive adsorption model
One extreme view of the adsorption of two proteins to an ion-exchanger is to assume that the adsorption sites for the two proteins are mutually independent, that is the adsorption of one type of protein to the ion-exchanger in no way affects the adsorption of the other species and there is therefore no competition between the proteins for the adsorption sites. If there is no competition between the proteins for adsorption, the adsorption characteristics of each protein will be the same as if the other protein were not present, i.e.

$$q_1^* = \frac{c_1^* q_{m1}}{c_1^* + K_{d1}} \quad \text{and} \quad q_2^* = \frac{c_2^* q_{m2}}{c_2^* + K_{d2}} \tag{2}$$

where the subscripts $_1$ and $_2$ indicate adsorbate species 1 and 2.

Totally competitive adsorption model
The other extreme approach to the analysis of two-component adsorption is to assume that there is total competition between proteins for adsorption to the ion-exchanger. Although the exchanger shows different maximum capacities for the two proteins (q_{m1} and q_{m2}), a

competitive model can be developed by involving a fractional occupancy of the adsorption capacity for each type of protein. Such an approach yields:

$$q_1^* = \frac{q_{m1} \, c_1^*}{K_{d1} + c_1^* + \frac{K_{d1}}{K_{d2}} c_2^*} \quad \text{and} \quad q_2^* = \frac{q_{m2} \, c_2^*}{K_{d2} + c_2^* + \frac{K_{d2}}{K_{d1}} c_1^*} \tag{3}$$

The equilibrium position of a batch system can be determined by solving equations (3) simultaneously with the mass balance equations:

$$Vc_{o1} = V c_1^* + v q_1^* \quad \text{and} \quad Vc_{o2} = V c_2^* + v q_2^* \tag{4}$$

using the values of K_{di} and q_{mi} determined in single component adsorption isotherm measurements.

EXPERIMENTAL

Materials
BSA and lysozyme (EC 3.2.1.17) were obtained from Sigma Chemical Company Ltd. (Poole, England), catalogue numbers A-3912 and L-6876 respectively. BSA has a relative molecular mass of 66,300 daltons and an isoelectric point (pI) of pH 4.7, whilst lysozyme has a relative molecular mass of 14,500 daltons and a pI of pH 11.1. All solutions were buffered with 0.1M sodium acetate/acetic acid, pH 5. This pH was chosen as it gave strong adsorption of both proteins and yet was not too far from neutrality. S Sepharose FF® was from Pharmacia LKB Biotechnology AB, Uppsala, Sweden.

Determination of Protein Concentration in the Liquid Phase
In experiments in which only one protein was present in solution, it was possible to determine protein concentration by measuring the optical density at 280 nm. In experiments in which both BSA and lysozyme were present in solution together, quantitation of the concentrations of the individual proteins was achieved by analytical separation of the proteins by molecular exclusion chromatography using a Fast Protein Liquid Chromatography system.

Batch Equilibrium Adsorption Studies
For single component experiments, a known volume of a 50/50 (v/v) suspension of S Sepharose FF in buffer was added to each of a series of flasks containing known volumes of buffered protein solution at different concentrations. The flasks were incubated overnight in a shaking water bath at 25°C to allow equilibrium to be established. The ion-exchanger was then allowed to settle under gravity for approximately 30 minutes and the resulting supernatant was filtered before determining the equilibrium concentration of protein in the soluble phase by UV spectrophotometry. The amount of protein adsorbed to the S Sepharose FF was then calculated by mass balance.

Two component adsorption experiments were performed in the same manner but with the difference that each flask contained a mixture of the two proteins. The amounts of BSA and lysozyme used in each flask were always equal on a mass basis. At equilibrium the concentration of each protein present in the liquid phase was determined by FPLC, and the amounts of each protein that were adsorbed to the ion-exchanger was calculated by mass balance.

The adsorption capacities quoted in all the results are based on the volume that the adsorbent would occupy when packed in a bed.

Packed Bed Experiments
All column experiments were performed with 2 ml (settled volume) of S Sepharose FF packed in a chromatography column, 1 cm diameter . All experiments were performed at a volumetric flow rate of 1 ml/min (superficial velocity 1.27 cm/min). Optical density at 280 nm of the outlet stream was recorded and fractions of 2 ml were collected at the column exit as required for FPLC analysis as described above.

Determination of Adsorbed Protein Profiles in Packed Beds

A series of packed bed experiments, in which the bed was loaded with a two-component protein mixture for various times, was performed. At the end of the loading period liquid was removed from the bed by passing 5 ml of air through the bed with a syringe in the reverse direction to that of the original liquid flow. The ion-exchange bed was then extracted from the column by pushing the lower endpiece upwards with a threaded rod. As the adsorbent emerged from the column, slices of approximately 2-3 mm (giving 9-12 slices in total from each bed) were removed with a scalpel and placed into weighed bijou bottles. The bottles were sealed and reweighed. The amount of ion-exchanger present in each slice was calculated from these weights. The adsorbed proteins were eluted from the ion-exchanger in each sample with 5 ml of 1M sodium chloride. The concentrations of BSA and lysozyme present in this solution were then determined by FPLC. The amount of each protein that had been adsorbed on each slice of S Sepharose FF could then be calculated.

Consecutive Application of Single Protein Solutions to Packed Beds

Packed bed experiments, in which a feed solution containing only one of the proteins was applied, were performed. When the protein concentration of the outlet stream (c), as determined from optical density measurements equalled, or was approaching, that of the inlet stream (c_o), the incoming feed stream was switched to a solution containing only the other protein. Fractions were collected at the column exit for analysis by FPLC.

RESULTS

Adsorption Isotherms

The isotherms for the adsorption of pure BSA or lysozyme to S Sepharose FF in 0.1M sodium acetate buffer, pH 5 are shown in Figure 1.

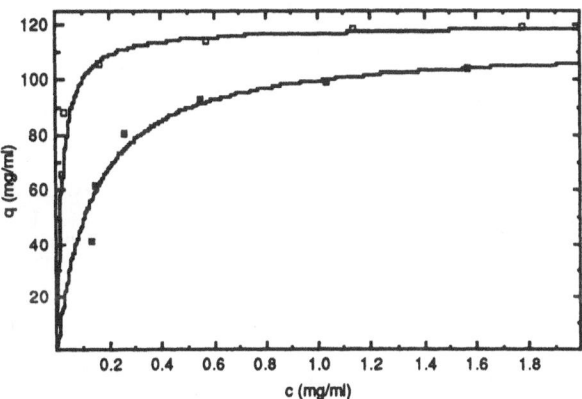

Figure 1: Adsorption isotherms for the binding of Lysozyme (□) and BSA (■) to S Sepharose FF in 0.1M acetate buffer, pH 5 at 25°C. The data are plotted as mg protein adsorbed per ml S Sepharose FF against mg/ml protein in solution.

The experimental data for both proteins fitted well to a Langmuir isotherm and the characteristic parameters K_d and q_m are shown in Table 1. The maximum capacities for the two proteins are similar when compared on a mass basis but the dissociation constants are only similar when compared on a molar basis.

The results of the two-component batch equilibrium adsorption experiments are plotted in Fig 2. From each adsorption experiment a pair of equilibrium adsorption results was obtained. Each result represents the concentration of protein in solution that was in equilibrium with an adsorbed amount of the same protein. The points pair with each other in a simple manner; the point at the lowest soluble BSA concentration pairing with that at the lowest soluble lysozyme concentration and so on over the entire range.

TABLE 1

Values of K_d and q_m for the adsorption of pure BSA and Lysozyme to S Sepharose FF.

	K_d (mg.ml^{-1})	K_d (M)	q_m (mg.ml^{-1})	q_m (mol.l^{-1})
Lysozyme	0.019	1.3x10^{-6}	120	8.4x10^{-3}
BSA	0.133	2.0x10^{-6}	113	1.7x10^{-3}

The adsorption capacities of the adsorbents are based on the volume that the adsorbent would occupy when packed in a bed.

The experimental data are compared to the results predicted by the two models of two-component adsorption. The values of q_{mi} and K_{di} used in the predictions were those determined above in the single component experiments. The non-competitive model (Fig 2a) gave a fairly accurate prediction of the lysozyme adsorption results but greatly over-predicted the amount of BSA that would be adsorbed to the ion-exchanger. Conversely the totally-competitive model (Fig 2b) gave a good prediction of the amount of BSA adsorbed but under-predicted the amount of lysozyme adsorbed.

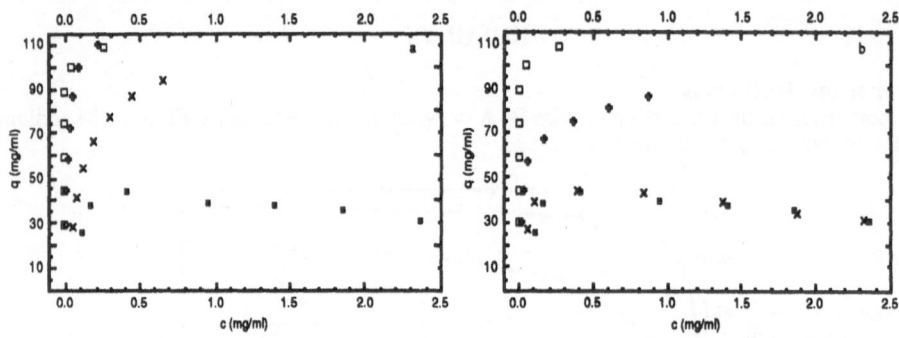

Figure 2. Batch adsorption of a mixture of lysozyme and BSA to S Sepharose FF in 0.1M acetate buffer, pH 5 at 25°C. The boxes represent experimental results for lysozyme (□) and BSA (■). The crosses represent the predicted values for lysozyme (✦) and BSA (✕). The data are plotted as mg of a particular protein adsorbed per ml S Sepharose FF (q) against mg/ml of that protein in solution (c). (a) The experimental results are plotted with the results calculated by the non-competitive model. (b) The experimental results are plotted with the results calculated by the fully-competitive model.

Frontal Analysis

The development of the breakthrough profiles for BSA and lysozyme when a solution containing a mixture of each protein at a concentration of 1 mg/ml was passed through a bed of S Sepharose FF are shown in Fig 3a. This figure shows that the breakthrough of BSA occurs before that of lysozyme, with the breakthrough profiles of both proteins having similar slopes. The concentration of BSA is seen to rise above that of the inlet concentration ($c/c_o > 1$) before it falls back towards it. This profile indicates that lysozyme is able to displace, and thereby elute, a certain amount of adsorbed BSA. The amounts of each protein which had bound to the beds was determined directly from the breakthrough profiles. The amount of BSA that had adsorbed was calculated to be 46 mg per ml S Sepharose FF, whilst the amount of lysozyme that had bound was 100 mg/ml. It is possible to calculate from the two models of multicomponent adsorption the amounts of protein that would be expected to adsorb to the ion-exchanger at equilibrium.

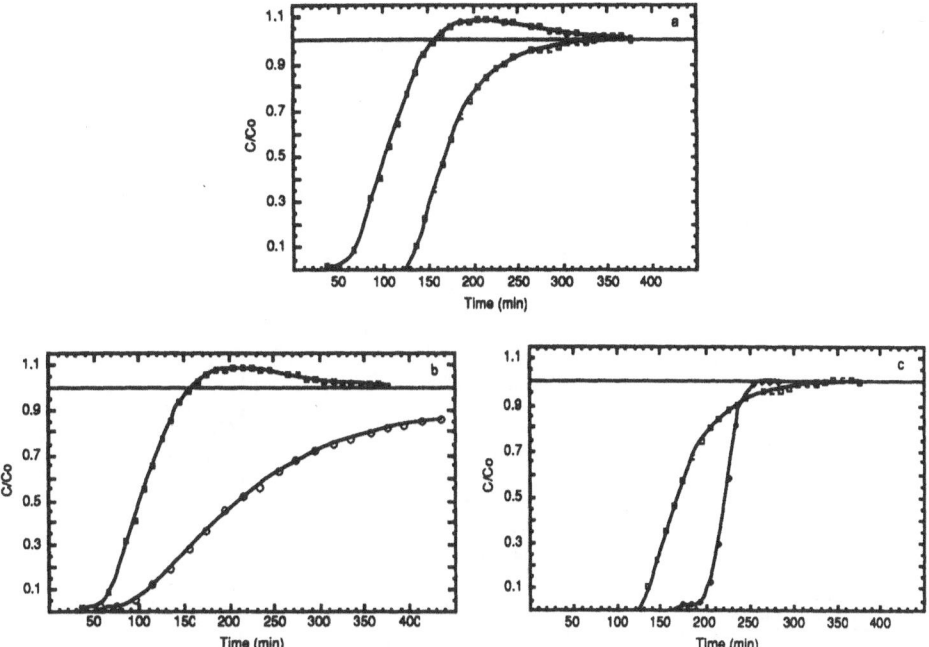

Figure 3.Breakthrough profiles for the adsorption of lysozyme and BSA to S Sepharose FF in packed beds. Fig 2(a): The breakthrough profiles of lysozyme (□) and BSA (■) when a solution containing both proteins, each present at a concentration of 1 mg/ml, was passed through the bed. Fig 2(b): Breakthrough curves for BSA from experiments in which pure BSA, at a concentration of 1 mg/ml, was applied to a bed (○) and when the mixture of BSA and lysozyme was applied (■). Fig 2(c): Breakthrough curves for lysozyme from experiments in which pure lysozyme, at a concentration of 1 mg/ml, was applied to a bed (●) and when a mixture of BSA and lysozyme was applied (□).

Performing these calculations for BSA gives an adsorbed concentration of 100 mg/ml for the non-competitive model and 14 mg/ml for the totally competitive model. The experimentally determined figure of 46 mg/ml of BSA adsorbed to the packed bed was not accurately predicted by either model but is closer to the value of the fully competitive approach. In the case of lysozyme the amounts of protein predicted to bind are 118 mg/ml for the non-competitive model and 103 mg/ml for the fully-competitive model. The experimentally determined figure of 100 mg/ml is almost completely consistent with the fully-competitive model.

In order to compare more easily the two-component breakthrough profiles with those obtained from the single component experiments for each protein, the profiles of multicomponent and single component experiments have been plotted in the same figures, those for BSA in Fig 3b and those for lysozyme in Fig 3c. Fig 3b clearly shows that the breakthrough profile of BSA in the presence of lysozyme is shifted considerably towards the origin compared to the position of the breakthrough curve when pure BSA is applied to the column. This is a reflection of the fact that significantly more BSA was able to bind to the packed bed in the absence of lysozyme than in the multicomponent experiment. The amount of BSA that was calculated to have adsorbed in the single component experiment was 130 mg of BSA per ml of S Sepharose FF compared to the 46 mg BSA per ml S Sepharose FF that was bound in the two-component experiment. The slope of the two-component BSA breakthrough profile is seen to be much sharper than that observed in the single component experiment.

The breakthrough profile of lysozyme obtained in the two-component experiments is shifted towards the origin by a much smaller amount than was the case for BSA (Fig 2c). The position of the two-component curve indicates that although less lysozyme bound to S Sepharose FF in the presence of BSA than was the case when lysozyme alone was present, the difference between the two-component and single component experiments is not as great as that for BSA. 100 mg of lysozyme per ml S Sepharose FF were bound in the two-component experiment compared to 125 mg/ml in the single component experiment. Also in contrast to the result for BSA, the gradient of the two-component lysozyme breakthrough curve is shallower than that obtained in the respective single component experiment.

The Adsorbed Protein Profile in Packed Beds

The development of the adsorbed protein profiles of BSA and lysozyme actually within the packed beds is shown in Fig 4.

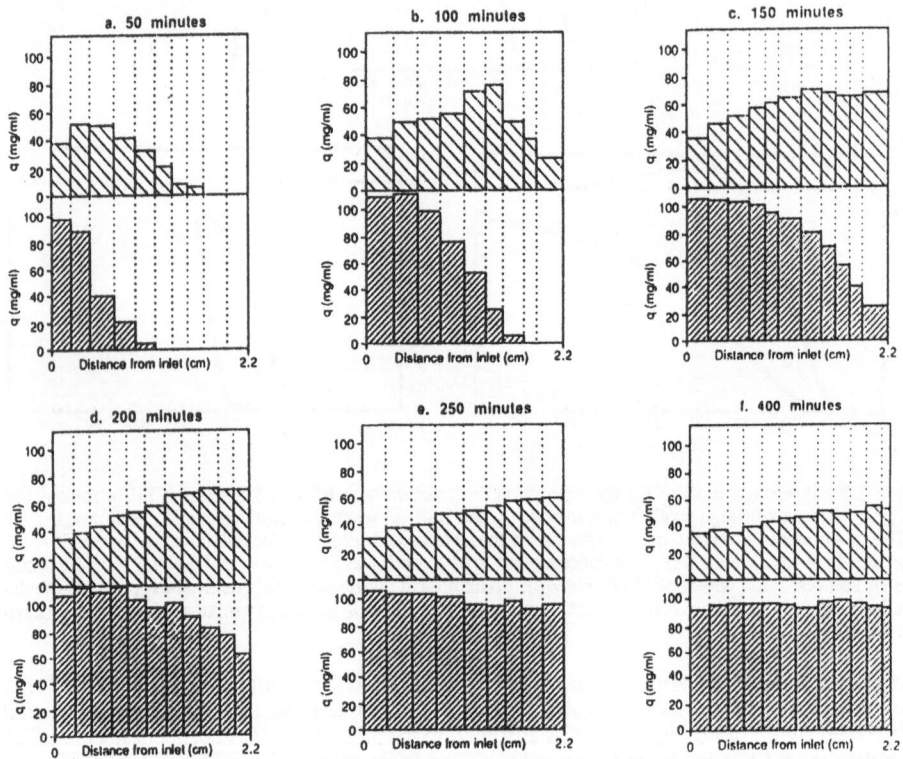

Figure 4. Adsorbed protein profiles for BSA and lysozyme within packed beds. The beds used were 1 cm diameter and 2.2 cm high and were loaded at 1 ml/min with a solution containing both proteins at a concentration of 1 mg/ml. The BSA profiles are shown in the upper section of each figure in light shading. The lysozyme profiles are shown in the lower section of each figure in dark shading. The results are plotted as the amount of protein adsorbed per ml of S Sepharose FF in each slice against the distance of each slice from the column inlet. The amount of S Sepharose FF in each slice is represented by the different widths of each slice plotted. Each figure shows the profile observed after loading the beds for the following times: (a) 50 minutes; (b) 100 minutes; (c) 150 minutes; (d) 200 minutes; (e) 250 minutes; (f) 400 minutes.

The diagrams show the concentrations of BSA and lysozyme adsorbed on each slice of S Sepharose FF that was taken from a packed bed. BSA can be seen to have penetrated further into the bed than lysozyme in the experiments where the column was loaded for 50 and 100 minutes. The adsorbed BSA profile developed a peak which was observed at a maximum in the 100 minute experiment. After this time the peak broadened and developed into a plateau, the concentration of BSA adsorbed on the bed falling as the system approached equilibrium, with the slices closest to the column entry nearing equilibrium first. In the final experiment, in which the column was loaded for 400 minutes, the amounts of lysozyme and BSA adsorbed, when averaged over the whole bed, were the same as the values obtained from analysis of the breakthrough curves.

Consecutive Application of Single Protein Solutions to Packed Beds of S Sepharose FF

Two experiments were performed in which a packed bed of S Sepharose FF was loaded with one protein and then the inlet stream switched to a pure feed of the other protein. The results of these experiments are shown in Fig 5.

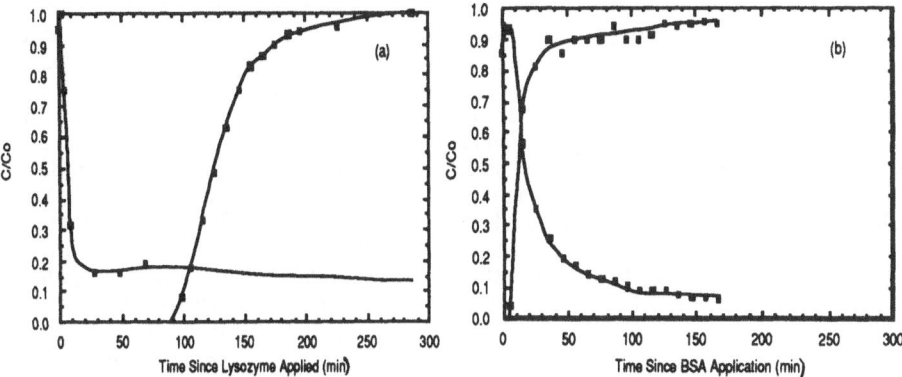

Figure 5. Adsorption of proteins to packed beds already loaded with another protein. In each experiment the bed was initially loaded with a solution of one protein at a concentration of 1 mg/ml until near to equilibrium with the feed stream. The feed stream was then switched to a solution containing the other protein at a concentration of 1 mg/ml. Concentrations of the two proteins in the exit stream were determined by FPLC analysis and are plotted from the time the feed stream was switched. (a) The profile of BSA (■) and lysozyme (□) in the exit stream of a bed initially loaded with BSA and subsequently loaded with lysozyme. (b)The profile of BSA (■) and lysozyme (□) in the exit stream of a bed initially loaded with lysozyme and subsequently loaded with BSA.

In the case of loading a bed with BSA and switching the feed to lysozyme, (Fig 5a), the concentration of BSA in the exit stream rapidly fell to a low level. 74 mg lysozyme per ml S Sepharose FF bound to a bed which initially contained over 125 mg/ml of adsorbed BSA and of which only 29 mg/ml of BSA was eluted from the column. In the complementary experiment, in which a bed was loaded with lysozyme and then the feed was switched to BSA (Fig 5b), the amount of BSA that was adsorbed to the ion-exchanger (17 mg BSA per ml S Sepharose FF) was less than the amount of lysozyme eluted (23 mg/ml).

DISCUSSION

The Langmuir isotherm has been successfully used to describe the adsorption of both BSA and lysozyme to the cation-exchanger S Sepharose FF in single component experiments. This is consistent with our studies of the adsorption of BSA to anion-exchangers in which a Langmuir isotherm was also found to be appropriate [3]. The equilibrium capacity (q_m) of S Sepharose FF for each protein was found to be similar on a mass basis although due to the greater relative molecular mass of BSA this protein has a smaller q_m value than lysozyme in molar terms. The dissociation constant, K_d, is a measure of the strength of the interaction between the protein and the ion-exchanger. In the case of BSA and lysozyme adsorbing to S Sepharose FF, lysozyme has a smaller dissociation constant than BSA on both a mass basis and, although the difference is less pronounced, in molar terms. As a result of the smaller dissociation constant and the fact that the molar concentration of lysozyme used in these studies was several times greater than that of BSA, it would be expected that if any competitive adsorption occurred, lysozyme would act as the more strongly binding component. Such an expectation is supported by predictions from the fully-competitive model, using the values of K_{di} and q_{mi} from the single component studies.

The results of both the two-component batch adsorption and packed bed experiments did show that greater amounts of lysozyme bound than BSA. However, neither of these models of multicomponent adsorption correctly predicted the amounts of each protein that bound in either agitated tank or packed bed experiments. Since the nature of the adsorption mechanisms of different proteins to an ion-exchanger is considered to be based on interactions with the same charged groups on the ion-exchanger surface, it was expected that the experimental results would be in close agreement with the fully-competitive model. In the batch adsorption studies, only the adsorption of BSA was similar to the results of the fully-competitive model, with this model greatly under-predicting the amounts of lysozyme that were bound. In contrast, the non-competitive model gave predictions which agreed closely with the amounts of lysozyme that were bound in these experiments, whilst over-predicting the amount of bound BSA. In the packed bed experiments the results of the fully competitive model agreed well with the amounts of lysozyme that were bound, but the amount of BSA bound was underestimated by the model. These results however, were closer to the fully-competitive model than the predictions of the non-competitive model, which over-predicted the amounts of both proteins which should bind, with the discrepancy between predicted and observed results being greatest in the case of BSA.

However, despite the lack of precise agreement between the experimental results and those calculated from the fully-competitive model, the results of packed bed experiments provided substantial evidence that competitive adsorption was occurring. The breakthrough profiles for both proteins were noticeably different in experiments in which a mixture of the two proteins was applied to a packed bed than was the case in single component experiments. The two-component BSA profile was seen to rise above a c/c_0 value of 1, that is the concentration of BSA in the exit stream was greater than that in the inlet stream. This type of profile is caused by a proportion of the more weakly binding component (BSA) being eluted from the adsorbent by the more strongly binding component (lysozyme). Elution of BSA by lysozyme also results in the sharper BSA breakthrough profile observed in the two-component experiments. The two-component breakthrough profiles observed are similar to those seen in studies of the multicomponent adsorption of smaller molecules in which competition between adsorbate molecules was occurring [8]. Elution of BSA by lysozyme was also observed in the experiments in which partially loaded beds of S Sepharose FF were extracted and the profile of adsorbed protein determined.

In the experiments in which packed beds of S Sepharose FF were first loaded with one protein and then the feed solution switched to a solution of the other protein, it was expected from consideration of a competitive model of adsorption, that lysozyme would elute and replace a proportion of the adsorbed BSA. Conversely it was expected that BSA, the more weakly binding component, would replace less adsorbed lysozyme. The results of applying lysozyme to a bed near to saturation with BSA did not entirely agree with this expectation. The amount of BSA which was eluted from the bed was much less than the additional amount of lysozyme bound. In the complementary experiment, applying BSA to a bed saturated with lysozyme, BSA was found to be less effective at eluting and replacing adsorbed lysozyme than was expected from equilibrium considerations which suggests that the dynamics of the process may be important

The results clearly show that the adsorption of BSA and lysozyme to the ion-exchanger S Sepharose FF is, to some extent, competitive in nature. However, the discrepancies between the experimental results and those predicted by either model of multicomponent adsorption indicate that the adsorption process is more complex than the models described. One of the primary assumptions of the fully-competitive model is that all adsorption sites are equally accessible to all adsorbate molecules. However it is likely that due to the smaller size of lysozyme in comparison to BSA, lysozyme is able to penetrate regions of the ion-exchanger particles which are too restricted for BSA to enter. Any lysozyme adsorption which occurred at sites which are inaccessible to BSA would be non-competitive and would result in the amount of adsorbed lysozyme being under-predicted by the fully-competitive model.

It should be remembered that predictions from the fully-competitive model refer to the amounts of protein that would be bound to the ion-exchanger at equilibrium with the liquid phase. It is possible that equilibrium may not have been achieved in some of the adsorption experiments described here and this may be a further reason for the discrepancies between the predictions and the experimental results. The mechanism for the displacement of an adsorbed molecule by another is unknown but it is possible that an adsorbed molecule may have to desorb from the ion-exchanger as a result of the reversible nature of the interaction before another molecule can be bound. The dissociation constants of the protein/adsorbent interaction are small, suggesting that the desorption rates may indeed be slow.

A further adsorption mechanism could be due to protein-protein interactions within the ion-exchanger particles. Electrostatic interactions between lysozyme and BSA in free solution have been reported previously [9, 10]. The large amounts of protein adsorbed within the ion-exchanger particles and the possibility that they are adsorbed in specific orientations, may promote such interactions, with the result that protein is adsorbed within the ion-exchanger but to other protein molecules rather than ion-exchanger functional groups. The BSA molecule has an ellipsoidal shape and the charge distribution along the major axis is known to be asymmetric, such that one end of the molecule carries predominantly positively charged groups and the other end negatively charged groups. Studies of BSA adsorption to a silica-based ion-exchanger in which the available surface area was known have indicated that BSA molecules were adsorbed in an "end-on" orientation, that is with the major axes perpendicular to the adsorbent surface [11]. Adsorption of BSA to S Sepharose FF in such an orientation with the positively charged region of the molecule interacting with the negatively charged sulphonic groups on the surface of the ion-exchanger, would present the negatively charged region of the BSA molecule pointing out towards the bulk liquid. This may have provided sites with which the strongly positively charged lysozyme molecules could have interacted electrostatically, resulting in adsorption. Any lysozyme adsorbed in this manner would be to sites which would not be available to BSA and would be in addition to those available during single component adsorption of lysozyme. Under these circumstances the amount of lysozyme predicted to adsorb by the fully-competitive model would indeed be an underestimate of the experimental results.

ACKNOWLEDGEMENT

The authors would like to thank the Science and Engineering Research Council for financial support. They are also grateful to Pharmacia LKB Biotechnology AB, Uppsala, Sweden, for the provision of experimental materials and equipment.

NOTATION

c liquid phase concentration of protein
c_0 initial, or inlet liquid phase protein concentration
K_d dissociation constant for the protein/ion-exchanger complex
q concentration of protein adsorbed to the ion-exchanger
q_m maximum protein capacity of the ion-exchanger
v volume of ion-exchanger
V volume of liquid
* value when system is at equilibrium

REFERENCES

1. Janson, J.C. and Hedman, P., *Biotechnol. Prog .*, 1987, **3**, 9.
2. Skidmore G.L., and Chase, H.A. In *Ion Exchange for Industry* (Ed. Streat, M.) Ellis Horwood, Chichester, 1988 p. 520.
3. Velayudhan A. and Horvárth, C., *J. Chromatogr..*, 1988, **443**, 13.
4. S.C. Nigam, A. Sakoda and H.Y. Wang, *Biotechnol. Prog.*, 1988, **4**, 166.
5. Jacobson, J.M., Frenz, J.H. and Horvath, C., *Ind. Eng. Chem. Res.*, 1987, **26**, 43-50.
6. Annesini, M.C. and Lavecchia, R., *Chem. Biochem. Eng. Q.*, 1987, **1**, 89.
7. Graham, E.E, Pucciani , A. and Pinto, N.G., *Biotechnol. Prog.*, 1987, **3** , 141.
8. Mansour, A., von Rosenberg, D.U. and Sylvester, N.D., *A.I.Ch.E.J.*, 1982, **28**, 765.
9. Steiner, R.F., *Arch. Biochem. Biophys.*, 1953, **46**, 291.
10. Steiner, R.F., *Arch. Biochem. Biophys.*, 1953, **47**, 56.
11. van der Wiel, J.P. and Wesselingh, J.A.. In *Adsorption: Science and Technology* (Eds. Rodriques, A.E., LeVan, M.D. & Tondeur, D.) Kluwer Academic Publishers, Dordrecht, 1989 p. 427.

Part 4

Product Finishing

THE MANUFACTURE OF PROTEINS FOR

HUMAN THERAPEUTIC USE

C.R. HILL

Celltech Limited, 216 Bath Road,
Slough, Berks. SL1 4EN, U.K.

INTRODUCTION

A key objective in the development of a process to manufacture any
pharmaceutical product is to ensure the safety and efficacy of that
compound.   Ultimately this can only be achieved through appropriately
designed human clinical studies.   However, it is essential that before a
protein pharmaceutical is introduced to humans, both it and the process
used in its manufacture are subjected to a rigorous programme of
characterisation and safety studies.   Guidelines for the testing of
protein pharmaceutical products derived by recombinant DNA and hybridoma
technology are given in documents prepared by the various National
Regulatory Agencies (1 - 4).   In this paper I will describe approaches
that can be taken to address some of the safety issues raised in the
guidelines.

THE NEED TO ASSAY

In comparison with small molecular weight drugs, protein pharmaceuticals
are very complex molecules.   Even a small protein drug such as human
growth hormone has a molecular weight of 22,000 and an intact
immunoglobulin has a molecular weight of 150,000.   The protein molecule
may contain a variety of carbohydrate groups and a number of disulphide
bonds which can theoretically combine in a large number of different
combinations.   A large number of other post-translational events may
also occur (5).

The extent and type of such events can be influenced by the process used
to manufacture the product.   The cell type used to express the product,
the fermentation process and purification methods used to recover the

product and subsequent formulation and storage prior to use can all
influence the quality and structure of the protein (6-9).   The complex
nature of protein molecules means that it is not possible to precisely
define the composition and conformation of the product.   The quality
control of a protein pharmaceutical must therefore be defined by
extensive characterisation of both the product itself and the specific
process used in its manufacture.

There is a wide range of potential impurities and contaminants that may
be associated with a recombinant protein pharmaceutical.   These may be
derived from any of the raw materials used in its manufacture or from the
product itself (Table 1).   For example, protein impurities may be

TABLE 1   Sources of Potential Impurities and Contaminants

Host Organism

- Endogenous viruses
- cell proteins
- DNA

Manufacturing Process

- Media components
- Adventitious viruses
- Mycoplasma
- Purification reagents
- Endotoxin
  Formulation excipients

Product Variants

- Aggregates
- Proteolytically cleaved product
- Deamidated variants
  Oxidation products
- Amino acid substitutions

derived from the host organism, media used in the fermentation process or
from excipients used in the formulation.   A major issue to be considered
if affinity chromatography is used during the purification of a
therapeutic protein is that of leakage of the ligand from the column.
This may result in contamination of the product with affinity ligand, be
it a protein (e.g. Protein A or an immunoglobulin) or small molecule
(e.g. textile dye).   Impurities derived from the product can include
proteolytically cleaved or aggregated protein and deamidated (10) or
amino-acid substituted forms (11).   A major safety issue for any product
derived from a continuous mammalian cell line is that of virus or DNA
contamination.   The main concerns centre on the possibility of
transforming DNA sequences and oncogenic infectious virus, originating
from the host cell line, contaminating the product.   The evaluation of
virus contamination is a challenging technical problem, particularly as
there always exists the possibility that a previously unforeseen agent
may appear (e.g. HIV).

Development of appropriate assays forms a key part of any process
development programme.   A large number of different assay types are
required (12,13).   These range from biological assays required to
measure potency and physicochemical and biochemical assays used to assess
purity to microbiological and animal tests to establish the sterility and
lack of pyrogenicity and toxicity of the compound.

In this paper I will discuss some of the approaches that are taken to
product and process characterisation and show how these can be used to
provide a high degree of assurance that the product can safely be
administered to humans.   Throughout I will illustrate this by referring
firstly to the specific issue of potential viral contamination of
recombinant products and secondly to the use of affinity chromatography
to purify products for human therapeutic use.   Both of these topics will
be dealt with in more detail in papers presented later in this session.

ADVENTITIOUS AND ENDOGENOUS VIRUSES

Potential viral contamination of recombinant proteins may arise either
from the presence of adventitious viruses or from the expression of
endogenous viruses.

434

TABLE 2   Approaches to Controlling Potential Virus Contamination

1.   Cell Bank Characterisation

2.   In-Process Testing

3.   Inactivation Procedures

4.   Raw Materials

5.   Process Validation

6.   Containment

7.   Post Market Surveillance

There are a number of approaches that can be taken to ensure the removal
of potential viral contaminants (Table 2).   An extensive programme of
testing for both adventitious and endogenous viruses is required.   This
should include appropriate testing of cell banks, raw materials and
samples taken throughout the process.   The primary aim of the testing
programme should be to identify any infectious virus which may compromise
the safety of the product.   Therefore the strategy that should be taken
when designing a testing programme is to test at those points in the
process where the risk of a virus being present is greatest.   For
example, it is appropriate to test the fermentation harvest for
endogenous retrovirus as it is at this point that the cells have reached
their maximum population density and hence there is the greatest
opportunity for the expression of retrovirus.

RETROVIRUS
There are a number of assays that can be used to detect retroviruses.
These are summarised in Table 3.   Many of these are technically
difficult to perform and relatively insensitive.   For example, in some
cases virus-like particles can be seen in cells by transmission electron
microscopy (15) whereas no infectious virus can be detected (16,17).
Furthermore, the sensitivity of biological virus assays is poor.   For

TABLE 3   **Assays Used to Detect Retrovirus**

Type	Test
Morphogenic	Transmission electron Microscopy
Biochemical	Reverse Transcriptase
Immunological	Radioimmunoassay
Biological	Plaque and focus forming co-cultivation
Molecular Biological	Hybridisation

Adapted from Lubiniecki (10).

TABLE 4   **Probability that a Test Sample contains an Infectious Particle**

Sample Volume (ml)	Virus Concentration (virus/litre)	Probability no virus in sample
1	10	0.99
1	100	0.90
1	1000	0.37
1	10000	0.00
4	10	0.96
4	100	0.67
4	1000	0.02
4	100000	0.00

sampling and statistical reasons it is difficult to detect low levels of virus contamination. The probability that a sample taken for testing from a bulk product will contain an infectious virus particle can be approximated by a Poisson distribution (Table 4). For example, if a 1 ml test sample is taken from a solution containing 100 infectious particles per litre there is a 90% chance that the sample will not contain a virus particle. If the test sample volume is increased to 4 ml there is still a 67% chance that the sample will not contain an infectious virus particle. Furthermore, it will never be possible to detect unknown agents using specific virus assays. For these reasons it is prudent not to rely solely on virus assay to establish the safety of the product.

PROCESS VALIDATION

A complementary approach to that of testing product and raw materials for viruses is to measure the capacity of the manufacturing process to remove or inactivate virus.

The capacity of the process to remove or clear virus from the product can be determined by applying samples of crude product to columns scaled down proportionately from the manufacturing process. The amount of virus retained with the product after purification can then be measured and a clearance factor calculated by dividing the total amount of virus applied to the column by that recovered with the product. By repeating this procedure for each step a cumulative clearance factor for the manufacturing process can be obtained. Examples of studies of this type can be found in references 18-21, 25.

It is essential when designing clearance studies to pay particular attention to the scaling-down of the manufacturing process. The scale-down factor should ideally be between 10 and 100 fold. Column height, linear flow rates and all buffers should be the same as those used for the manufacturing process. It must also be shown that the scaled-down column has the same performance characteristics as the full-scale column.

Process validation studies should also include an evaluation of those steps that could result in inactivation of viruses. For example, the use of a buffer with a pH below 4.0 to elute an antibody from a Protein A

affinity column would result in a significant inactivation of
retroviruses.   The uncertainty arising from the possibility of
contamination from previously unknown agents has led to a greater
emphasis on the inclusion of steps in a process designed specifically to
inactivate viruses.   This has been particularly the case in the plasma
fractionation industry.   The impact of Human Immunodeficiency Virus has
led to a major programme of work to develop inactivation procedures for
viral contaminants that can be applied to protein products such as Factor
VIII derived from plasma.   For recent reviews of this subject see
references 22-24.   The main approach has been to develop formulations
that allow heating of the final product in a dry form to temperatures
between 60-80°C for periods up to 72 hours without destroying the
activity of the products.   In some cases heating in the wet state has
been employed and the inclusion of solvent and detergent mixtures (25)
designed to disrupt the virus envelope has also been employed.   The
experience gained in the plasma fractionation industry is now being
applied to the development of virus inactivation procedures for cell
culture derived protein pharmaceuticals.

AFFINITY CHROMATOGRAPHY OF THERAPEUTIC PROTEINS
The demand for very high purity proteins for therapeutic use has resulted
in recent years in the development of a number of processes where
immunoaffinity chromatography is a key step (27-29).   Other protein
ligands, such as Protein A, are also being used to purify proteins for
human therapeutic use.

Affinity chromatography can give very high levels of protein purity in
high yield.   In addition affinity chromatography steps will generally
give high step clearances of impurities such as DNA and viruses (26).
However, a major issue to be addressed during the development of an
affinity step is the extent to which protein ligand leaches from the
column thus contaminating the product.   There are several approaches
that can be taken to characterising an affinity chromatography matrix.
These are summarised in Table 5.

It is essential that an appropriate assay to detect leakage products is
developed (30-32).   It is important to validate the assay for the
detection of intact ligand but also to show that the assay can measure

TABLE 5   Approaches to Characterisation of an Affinity Chromatography
            Step

1.   Ligand Assay Development

2.   Process Development

3.   Repetitive use studies

4.   Ligand Stabilisation

fragments of the immobilised ligand if it is anticipated that these may
be shed from the column.   An appropriate insoluble matrix and coupling
chemistry must be identified, as the extent of ligand leakage can vary
over a wide range depending on the fabrication of the affinity matrix
(33).   In general an affinity column would be followed by at least one
further step such as an ion-exchange column.   This can significantly
reduce the level of ligand that may contaminate the product
(32).

If an affinity column is to be used for more than a single cycle it is
important to demonstrate that this does not impair the performance of the
column.   This can be done by carrying out repetitive use studies with
scaled-down columns (34) and also by monitoring each cycle of use at the
full manufacturing scale.   In addition to measuring the yield and purity
of the product after each cycle it is important to demonstrate that the
clearance of impurities such as DNA and viruses is not reduced following
a number of cycles of column re-use.   In order to reduce ligand leakage
to a minimum a number of methods have been published in which
stabilisation of the immobilised ligand by covalent modification is
described (35-37).

Finally it is essential that the principles of current Good Manufacturing
Practice are applied during the manufacture of any affinity reagent to be
used in the manufacture of a therapeutic product.   Furthermore, if the
reagent is to be supplied to a third party comprehensive documentation
including a detailed Certificate of Analysis must be made available in
order to demonstrate the consistency and quality of the product.

CONCLUDING REMARKS

In this chapter I have discussed some of the key safety issues for the development of protein pharmaceuticals derived from continuous mammalian cell lines. I have emphasised the importance of applying a number of techniques during the assessment of any risk factor. For example, approaches to assessing potential risks from virus contamination include cell bank characterisation, in-process testing and process validation. During the design of strategies to assess product safety it is essential to apply sound scientific principles and to concentrate efforts at those points where risks may be greatest. This information, taken together with manufacture of the product under the auspices of current Good Manufacturing Practice can provide a high degree of assurance that the product can safely be administered to humans.

REFERENCES

1.   Office of Biologics Research and Review (1987).   Points to consider in the manufacture and testing of Monoclonal Antibody products for human use.   Food and Drug Administration, Bethesda, Maryland, 20892, USA.

2.   Australian Department of Community Services and Health (1987). Guidelines for the Preparation and Presentation of Applications for General Marketing of Monoclonal Antibodies Intended for Use in Humans, GPO Box 9848, Canberra, ACT 2601, Australia.

3.   Committee for Proprietary Medicinal Products (1988).   Guidelines on the Production and Quality Control of Medicinal Products Derived by Recombinant DNA Technology.   Trends Biotechnol. $\underline{5}$, G1 – G4.

4.   Committee for Proprietary Medicinal Products (1988).   Guidelines on the Production and Quality Control of Monoclonal Antibodies of Murine Origin Intended for Use in Humans.   Trends Biotechnol. $\underline{6}$, G5 – G8.

5.   Wold, F. (1981) In Vivo chemical modification of proteins.   Ann.
      Rev. Biochem. 50, 783–814.

6.   Utsumi, J., Mizono, Y., Hosoi, K., Okana, K., Sawada, R.,
      Kajitani, M., Sakai, J., Naruto, M. and Shimizu, H. (1989).
      Characterisation of four different mammalian–cell–derived
      recombinant human interferon–Beta's: identical polypeptides and
      non–identical carbohydrate moieties compared to natural ones.
      Eur. J. Biochem. 259, 6790 – 6797.

7.   Moellering, B.J., Tedesco, J.L., Townsend, R.R., Hardy, M.R.,
      Scott, R.W. and Prior, C.P. (1990) Electrophoretic differences in
      MAb expressed in three media. BioPharm 3, 30–38.

8.   Sadana, A. (1989) Protein inactivation during downstream
      separation, Part II: The Parameters. BioPharm, 2, 20–23.

9.   Underwood, P.A. and Bean, P.A. (1985) The influence of methods of
      production, purification and storage of monoclonal antibodies upon
      their observed specificities.   J. Immunol. Meths. 80, 189–197.

10.   Di Augustine, R.P., Gibson, B.W., Aberth, W., Kelly, M., Ferrua,
      C.M., Tanooka, Y., Brown, C.F. and Walker, M. (1987) Evidence for
      isoaspartyl (Deamidated) forms of mouse epidermal growth factor.
      Anal. Biochem. 165, 420–429.

11.   Bogosian, G., Violand, B.N., Dornward–King, E.J., Workman, W.E.,
      Jung, P.E. and Kane, J.F. (1989). Biosynthesis and incorporation
      in protein of norleucine by Escherichia Coli. J. Biol. Chem. 264,
      531–539.

12.   Anicetti, V.R., Keyt, B.A. and Hancock, W.S. (1989) Purity
      analysis of protein pharmaceuticals produced by recombinant DNA
      technology. Trends Biotechnol. 7, 342–349.

13.   Garnick, R.L., Solli, N.J. and Papa, P.A. (1988) The role of
      quality control in biotechnology: An analytical perspective.
      Anal. Chem. 60, 2546–2557.

14. Lubiniecki, A.S. (1988) Safety of recombinant biologics: Issues and emerging answers. Animal Cell Biotechnology, 3, 3-12.

15. Stavrov, D., Bilzer, T., Tsangaris, T., Durr, E., Steinecke, M. and Anzil, A.P. (1983) Presence and absence of virus particles in hybridomas secreting monoclonal antibodies against gliomas. J. Cancer Res. Clin. Oncol. 106, 77-80.

16. Lubiniecki, A.S. and May, L.H. (1985) Cell bank characterisation for recombinant DNA mammalian cell lines. Dev. Biol. Stand. 60, 141-146.

17. Jackson, M.L., Nakamura, G.R., Lubiniecki, A.S. and Patzer, E.J. (1987) Attempts to detect retrovirus in continuous cell lines: Radioimmunoassays for hamster P30 protein. Dev. Biol. Stand. 66, 541-553.

18. Levy, J.A., Lee, H.M., Kawahata, R.T., Spitler, L.E. (1984) Purification of monoclonal antibodies from mouse ascites eliminates contaminating infectious mouse type C viruses and nucleic acids. Clin. Exp. Immunol. 56, 114-120.

19. Mitra, G., Wang, M.F., Mozen, M.M., McDougal, J.S. and Levy, J.A. (1986) Elimination of infectious retrovirus during preparation of immunoglobulins. Transfusion 26, 394-397.

20. Schreiber, A.B., Hrinda, M.E., Newman, J., Criss Tarr, G., D'Alisa, R. and Curry, W.M. (1989) Removal of viral contaminants by monoclonal antibody purification of plasma proteins. Curr. Stud. Hematol. Blood Transfus., Karger Basel, 56, 146-153.

21. Piszkiewicz, D., Sun, C-S., Trondreau, S.C. (1989) Inactivation and removal of human immunodeficiency virus in monoclonal purified antihemophilic factor (human) (Hemofil M) Thromb. Res. 55, 627-634.

22. Pierce, G.F., Lusher, J.M., Brownstein, A.P., Goldsmith, J.C. and Kessler, C.M. (1989) The use of purified clotting factor concentrates in hemophilia - Influence of viral safety, cost and supply on therapy. J. Am. Med. Assn. 261, 3434-3438.

23.   Lelie, P.N., Reesink, H.W. and Lucas, C.J. (1987) Inactivation of
       12 viruses by heating steps applied during the manufacture of a
       Hepatitis B Vaccine. J. Med. Vir. 23, 297–301.

24.   Foster, P. (1990) This volume.

25.   Edwards, C.A., Piet, M.P.J., Chin. S. and Horowitz, B. (1987) Tri
       (n–butyl) phosphate/detergent treatment of licensed therapeutic
       and experimental blood derivatives. Vox. Sang. 52, 53–59.

26.   Brady, D., Bonnerjea, J. and Hill, C.R. (1990) Purification of
       Monoclonal Antibodies for human clinical use: Validation of DNA
       and Retroviral clearance. This volume.

27.   Jack, G.W. and Wade, H.F. (1987) Immunoaffinity chromatography of
       clinical products. Trends Biotechnol. 5, 91–95.

28.   Hochuli, E. (1988) Large–scale chromatography of recombinant
       proteins. J. Chromatog. 444, 293–302.

29.   Weinstein, R.E. (1989) Immunoaffinity purification of Factor
       VIII. Ann. Clin. and Lab. Sci. 19, 84–91.

30.   Dertzbaugh, M.T., Flickinger, M.C. and Lebherz, W.B. (1985) An
       enzyme immunoassay for the detection of Staphylococcal Protein A
       in affinity-purified products. J. Immunol. Meths. 83, 169–177.

31.   Nelson, L.C., Peterson, M.L., Frie, S., Vetterlein, D., Gregory,
       T. and Chen, A.B. (1988) Enzyme–linked immunosorbent assays
       (ELISA) for the determination of contaminants resulting from the
       immunoaffinity purification of recombinant proteins. J. Immunol.
       Meths. 113, 113–122.

32.   Bloom, J.W., Wang, M.F., Mitra, G. (1989) Detection and reduction
       of Protein A contamination in immobilised Protein A purified
       monoclonal antibody preparations. J. Immunol. Meths. 117, 83–89.

33.   Fugistaller, P. (1989) J. Immunol. Meths. 124. 171–177.

34.   Francis, R., Bonnerjea, J. and Hill, C.R. (1990) Validation of the
       re-use of Protein A Sepharose for the purification of monoclonal
       antibodies. This volume.

35.   Gyka, G., Ghetie, V., Sjoquist, J. (1983) Crosslinking of
       antibodies to Staphylococcal Protein A matrices. J. Immunol.
       Meths. 57, 227-233.

36.   Hagen, M., Strejan, G.H. (1987) Antigen leakage from
       immunosorbents - Implications for the detection of site-directed
       auto-anti-idiotypic antibodies. J. Immunol. Meths. 100, 47-57.

37.   Kando, A., Kishimura, M., Katoh, S. and Sado, E. Improvement of
       proteolytic resistance of immunoadsorbents by chemical
       modification with polyethylene glycol. Biotech. and Bioeng. 34,
       532-540.

IMMUNOAFFINITY PURIFICATION OF GUAR ALPHA-GALACTOSIDASE USING MILD ELUTION CONDITIONS

M.J. BERRY; J. PIRON; M.M. GANI; P. PORTER;
Unilever Research, Colworth House, Sharnbrook, Bedford U.K.

ABSTRACT

The value of guar galactomannan may be improved by enzymic modification using guar α-galactosidase. The reaction requires a preparation of α-galactosidase which is completely free from β-mannanase contamination. α-galactosidase was purified to this specification from guar seeds using immobilised monoclonal antibodies raised against the enzyme. Of three antibodies studied, each interacted with α-galactosidase less strongly than would be typical-making complete separation of κ-galactosidase from other enzyme activities difficult. However, it was found that the antibody-enzyme interaction was stronger in high ionic strength (1M NaCl) or mild acidity (pH 5.5). This increased binding was sufficient to effect complete separation from β-mannanase. The increased antibody affinity in buffers that are dissimilar from physiological conditions was a very unexpected result. It serves as a good example of the importance of optimising loading buffers to maximise the resolution obtained with immunoadsorbents.

INTRODUCTION

α-Galactosidase can be used to improve the properties of guar gum, in particular its readiness to form gels. Guar gum that has been modified by this enzyme may be used as a substitute for the more expensive locust bean gum. Locust bean gum is used as a food stabiliser in low fat spreads, ice cream and pet foods. The required modification of guar galactomannan is achieved by the removal of $(1 \rightarrow 6)$ $-\alpha$-D-galactose side chains. This requires a preparation of α-galactosidase enzyme which is free from β-mannanase, an enzyme which hydrolyses the $(1 \rightarrow 4)$ $-\beta$-D-mannan backbone into oligosaccharides, thus reducing its functionality. (1)

In the guar seed, α-galactosidase and β-mannanase are both present in the aleurone layer. On germination, both enzymes are activated. This results in complete digestion of guar gum (the storage galactomannan in the endosperm) with the release of energy (2). The challenge for foods companies has been to develop a process for the preparation of α-galactosidase which is completely free from β-mannanase contamination. (3) Homogeneous α-galactosidase is not necessarily required.

Previously published methods for the purification of α-galactosidase from guar seeds have required two chromatography steps (4) and in our hands have been very time consuming. The aim of this study was to develop a rapid single step process for purifying α-galactosidase from guar extract using immunoadsorbents. This department has used immunoaffinity chromatography to purify a number of proteins such as enzymes (5), human serum proteins, and viral antigens. In our experience, the degree of success of different immunoaffinity chromatography processes has been determined by the quality of the antibody that was used as the immunoligand. Sometimes a monoclonal antibody (MCA) was available which had been specially selected from a library of candidate MCAs for its binding characteristics. When such a reagent was used as the immunoligand, outstanding purification could be achieved.

In this study, a selection of monoclonal antibodies were raised against α-galactosidase. Two screening assays were developed for evaluating hybridomas for their potential to produce anti α-galactosidase antibodies. The use of more than one assay usually increases the number of positive hybridomas identified (6). Hybridomas of interest were reproduced in serum free media. Purified antibodies were evaluated for their potential in immunoaffinity chromatography.

MATERIALS AND METHODS

Development of Screening Assays

Two ELISAs were developed: Assay 1 in which monoclonals were captured by immobilised α-galactosidase, and then detected by a rabbit anti-mouse conjugate, and Assay 2 in which monoclonals were captured by immobilised anti-mouse IgG antibody and then detected with an α-galactosidase conjugate. (α-galactosidase was conjugated to calf-intestine phosphatase because the former enzyme generates too low a signal for detection in ELISA). The two screening assays were developed using positive mouse serum ie serum from mice that had been repeatedly immunised with pure α-galactosidase enzyme.

Immunisation of Mice

Male Balb/c mice, 8 to 12 weeks of age, were each immunised sub-cutaneously and intraperitoneally with approximately 50 μg of α-galactosidase which was emulsified with an equal volume of Freund's Complete Adjuvant. Booster injections with Freund's Incomplete Adjuvant were given on day 30 followed by further booster injections without adjuvant on days 37 and 44. After the fourth immunisation, sera from these mice were tested for the production of antibody against α-galactosidase using screening assays 1 and 2. The mouse whose serum exhibited the highest anti-α-galactosidase activity was selected for cell hybridisation. This mouse was given three further daily intravenous and intraperitoneal injections immediately prior to fusion.

Fusion and Selection of Hybridomas

Spleen cells were fused to SP20 myeloma cells with PEG 3000 (Fluka AG Batch No 251778).

Supernatants from established hybridoma cell lines were tested for anti-α-galactosidase activity with assays 1 and 2. Cell populations of interest were cloned at least twice using limiting dilution technique (0.5 cells/ml) to ensure clonality. These single clones were allowed to grow before being re-evaluated for anti α-galactosidase activity in assays 1 and 2.

Production of Pure IgG

Three hybridomas were found to be good producers of specific antibody. These were grown up in serum free media in fermenters. Cultures were harvested after cell death. Cell debris was removed by a microfilter with a pore-size of 6.4 μm (Pall Ltd). The clarified culture was concentrated by ultrafiltration across membranes with a molecular-weight cut-off of 30,000 (Amicon Ltd). Antibody was purified by a combination of gel filtration chromatography (Sephacryl S200, Pharmacia) and ion-exchange chromatography (DEAE Sephacel, Pharmacia).

Immobilisation of Mouse IgG on Sepharose 4B

Purified mouse antibody preparations were dialysed against coupling buffer (0.1M Na HCO$_3$, 0.5 M NaCl, pH 8.3) and concentrated in a stirred cell concentrator. (Amicon Ltd). Antibody preparations were concentrated to approximately 5 mg/ml. The concentrated antibody preparations were mixed with an equal volume of cyanogen bromide derivatised Sepharose 4B (Pharmacia), to which antibody was covalently bound according to manufacturer's instructions.

An affinity column of approximately 5mls wet volume was made with each of the three monoclonal antibodies.

Purification of α-Galactosidase from Guar Extract

An extract made from guar seeds was exhaustively dialysed against phosphate buffered saline, pH 7 (PBS); then loaded onto an affinity column at 20 mls/hour. Columns were washed with PBS. Fractions collected were analysed for α-galactosidase and β-mannanase content by standard colourimetric assays - (4) and (7) respectively.

Optimisation of α-Galactosidase Purification

To maximise the interaction between the antibodies and κ-galactosidase, thus optimising the potential for purification, several different loading buffers were evaluated. Buffers covering a range of pH values and ionic strengths were used.

Isoelectric Focussing of Proteins

The isoelectric points (pI) of α-galactosidase enzyme and the three monoclonal antibodies were determined by focussing the proteins on a 5% polyacrylamide gel (LKB- Pharmacia). This gel was also used to determine the purity of monoclonal antibody preparations before coupling. Since the only high molecular-weight component in our serum free media is transferrin, the absence of transferrin on an isoelectric focussing gel is indicative of a high-purity preparation. In addition, isoelectric focussing was used, to check that each antibody product was clonal:- a clonal product typically focusses to a single cluster of equidistant bands.

RESULTS

Performance of Screening Assays

Assays 1 and 2 produced characteristic sigmoid curves across a range of dilutions of positive mouse serum. There was no detectable signal with negative mouse serum. This confirmed that the two assays were specific for anti α- galactosidase antibody.

Immobilisation of Monoclonal Antibody

The efficiency of immobilisation was estimated by measuring the O.D. 280 of the supernatant solution after reaction with the gel. The percentage protein bound was greater than 95% in each case.

Purification of κ-Galactosidase from Guar Extract in PBS

When the starting material was loaded in PBS and the columns were washed in PBS, none of the columns charged with antibody bound α-galactosidase tightly; adsorption of the enzyme did not occur. However, the passage of α–galactosidase through all three of the columns was significantly retarded, and the enzyme eluted last as an identifiable peak. β–mannanase was not retarded on these columns, and so moderate separation of the two activities was achieved. A typical separation profile for a column charged with 2595.9 is shown in Fig 1. A column of the same dimensions made with the same batch of Sepharose 4B, but without bound antibody, did not retard α–galactosidase (Fig 2). A similar result was obtained for a column to which non-specific antibody was bound. This demonstrated that retardation was due to a specific interaction between 2595.9 and enzyme. The degree of separation achieved by the immunoadsorbents made with each of the three MCAs was expressed quantitatively as the selectivity. (Table 1).

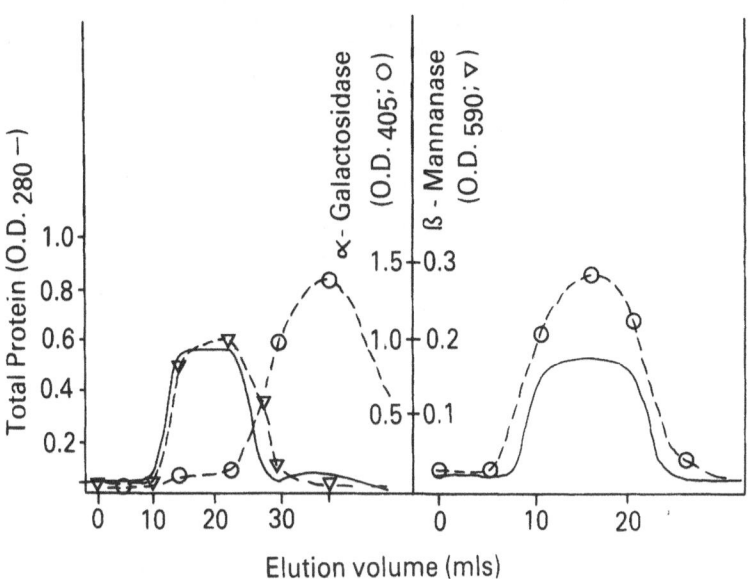

Figures 1 and 2. Interaction of α–galactosidase with columns in PBS. The column in Figure 1 (left) was functionalised with monoclonal antibody specific for α–galactosidase. The column in Figure 2 (right) was not functionalised.

TABLE 1

Selectivity of three different monoclonal antibodies (MCA)

MCA	Selectivity
2595.9	4.0
2594.7	2.1
2597.1	2.3

Selectivity $= (V_2 - V_0)/(V_1 - V_0)$

where V_2 = elution volume of α–galactosidase
V_1 = elution volume of β–mannanase
V_0 = void volume of column

Improvement of Resolution by Loading in a High Salt Buffer

When the starting material was loaded onto the column in a high salt buffer (PBS + 1M NaCl, pH 7) the affinity for α-galactosidase was enhanced so that the enzyme was adsorbed onto the column rather than merely retarded by it. Washing the column with this buffer removed all the β-mannanase present. α-galactosidase could then be eluted in PBS. This produced an α-galactosidase fraction which contained no detectable β-mannanase. Recovery of α-galactosidase in this purified fraction was almost 100% providing that the column's capacity was not saturated. A typical purification profile for 2595.9 is shown in Fig 3. Similar profiles were obtained with the other two antibodies.

Enhanced binding, sufficient to cause adsorption rather than retardation of α-galactosidase, also occurred if the starting material was loaded in 0.02M phosphate, pH 5.5, in the absence of high salt.

A column of the same dimensions made with the same batch of Sepharose 4B, but without bound antibody, did not adsorb α-galactosidase when loaded in the high salt buffer. (Fig 4).

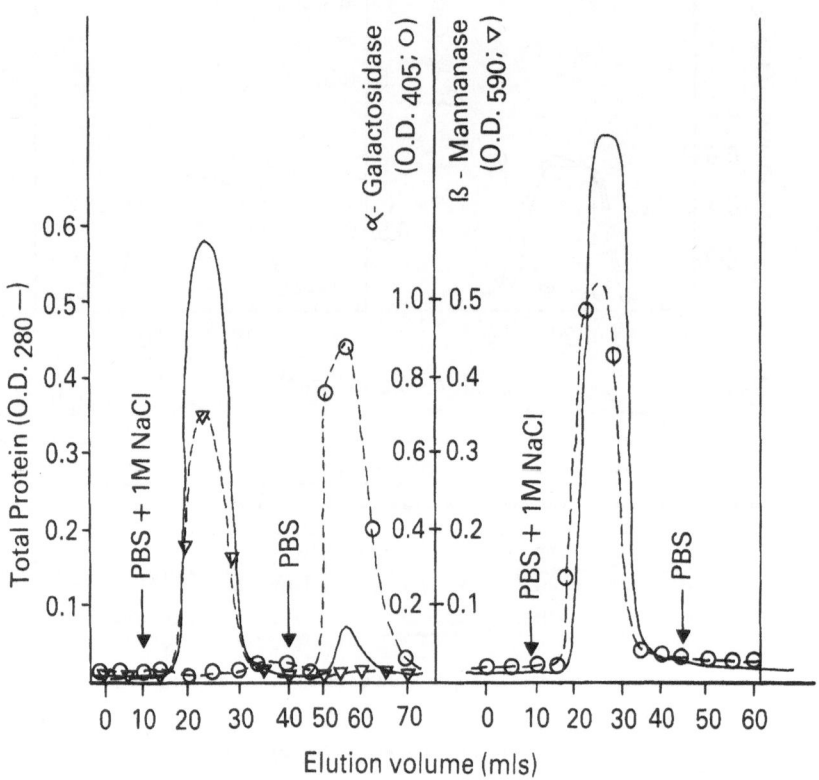

Figures 3 and 4. Interaction of α-galactosidase with columns in high salt buffer. The column in figure 3 (left) was functionalised with monoclonal antibody specific for α-galactosidase. The column in Figure 4 (right) was not functionalised.

A similar result was obtained for a column to which non-specific antibody was bound. This confirmed that the effect of the modified buffers was to increase the specific interaction between the enzyme and 2595.9, rather than to introduce a non-specific interaction between the enzyme and the gel.

1M NaCl did not cause observable denaturation of antibody or enzyme.

Isoelectric Focussing of Enzyme and Antibodies

The enzyme focussed at approximately pH4. A pI of 3.7 has previously been reported for guar α-galactosidase in the literature (4).

All the antibodies focussed to a cluster between pH 6.5 and pH 7. This microheterogeneity is typical and is thought to be due to differences in carbohydrate modifications (8). The fact that each antibody preparation focussed to a single cluster was taken as evidence that the antibodies were produced from single clones. (8). No contaminatory transferrin was detected in any of the antibody preparations.

DISCUSSION

It was important to establish that the two screening assays were specific for anti-α-galactosidase antibodies. This confirms that monoclonal antibodies isolated using these assays will produce specific immunoadsorbents. Three specific monoclonals were isolated in this study; it would be straight forward to use these assays to isolate more antibodies and build up a library of anti- α -galactosidase monoclonals. The comparison of selectivities of different immunoadsorbents is a good example of a criterion which may be used to select the best monoclonal from a library, rather than seizing upon the first antibody which is found to work adequately. Selectivity is directly related to the binding constant of the antibody (9) and therefore should be independent of the amount of antibody immobilised on the gel, and independent of column geometry. Hence the identification of the column made with 2595.9 as the one having the highest selectivity is an important result.

All three immunoadsorbents exhibited retardation of α-galactosidase in PBS rather than tight binding. This is in marked contrast to other immunoadsorbent processes designed in our laboratory and by others (5,10). Typically, immunoadsorbents bind their antigen so tightly that an optimisation of elution buffer is required to effect release without inactivating either antibody or antigen. (5,10). However, immunoadsorbents specific for single subunit proteins (such as α-galactosidase) may be expected to exhibit weaker binding than immunoadsorbents which bind dimers. This is because monomeric antigens have no potential for multivalent binding unless a polyclonal antibody preparation is used. Multivalent binding is known to be capable of enhancing the 'functional association constant' of an antibody for its antigen from about 10^7 to 10^{11} M^{-1} (11). Nonetheless, we have found that immunoadsorbents raised against other monomeric proteins (such as human transferrin) bind sufficiently tightly in PBS to cause complete immobilisation of antigen rather than retardation. Since all three antibodies isolated in this study bind α-galactosidase with lower affinity than would be typical, it seems probable that there is some generic effect brought about by a peculiar immunogenicity of this enzyme.

Adsorption (ie immobilisation) of α-galactosidase, and therefore improved resolution, was obtained by optimising the loading buffer. The improved binding of an immunoadsorbent in buffers that are dissimilar from physiological conditions has not been previously reported to our knowledge. That this anomalous behaviour was due to a property of the antibody was verified by two control experiments. Firstly, it was determined that α-galactosidase did not bind to a 'naked' column (with no antibody attached) or to a column with a non-specific antibody attached (Fig 4). Secondly, the antibody was found to exhibit similar binding characteristics in ELISA. (results not shown).

One possible hypothesis to explain this anomalous behaviour runs as follows. The antibodies each have a pI range from 6.5 to 7 and α-galactosidase has a pI of 3.7. Therefore in PBS, pH7, the antibody has a slight net negative charge and the enzyme has a large net negative charge. Mutual charge repulsion reduces the probability of close approach between antibody and antigen; thus reducing the opportunity for formation of antigen-antibody complexes. However, at high ionic strength, the charge on each protein is masked by counterions, therefore abolishing mutual repulsion. Close approach of antibody and antigen is now a more common event: this allows more antibody-antigen complexes to form. Thus at pH 7, high salt has the effect of increasing the antibody-antigen association constant.

The results obtained at pH 5.5, where enhanced binding occurs in the absence of high salt, can also be explained by considering the net charges on the antibody and antigen. At pH 5.5, the antibody will be positively charged (pI 6.5 - 7.0) while the antigen will be negatively charged (pI 3.7). Therefore at this pH, the net charges on the two proteins are opposite. This increases the probability of close approach and subsequent complex formation.

Another possible interpretation is that hyrophobic interactions make a major contribution to this particular antibody-antigen bond. Consequently binding is enhanced in high salt buffer or at a pH where neither of the two participating proteins are excessively charged.

Whatever the explanation, the findings in this study represent an extreme example of a generally applicable principle: the importance of optimising the loading buffer for immunoadsorbents. Whilst a lot of attention has been paid to elution buffers for immunoaffinity chromatography (10), relatively little research has been applied to loading buffers. In our opinion, the composition of loading buffer is equally important and should be tailored to maximise specific antigen binding and minimise non-specific binding to either antibody or column. Without this optimisation, it is unlikely that the full potential of the technique will be realised.

The binding properties of the antibodies described in this paper lend themselves to scale-up since the mild elution buffers which are used should result in a long column life-time. Current work involves immobilising the antibodies on a silica solid phase which has been customised for immunoaffinity chromatography. (Sorbsil C1000, Crosfield Chemicals).

ACKNOWLEDGEMENT

We thank Dr R. Jeffcoat for suggesting this project

REFERENCES

1. McCleary, B.V., Effect of galactose content on the solution and interaction properties of guar and carob galactomannans. Carbohydrate Research, 1981, 92, 269-285.

2. Hughes, S.G., Overbeeke, N., Robinson S., and Smeets, F.L.M.
    Messenger RNA from isolated aleurone cells directs the synthesis
    of an alpha–galactosidase found in the endosperm during
    germination of guar seed. Plant Molecular Biology 1988, 11,
    783–789.

3. Overbeeke, N., Fellinger, A.J., and Hughes, S.G. Production of
    guar alpha–galactosidase by hosts transformed with recombinant DNA
    methods. International Patent, 1987, WO 87/07641.

4. McCleary, B.V., Enzyme interactions in the hydrolysis of
    galactomannan in germinating guar, Phytochemistry 1983, 22,
    649–658

5. May, K., Gani M.M. and Senior, S.J., Pure alkaline phosphatase its
    preparation and use. European Patent 1985, PN 0151 320

6. Gani, M.M., Coley, J., and Porter, P., Epitope masking and
    immunodominance – complications in the selection of monoclonal
    antibodies against HCG Hybridoma 1987, 6, 637–641.

7. McCleary, B.V., A simple assay procedure for β–D–mannanase.
    Carbohydrate Research 1978, 67, 213–221.

8. Hoffman, D.R., Structure and synthesis of immunoglobulins. In
    Biological and Biomedical Applications of Isoelectric Focussing,
    ed. Catsimpoolas, N, and Drysdale, J., Plenum, New York 1984, pp
    121–149.

9. Eilat, D and Chaiken, I.M., Expression of multivalency in the
    affinity chromatography of antibodies. Biochemistry 1981, 18,
    790–795.

10. Bureau, D and Daussat, J., Immunoaffinity chromatography of
    proteins. A gently and simple desorption procedure. J. Imm.
    Methods 1981. 41, 387–392.

11. Karush, F., Multivalent binding and functional affinity. Contemp.
    Top Mol Immunol 1976, 5, 217–228.

PURIFICATION OF MOUSE IgG1 ON PROTEIN A
AND THE MEASUREMENT OF CONTAMINATING PROTEIN A

KAMI BEYZAVI/HELEN C. WOOD

Bioprocessing Limited,
No. 1 Industrial Estate, Consett, Co. Durham DH8 6TJ.

ABSTRACT

A new affinity matrix (PROSEP-A High Capacity) has been developed for the purification of immunoglobulin G (IgG). The matrix consists of protein A, covalently bound to controlled pore glass. The binding capacity of PROSEP-A High Capacity for mouse IgG1 in various binding buffers of different pH and ionic strength has been determined. The binding capacity was in the region of 10 to 21mg/ml depending on the pH and ionic strength of the binding buffer. The association constants of PROSEP-A and protein A bound to sepharose for mouse IgG1 were determined and values of $1.15 \times 10^5 (M^{-1})$ and $0.97 \times 10^5 (M^{-1})$ were obtained respectively. The equilibration time for binding mouse IgG1 to PROSEP-A High Capacity was determined to be about 20 minutes and a sharp breakthrough curve was observed for the adsorption of mouse IgG1 to PROSEP-A High Capacity. A sensitive enzyme immunoassay was developed to measure the amount of protein A contaminating the purified antibody. The assay has a sensitivity of 1ng protein A per milligram IgG1 and the leakage of protein A from PROSEP-A High Capacity in the presence of mouse IgG1 was between 2-5 ng protein A per milligram antibody.

INTRODUCTION

Over the last few years, the commercial interest in monoclonal antibodies has rapidly increased. They offer exciting potential as diagnostic and therapeutic substances and also serve as affinity ligands for purifying other high value proteins of biopharmaceutical importance. Some applications

require only milligram quantities of monoclonal antibody
whereas others such as cancer therapy, use much greater
quantities; doses may be in grams per patient per year.

Cell culture currently is the only viable large-scale
production source for murine derived monoclonal antibodies and
the majority of monoclonal antibodies produced on a large scale
will be murine-derived as this is a well established technology
[1]. The culture medium used is complex, therefore, a precise
protein specific method must be used to purify the antibody.
In addition, to satisfy the potential market requirements,
multi kilogram quantities of highly purified material must be
produced in a cost effective manner.

The ability of protein A to bind to the Fc portion of
immunoglobulins is widely used as a basis for separation of
IgG's from other proteins by affinity chromatography. Protein
A is a component of the cell wall of Staphylococcus Aureus that
is able to bind various mammalian immunoglobulins [2]. Its
reactivity with mouse immunoglobulin sub-classes has been
described [3] and immobilised protein A on various matrices is
used extensively for the purification of monoclonal antibodies
from both ascites fluid and cell culture supernatants [4].

Usefulness of protein A has been hampered by the fact that
some species and sub-classes of IgG do not react with the
molecule. Protein A has been reported to bind weakly to mouse
IgG1 [5]. In this paper we describe the purification of mouse
IgG1 using a novel solid support, PROSEP-A High Capacity
(PROSEP-A HC), which is protein A immobilised onto porous
glass. We have also developed an EIA to measure the protein A
contaminating the purified antibody.

MATERIALS AND METHODS

PROSEP-A HC and Sepharose/Protein A are standard products of
Bioprocessing Ltd. Mouse IgG1 was a gift from Dr. D. Price,
Unipath/Oxoid Ltd. Chicken anti-protein A was obtained from
Dr. D. Pepper, SNBTS, Edinburgh. Horseradish peroxidase type
VI, urea hydrogen peroxide and tetramethyl-benzidine (TMB) were
supplied by Sigma Chemical Company, Poole, Dorset, UK. All
other chemicals of AnalaR grade were purchased from BDH Ltd.,
Poole, Dorset, UK. Cecil CE594 U.V. & Visible spectro-
photometer, Titerteck Uniskan I, Flow Laboratories and
Pharmacia Phast System electrophoresis were used in this
project.

Mouse IgG1 preparation: The Mouse IgG1 used in this project
was grown in serum free media and supplied to us as 50%
ammonium sulphate precipitate. The precipitate was dissolved
in 0.1M borate/0.15M NaCl buffer, pH 8.5 and dialysed against
three changes of this buffer over 72 hours. The final antibody
solution was 10mg/ml by absorbance at 280nm and better than 95%
pure when analysed by silverstained polyacrylamide gel
electrophoresis (Pharmacia).

Binding Capacity: The binding capacity of PROSEP-A HC and
Sepharose/Protein A for mouse IgG1 were determined by packing
about 2ml of the matrix into a plastic column, 7.5mm diameter x
45mm height, and equilibrating the matrix with the binding
buffer. 50mg of the antibody in the binding buffer was applied
to the column under gravity. The column was washed with five
column volumes of the binding buffer and the bound antibody was
eluted using 0.1M glycine/HCl at pH 2. The amount of antibody
eluted was determined by measuring the absorbance of the
solution at 280nm.

Isotherm: The matrix in batches of 0.5ml was mixed overnight
on a roller-mixer at room temperature with the antibody in 0.1M
borate/0.15M NaCl pH 8.5 at the desired concentration. At the
end of the mixing period, the antibody remaining in the
solution was determined using Pierce BCA protein assay kit.
The amount of antibody in the solution before mixing with the
matrix was also determined using the same assay.

Equilibration Time: 1ml of the matrix was mixed with 10ml of
mouse IgG1 at 1mg/ml. Samples of the supernatant were analysed
for antibody at different time intervals. The Pierce BCA assay
was used for antibody analysis.

Breakthrough Curve: 2ml PROSEP-A HC was packed into a column
(7.5mm diameter, 45mm height). Mouse IgG1 at 1.5mg/ml was
applied to the column at 0.25ml/minute. Five ml fractions of
the flowthrough were collected and the absorbance of the
solution at 280nm was measured.

Enzyme Immunoassay for Protein A: The chicken anti-protein A was enzyme labelled with horseradish peroxidase using the method described by Nakane and Kawaoi [6].

Assay procedure: The Nunc 96 well plates were coated with 100ul of chicken anti-protein A in 0.1M bicarbonate-carbonate buffer at pH 9.6 for 4 hours at room temperature. The plates were washed three times with phosphate buffered saline containing 0.1% tween 20. 100ul of each sample was added to the appropriate wells and the plates incubated overnight at 4°C. The plates were then washed a further three times with phosphate buffered saline containing 0.1% tween 20 and 100ul of enzyme conjugate added. The plates were incubated for two hours at room temperature. The washing procedure was repeated and 200ul of TMB substrate in phosphate citrate buffer pH 6.0 + 0.1 was added. The reaction was stopped after 20 minutes by the addition of 50ul of 2.5M H_2SO_4 and the colour was read on an EIA reader at 450 nm.

RESULTS:

PROSEP-A HC and Sepharose/Protein A: results comparing the binding capacities of PROSEP-A HC and Sepharose/Protein A for mouse IgG1 in two different buffers are shown in table 1. The binding capacities of both matrices are affected by the concentration of sodium chloride in the buffer. The binding capacity of PROSEP-A HC is about 9 times higher than Sepharose/Protein A in the presence of 0.15M sodium chloride. This difference in capacity decreases to a factor of two in the presence of 3M sodium chloride.

TABLE 1

Comparison of binding capacities of PROSEP-A High Capacity and Sepharose/Protein A for mouse IgG1.

MATRIX	NaCl CONCENTRATION IN BUFFER	CAPACITY (mg/ml)	MOLAR RATIO (IgG/PROTEIN A)
PROSEP-A HC	0.15M	13.4	9.6
PROSEP-A HC	3.0M	21.4	15.3
Sepharose/Protein A	0.15M	1.5	2.1
Sepharose/Protein A	3.0M	9.9	14.1

Binding Buffer : 0.1M borate at pH 8.5 containing NaCl

pH effect: The results showing the effect of pH of the buffer on the binding capacity of PROSEP-A HC for mouse IgG1 are shown in table 2.   Initially there is a slight decrease in the binding capacity by excluding sodium chloride from the buffer and then further decreases are observed by reducing the pH of the buffer.   At pH 5 the binding of mouse IgG1 to PROSEP-A HC becomes negligible.   This pH can be used to elute the antibody.

TABLE 2

Effect of pH on the binding of IgG1 to PROSEP-A High Capacity

Buffer	pH	Capacity (mg/ml)
0.1M borate/0.15M NaCl	8.5	13.4
0.1M borate (No NaCl)	8.5	12.4
0.1M Na_2HPO_4/NaH_2PO_4	8.0	11.7
0.1M Na_2HPO_4/NaH_2PO_4	7.7	11.1
0.1M Na_2HPO_4/NaH_2PO_4	7.0	10.2
0.1M Na_2HPO_4/NaH_2PO_4	6.5	7.8
0.1M Na_2HPO_4/NaH_2PO_4	6.0	5.3
0.1M citrate/Na_2HPO_4	5.0	0.4
0.1M glycine/HCl	2.0	0

Isotherm: Figure 1 shows the isotherms for both PROSEP-A HC and Sepharose/Protein A.   The values for the association constants between protein A bound to sepharose or PROSEP and mouse IgG1 were determined using this isotherm.   The values for association constant were determined at 50% saturation [7] which were $1.15 \times 10^5 M^{-1}$ and $0.97 \times 10^5 M^{-1}$ for PROSEP-A HC and Sepharose/Protein A respectively.

Equilibration and breakthrough curve: Figure 2 shows percentage equilibration reached against mixing time.   80% of

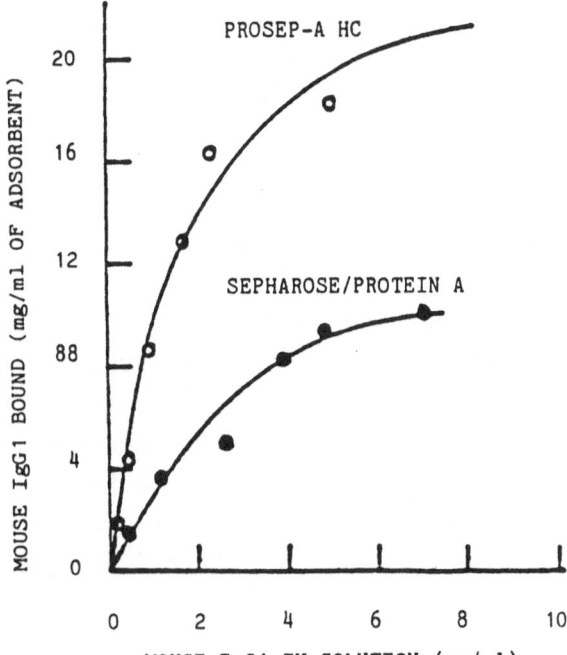

Figure 1. Adsorption Isotherm

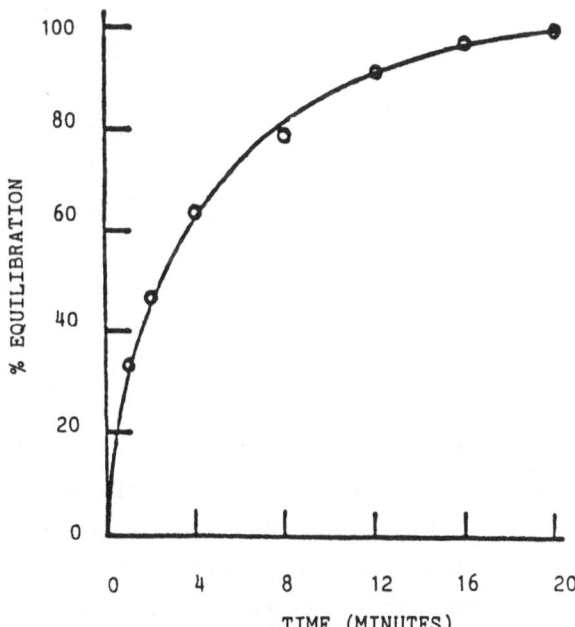

Figure 2. Equilibration time for binding IgG1 to
PROSEP-A HC

equilibration is reached in the first 10 minutes and 100%
equilibration is reached after mixing PROSEP-A HC with mouse
IgG1 for 20 minutes. The breakthrough curve for the adsorption
of mouse IgG1 to PROSEP-A HC at a flow rate of 0.25ml/min is
shown in figure 3.

Protein A assay: The assay was run in various buffers with and
without mouse IgG1 and human polyclonal IgG. Results are shown
in figure 4. The assay showed the highest sensitivity when run
in the neutralised citrate buffer. The presence of mouse IgG1
had no effect on the sensitivity of the assay. The assay had
the lowest sensitivity when run in the presence of human
polyclonal IgG. The leakage of protein A from PROSEP-A HC was
determined in neutralised citrate buffer and in the presence of
mouse IgG1. The values of between 2 to 5 ng protein A per mg
IgG1 were obtained.

DISCUSSION

PROSEP-A HC has shown a very high binding capacity for mouse
IgG1 which is a sub-class of mouse IgG with low binding
affinity to conventional protein A adsorbents such as
Sepharose/Protein A. The use of buffers with high pH and high
salt concentrations e.g. pH 9 and 3M NaCl, has been adopted by
many workers for the effective binding of mouse IgG1 to protein
A adsorbent. With PROSEP-A HC mouse IgG1 can be bound
efficiently under physiological conditions. Therefore PROSEP-A
HC can be used to purify mouse IgG1 directly from cell culture
supernatant or ascitic fluid without any pretreatment.
Furthermore, the use of buffers with high salt concentrations
during purification of antibodies on protein A increases the
non-specific binding of contaminating proteins to protein A by
hydrophobic interaction. This has been shown in our
laboratories for both PROSEP-A HC and Sepharose/Protein A. The
use of high salt concentration is avoided when PROSEP-A HC is

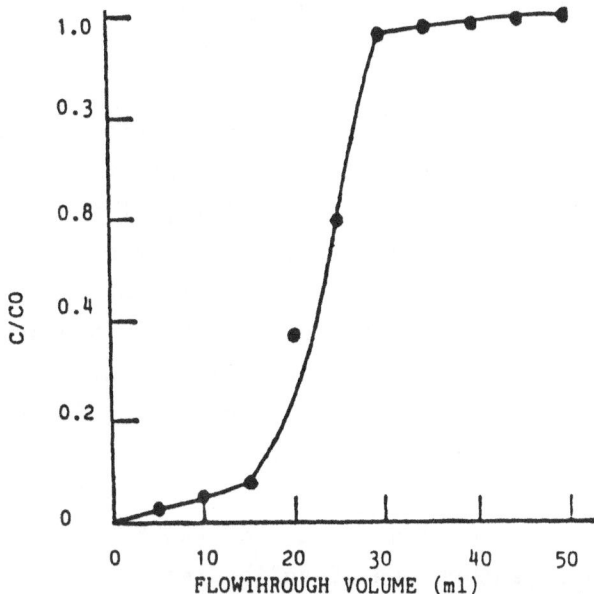

Figure 3. Breakthrough curve for adsorption of IgG1 to
PROSEP-A HC
 C: Concentration of IgG1 in flowthrough
 CO: Initial IgG1 concentration

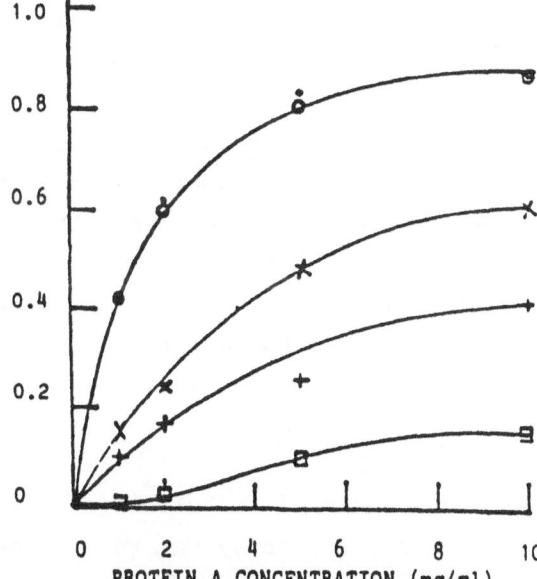

Figure 4. Enzyme Immunoassay for protein A
A: Protein A in neutralised citrate buffer with or without mouse
IgG1 at 1mg/ml. B: Protein A in borate buffer pH 8.6. C: Protein
A in citrate buffer pH 3.0. D: Protein A in neutralised citrate
buffer with human polyclonal IgG at 1mg/ml.

used with buffers of physiological salt concentration and hence the high purity of the final product is ensured.

Similar values for the association constants of PROSEP-A HC and Sepharose/Protein A were obtained. This implies that there is no change in the affinity of protein A bound to PROSEP for mouse IgG1. The higher capacity of PROSEP-A HC is due to a better orientation of immobilised protein A. Also since the affinity constants of PROSEP-A HC and Sepharose/Protein A in binding buffers with 0.15M NaCl concentration are the same, similar eluting conditions can be used for both matrices when purifying mouse IgG1.

Uniform microstructure of PROSEP matrix results in the efficient adsorption of antibodies which is shown by the sharpness of the breakthrough curve. Total antibody eluted from the column was 25mg (12.8mg/ml) which agrees closely with the results in Table 1.

The assay developed for the protein A determination is affected by the nature of the buffer and also the presence of human polyclonal antibody. The highest sensitivity was obtained using neutralised citrate buffer. The presence of mouse IgG1 in this buffer does not have any effect on the sensitivity of the assay. However, the presence of human polyclonal IgG reduces the sensitivity of the assay considerably. This is because of the high affinity of human IgG for protein A. Other antibodies or sub-classes with higher affinity for protein A than IgG1 will also effect the sensitivity of this assay.

For the purpose of determining the leakage of protein A from PROSEP-A HC, the assay was run in neutralised citrate buffer and in the presence of mouse IgG1 at 1mg/ml. The leakage of protein A was very low, in the region of 2-5 ng protein A per milligram of mouse IgG1 eluted from PROSEP-A High Capacity.

REFERENCES

1. Duffy, S.A., Moellering, B.J., Prior, G.M., Doyle, K.R. and Prior, C.P., Recovery of Therapeutic-Grade Antibodies: Protein A and Ion-Exchange Chromatography, Pharmaceutical Tech. Inter., 1989, September, 46-52.

2. Lindmark, R., Thoran-Tolling, K and Sjoquist, J., Binding of Immunoglobulins to Protein A and Immunoglobulin Levels in Mammalian Sera, J. Immunol. Methods, 1983, 62, 1-13.

3   Mackenzie, M.R., Warren, N.L. and Mitchell, G.F., The Binding of Marine Immunoglobulins to Staphylococcal Protein A, J. Immunol., 1978, 120, 1493-1496.

4. Scott, R.W., Duffy, S.A., Moellering, B.J. and Prior, C., Purification of Monoclonal Antibodies from Large-Scale Mammalian Cell Culture Perfusion Systems, Biotech. Progress, 1987, Vol. 3, No. 1, 49-56.

5. Ey, P.L., Prowse, S.J. and Jenkin, C.R., Isolation of Pure IgG1, IgG2a and IgG2b Immunoglobulins from Mouse Serum Using Protein A-Sepharose, Immunochemistry, 1978, 15, 429-436.

6. Nakane, P.K. and Kwaoi, A., Peroxidase-Labelled Antibody: A New Method of Conjugation, J. Histochem. Cytochem., 1974, 22, 1084-1099.

7. Chase, H.A., Scale-up of Immunoaffinity Separation Processes, J. Biotech., 1984, 1, 67-80.

ANALYSES OF PURITY AND HOMOGENEITY IN PRODUCTION PROCESSES

KIRSTEN BIEDERMANN, PIA KNAK JEPSEN, JYTTE PEDERSEN and BJARNE
RØNFELDT NIELSEN
Department of Biotechnology, The Technical University of Den-
mark, building 223, DK-2800 Lyngby

ABSTRACT

An analytical procedure based on isoelectric focusing was developed
to identify contaminants in a nuclease preparation. Two categories
of contaminants were visualized: protein impurities from the
process background and modified forms of the enzyme. The potential
of the analysis was tested on a fermentation with $\underline{E.}$ $\underline{coli}$ producing
and secreting the enzyme, and using different purification pro-
cedures. Introduction of simple analyses to validate quality of
a product at any stage in the developmental phase of a production
can help to outline a production strategy.

INTRODUCTION

Purity and homogeneity are factors of increasing importance for
validating the production of proteins. Microorganism and cell
lines are the future source for production of proteins such as
enzymes and polypeptides. Proteins destined for pharmaceutical
therapy have to fulfil a requirement for a high level of purity,
which may include removal of modified forms of the product.
Accessibility of sensitive analytical methods is therefore of
importance in the developmental phase of a production process in
order to evaluate each step of the system. However this is not
only true for production of drugs, but is also of great help for
the design of any production process in order to optimize it with
regard to the yield and purity required for the application of
the particular product.

Impurities, which have to be removed, can be divided into two
categories: 1) impurities originating from the host and culture
medium chosen, and from particular separation processes such as
ligands from affinity chromatography, 2) impurities which are

modified forms of the product. They can be formed in the cell during the process by proteolytic cleaving (1), incorrect folding or incorrect paired disulphide bonds, or be caused by adverse culture conditions. They may also arise during purification as a result of limited proteolytic degradation of the product, or of conformational changes due to contact with specific reagents in a separation step. Chemical modification can also occur during storage, such as deamination of asparagine and glutamine residues. Oxidation of methionine to methionine sulfoxide has also been seen in recombinant products such as insulin-like growth factor and human growth hormon. The first category will subsequently be called impurities or contaminants and the second category modified forms or variants of the product.

Direct determination of purity and homogeneity of proteins is most frequently performed by polyacrylamide gel electrophoresis, SDS-PAGE, which separates proteins with regard to size. However a more sensitive method is isoelectric focusing, IEF, which separates proteins due to charge differences (2,3). Furthermore the separated proteins can, after their separation, be characterized if they are blotted onto nitrocellulose. Identification can be carried out with antibodies or by using biological activity. In this way simple and fast analytical procedures can be used without any particular pretreatment of the samples, which is normally a bottle neck for direct characterization of proteins by methods such as reverse phase HPLC.

Analyses, based on isoelectric focusing to determine purity and homogeneity are demonstrated here on the microbial production of an enzyme nuclease. The enzyme originates from Serratia marcescens and it can be produced and secreted by Escherichia coli. The enzyme is an endonuclease and degrades DNA and RNA unspecifically. It has a molecular weight of 30 kDa as determined by SDS-PAGE, and an isoelectric point of 6.9 (4). Strains differing in their ability to produce a homogeneous product were included in the examination. The efficiency of different separation methods to remove impurities and modified forms of the product were tested.

MATERIAL AND METHODS

Analytical methods
Isoelectric focusing was performed in 1% agarose gel on an LKB multiphor 211 apparatus. The pH gradient was formed with carrier ampholytes (Serva) pH 4-10 and 3-10 in a ratio of 1:3. After focusing, i.e. when the proteins had migrated to their respective

isoelectric points, the gel was fixed and stained unspecifically with Coomassie Blue modified with Crocein Scarlet (Serva) and $CuSO_4$ (3).

Proteins in similar gels were transferred to nitrocellulose by capillary blotting (5). Free sites of attachment were blocked with 0.05% Tween 20. Immunochemical detection was performed with rabbit antiserum against nuclease, for identifying nuclease, and variants of nuclease and rabbit serum against E. coli proteins for identifying impurities. The antibodies bound to the proteins were visualized after reaction with a secondary antibody, swine anti rabbit IgG, conjugated with alkaline phosphatase (6).

Detection of the enzymatic activity of nuclease was performed by a zymogram technique obtained by layering the nitrocellulose on an agarose gel for one hour with the substrate DNA incorporated, DNase-test-agar (Difco). The enzymatic activity was visualized as clearing zones on the gel after non-degraded DNA had been precipitated with 1 N HCl.

Two-dimensional gel electrophoresis was carried out in the first dimension by IEF as described. The second dimension was performed vertically by SDS-PAGE using 0.1% SDS and 11% acrylamide (7,2,3).

Biological Materials

Fermentation: E. coli strain MT 102 (a derivative of MC 1000) carrying a plasmid p403-SD2, encoding nuclease and its signal-peptide was cultivated in a 15 litre fermentor. The cells were cultivated at 35°C with an aeration rate of 1.0 vol/vol/min, without stirring, for 48 h, in minimal medium enriched with thiamine, casaminoacids, tryptone, yeast extract and glucose. Cell density was measured through optical density at 450 nm and nuclease activity through its ability to degrade DNA (4). Periplasmic nuclease was released by sonication.

Protease inhibitors pepstatin A and PMSF were added to samples from the fermentation to be used for IEF to a final concentration of $10^{-7}M$ and $10^{-3}M$ respectively. Cells and medium were separated by centrifugation. The medium samples were, if required, concentrated by dialysis against 20% polyethyleneglycol (20.000). The cells could be applied directly to the gels, but better results were obtained if they were pretreated for 30 min with 0.6 mg per ml lysozyme. For Coomassie Blue staining the samples contained approximately 15o to 6oo U of nuclease. The samples analysed by immunoblotting and zymograms contained 10 to 50 U of nuclease. The specific activity of purified nuclease has been determined to 4.0×10^4 U/mg (4).

Shake flask experiments were performed with different E. coli strains (MT 102, W 3110, CH 30 and CSH 50) carrying the same plasmid p403-SD2 and a S. marcescens strain (W 280) (9), in order to evaluate the ability of the different host cells to produce homogeneous nuclease. Cells and medium for isoelectricfocusing were prepared as described above.

Purification: Cells and medium were separated by tangential flow microfiltration (Pellicon Cassette system, Millipore). The closed filtration system consisted of a tank, a pump and the filtration unit. Periplasmic nuclease was released from the cells left in the system by an osmotic shock treatment. This was performed in three steps: conditioning of the cells in 20% sucrose (w/v) in 0.03 M Tris-HCl pH 8.3 containing 1 mM EDTA, shock treatment of the cells and wash of the speroplast, both with cold water. Each step was performed by alternately recirculating the cells in the appropiate solution and subsequently filtering off the solution by opening the outlet valve. The medium and the periplasmic extract, obtained from the shock treatment and speroplast wash, were combined and concentrated by ultrafiltration (10). The enzyme was precipitated with ammonium sulphate (80% saturation) and purified by ion exchange chromatography on DE-52 (Whatman) and CM-Sepharose (Pharmacia) (4). Preparative IEF was performed in a multi-chamber apparatus consisting of 32 separation chambers (Kem-En-Tec) (11) and immunoaffinity chromatography was performed on CNBr-activated agarose (Pharmacia) coupled with our own monoclonal antibody (NUC 1-4) (12).

RESULTS

After 24 h of fermentation the cell density was measured as 7 (OD_{450} nm) and the amount of nuclease was measured as 150 mg per litre with 85% located in the periplasm of the cells. After 48 h of fermentation 85% of the enzyme was located in the medium.

Analyses by isoelectric focusing of cells and medium from the fermentation with E. coli cells producing nuclease are shown in Figure 1. The samples examined are taken after 24 and 48 h of fermentation, i.e. when the majority of the enzyme is respectively periplasmic and extracellularly located. Samples from medium and cells look similar on the gels. Nuclease separated as the predominant band at pH 6.9 vizualized by Coomassie Blue staining, immunoblotting and at the zymogram. The other bands seen by Coomassie Blue staining are impurities. However, some of them react with antiserum against nuclease and also have nuclease activity which is seen on the zymogram. These results confirm that heterogeneities are already formed in the cells during cultivation. One variant has a pI of 8.3 and the pIs of the others are seen to be more acidic than the pI of nuclease. SDS-PAGE and the same blotting techniques did not detect similar active variants indicating that the molecular weights of the modified forms of nuclease are the same or nearly the same as nuclease. We also analysed E.coli without plasmid for nuclease activity with the same techniques, and no clearing zones was seen.

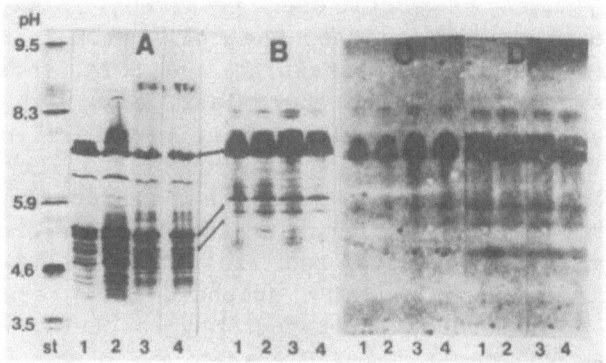

Figure 1. Isoelectric focusing of medium and cells from the E. coli fermentation. A. Coomassie Blue staining (600 U); B. immunoblot (10 U); C. zymogram (10 U); zymogram (50 U). 1,2. medium after cultivation 24 and 48 hours; 3,4. cells after cultivation for 24 and 48 hours.

In order to discover whether the observed variants of the enzyme nuclease were related to the host strain, we analysed different E. coli strains with the same plasmid and also a S. marcescens strain which naturally produces extracellular nuclease. It was observed that all strains produced modified forms of the enzyme with pIs in the range 4.5-8.3, but that the number varied. It was also observed that the nature of other proteins in the medium, and the cells from the different strains, varied. S. marcescens itself produced a larger number of variants of nuclease than any of the E. coli strains, and additionally this strain produces extracellular proteases. We had previously chosen E. coli MT102(p403-SD2) for our fermentation and purification studies because this strain is able to produce larger quantities of nuclease than the other transformed E. coli strains, and larger amounts are secreted into the medium. The analyses showed that this strain also produces less variants of nuclease when compared to the other strains. The analyses described are able to detect variants of nuclease but they do not give a quantitative picture of the amount of variants relative to nuclease (Figure 1).

The purification of the enzyme was analysed by the same analytical procedures, to discover whether more variants of nuclease were formed during the purification, or to find out which step in a purification procedure removed impurities and variants. The analyses were now expanded with an immunoblot using antiserum against E. coli proteins from the host strain in order to ensure detection of all impurities. The purifications by salt precipitation, anion and cation exchange chromatography are shown in Figure 2. Impurities and almost all the variants having pIs different from the pI of nuclease are removed. In particular the last cation exchange chromatography removes the basic variant with pI 8.3. The figur also shows that the variant having pI 8.3 has increased during the operations.

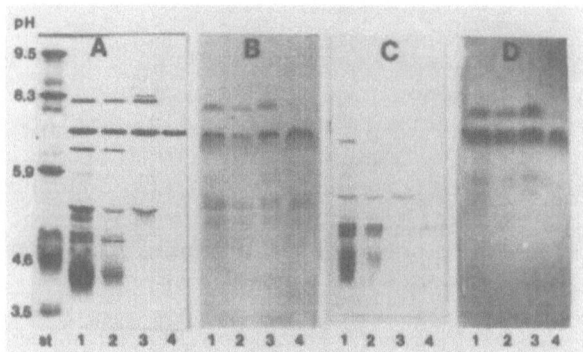

Figure 2. Isoelectric focusing of the purification of nuclease. 1. medium and periplasmic extract concentrated by ultrafiltration; 2. salting out; 3. anion exchange chromatography; 4. cation exchange chromatography; A. Coomassie Blue staining (200 U); B. immunoblot with antiserum against nuclease (10 U); C. immunoblot with antiserum against E. coli proteins (10 U). D. zymogram (10 U).

We also examined the separation efficiency of preparative IEF on a sample after salt precipitation and anion exchange chromatography. After focusing, the content of each of the 32 separation chambers was analysed by IEF. All variants and all impurities with pIs lower than nuclease were collected in the chambers having acidic pH. Nuclease had accumulated in 6 out of the 32 chambers around chamber 10 with pH 6.9, however the variant with pH 8.3 was not separated from the enzyme. The nuclease activity of the variants with acidic pIs could now be measured to approximately 7% of the total activity fractionated in the apparatus. However we do not know if the specific activities of the acidic variants is the same as that of nuclease with pI 6,9.

Though IEF reveals small variations in proteins due to charge differences, it will not reveal whether there are more molecules with the same pI hidden in a protein band after focusing. Immunoblotting of our gels with antiserum against E. coli proteins had shown that some E. coli contaminants focused at the same pH as nuclease. Two-dimensional gel electrophoresis and silver staining of the proteins is an efficient method to identify any component present in a sample. We analysed a sample from the preparative IEF (chamber 10) containing nuclease and the variant with pI 8.3. It appeared that nuclease was contaminated with two proteins with higher molecular weights than nuclease. The same analysis showed that the variant at pH 8.3 has a slightly lower molecular weight than nuclease.

The purity of nuclease was also analysed after immunoaffinity chromatography. The monoclonal antibodies used as ligands were specially screened for their ability to dissociate nuclease at neutral pH, where the enzyme is stable, by introducing magnesium ions in the elution buffer. Nuclease in the concentrated medium and periplasmic extract was purified directly onto the affinity column, Figure 3. It is seen that the impurities and the modified forms of nuclease are removed in one step.

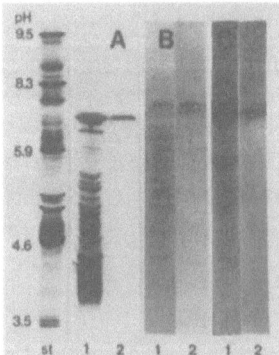

Figure 3. Isoelectric focusing of the purification of nuclease by immunoaffinity chromatography with monoclonal antibodies. A. Coomassie Blue staining (150 U); B. immunoblot (10 U); C. zymogram (10 U). 1. concentrated medium and periplasmic extract; 2. immunoaffinity chromatography.

DISCUSSION

In production processes on-line measurements of proteins are followed by methods which monitor total protein, such as Bradford or ultraviolet absorption. More product related-methods are estimation of retention time by reverse phase HPLC or affinity chromatography. However none of these methods gives a true picture of contaminants and modified forms of the proteins, and they can only be trusted if additional characterization of the protein pools has taken place. This may be electrophoresis, amino acid analysis, N- and C-terminal sequencing and peptide mapping. The latter methods are expensive and laborious and should therefore only be employed in the characterization of the final product. Often variants of the product will not even be detected by these methods, in particular, if they are present in small amounts. Electrophoretic techniques, including IEF, are powerful methods and their sensitivities can be increased to detect less than a picogram of a protein (13).

The analytical procedures based on IEF were shown in this study to be able to detect purity and homogeneity of an enzyme nuclease, both when cells were producing the enzyme, and during its purification. When the analyses were combined with immuno-blotting and zymogram, where the biological activity of the enzyme was utilized, the detection sensitivity increased, but information about the nature of the contaminants was also obtained. The analyses are cheap when compared with other analytical methods, and they are fast and easy to perform. If the analyses are supplemented with two-dimensional gel electrophoresis, both size and charge differences of the proteins are detected.

Cells and medium from the fermentation with E. coli could be analysed directly in our study. We also found that S. marcescens produced more impurities and variants of the enzyme than any of the E. coli strains examined. Such information is valuable before a production strategy can be established. It is important to know the physical and chemical properties of the product to be purified, but it is also helpful to have information about the impurities which have to be separated from the product. Therefore strains or host/plasmid combinations should not only be screened with regard to productivity, but also their specific impurities and their ability to process the product correctly should be taken into consideration. Fermentation processes are optimized with regard to yield, but it is important to remember that the conditions chosen do not necessarily give the most homogeneous product.

Nuclease was purified using traditional methods such as salt precipitation and ion exchange chromatography as well as more advanced methods such as preparative IEF and immunoaffinity chromatography. The analytical techniques detected any contaminants not separated from the homogeneous enzyme. Two-dimensional electrophoresis detected several contaminants which had the same pI as the enzyme. In such cases separation methods based on charges can not be used. Gel filtration can sometimes be successful but another possibility is affinity chromatography. A more advanced method to circumvent these problems is to produce the protein with an extension which alters its charge or affinity. The extension is cleaved off after the hybrid protein has been purified and the purification is repeated.

The variants of the enzyme nuclease had pI-values ranging from 4.5 to 8.3. Possible explanations for formation of modified forms of a product which is secreted can be: incorrect cleavage of the signal peptide which leads the molecule across the cytoplasmic membrane, and limited proteolytic attack at the N- or C-termini of the molecule when the folding process takes place. Recently we isolated the variant with pI 8.3 and it was characterized by means of plasma desorption mass spectrometry. It was found that the first three amino acids at the N-terminus were missing. At the present time we do not know the nature of the other variants relative to nuclease nor the reason for their appearance during cultivation.

CONCLUSIONS

In the developmental phase of the production of a protein it is important to develop simple analyses to identify impurities and modified forms of the product. For this purpose isoelectric focusing is an effective technique. The nature of the separated proteins can be further analysed by immunoblotting, techniques utilizing the biological activity of the proteins, or electrophoresis in the second dimension, utilizing other separation criteria. Knowledge about the origins and characteristics of the impurities and variants of the product is valuable in order to prevent their appearance or ensure their removal.

REFERENCES

1. Shibui, T., Uchida, M., Nagahari, K. and Teranishi, Y., High level secretion of Human Cardiodilatin by Escherichia coli. Agric. Biol. Chem., 1988, **52**, 1145-1150.

2. Hames, B. D. and Rickwood, D., Gel Electrophoresis of Proteins a Practical Approach, IRL Press, 1987.

3. Pharmacia Fine Chemicals, Isoelectric Focusing Principles and Methods, Ljungföretagen AB, Örebro, 1982-1.

4. Biedermann, K., Jepsen, P.K., Riise, E. and Svendsen, I., Purification and characterization of a Serratia marcescens nuclease produced by Escherichia coli. Carlsberg Res. Commun., 1989, **54**, 17-27.

5. Andrews, A. T., Electrophoresis, Clarendon Press, Oxford, 1986, pp. 59-77.

6. Blake, M.S., Johnstone, K.H., Russell-Jones, G.J. and Gotschlich, E.C., Anal. Biochem., 1984, **136**, 175-179.

7. Laemmli, U.K. Cleavage of structural protins during the assembly of the head of Bacteriophage T4. Nature, 1970, **227**, 680-685.

8. Scopes, R., Protein Purification Principles and Practice, Springer-Verlag, 1984, pp. 185-200.

9. Jepsen, P.K., Riise, E., Biedermann, K., Kristensen, P.C.R. and Emborg, C., Two-level factorial screening for influence of temperature, pH and aeration on production of Serratia marcescens nuclease, Appl. Environmental Microbiol., 1987, **53**, 2593-2596.

10. Biedermann, K. and Jepsen, P. K., Release of periplasmic enzymes from Eschericha coli using tangential flow filtration. Biotechnol. Techniq., 1989, **3**, 39-44.

11. Kyhse Andersen, J.,Multi-chamber apparatus for preparative isoelectric focusing. Electrophoresis, 1989, **10**, 6-10.

12. Andresen, L. O., Koch, C., Sørensen, B. B. and Emborg, C., Production of monoclonal antibodies against the enzyme nuclease from Serratia marcescens and evaluation of the individual antibodies for their use in EIA and affinity chromatography. Biotechnol. Techniq., 1989, **3**, 407-410.

13. Nespolo, A.. Bianchi, G., Salmaggi, A., Lazzaroni, M., Cerrato, D. and Taljoli, L. M., Immunoblotting techniques with picogram sensitivity in cerobrospinal fluid protein detection. Electrophoresis, 1989, **10**, 34-40.

PURIFICATION OF MONOCLONAL ANTIBODIES FOR HUMAN CLINICAL USE: VALIDATION OF DNA AND RETROVIRAL CLEARANCE

DAVE BRADY, J. BONNERJEA AND C.R. HILL
Celltech Ltd., 216 Bath Road,
Slough, Berks. U.K. SL1 4EN.

ABSTRACT

The production of therapeutic grade monoclonal antibodies using mammalian cell culture requires the development of a purification process that consistently yields a safe and efficacious product. Process validation studies provide the assurance that the purification process is capable of reducing the levels of potential contaminants to safe and acceptable levels.

The monoclonal antibody purification process used at Celltech has been validated for its ability to remove DNA and retrovirus by means of challenge studies. Moloney Murine Leukaemia Virus and ^{32}P-labelled murine hybridoma DNA were used to challenge individual chromatography steps that had been scaled-down to mimic the full-scale manufacturing process. These studies have demonstrated that the risk factors associated with contaminating DNA and virus can be substantially reduced by the purification procedure described here.

INTRODUCTION

Since the first description of the hybridoma technique for the preparation of monoclonal antibodies [1] a large amount of research and development effort has been directed towards developing applications of monoclonal antibodies as human therapeutic and diagnostic agents.   A primary objective when developing a human therapeutic product is to ensure the purity, potency, safety and efficacy of the compound. The guidelines for therapeutic monoclonal antibody standards provided by the various national licencing authorities [2 - 4] provide a framework for the development of an appropriate strategy.

A major issue regarding the purity and safety of the compound is the level of potential contaminating substances derived from the host cell line and the raw materials used in its manufacture including the media components and purification reagents. The purity and safety issues related specifically to contamination by viruses and DNA are of particular importance.

Many hybridoma cell lines contain endogenous retrovirus observed as particles by electron microscopy. A combination of approaches is required to ensure the safety of the product. These include extensive testing of the cell bank for a wide range of different viruses [2 - 4] and specific testing of raw materials, in process samples and final product.

An essential component of manufacturing process safety validation is to demonstrate the capacity of the purification process to remove or inactivate retrovirus [5]. This can be achieved by designing a scaled-down process that mimics the manufacturing process as closely as possible. The scaled-down purification step can then be challenged with a known amount of retrovirus and its removal from the purified product can be determined giving a clearance capacity for that particular step. This procedure can then be repeated for the other steps in the process to give an overall clearance capacity for retrovirus.

Similar approaches can be taken to assure the removal of host-cell DNA from the product [5]. A level of 10 pg/dose or less is suggested as a target by the regulatory authorities in their guidelines [2 - 4]. Assay of the final product with a sensitive hybridisation assay is essential. However, in some cases relatively large doses of antibody are administered, often exceeding 50 mg of protein. Although hybridisation assays are very sensitive, generally being capable of measuring 1 - 5 pg of DNA/5 mg of antibody, it is often difficult to accurately measure less than 10 pg of DNA in the presence of greater than 50 mg of antibody.   Process clearance studies similar to those described above for virus can then be used to give greater assurance that DNA has been reduced to an appropriate level in the therapeutic product.

A large number of methods have been described for purification of antibodies [6].   Two commonly used methods are Protein A affinity chromatography and anion-exchange chromatography.   In this report we describe the results of a study in which clearance of a retrovirus and DNA for these two steps has been determined.

MATERIALS AND METHODS

Retrovirus

Moloney murine leukaemia virus (MLV), a type C retrovirus containing single stranded RNA, was selected as a model retrovirus.   This is typical of the type of virus particle often observed in hybridoma cells by electron microscopy.   The retrovirus was prepared by passage in NIH 3T3 cells grown in Dulbecco modified MEM supplemented with 10% foetal calf serum.   Viral titres were determined using mink S^+ L^- cells and the results expressed in foci forming units.   The limit of detection of this assay is 10 ffu/ml.   The viral stock used for the study has a titre of 1 x 10^9 ffu/ml.

Control experiments were carried out on all samples tested to ensure that they were not cytotoxic for the mink S^+L^- detection cell line.

DNA Extraction and Nick Translation

Murine DNA was extracted from an overgrown murine hybridoma cell culture.   Cell debris was removed from the culture broth by centrifugation (1,200 rpm, 5 min) and the supernatant was precipitated with isopropanol (2 volumes).   Protein was removed by digestion with Proteinase K, followed by extraction with phenol and chloroform.   The nucleic acid was precipitated by addition of ethanol and salt and treated with ribonuclease A (50 µg/ml, 37°C/hr). After re-extraction with phenol and chloroform to remove the ribonuclease the DNA was precipitated from ethanol.

The purified murine DNA was nick translated using an Amersham Nick Translation Kit (Product number N-5500). A reaction mixture containing murine DNA (1 µg), 2 nmoles dCTP, dGTP and dTTP, 100 µCi alpha labelled dATP-32 (>6000 Ci/nmol), 5 units of DNA polymerase, was prepared and incubated at 16°C for 2 hours. The reaction was then stopped by the addition of EDTA and unincorporated counts separated by gel filtration using Sephadex G50.

Protein A Affinity Chromatography

Retrovirus Clearance: A column of Protein A Sepharose (Pharmacia) (2 ml) was prepared and equilibrated at room temperature. A sample of monoclonal antibody was prepared (33.4 ml, 17 mg IgG) and spiked by the addition of MuLV (1 ml, 3.89×10^8 ffu). This was applied to the column which was then washed and the antibody eluted with sodium citrate buffer, pH 2.5. The pH of the eluted antibody was adjusted to 7.0 by the addition of Tris/HCl and samples were taken for virus assay.

The elution of antibody from a Protein A affinity column is dependent on pH [7] and can vary over a range from pH 7.0 to pH 2.5. In order to study the effect of elution pH on the extent of retrovirus clearance achieved, two further experiments were carried out in which the column was eluted with sodium citrate buffer, pH 4.0 and sodium phosphate buffer, pH 7.0.

Murine DNA Clearance: A column of Protein A Sepharose (Pharmacia) (2 ml) was prepared. A sample of monoclonal antibody (32 ml, 17 mg) was spiked with radiolabelled murine DNA (1 µg, 1.5×10^7 cpm) and applied to the column. The column was washed and the antibody eluted with sodium citrate buffer pH 2.5.

The pH of the eluted antibody was adjusted to 7 by the addition of Tris/HCl and samples were taken for scintillation counting.

DEAE Anion-Exchange Chromatography

Murine Retrovirus Clearance: A column of DEAE Sepharose (Pharmacia) (2 ml) was prepared and equilibrated with 10 mM Tris pH 8.0. A sample of antibody was prepared in equilibration buffer (18 ml, 18 mg) and spiked with MuLV (1 ml, 1.9 x 10^8 ffu). The column was washed and the antibody eluted with an increasing linear salt gradient to 200 mM sodium chloride in equilibration buffer. Samples of the eluted antibody were taken for retrovirus assay.

Murine DNA Clearance: A column of DEAE Sepharose (Pharmacia) was prepared as above. A sample of monoclonal antibody was spiked with radiolabelled murine DNA (1 µg, 9.1 x 10^7 cpm) and applied to a column of DEAE Sepharose prepared and run as described above. Samples of eluted antibody were taken for scintillation counting.

RESULTS AND DISCUSSION

Murine Retrovirus Clearance

The results for murine retrovirus clearance by Protein A affinity and anion-exchange chromatography are given in Table 1. The largest step clearance is obtained by Protein A affinity chromatography with elution at pH 2.5. No virus was detected in the antibody eluted at this pH. The limit of sensitivity of the virus assay is 10 ffu/ml. As the antibody was eluted from the column in 11.5 ml a value of 115 ffu is recorded as the maximum amount of virus that may be eluted with the antibody. This figure is used for calculation of the step clearance factor of 6.5 logs. However, as no virus was detected in the sample the actual step clearance is likely to be greater than 6.5 logs.

The step clearance factor for Protein A affinity chromatography progressively decreases as the elution pH increases to pH 7.0. Retrovirus are enveloped viruses and are known to be sensitive to inactivation at low pH.

This was confirmed by a series of control experiments. When the murine retrovirus was incubated at room temperature in column equilibration buffer for the duration of the experiment a reduction of just 1 log of virus titre was observed. However, when the retrovirus was incubated in elution buffer at pH 4.0 and pH 2.5 a reduction of 7.0 and >7.7 logs respectively of virus titre was seen.

TABLE 1
Murine Retrovirus Clearance by Protein A Affinity and
Anion-Exchange Chromatography

Column Step	Total Retrovirus Applied(1) (ffu)	Total Retrovirus Recovered(2) (ffu)	Step Clearance(3) (Logs)	Cumulative Clearance(4) (Logs)
Protein A Affinity				
Elution pH 2.5	3.9×10^8	<115	>6.5	
Elution pH 4.0	4.7×10^8	345	6.1	
Elution pH 7.0	1.5×10^8	5×10^3	4.5	
Anion-Exchange	1.9×10^8	< 78	>6.4	>12.9 - 10.9 (5)

Footnotes:

1. The total retrovirus applied to the columns.

2. The total retrovirus recovered with the eluted antibody.

3. The step clearance factor is calculated as follows:

   Total retrovirus applied
   Total retrovirus recovered

   e.g. for elution at pH 2.5

   $$\text{Virus clearance} = \frac{3.9 \times 10^8}{<115} = >3.4 \times 10^6$$
   $$= >6.5 \text{ logs}$$

4. Calculated by multiplying step clearance for column 1 by step clearance for column 2 (adding logs).

5. The range is dependent on the elution pH used for the column preceding the anion exchange column.

For anion-exchange chromatography a reduction of >6.4 logs virus titre was obtained. The cumulative clearance factor for Protein A affinity chromatography followed by anion-exchange chromatography is >12.9 - 10.9 logs. The range depends on the elution pH used for the Protein A column, the greatest clearance being achieved with the pH 2.5 elution buffer.

A target value for clearance of retrovirus clearance by a purification process is generally 10 - 15 logs. The results presented here show that this can be achieved in just two purification steps. These results, taken together with an appropriate programme of virus testing on the cell banks and schedule of in-process testing, provide a high degree of assurance that the final product is free of retrovirus contamination.

Murine DNA Clearance

The results for step clearance of murine DNA by Protein A affinity and anion-exchange chromatography are given in Table 2. For Protein A affinity chromatography a clearance of 5.7 logs DNA has been achieved. Some radiolabelled DNA was eluted with the antibody although this represents less than 0.002% of the total counts per minute applied to the column. A clearance of 3.4 logs DNA has been achieved for the anion exchange column and here again some of the radiolabelled DNA has been retained and eluted with the monoclonal antibody.

The cumulative clearance of murine DNA by Protein A affinity chromatography followed by anion-exchange chromatography is given in Table 4. An overall clearance of 9.1 logs DNA has been achieved by the 2 steps. In other words, if 10 mg DNA/mg antibody was applied to the Protein A affinity column the purified product obtained from the anion-exchange column should contain less than 10 pg DNA/mg antibody.

TABLE 2
Murine DNA Clearance by Protein A Affinity and
Anion-Exchange Chromatography

Column Step	Total DNA Applied (cpm)	Total DNA Recovered (cpm)	Step Clearance (Logs)	Cumulative Clearance (Logs)
Protein A Affinity	1.5×10^7	3.0×10^2	5.7	
Anion-Exchange	9.1×10^7	3.8×10^4	3.4	9.1

CONCLUSION

Purification process contaminant clearance studies are an essential part
of process validation for the manufacture of therapeutic proteins. They
give a valuable insight into the ability of the purification process to
remove potential contaminating substances. Taken together with
complementary testing of cell banks, raw materials, in-process samples
and final products they provide a high degree of assurance that potential
contaminants such as retrovirus and DNA are eliminated from the purified
product.

REFERENCES

1. Kohler, G. and Milstein, C., Continuous culture of fused cells
    secreting antibody of predefined specificity. Nature (Lond.) 1975,
    256, 495-497.

2. Committee for Proprietary Medicinal Products. Guidelines on the
    production and quality control of monoclonal antibodies of murine
    origin intended for use in humans. Trends Biotechnol., 1988, 6,
    G5-G8.

3. Australian Department of Community Services and Health, Guidelines for the preparation and presentation of applications for general marketing of monoclonal antibodies intended for use in humans, G.P.O. Box 9848, Canberra, ACT 2601, Australia.

4. FDA Office of Biologics Research and Review, Points to consider in the manufacture and testing of monoclonal antibody products for human use, 1987, Bethesda, Maryland 20892, U.S.A.

5. Levy, J., Lee, H.M., Kawahata, R.T. and Spitler, L.E., Purification of monoclonal antibodies from ouse ascites eliminates contaminating mouse type C viruses and nucleic acids, Clin. Exp. Immunol., 1984, 56, 114–120.

6. Kenney, A.C., Large-scale purification of monoclonal antibodies. In Monoclonal Antibodies: Production and Application, Alan R. Liss Inc., 1989, pp. 143-160

7. Ey, P.L., Prowse, S.J. and Jenkin, C.R. Isolation of pure IgG_1, $IgG2a$ and IgG_{2b} immunoglobulins from mouse serum using protein A Sepharose, Immunochemistry, 1978, 15, 429.

ANALYSIS OF RECOMBINANT HUMAN IFN-GAMMA PRODUCED IN CHO CELLS:EFFECTS OF GLYCOSYLATION AND PROTEOLYTIC PROCESSING ON PRODUCT HETEROGENEITY.

ELISABETH CURLING, PAUL HAYTER, ANTHONY BAINES, ALAN BULL, PHILIP STRANGE AND NIGEL JENKINS.
Biological laboratory, University of Kent
Canterbury, Kent. CT2 7NJ.

ABSTRACT

Recombinant human IFN-gamma (Hu-IFN-γ) secreted by CHO cells was analysed in static and stirred batch cultures. After immunoprecipitation and SDS-gel analysis twelve molecular weight variants could be separated. After complete deglycosylation or tunicamycin treatment, only three IFN forms were seen, indicating that (a) variation in glycosylation, and (b) processing of the IFN-γ polypeptide, was occurring. In common with natural IFN-γ, both fully-glycosylated and partially-glycosylated IFN-γ were secreted. When the cells were grown in suspension culture the proportion of non-glycosylated IFN rose steadily, with a corresponding decrease in the fully-glycosylated form which was independent of the glucose concentration present in the medium. The shift in glycosylation may be in response to changing culture environment in line with the decline in specific growth and IFN production rates after 50 hours of culture.

INTRODUCTION

Human IFN-γ is a lymphokine normally secreted by T lymphocytes during an immune response and has potent antiviral activity [1]. The gene has been cloned and expressed in E. coli [2] and CHO cells [3]. Typically, recombinant glycoproteins expressed in CHO cells most closely resemble the naturally-occurring form in terms of post-translational modifications including glycosylation [4]. Both natural [5] and CHO-derived [3] human

IFN-γ express complex bi-antennary N-linked oligosaccharides attached to Asn-$_{100}$ and/or Asn-$_{25}$ on a mature polypeptide of 146 amino acids (17.1 kDa), to yield a mixture of fully and partially glycosylated product. Processing of the core polypeptide can also occur with the formation of up to six molecular weight variants [5]. An intact carboxy-terminus may [6] or may not [7] be necessary for full biological activity and a glycosylation pattern which differs from that seen in the natural form can lead to increased clearance rates in vivo.[8].

In order to study the fidelity of human IFN-γ production in CHO cells, the variants were analysed by labelling with ^{35}S-methionine labelling, and by the use of specific glycosidases. The relative proportions of each major variant identified was then monitored under various culture conditions, to assess the influence of culture environment on IFN-heterogeneity.

MATERIALS AND METHODS

Cell line

A CHO-K1 mutant which lacks the dihydrofolate reductase (DHFR) gene was co-transfected with the Hu-IFN-γ and DHFR genes. The IFN-γ gene copy number was co-amplified using methotrexate selection up to a concentration of 0.1μM.

Reagents

Protease inhibitors (used during immunoprecipitation), tunicamycin, ^{14}C-labelled low molecular weight markers, Protein-A Sepharose, methionine-deficient MEM medium and silver nitrate were obtained from Sigma (Poole, U.K.).

L-$^{[35]}$S-methionine was purchased from Amersham (Amersham, Bucks). N-glycanase (peptide-glycopeptidase F) was obtained from Genzyme Biochemicals Ltd (Maidstone, Kent). Monoclonal antibody 20D7 anti-IFN-γ was supplied by Celltech Ltd (Slough, U.K.).

Cell culture

CHO cells were grown in 500ml spinner culture, or in 2 litre stirred fermenters modified for use with animal cells (Bioengineering Ltd, Switzerland) where a pH of 7.2 and a dissolved oxygen tension of not less than 40% was maintained. Cultures were seeded at 10^5 cells/ml in serum-free medium and daily samples were taken for IFN and metabolite analysis.

To produce ^{35}S-labelled IFN, 10^5 cells/well were incubated in 96 well plates (Nunc; Becton Dickinson, Oxford) for 3 hours at 37C using 8μCi ^{35}S-methionine/100μl methionine-free MEM.

Immunoprecipitation of IFN-γ

Secreted IFN-γ was immunoprecipitated by incubation of pre-cleared samples with purified antibody 20D7 (10μg) for 3 hours on ice followed by the addition of 6μl of Protein A Sepharose for 60 minutes at 4C, as described elsewhere [9].

Glycosylation analysis

To remove all the N-linked oligosaccharides from the IFN polypeptide, immunoprecipitated and boiled samples containing IFN were treated overnight with N-glycanase according to the manufacturers instructions. Control N-glycanase-treated samples were treated in the same way but omitting enzyme. To inhibit glycosylation, cells were pre-treated for 1 hour before and 3 hours during methionine-labelling with 10μg/ml tunicamycin.

SDS-gel electrophoresis and densitometry

Immunoprecipitated samples were dissolved in sample buffer and boiled to dissociate IFN from the Sepharose beads. The samples were then loaded onto 14% polyacrylamide SDS gels an run at a constant current of 15mA. Subsequently the gels were dried for autoradiography on pre-flashed X-ray film or silver-stained as described previously [10]. The resulting bands were quantified by conversion into absorbance units using a scanning densitometer (BioRad Model 1050). From the scans obtained, the area under each peak (corresponding to each band) was integrated to allow the relative proportions of each IFN variant to be calculated.

RESULTS

Analysis of IFN variants.

To analyse the products of a CHO cell line transfected with the Hu-IFN-ɣ construct, cells were grown in the presence of ^{35}S-methionine to obtain radiolabelled IFN-ɣ. Using immunoprecipitation followed by SDS PAGE electrophoresis and autoradiography, a total of twelve electrophoretic variants of IFN (precipitated by the antibody 20D7) were separated as shown in Figure 1a. To determine which of these variants were glycosylated, samples were treated with N-glycanase which removes all high mannose, hybrid, bi-, tri, and tetra-antennary complex oligosaccharides attached to asparagine residues. Three non-glycosylated IFN- forms were found after this treatment as shown in Figure 1b. The peak with an Mr of 17 kDa corresponded to the sequence-derived molecular weight of the IFN-polypeptide [2]; the two major peaks with an Mr of 16.2 and 15.2 are probably proteolytically-processed forms of the intact IFN-ɣ protein. To confirm that all three forms were non-glycosylated, the cells were treated before and during labelling with tunicamycin which inhibits the first stage of glycosylation by preventing the transfer of GlcNAc-1-phosphate to the dolichol phosphate lipid carrier. As Figure 1c shows, the same three peaks were seen, with a small fourth peak present.

All the other peaks with an Mr > than 17 kDa expressed sialated oligosaccharides since treatment with neuraminidase to remove terminal sialic acid residues reduced the Mr of each by a factor corresponding to the fully and partially-glycosylated forms described previously (results not shown). Thus, it appeared that the observed heterogeneity in IFN-ɣ was due to a combination of glycosylation and proteolytic processing events. Attempts to suppress the proteolytic cleavage of the IFN polypeptide by adding a range of protease inhibitors (listed in Materials and Methods) to the cell cultures during radiolabelling were unsuccessful, suggesting that cleavage was occurring prior to secretion (Results not shown).

Figure 2 summarises the interpretation of the

autoradiograph bands into relative molecular weights and includes the positions of the major glycosylation variants in both their intact and proteolytically-cleaved forms.

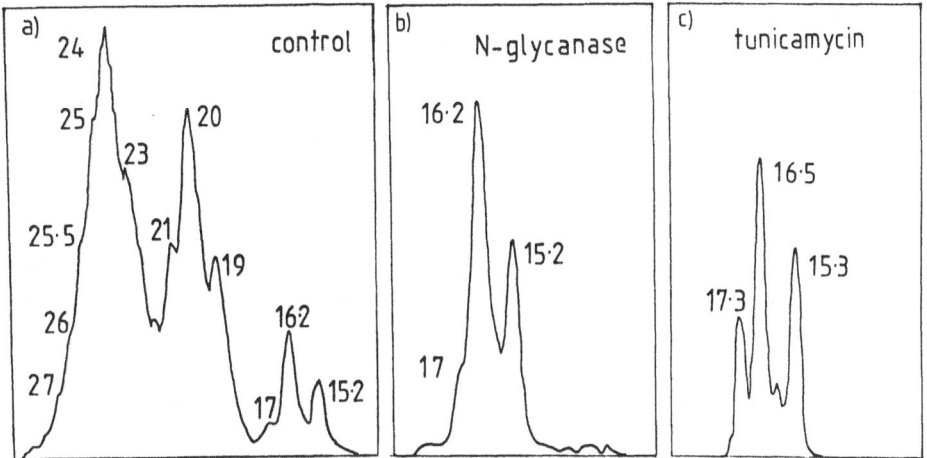

FIGURE 1. Analysis of recombinant IFN-γ secreted by CHO cells. (a) Control IFN-γ sample; (b) Samples treated with N-glycanase to remove all oligosaccharide side chains (c) Samples where cells were treated with tunicamycin both before and during [35]-S methionine labelling to inhibit glycosylation of IFN.

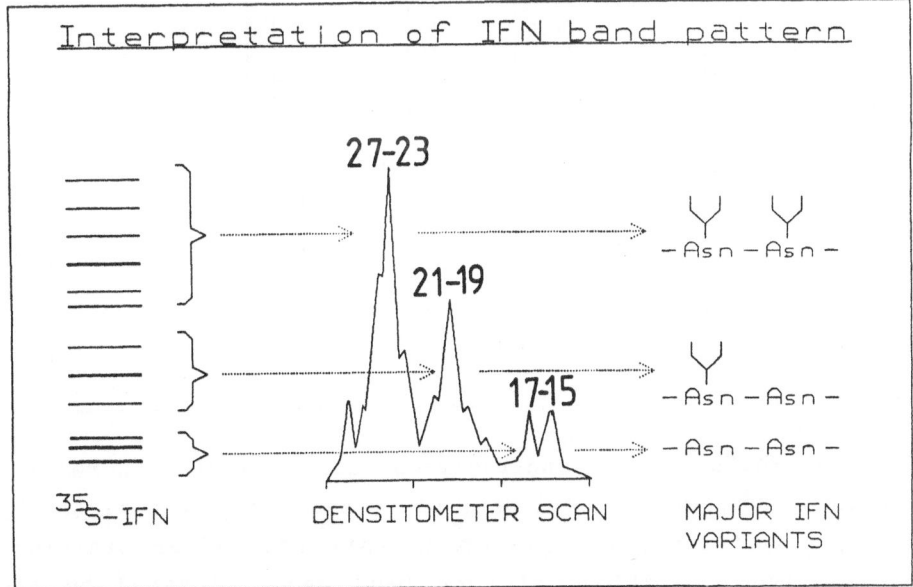

FIGURE 2. Summary of glycosylation heterogeneity in CHO IFN-γ.

Glycosylation patterns during stirred batch culture

The effect of external environment on the degree of
glycosylation seen in the IFN-γ product was analysed. Daily
samples from 500ml spinner cultures of the CHO cell line were
harvested, immunoprecipitated, separated on 14% polyacrylamide
SDS gels and then silver-stained. The relative abundance of
each IFN band was quantified by integration of area under each
peak obtained using scanning densitometry as shown in Figure 3.
A marked change in the pattern was seen with the proportion of
non-glycosylated IFN-γ increasing with time, matched with a
concomitant decrease in the relative proportions of fully-
glycosylated IFN from 61% on day 2 to 29% on day 5. The singly
glycosylated form slightly increased in proportion with time.

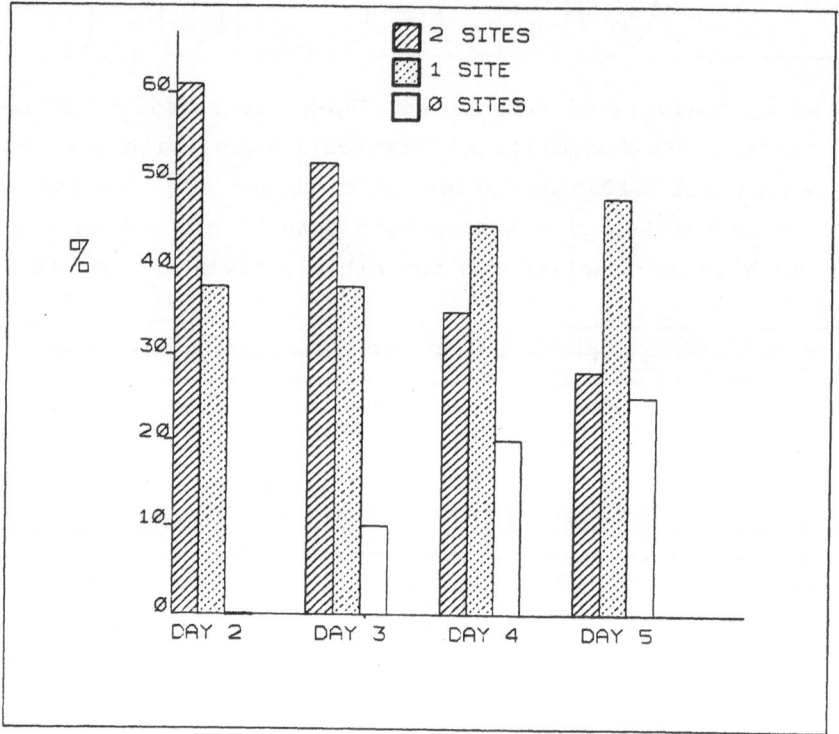

FIGURE 3. Shift towards under-glycosylated IFN with increasing
time in a 500ml spinner culture. Daily samples were analysed by
immunoprecipitation, gel electrophoresis and silver staining.
The relative proportions of each band were estimated by the
integration of the IFN-γ bands scanned by densitometry.

To see if the same shift in glycosylation was seen in a two litre stirred batch system, daily samples were taken for IFN glycosylation analysis 195 hours in a fermenter culture under pH and oxygen control. In addition, specific growth and IFN production rates were calculated. During exponential growth, the specific growth rate was $0.03h^{-1}$ and the specific IFN production rate 200 $IU/10^6$ cells/hr; both rates declined steadily after 50 hours of culture. The results in Figure 4 show a similar increase in the relative levels of the non-glycosylated IFN-ɣ variant which drops slightly at the last time point when the cell viability decreases. This suggests that the the non-glycosylated form is more susceptible to proteolytic degradation compared to the glycosylated forms of IFN-ɣ. Laboratory scale two litre stirred cultures were also initiated with a higher initial glucose concentration to see if

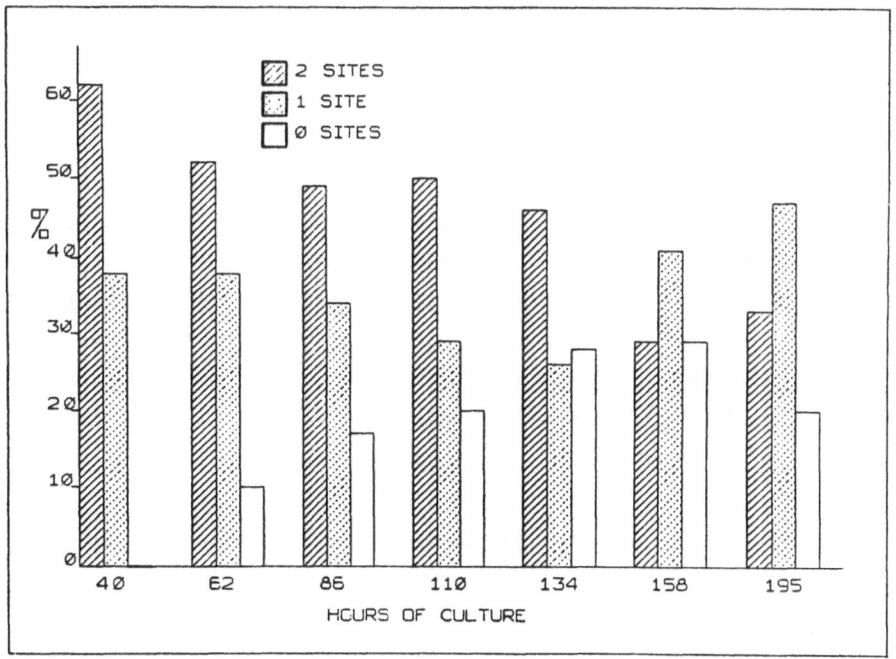

FIGURE 4. Fermenter batch culture of CHO cells producing IFN-ɣ Daily samples from a 2 litre stirred batch culture maintained for 195 hours were analysed as described in FIGURE 3.

glycosylation was limited by low glucose levels (results not shown). No effect of a 20mM compared to a 10mM starting concentration of glucose was detected.

DISCUSSION

The recombinant IFN-γ produced by a transfected CHO cell line has been analysed and was found to be heterogeneous with up to twelve molecular weight forms secreted. The variants were a product of (a) inconsistent complex bi-antennary oligosaccharide expression (at positions Asn-$_{28}$ and Asn-$_{100}$) leading to the secretion of incompletely glycosylated or non-glycosylated IFN-γ , and (b) proteolytic processing at two separate sites on the polypeptide. In addition, there may be some variation in the oligosaccharide structure seen in the fully glycosylated form as there were six variants, compared to three of the partially-glycosylated form. Cell-free translation of IFN-γ mRNA in the presence of microsomes prepared from the CHO cell line described here also showed variation in the Mr of the fully glycosylated form, but no truncated forms of the polypeptide were detected [11].

Previous analysis of CHO derived IFN-γ variants using mass spectroscopy [12] detected a major variant of 15.2 kDa, which corresponds to the lowest molecular weight band shown in Figure 1. Others [13] have detected a 16.6 kDa form of E.coli-derived recombinant IFN-γ , which coincides with the second truncated variant described here. Interestingly, both sites of proteolysis consist of a cluster of basic residues, but the precise protease enzyme(s) responsible are unknown.

To discover if the pattern of glycosylation remained constant with time, the IFN-γ from CHO cells grown in stirred batch culture was immunoprecipitated and resolved by SDS PAGE electrophoresis. Daily samples revealed that a gradual shift away from the fully glycosylated form occurred, presumably in response to the changing external environment, since we have evidence that IFN-γ retains a constant glycosylation pattern when incubated over 96 hours with cell-free spent medium from batch cultures. (Data not shown).The non-glycosylated form was only detectable after three days of culture due to the limits of silver stain sensitivity to the low level of protein present; but from autoradiographs of labelled IFN it can be seen (Figure 1) that at the start of cultures up to 5% of IFN secreted is non-glycosylated.

For recombinant products commonly expressed in CHO cells which depend on glycosylation fidelity for biological activity and extended half-life in vivo (notably erythropoietin [14]), it may be important to monitor levels of carbohydrate expression when designing media and bioreactor systems. Analysis of other recombinant proteins produced in heterologous cells needs to be done to discover whether the data presented here is typical for any other recombinant proteins currently produced in animal cells.

ACKNOWLEDGEMENTS

We wish to thank Wellcome Biotechnology for the CHO cell line; Celltech for antibody 20D7 and the SERC Biotechnology Directorate, Celltech, Glaxo, Porton International, Smith-Kline Beecham and Wellcome Biotechnology Ltd for support of the research programme of which the work described here was part.

REFERENCES

1. Johnson, H.M., Mechanism of interferon gamma production and assessment of immunoregulatory properties. Lymphokines, 1985, 11, 33-46.

2. Gray, P.W., Leung, D.W., Pinnica, D., Yelverton, E., Najarian, R., Simonsen, C.C., Derynck, R., Sherwood, P.J., Wallace, D.M., Berger, S.L., Levinson, A.D. and Goeddel, D.V., Expression of human immune interferon cDNA in E.Coli and monkey cells. Nature, 1982, 295, 503-508.

3. Mutsaers, J.H.G.M., Kamerling, J.P., Devos, R., Guisez, Y., Fiers, W. and Vliegenthart, J.F.G., Structural studies of the carbohydrate chains of human γ-interferon. Eur. J. Biochem. 1986, 156, 651-654.

4. Utsumi, J., Mizuno, Y., Hosoi, K., Okano, K., Sawada, R., Kajitani, M., Sakai, I, Naruto, M. and Shimizu, H., Characterization of four different mammalian-cell-derived

recombinant human interferon-Beta-1S:-identical polypeptides and non-identical carbohydrate moieties compared to natural ones. Eur. J. Biochem. 1989, **181**, 545-553.

5. Rinderknecht, E., O'Connor, B.H. and Rodriguez, H., Natural human interferon-γ., Complete amino acid sequence and determination of sites of glycosylation., J. Biol. Chem. 1984, **259**, 6790-6797.

6. Seelig, G.F., Wijdenes, J., Nagabhashan, T.L. and Trotta, P.P.,Evidence for a polypeptide segment at the carboxyl terminus of recombinant human gamma interferon involved in the expression of biological activity., Biochemistry, 1988, **27**, 1981-1987.

7. Sakaguchi, M., Honda, S., Ozawa, M. and Nishimura, O., Human interferon-γ lacking 23 COOH-terminal amino acids is biologically active., FEBS Letts., 1988, **230**, 201-204.

8. Ashwell, G. and Hartford, J., Carbohydrate-specific receptors of the liver., Ann. Rev. Biochem., 1982, **51**, 531-554.

9. Curling, E.M.A., Hayter, P., Baines, A.J., Bull, A.T., Strange, P. and Jenkins, N., Recombinant human interferon-gamma: differences in glycosylation and proteolytic processing lead to heterogeneity in batch culture., (Submitted for publication).

10.Johnson, A. and Thorpe, R., Immunochemistry in practice, Blackwell Scientific Publications, Oxford, 1987, pp 150-151.

11.Bulleid, N., Curling, E.M.A., Freedman, R.B. and Jenkins, N.J., Source of heterogeneity in secreted interferon-gamma: a study on products of translation in vitro., Biochem. J. (In Press).

12.Morris, H.R. and Greer, F.M., Mass spectroscopy of natural and recombinant proteins and glycoproteins., TIBTECH, 1988, **6**, 140-147.

13.Honda, S., Asano, T., Kajio, T., Nakagawa, S., Ikeyama, S, Ichimori, Y., Sugino, H., Nara,K., Kakinuma, A. and Kung, H-F., Differential purification by immunoaffinity chromatography of two carboxy-terminal portion-deleted derivatives of recombinant human interferon-γ from Escherichia coli., J. Interferon. Res.,, 1987, **7**, 145-154.

14.Kagawa, Y., Takasaki, S, Utsumi, J, Hosoi, K., Shimizu, H., Kochibe, N. and Kobata, A., Glycosylation at specific sites of erythropoetin is essential for biosynthesis, secretion and biological function., J. Biol. Chem., 1988, **33**, 17508-17515.

VALIDATION OF THE RE-USE OF PROTEIN A SEPHAROSE FOR THE PURIFICATION OF MONOCLONAL ANTIBODIES

R. FRANCIS, J. BONNERJEA AND C.R. HILL
Celltech Ltd., 216 Bath Road,
Slough, Berks., U.K. SL1 4EN.

ABSTRACT

The use of Protein A affinity chromatography at an early stage in the purification of a monoclonal antibody results in a high purity product with high yield. However, in order to operate the process in an economic manner it is often necessary to use a Protein A affinity column for a number of consecutive cycles to purify a single batch of antibody. An important aspect of process safety validation is the impact of repeated use of the column on the stability of the affinity ligand and hence on column performance and quality of the purified monoclonal antibody. A programme of work has been undertaken to study the effect of repeated use of a single column of Protein A Sepharose FF on the yield and quality of monoclonal antibody.

An automated chromatography system was used to carry out 25 sequential purification cycles. For each cycle the yield of antibody was measured by A_{280} readings and the purity was determined by SDS-PAGE, isoelectric focussing and gel permeation HPLC analysis. The level of Protein A residues in the affinity purified monoclonal antibody was also determined for each cycle by ELISA.

INTRODUCTION

Monoclonal antibodies are currently under intensive investigation for
their therapeutic potential in a wide range of diseases. A critical
requirement for those studies is the development of a purification
process that consistently yields a safe and efficacious product. A wide
range of purification procedures have been applied to the preparation of
monoclonal antibodies[1]. A commonly used method is affinity
chromatography with immobilised Protein A[2]. The use of Protein A
affinity chromatography, particularly at an early stage in the
purification process, can result in the preparation of a high purity
product at high yield. However, the cost of such affinity matrices
often necessitates their use for a number of consecutive cycles in order
to operate the process in an economic manner. A single column of
Protein A affinity matrix may be re-used for a number of cycles during
the purification of a single batch of a therapeutic antibody. It is
important to note, however, that the affinity matrix must be dedicated to
the purification of a single antibody and not to use it for purification
of other antibody proteins.

An important consideration for the development of an affinity step
is the impact of repeated use of the column on the stability of the
affinity ligand and hence the column performance and leakage of ligand
into the purified product. In the study described here a column of
Protein A Sepharose was used for 25 consecutive cycles of antibody
purification. For each cycle the yield of antibody was measured and the
purity determined by SDS-PAGE, isoelectric focussing and gel permeation
HPLC analysis. The level of Protein A residues in the purified antibody
was also determined for each cycle of purification.

MATERIALS AND METHODS

Protein A Affinity Purification of Monoclonal Antibody

A column of Protein A Sepharose (Pharmacia) (2 ml) was prepared and
washed with equilibration buffer. A sample of monoclonal antibody was

applied to the Protein A column at a linear flow rate of 50 cm/hr.   The column was washed with equilibration buffer until A_{280} column eluates reached base line.   The unbound fraction was retained for analysis. Bound antibody was eluted by the application of 0.1 M sodium citrate, 0.1 M sodium chloride, pH 4.0, this fraction was also retained for analysis.   After every second purification cycle the column was washed with 0.1 M citric acid pH 2.5 after elution of antibody.   The column operation was automated using a system of remotely actuated values (Pharmacia C3 controller) and run for 25 consecutive cycles of antibody purification.

Analytical Methods

The unbound and eluted fractions were analysed using a range of analytical techniques outlined below.

Protein Concentration: Absorbance at 280 nm was used to determine the protein concentration in the sample using an extinction coefficient of $e^{1\%}$ 1 cm = 14.3.   The test sample was suitably diluted to give an absorbance reading of approximately 1.2.

SDS–PAGE Gel Analysis: Samples were denatured by treatment with SDS (sodium dodecyl sulphate) and the disulphide bonds disrupted by treatment with 2-mercaptoethanol.   Polypeptides were separated according to sub-unit molecular size by electrophoresis through 7.5 – 15% (w/v) SDS–PAGE gels and visualised by staining with Coomassie Blue.   Peptide bands were quantified by laser densitometry.   The detection limit of the staining procedure is approximately 0.1 microgrammes per peptide band. When a 10 mg sample was applied, single discrete contaminants are detectable at a level of approximately 0.45% of the total protein.

Isoelectric Focussing   Samples were applied to preformed gels (Pharmacia, Phast gels, pH 3 – 9) and focussed for 200 volt hours. Protein bands were visualised by staining with silver stain.

Gel Permeation HPLC   A Gilson HPLC system was used.   Samples were injected onto a Zorbax GF-250 column and eluted in 50 mM sodium phosphate buffer (pH 6.85).   Protein elution was monitored at 280 nm.   Single discrete contaminants were detectable at 0.5% of the total protein.

Protein A Residues   A non competitive ELISA for _Staphylococcus_
aureus Protein A has been developed.   A solid phase capture antibody
binds Protein A which is detected using biotinylated rabbit anti–Protein
A and streptavidin–horseradish peroxidase.   Detection limit of the assay
is 0.2 ng Protein A/ml.

RESULTS AND DISCUSSION

Elution Profile and Antibody Yields

The elution profiles for each cycle was recorded.   The shape of the
eluted antibody peak did not change throughout the 25 cycles indicating
that gross deterioration of the column did not occur.   In particular the
absence of trailing edges on the eluted peak after 25 cycles indicated
that the kinetics of binding to and elution of the antibody from the
immobilised Protein A was not significantly altered as cycle number
increased.   The yield of monoclonal antibody for every fifth cycle is
given in Table 1.   A yield of approximately 81% purified antibody was
obtained for each of the 25 cycles.   As the column was loaded to 80% of
its dynamic capacity for each cycle, the constant high yield indicates
that no major deterioration of the column performance has occurred over
the 25 cycles tested.

TABLE 1
Yield of Purified Monoclonal Antibody

Cycle Number	Monoclonal Antibody (1) Recovered (mg)	Yield (2) (%)
2	16.8	88
5	14.3	75
10	14.4	76
15	16.8	88
20	15.2	80
25	14.8	78

(1)   Determined by A_{280} measurement of eluted antibody.

(2)   19.0 mg monoclonal antibody applied to the column for each
      purification cycle.

Purity of Eluted Antibody

<u>SDS-PAGE Analysis</u>:   Samples of eluted antibody from selected cycles were analysed by SDS-PAGE under reducing conditions (Figure 1).   The major proteins present in the sample applied to the column are bovine serum albumin and antibody heavy and light chain (Figure 1, lane 2).

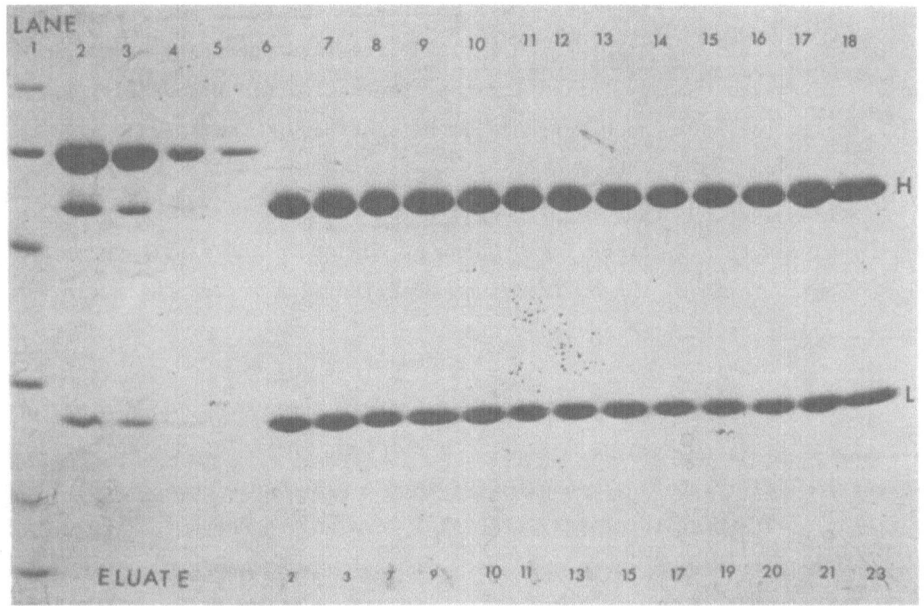

Figure 1.   SDS-PAGE Analysis under reducing conditions of eluted antibody
from cycles 2 - 23

Lane

1	Molecular Weight markers (94, 67, 43, 30, 21 and 14.4 kd).
2-5	Column load sample (loadings 20, 15, 7.5, 3 µg).
6-18	Eluted antibody (20 µg) from cycles 2-23.

The SDS-PAGE analysis of samples from cycles 2 to 23 are shown in lanes 6 – 18. After purification it can be seen that the antibody fraction contains predominantly heavy and light chains. The relative proportion of the heavy and light chains does not change with increasing cycle number. Furthermore, both chains remain fully intact with no sign of peptide cleavage as the cycle number increases.

Isoelectric Focussing: The eluted antibody from every fifth cycle was analysed by isoelectric focussing and silver stained. No change in the focussed pattern was observed in eluted antibody from cycles 1 – 25. The pattern obtained by isoelectric focussing was typical of that seen for monoclonal antibodies. The antibody separated into 5 major bands with 2 minor basic bands and 1 minor acidic band. The isoelectric focussing pattern of the purified antibody remained unchanged over the number of cycles tested, indicating that the matrix did not develop a selectivity for any change variants of the antibody.

Gel Permeation HPLC: Antibody eluates from all purification cycles were analysed by gel permeation HPLC. No change in the elution profile and retention time of the antibody was observed with increasing cycle number and no aggregated antibody was detected.

Protein A Residues: The level of Protein A residues in each of the eluted antibody samples was measured by ELISA. The ELISA system has a detection limit of 0.2 ng Protein A/ml in the presence of monoclonal antibody. The results of analysis for 15 cycles are shown in Figure 2.

The highest level of Protein A was detected in the antibody purified on the first cycle of use. By cycle 4 the level has fallen to 5-7 ppm (ng Protein A/mg monoclonal antibody) and remains constant at this level for all subsequent cycles. The higher levels detected in antibody purified on cycles 1 – 3 is possibly due to incomplete removal of non-covalently coupled Protein A from the unused Protein A Sepharose matrix prior to its first use. In studies in which we have measured the leakage of protein ligands from immunopurification columns, we have observed both a similar trend and similar levels of ligand leakage to that reported here for Protein A Sepharose.

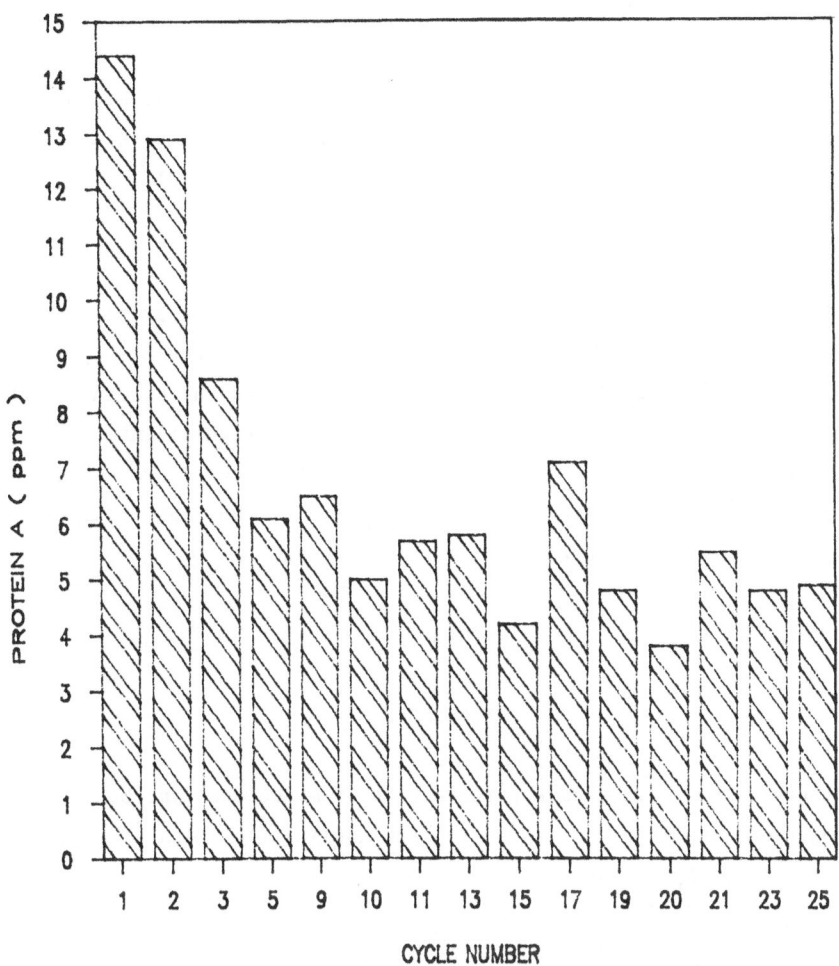

<u>Figure 2</u>   Histogram of Protein A (ng/mg antibody) Measured in Eluted
Antibody from Cycles 1 – 25

A major issue to be addressed when developing an affinity
purification step is the level of leakage of the affinity ligand into the
purified product.   A recent study has shown that this can vary widely
depending on the source of Protein A matrix used(3).   The highest level
of leakage reported by Fuglistaller (3) was 88 ng Protein A/mg antibody
although more typically the values fell within the range of 2 – 10 ng
Protein A/mg antibody.   In this study we have detected similar levels
(less than 10 ng/mg antibody) in the purified antibody eluted from the

affinity column. In general further chromatography steps would follow an affinity column when used to purify a therapeutic protein. Work carried out at Celltech has shown that anion-exchange chromatography following Protein A affinity chromatography will result in a reduction in the level of Protein A in the product of at least 5 - 10 fold.

CONCLUSION

A study has been carried out in which the performance of a Protein A Sepharose affinity column has been evaluated for 25 consecutive cycles of monoclonal antibody purification. The performance of the column and the quality of the purified antibody was measured. No significant deterioration in either the column performance as measured by peak shape and antibody yield or antibody quality as judged by SDS-PAGE, isoelectric focussing, GP-HPLC and Protein A ELISA was detected.

REFERENCES

1. Kenney, A.C., Large-scale purification of monoclonal antibodies In Monoclonal Antibodies: Production and Application, Alan R. Liss Inc., 1989, pp143-160.

2. Ey, P.L. Prowse, S.J. and Jenkin, C.R., Isolation of pure IgG_1, IgG_{2a} and IgG_{2b} immunoglobulins from mouse serum using protein A Sepharose. Immunochemistry, 1978, 15, 429.

3. Fuglistaller, P. Comparison of immunoglobulin binding capacities and ligand leakage using eight different protein A affinity chromatography matrices. J. Immunol. Meths., 1989, 124, 171-7.

PROCESS DESIGN FOR THE INACTIVATION OF VIRUSES IN THE MANUFACTURE OF PHARMACEUTICAL PROTEINS

Ronald V McIntosh and Peter R Foster
SNBTS Protein Fractionation Centre
21 Ellen's Glen Road
EDINBURGH
EH17 7QT

ABSTRACT

The transmission of viruses such as HIV-1 has made the inactivation of viral contaminants a major development area in the manufacture of pharmaceutical proteins from human blood plasma. The risk of infection from biological materials is also an important issue in the manufacture of cell culture derived products.   The experience gained in developing different   viral   inactivation strategies   in the production of blood products can be used to help design similar procedures for the production of pharmaceuticals from biosynthetic feed stocks.   Ideally the inactivation process should be effective against a broad spectrum of viruses; it should not increase the toxicity or decrease the efficacy of the product; it should be reproducible and ensure that the   treated product   remains free from subsequent contaminantion. Methods which   can be applied when products are sealed in the final container are most likely to fulfil all of these conditions.

INTRODUCTION

Comparatively   recent   developments   in molecular biology and immunology have increased awareness in the potential of proteins as macromolecular drugs (1). However proteins have been established in clinical use for many years and are an essential feature of modern medicine.   In contrast to recombinant   materials the   established pharmaceutical proteins are derived from natural   sources principally blood plasma provided by human donors; for a review, see Stryker et al (2).   The fractionation of human plasma into its different constituents has enabled high quality clinical products to be manufactured world wide on an industrial scale.   These products include Albumin solutions (for volume replacement and the treatment of shock), Immunoglobulin preparations (for the prevention and   treatment of infectious diseases and immune disorders) and Coagulation Factor concentrates (for the prevention and treatment of bleeding disorders).

The experience gained in the preparation of products from human plasma provides an ideal framework on which to build further advances in downstream processing which can benefit both the natural (human plasma) and biosynthetic (recombinant/cell culture) routes to macromolecular drugs. An example of this approach is the development of virus inactivation procedures to allow the preparation of biologicals free from the transmission of infectious agents. The transmission of viruses such as HIV-1, Hepatitis B and non-A, non-B Hepatitis by the use of untreated human blood products (3) illustrates the potential hazard associated with the use of therapeutic products derived from biological sources. The use of cell culture systems to synthesise therapeutic biologicals such as tissue plasminogen activator and monoclonal antibodies is often promoted as reducing the risk of infection by intensive screening of the cell line used in the production process. However the use of a cell culture system does not of itself remove the risk of transmitting disease. For example, mouse hybridoma cells are known to be infected with retroviruses, many secreting type c particles (4) and monoclonal antibodies which are produced in ascites fluid could be contaminated by the many viruses that are known to infect rodents (5). The Licencing Authorities are well aware of these problems and in guidelines concerning the therapeutic use of monoclonal antibodies have recommended extensive testing for likely viral contaminants (6). However even the most comprehensive of test systems cannot detect as yet unidentified or undiscovered viruses which is just how HIV infection arose in blood products. There is a need therefore, in addition to identification and screening procedures, to develop also inactivation technologies as part of an overall anti viral strategy.

AIMS OF VIRUS INACTIVATION

Performance

The ideal virus inactivation method would inactivate all types of viral contaminants. Therefore the selected method(s) must be active against the broadest possible spectrum of virus types.

Toxicity

Products which have been treated to inactivate viral contaminants should contain no residues from potentially toxic reagents used in the process, nor should the process have altered the product so as to have made it toxic for example by the formation of neo-antigens.

Efficacy

The treated product should maintain its normal biological activity, in-vivo performance and half-life.

Efficiency

The inactivation process should be cost effective ie. an acceptable yield with an inexpensive process.

Reproducibility

The inactivation process to be used should be able to be calibrated, validated, monitored and controlled such that consistent routine performance can be assured.

Prevention of Reinfection

In an ideal process, once the product has completed the virus inactivation stage, it should not come into contact with untreated product ie. there should be no possibility of accidental recontamination.

CURRENT VIRUS INACTIVATION METHODS

There are several virus inactivation methods currently being used in the preparation of pharmaceutical proteins from human plasma.

Heating In Solution

Albumin products have been heated in solution at $60^{\circ}C$ for 10 hours ("pasteurisation") for over 40 years with an outstanding safety record against a range of potentially infective agents. Edsall (7) has reviewed the development of this process. Albumin can be physically stabilised relatively easily by low concentrations of non toxic additives (sodium caprylate, sodium acetyltryptophanate) for heating in solution. As the final dose form is a solution the heating can be performed as a terminal step in the sealed final product container which eliminates all possibility of accidental infection by contaminated equipment or other batches. The use of a heat inactivation method in this way is also advantageous in that it is relatively simple to validate that each container in a batch of albumin product has been subjected to the correct time/temperature combination and a record can be kept for each pasteurisation treatment.

There are no simple stabilising formulations for the heating in solution of products which unlike albumin (used simply as a blood volume expander) contain labile biological activities. Stabilisation of human factor VIII activity for example against pasteurisation at $60^{\circ}C$ for 10 hours can only be achieved by the addition of high concentrations of carbohydrate and glycine (8,9) and although this still allows a significant degree of virus inactivation (10) the loss of factor VIII activity can be high. However such formulations cannot be used for final products and extra processing is required to remove the stabilisers after heat treatment. Therefore the added safety and process security of applying the heat treatment as a terminal step are lost.

Heating in the Lyophilised State

Coagulation factors such as factor VIII are lyophilised final products as most pharmaceutical proteins would be. The development of appropriate formulations and freeze drying cycles has enabled lyophilised coagulation factor preparations to be heated at high temperatures for prolonged periods (80°C, 72hrs) in their final product form (11,12) with good recoveries of factor VIII activity across the heating step. These products have been shown to be potentially very safe from virus infection (13,14) and because the heat treatment can be applied as a terminal step they have the additional safety advantage of there being no possibility of recontamination.

Lyophilised products have also been heated at a raised moisture content by introducing steam (15) and although there is no published description of the method, clearly it could not be carried out as a terminal process in a sealed container.

Chemical Treatments

Several chemical treatments have been used as virus inactivation procedures in the manufacture of plasma products. These include the heating of a bulk freeze dried intermediate coagulation factor preparation in a slurry with n-heptane (16) about which little or no information has been published and the treatment of bulk intermediate solutions with β-propiolactone in combination with UV irradiation (17) which has not proved suitable for factor VIII preparations due to excess yield losses. The treatment of bulk intermediate solutions with an organic solvent, tri (n-butyl) phosphate (TNBP) together with a detergent usually Tween 80 or sodium deoxycholate (18) has been evaluated extensively in blood products (19) showing good yield and virus safety characteristics. However this method relies for its action on the majority of viruses transmitted by plasma products having a lipid envelope sensitive to the solvent. A non-lipid enveloped virus was shown to be completely resistant to TNBP/detergent treatment (18) which limits the range of viruses against which this method would be effective. A more fundamental inactivation treatment such as heat might be expected to be effective to a greater or lesser extent across a broader spectrum, if not all, virus types.

All the chemical processes discussed above use potentially toxic reagents whose removal involves considerable extra processing which in turn means that they cannot be applied as terminal steps. A further complication in using bulk intermediate chemical treatments as opposed to processes which can be applied to individual doses of final product is in their validation. It is not a simple matter to ensure for a bulk batch treatment method that the same degree of chemical contacting takes place from batch to batch or during process scale up. The effects of scale are also important considerations in accurately reproducing the inactivation process at laboratory level to calibrate its effect for example using model viruses. A non-invasive method applied to final products can be carried out on one vial or one thousand vials with equal validity.

CONCLUSIONS

The manufacture of products which are free from the transmission of infectious agents is an important development area for both natural (eg. human plasma) and biosynthetic (eg. cell culture) derived macromolecular drugs. In addition to identification screening and elimination of viruses the development of inactivation methods is an important part of an overall anti viral strategy.

From the experience gained in the manufacture of pharmaceutical proteins from human blood plasma there are several different inactivation methods to choose from which could be used on their own or in combination with a complementary method. We would conclude that the method of choice and the method which should always be included in any combination of inactivation steps is heat treatment of the product in its final sealed container. In this particular application, of macromolecular (protein) drugs, one could go further and state that the preferred method would be heat treatment of the lyophilised product in its final sealed container. Being able to present the product as a freeze dried preparation gives a stable final product form needed for complex labile macromolecules. Even at this relatively early stage in its development the heat treatment of lyophilised products has shown good virus inactivation potential and it carries a low yield penalty. Furthermore as a reproducible, controllable and terminal step, where all possibility of re-contamination is eliminated, it offers the highest standard of Good Pharmaceutical Manufacturing Practice of the methods available at present.

The feasibility of this approach for biosynthetic products has been demonstrated in the significant killing of two types of test virus by the heat treatment of an appropriately formulated and freeze dried murine monoclonal antibody preparation (20).

REFERENCES

1.   Blohm   D.,   Bodschweiler   C. and   Hillen. Pharmaceutical proteins. Angew. Chem. Int. Ed. Engl., 1988, 27: 207-225.

2.   Stryker M.H., Bertolini M.J. and Hao Y.L.   Blood Fractionation: Proteins. In Advances in Biotechnological Processes 4, Alan R Liss, New York, 1985, pp275-336.

3.   Roberts H.R. and Macik B.C, Factor VIII and Factor IX concentrates; clinical efficacy as related to purity. In Thrombosis and Haemostasis eds. Verstraete M., G. Vermylen, R. Lijuen, G. Arnout, Leuven University Press, Belgium, 1987, pp. 563-81.

4.   Weiss R.A.   Retrovirus produced by hybridomas. New Eng. J. Med., 1982, 307, 1587.

5.   Carthew P.,   Is rodent virus contamination of monoclonal antibody preparations for use in human therapy a hazard? J. Gen. Virol., 1986, 963-74.

6. Committee for Proprietary Medical Products: Ad hoc working party on Biotechnology/Pharmacy. Notes to applicants for marketing authorizations on the production and quality control of monoclonal antibodies of murine origin intended for use in Man. J. Biol. Stand., 1989, 17, 213-22.

7. Edsall J.T. Stabilization of serum albumin to heat and inactivation of the hepatitis virus. Vox. Sang., 1984, 46, 338-40.

8. Ng P.K and Dobkin M.B. Pasteurisation of antihemophilic factor and model virus inactivation studies. Thromb. Res. 1985, 39, 439-47.

9. MacLeod A.J, Welch A.G, Dickson I.H, Cuthbertson B. and Foster P.R. Stabilisation of proteins to heat. Research Disclosures, 1984, 244, 380.

10. Heimburger N and Karges H.E. Strategies to produce virus safe blood derivatives. In Virus Inactivation in Plasma Products ed. J.J Morgenthaler, Karger, Basel 1989 pp. 23-33.

11. Winkelman L., Owen N.E, Haddon M.E and Smith J.K. Treatment of a new high specific activity factor VIII concentrate to inactivate viruses. Thromb. Haemostas, 1985, 54, 19.

12. McIntosh R.V, Docherty N, Fleming D and Foster P.R. A high-yield factor VIII concentrate suitable for advanced heat treatment. Thromb. Haemostas., 1987, 58, 306.

13. Colvin B.T, Rizza C.R, Hill F.G.H, Kernoff P.B.A et al. Effect of dry-heating of coagulation factor concentrates at 80°C for 72 hours on transmission of non-A non-B hepatitis. Lancet, 1988, II, 814-16.

14. Ludlam C.A, Chapman D, Cohen B. and Litton P.A. Antibodies to hepatitis C virus in haemophilia, Lancet, 1989, II, 560-61.

15. Barret P.N, Wober G., Eibl J. and Dorner F. Efficiency of steam treatment procedures used for virus inactivation in human coagulation factors Thromb. Haemostas. 1987, 58, 371.

16. Heldebrant C.M, Gomperts E.D., Kasper C.K., McDougal J.S et al. Evaluation of two viral inactivation methods for the preparation of safer factor VIII and factor IX concentrates. Transfusion, 1985, 25, 510-15.

17. Stephan W. Inactivation of hepatitis viruses and HIV in plasma and plasma derivatives by treatment with β-propiolactone/UV irradiation. In Virus Inactivation in plasma products, ed. J.J Morganthaler Karger, Basel, 1989, pp122-27.

18. Horowitz B., Wiebe M.E., Lippin A. and Stryker M.N. Inactivation of viruses in labile blood derivatives I. Disruption of lipid enveloped viruses by tri (n-butyl) phosphate detergent combinations. Transfusion, 1985, 25, 516–22.

19. Edwards C.A., Piet M.P.J., Sing Chin and Horotwitz B. Tri (n-Butly) Phosphate/detergent treatment of licensed therapeutic and experimental blood derivatives. Vox. Sang. 1987, 52, 53–59.

20. Harbour C., MacLeod A.J., Reid K.G and McIntosh R.V. Viral inactivation strategies for monoclonal antibodies: dry heating. Biotech. Techniques 1989, 3, 261–64.

INVESTIGATIONS ON APPARENT HOMOGENEITY AND STABILITY OF RECOMBINANT HuIFN-ALPHA 1

REINHARD H. WONDRACZEK, AXEL STELZNER[1] and BRIGITTE GLÜCK[1]

Dept. Proteinbiotechnology and Dept. Virology[1], Central Institute for Microbiology and Experimental Therapy, Acad. Sci. GDR, Jena 6900, GDR

ABSTRACT

During the purification of recombinant HuIFN-alpha 1 from E. coli Tg1 two observations were extensively investigated:
1. the presence of two main peaks in preparative reversed-phase HPLC
2. the rapid decline of biological activity of the final product with a purity of + 99 %
The first finding may be interpreted as apparent diversity due to strong interactions between the protein and the chromatographic matrix leading to "forced conformations". The second one is the result of a rapid degradation which could be overcome by the addition of an appropriate stabilizer.

INTRODUCTION

The introduction of new technologies for the production of protein drugs in large amounts is leading to strict regulations concerning both the bioprocess and the target product. Like other drugs, the testing has to include the following main categories: purity, identity, safety, activity and stability.

The structure of the final product and its purity are strongly influenced by the producing microorganism and the purification process. Although the primary structure of the protein should be strictly determined through the nucleotide sequence

the encoding gene, structural variations may emerge as a consequence of expression errors, incomplete post-translational
processing, mutations, and denaturation during the purification process. This induced heterogeneity may not only render these proteins more difficult to purify than natural species, but may also make the purity test a very strong objective in product analysis.

Many methods serving this purpose are based on chromatography,
particularly on HPLC. As Regnier (1) has pointed out, the
mechanism of interaction between proteins and chromatography
media is not yet completely understood and may lead sometimes
to experimental results which are difficult to interpret.

Tests on activity and stability of the recombinant protein
comprise physico-chemical methods as well as biological and
immunological assays. They are essential for the formulation
of the drug.

In this paper we want to discuss some observations we had
to deal with in the course of the process development for a
highly purified rHuIFN-alpha 1 (D) for research purposes.

MATERIALS AND METHODS

Purification of rHuIFN-alpha 1

The details of the purification scheme will be published in
a forthcoming paper. Briefly, it includes the following steps:

1. suspension of frozen cells in 6M urea and subsequent
   high-pressure homogenization
2. aqueous two-phase partitioning followed by a series of
   consecutive fractionated precipitations
3. immuno-affinity chromatography (IAC)
4. weak cation-exchange chromatography (WCX) at pH = 4.5
   using TSK Fractogel CM 650 (F)
5. preparative reversed-phase HPLC (RPC)
6. Buffer exchange by size-exclusion chromatography (SEC)
   into 30 mM citrate, pH = 3.0 using TSK Fractogel HW 40 (F)

HPLC-investigations

A model LC-7A SHIMADZU biocompatible (titanium) HPLC system
was used for all analytical and preparative runs. Columns and
buffers are listed in Table 1. Gradient profiles are shown in
Figure 2. Water was purified by an ELGASTAT-UHQ (ELGA Ltd.).
Freshly distilled isopropanol was used. All buffer substances
were of "suprapur" quality (MERCK). Buffers and samples were
filtered through 0.22 /u DURAPORE membranes (MILLIPORE) prior
to application.

TABLE 1
Chromatographic conditions for HPLC studies

Method	Column	Eluents	
SEC	SynChropak GPC 300		200 mM K-phosphate, pH = 6.8
RPC	Vydac 214 TP 1010	A:	20 mM K-phosphate, pH = 2.5
		B:	Isopropanol
HAC	Shim-pack HAC	A:	10 mM K-phosphate, pH = 6.8
		B:	300 mM K-phosphate, pH = 6.8
HIC	Shim-pack HPC-2T	A:	20 mM Na-phosphate, pH = 7.0
		+	2 M Ammoniumsulfate
		B:	20 mM Na-phosphate, pH = 7.0
WAX	Shim-pack WAX-2T	A:	20 mM K-phosphate, pH = 6.0
		B:	1 M KCl in A

Analytical characterizations

Polyacrylamide gel-electrophoreses were carried out
according to Laemmli (2).

N-terminal amino acid sequence analyses were performed
using an automatic model 470 A gas phase sequencer (APPLIED
BIOSYSTEMS).

Biological activity was measured in terms of 50 %
inhibition of CPE of VSV on MDBK cells, calibrated with the
international standard HuLeIFN B69/19 (NIBSC, London).

Immunological activity was analyzed in an ELISA test
employing two monoclonals developed at our Institute.

Trace impurities such as host proteins (E. coli Tg1) and

mouse IgG (IAC column bleeding) were detected down to a level
of 100 ppm and 10 ppm, respectively, by ELISA tests developed
at our Institute.

RESULTS AND DISCUSSION

The product obtained after IAC and subsequent WCX appeared to
be sufficiently pure when analyzed by routine methods only
(compare figure 1), however, it still contained unacceptable
amounts of trace impurities (host protein > 1000 ppm and
IgG > 100 ppm). Therefore, further purification was attempted
by preparative reversed-phase HPLC. After this step, recovery
of biological activity was about 90 %, and trace impurities
were below the detection limit, however, the elution pattern
displayed a multiplicity of peaks (figure 2 A).

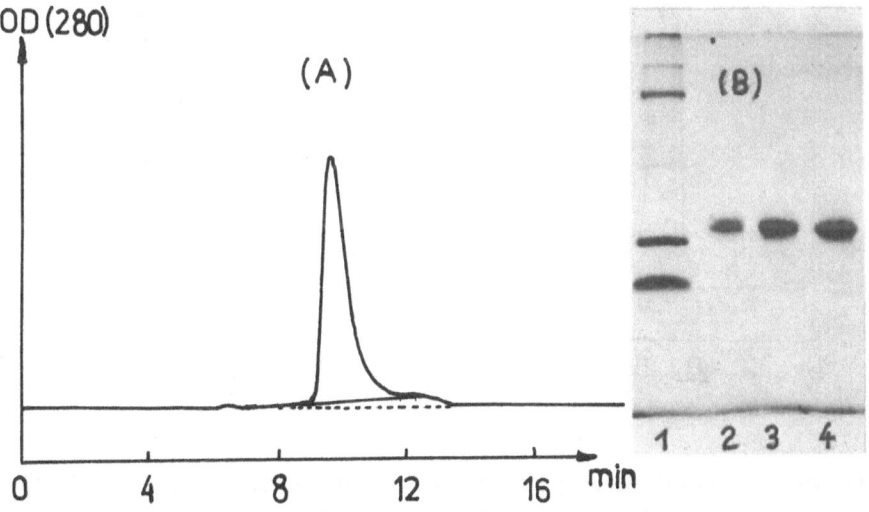

Figure 1.   Purity test of rHuIFN-alpha 1 obtained after IAC
            and subsequent WCX. (A) = analytical SEC (table 1),
            (B) = SDS-PAGE: Lane 1 = molweight markers,
            lane 2 = rHuIFN-alpha 1 (5 /ul), lane 3 = 10 /ul,
            lane 4 = 15 /ul

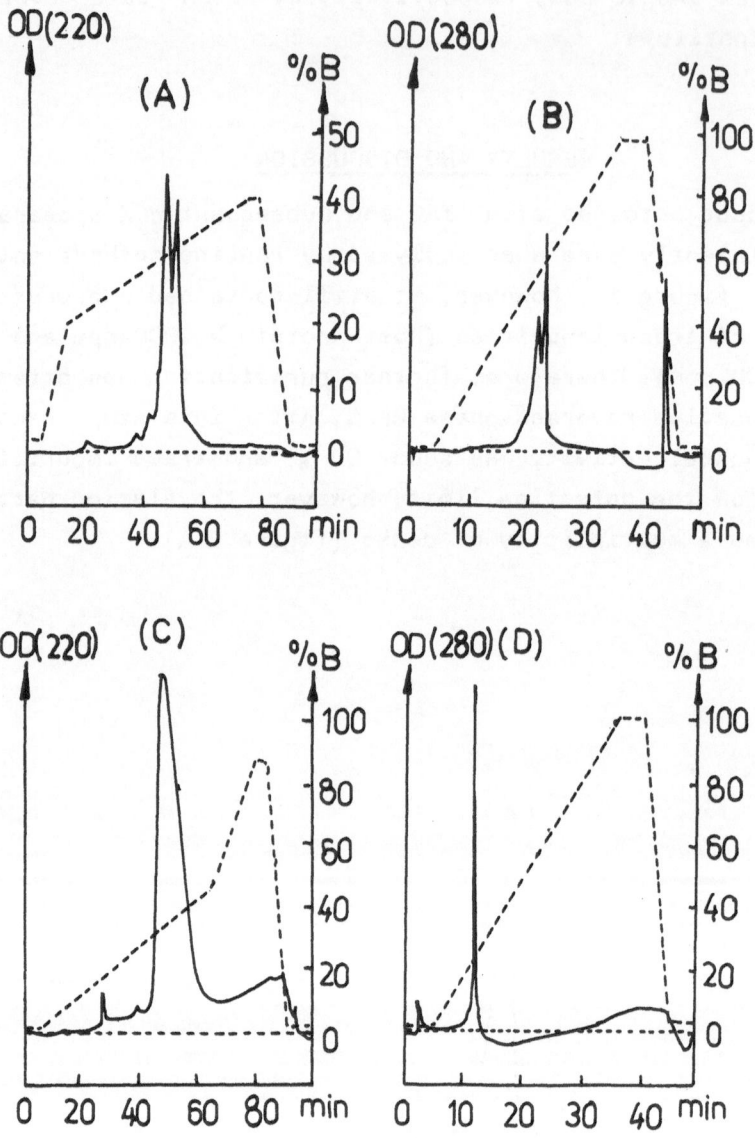

Figure 2. HPLC analyses of the product obtained after IAC
and subsequent WCX.
(A) = preparative RPC, (B) = analytical HIC,
(C) = analytical WAX, (D) = analytical HAC
Conditions are given in table 1

This observation seems to indicate a diversity of the product analogous to the findings reported by a Japanese group (3,4) for rHuIFN-alpha 2. On the other hand, conformational changes of proteins through strong interactions with the chromato-graphic matrix under denaturing elution conditions, as in the presence of strong acids and organic modifiers in RPC, have been extensively discussed lately (1, 5-8).

Basically, two hypotheses have to be taken into consideration:

1. Figure 1 shows only apparent homogeneity of the product; i.e. the resolution of the methods employed is too poor to detect sample diversity

2. Figure 2 reflects apparent diversity of "forced conforma-tions" induced by unfolding-folding equilibria under the special chromatographic conditions

Whereas HPLC separations based on hydrophobicity, such as re-versed phase (RPC = figure 2A), and hydrophobic interaction (HIC = figure 2B) always showed peak-multiplicity, other me-thods like hydroxyapatite (HAC = figure 2D) and weak anion-exchange HPLC (WAX = figure 2C) suggested homogeneous samples.

The two peaks in figure 2A could be easily separated. After buffer exchange into 30 mM citrate pH = 3.0 the two species were analyzed (Figure 3 and Table 2).

TABLE 2
Characterization of the two peaks separated by
preparative RPC

Data		"peak 1"	"peak 2"
Retention time (min.)		48.2	51.3
+ specific activity	Biotest	2.5±0.4	2.3±0.5
$(\times 10^{-8}$ IE/mg)	ELISA	2.8±0.7	2.7±0.7
trace impurities	IgG	< 10	< 10
(ppm)	host protein	< 100	< 100

+ specific activities were calculated from protein content (Bradford) and 40 and 10 parallel measurements in biotest and ELISA, respectively

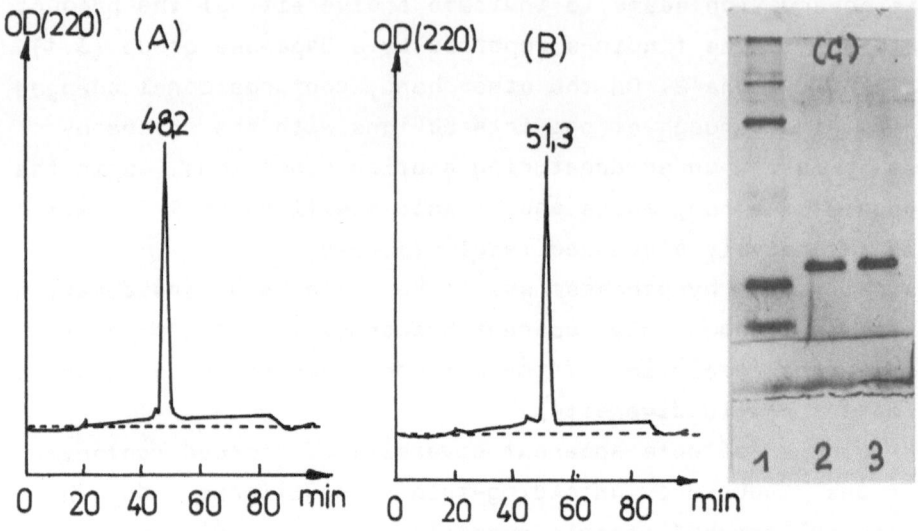

Figure 3. Characterization of the two species obtained by
preparative RPC.
(A) and (B) = RPC of "peak 1" and "peak 2",
respectively, after buffer exchange (30 mM citrate)
(C) = reducing SDS-PAGE: lane 1 = molweight markers
lane 2 = "peak 1", lane 3 = "peak 2"

N-terminal amino acid sequence and the C-terminal unit are:

```
"peak 1":  (C) D L P E T H S L D  .... E
"peak 2":  (C) D L P E T H S L D  .... E
            1  2 3 4 5 6 7 8 9 10  .... 166
```

Both species appear to have the primary structure published
for HuIFN-alpha 1 (D) (9). From the virtually identical specific activities of "peak 1" and "peak 2" it may be concluded
that they also have identical epitopes and receptor-binding
sites.

Summarizing these observations: The two proteins appear
to have the same primary structure, however, there exist small
differences in the tertiary/quarternary structure which are
responsible for changes in the "chromatographic contact region" (1) and, thus, distinct elution in RPC. Surprisingly, the
two conformations are quite stable and even preserved during
buffer exchange. The origin of the two species; i.e. whether
they are generated in RPC or are already present, and their

exact nature, are still to be investigated.

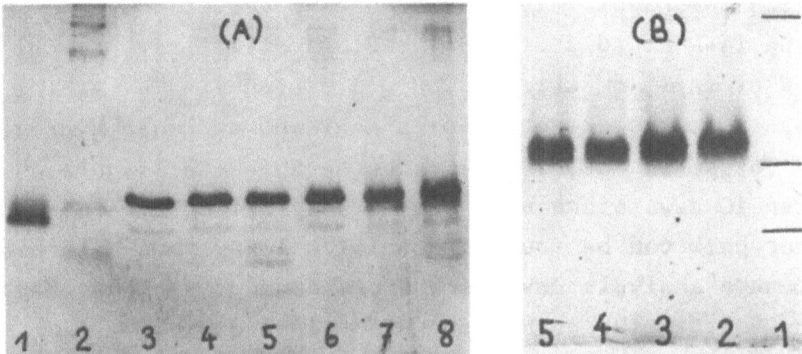

Figure 4. Investigations on stability of rHuIFN-alpha 1 by
SDS-PAGE for samples in solution (A) and stabilized
(A): 1 = non-reduced, 2 = markers, 3-5: reduced by
ß-mercaptoethanol, 6-8 = reduced by DTE, stored in
30 mM citrate pH = 3.0 for 12 h at 4°C (4+7) and
at 20°C (3+6) and for 24 h at 37°C (5+8)
(B): 1 = markers, 2+3 = samples in above buffer
with 10 % (v/v) gelafusal at 4°C for 12 h and 10
days, 4+5 = lyophilized samples stored for 10 and
100 days, respectively, at 4°C and re-dissolved in
water, all samples 20 times overloaded

TABLE 3

Recovery of biological activity after storage at 4°C
A = sample in the elution buffer of RPC, B = sample in 30 mM
citrate pH = 3.0, C = B + 2 mg/ml human serum albumin,
D = B + 10 % (v/v) gelafusal, E = D lyophilized. At least 10
parallel determinations of biological activity were used for
the average.

storage time	average recovery of biological activity (4)					
	peak 1-A	peak 2-A	B	C	D	E
0	100	100	100	100	100	100
3- 6 h	-	-	81±15	122±35	116±10	-
9-12 h	-	-	60±10	92±20	116±15	-
1- 5 d	105±10	110±15	60±15	103±20	116±15	-
10-15 d	110±20	100±10	33±10	84±15	83±10	115±20
100 d	80±20	75±15	7± 5	61±20	40±25	103±10

Table 3 and Figure 4 summarize the results of the stability studies. The following conclusions can be drawn for both

the "peak 1" and the "peak 2" species after preparative RPC:

1. Samples in the elution buffer of RPC (ca. 30 % isopropanol) retain their activity over several weeks at 4°C.

2. There is a rapid decline of biological activity of the product in 30 mM citrate, pH = 3.0 even at 4°C. At the same time, the emergence of a new band at lower hydrodynamic volume is observed by reducing SDS-PAGE (figure 4).

3. After 10 days storage in the citrate buffer at 4°C, a new minor peak can be found in analytical RPC, and N-terminal sequence analysis reveals the presence of another component besides the authentic HuIFN-alpha 1:

$$L-A-E-H-S-R-I-S-P- \ldots$$

This could correspond to degraded interferon where the labile Leu(17)-Leu(18) position has been cleaved.

4. By the addition of 10 % (v/v) GELAFUSAL, a commercially available plasma substitute, the loss of biological activity is substantially retarded, and no additional band appears in SDS-PAGE even after a prolonged storage time. Similar results were confirmed for lyophilized samples.

CONCLUSIONS

RP-HPLC is a very valuable technique for protein quality control. However, denaturation of the protein as a result of its interaction with the chromatographic sorbent can produce artifact peaks and complicate the interpretation of the results. Stability problems may render these analyses even more difficult. Therefore HPLC methods, particularly those based on hydrophobicity, that are to be used in purity and/or stability tests have to be validated prior to use (6,10). Conformational changes may be induced by RP-HPLC and turn out to be irreversible. Consequently, the proof of identity of the recombinant protein becomes even more difficult.

In our case we are not yet able to decide which of the two species comprises the native "authentic" HuIFN-alpha 1.

ACKNOWLEDGEMENTS

The authors are grateful to the following collaborators in our Institute: Dr. M. Tonew for essentially contributing to the establishment of the biological assay, Dr. Wollweber and Dr. Römer for the development of the ELISA tests, Dr. Fricke for IAC purifications, Dr. Vettermann and Dr. Kraft (ZIM, Berlin) for N-terminal sequence analyses.
Special thanks go to SHIMADZU (Europa) GmbH, Duisburg FRG for the generous support with HPLC instrumentation and columns.

REFERENCES

1. Regnier, F.E., The role of protein structure in chromatographic behavior. Science 1987, 238, 319-323

2. Laemmli, U.K., Cleavage of structural proteins during the
assembly of the head of bacteriophase T4. Nature 1970,
227, 680-685

3. Honda, S., Sugino, H., Nishi, K., Nara, K. and Kakinuma,
A., Purification of human leukocyte interferon A derived
from Escherichia coli: an aging process to prevent the
formation of oligomers. J. Biotechnology, 1987, 5, 39-51

4. Nakagawa, S., Honda, S., Sugino, H., Kasamoto, S., Sasaoki,
K., Nishi, K. and Kakinuma, A., Characterization of three
species of Escherichia coli-derived human leukocyte interferon A separated by reversed-phase high-performance liquid chromatography. J. Interferon Res., 1987, 7, 285-299

5. Lu, X.M., Benedek, K. and Karger, B.L., Conformational effects in the high-performance liquid chromatography of proteins. Further studies of the reversed-phase chromatographic behavior of ribonuclease A. J. Chromatography 1986,
359, 19-29

6. Garnick, R.L., Solli, N.J. and Papa, P.A., The role of
quality control in biotechnology: an analytical perspective. Anal. Chem. 1988, 60, 2546-2557

7. Thevenon, G. and Regnier, F.E., Reversed-phase liquid
chromatography of proteins with strong acids. J. Chromatography 1989, 476, 499-511

8. Drake, A.F., Fung, M.A. and Simpson, C.F., Protein conformational changes as the result of binding to reversed-
phase chromatography column materials. J. Chromatography
1989, 476, 159-163

9. Henco, K., Brosius, J., Fujisawa, A., Fujisawa, J.-I.,
Haynes, J.R., Hochstadt, J., Kovacic, T., Pasek, M.,
Schamböck, A., Schmidt, J., Todoroko, K., Wälchli, M.,
Negata, S. and Weissmann, C., Structural relationship of
human interferon alpha genes and pseudogenes. J. Mol.
Biol. 1985, 185, 227-260

10. Maldner, G., Requirements and tests for HPLC apparatus
and methods in pharmaceutical quality control.
Chromatographia 1989, 28, 85-88

Part 5

Integrated Processes

THE RATIONAL DESIGN OF LARGE SCALE PROTEIN SEPARATION PROCESSES

J.A. Asenjo

Biochemical Engineering Laboratory
University of Reading
P.O. Box 226, Reading RG6 2AP, England

ABSTRACT

This paper reviews and discusses recent developments in rational process design and their potential application in present and future biotechnology for large scale protein separation processes. It includes a review of the main issues involved in process design for protein separation and purification and a brief evaluation of rigorous computer methods in chemical engineering, recent developments in artificial intelligence, particularly the field of expert systems and the potential of using a combination of such techniques. The downstream process was divided into two distinct subprocesses: a first subprocess called 'recovery' after which the total protein concentration is 60 - 70 g/L and a second subprocess called 'purification'. The main deficiency of accurate information was found to be in that required for the selection of high resolution purification operations on a rational basis. An expert system for selection of optimal protein separation sequences will give the user a number of alternatives chosen based on extensive data back-up on proteins and unit operations.

INTRODUCTION

A protein purification process is usually composed of a sequence of separation and purification operations whose final aim is to obtain the required product at a prespecified level of purity. On a large scale it is necessary to obtain the highest possible recovery yield minimizing the resources utilized and hence the cost. This can be achieved if a minimum number of efficient separation steps are employed [1]. For the manufacture of chemical products one can consider two extreme cases.

1. The product is of high value in a virtually competition free market
2. The product is a high-volume chemical with many producers in a highly competitive market.

In the first case one would probably choose the first successful purification procedure found and the product will probably have a short market life expectancy. The overall economics of such a case are determined by getting into the market ahead of possible competitors. If the market lasts and increases, and competitors move in, products will eventually fall into the second category in which process optimization and economics play a *very important* role.

Modern biotechnology processes are virtually all still very close to case 1 but as its products become more competitive and therefore their use more widespread (e.g. hepatitis B vaccine) they will move closer to the second case. As this happens the importance of rational design tools in biotechnology will grow and its economic importance will evidently increase [2]. In order to achieve this we can recognize two different tasks:

1. Selection of Separation Sequence: giving maximum recovery yield (purity is determined by requirements) and using minimum resources (number of steps and cost of each step).

2. Design of Individual Unit Operations: using design equations and correlations as well as characteristics of operations, proteins and contaminants.

It is important to bear in mind that a process designed for small quantities is usually not optimal for large scale.

PURIFICATION STRATEGY

To design a large scale process to economically purify a protein or other biotechnology product, maintaining a high yield, yet obtaining a virtually pure product while also minimizing the cost, requires three main considerations: i) clearly defining the final product objective, ii) characterizing, as far as possible, the starting material and with these two pieces of information iii) defining possible separation steps.

The following five main heuristics or rules of thumb [1,3] provide a good basis for process selection:

Rule 1: 'Choose separation processes based on different physical, chemical or biochemical properties'.

Rule 2: 'Separate the most plentiful impurities first'.

Rule 3: 'Choose those processes that will exploit the differences in the physicochemical properties of the product and impurities in the most efficient manner'.

Rule 4: 'Use a high resolution step as soon as possible'.

Rule 5: 'Do the most arduous step last'.

In any event we have to keep a very open mind as far as any possible changes in the process as well as to keep it as simple as possible. The main steps in a large scale protein purification procedure are usually not more than four or five necessary ones and they consist of:

Recovery/Isolation
1.- Cell Separation
2.- Cell Disruption and Debris Separation (for
   intracellular proteins only)
3.- Concentration
Purification
4.- Pretreatment or Primary Isolation
5.- High Resolution Purification
6.- Polishing of Final Product

Rationalization of Procedure

An important point that needs consideration is that once the purification procedure is set and regulatory approval of the product is underway the purification procedure cannot be changed. Only a particular product obtained by a specific procedure obtains regulatory approval, hence once this is given, the purification procedure is fixed. This stresses the value of early rationalization of the purification process. It also means that for a protein to be used for therapeutical application or other human use which, if successful, relatively large quantities of the product will be required, even in the very early stages of protein purification one should only use in the laboratory such procedures that can be realisticaly used in large scale, ie. for which suitable large scale equipment either exists or might be developed in the foreseable future. Also, from early on, concepts related with maximization of yield in each step and in the whole separation sequence, minimizing the number of steps used and minimizing resources (economics), should be introduced [3].

LARGE SCALE OPERATIONS

The main operations used on a large scale for separation and purification of proteins are shown in Table 1.

TABLE 1
Separation and Purification Operations for Large Scale
Recovery and Purification of Proteins

Operation	physico-chemical property
Centrifugation	sedimentation velocity
Filtration	particle size
Microfiltration	particle size
Homogenization	intracellular nature (pressure gradient)
Bead milling	intracellular nature (liquid/solid shear)
Ultrafiltration	molecular size
Two phase extraction	partition coefficient
Precipitation	solubility (hydrophobic interaction)
Adsorption	van der Waals forces, H bonds, polarities, dipole moments
Ion-exchange	charge (titration curve)
Hydrophobic interaction	surface hydrophobicity
Affinity chromatography	biological affinity
Gel filtration	molecular size
Reversed phase liquid chromatography	hydrophilic and hydrophobic interactions

If the intracellular product is manufactured in E. coli, the heterologous proteins will accumulate in the form of insoluble inclusion bodies. This makes necessary the processing of the inclusion bodies into the native protein by denaturing and refolding. If the intracellular product is manufactured in yeast in many instances the protein is present in homogeneous particulate form, typically 30 - 60 nm particles such as VLP's or 'virus like particles'. Although the processing of intracellular particulate recombinant proteins is an important aspect of downstream processing there are not many satisfactory methods for large scale separation, denaturation and refolding of the particulate proteins. Recent developments in the use of reverse micelles for protein refolding and of aqueous two-phase systems for separation of VLP's from yeast homogenates appear particularly attractive.

Variables Affecting Scale-up
In virtually all chromatographic procedures, particularly in those based on adsorption type interactions, scale up is achieved by increasing the radius of the column while maintaining the column height. The effect of process parameters on resolution and throughput in chromatographic procedures have been reviewed [3]. The resolution of proteins is mainly determined by the elution strategy (desorption) and not by the length of the column. Relatively short columns are favoured in the range of 15 - 30 cm in height. The largest available column for protein purification appears to be a 1700 to 2500 liter one, 2 m in diameter with an adjustable bed height between 55

and 80 cm (Amicon). In such systems the appropriate design of flow distributors is a crucial concern to obtain reproducible scale-up. Radial flow chromatography, a radically new column design which has been developed by Sepragen, constitutes a more rational approach to scale up chromatography. The mobile phase flows radially and hence scaling up is achieved by increasing column length. Flow rates seem high and hence processing times appear to be shorter than with traditional column design; however, scale-up is very hardware intensive (virtual linear increase of equipment with size).

Development of Novel Large Scale Operations

The development of protein separation and purification techniques for efficient use on a large scale (eg. separation and renaturation of inclusion bodies or separation of VLPs) is in its infancy. Most operations in use today are just large versions of laboratory procedures and have not been developed specifically for large scale use. In the next 5 to 15 years we should see important new developments in the conception of procedures and apparatus specifically developed for large scale use. For instance an alternative to chromatographic affinity adsorption that can be used continuously in the large scale constitutes the use of two stages of adsorption and desorption where the affinity support phase is recycled between the adsorption and desorption stages. The two phases can be either liquid or solid (eg. a liquid fluorocarbon [4], or solid agarose [5]). Such a continuous system can also be used for ion-exchange adsorption. Another important development for large scale processing is aqueous two-phase partitioning systems where it is now becoming possible to manipulate the system to separate proteins based on hydrophobicity, charge, affinity and m.w., similarly to the way it is done in chromatography. Other examples of emerging technologies are reverse micelles, differential product release (see paper in this book), and fluidised bed adsorption (paper in this book).

PROCESS DESIGN AND OPTIMIZATION

Process design and selection of operations is a complex process where the design evolves from a preliminary stage to the final stage in a trial and error fashion, repeatedly revising and refining the initial assumptions and restrictions (1. flowsheet generation (qualitative/ semiquantitative), 2. quantitative design of units, 3. revise flowsheet (1.), then 2. and etc. until some objective is reached). An important aspect of process design involves the selection of operations and design of plant equipment. In the initial stages this process is more or less done using heuristics: using rules of thumb to arrive at a rapid (and reliable) specification of equipment type, size and maybe cost.

The design of a protein recovery and purification process shares many characteristics with other engineering design activities. To design a process or an operation requires the satisfaction of a number of constraints (purity, quality, process temperature, desired yield) using what is known about the materials (chemical and biochemical properties, thermodynamics and fluid dynamics of the process material) to end with a sequence of equipment interconnected in a particular order. It is important to stress that the type of reasoning behind the design process does not rely only upon strict mathematical models. Equations could only provide the information necessary to conclude that a particular equipment is appropriate, but the inclusion or not of the step is left for the designer in a job which is based mainly on judgement.

Properties of Proteins and Main Protein Contaminants

It is important to have detailed information of the physico-chemical characteristics of the final protein product and of the main contaminants. This will have an impact on the level of rigorousness that can be used in the analysis of the interaction of the protein and contaminants with the individual separation processes. These include: charge (titration curve), surface hydrophobicity, biospecificity, size (M.W.), isoelectric point, particle size, density, viscosity, difusivity, stability and maybe some other characteristics such as e.g. potential inhibitors in the presence of proteases.

An important parameter to use in evaluation of purification and high resolution polishing of proteins is the clearance factor [2]

$$\text{clearance factor} = \cfrac{\cfrac{\% \text{ purity (finish)}}{\% \text{ impurities (finish)}}}{\cfrac{\% \text{ purity (at start)}}{\% \text{ impurities (at start)}}}$$

This factor gives similar values to the purification factor (% purity (finish)/% purity (at start)) at low product protein concentrations but at high protein concentrations the purification factor is of little use whereas the clearance factor gives a measure of the reduction in the level of impurities. For instance, when purifying a protein from 99 % to 99.9 % purity where a 10 fold reduction in the level of impurities is obtained, the value of the clearance factor is 10.1 whereas the value of the purification factor is 1.01. This concept has been used for a number of years in the analysis and calculations for separation of multicomponent mixtures where it is known as the separation factor [7].

Characteristics of Operations and Design Elements

Quantitative information on the performance of individual operations including design correlations and data bases of properties of materials is vital. For instance, in the case of mechanical cell disruption the design information available has been described [8]. It will include flow rate, type of agitator or operating pressure, cell concentration and type, fraction of product release and size of the disrupter. For a chromatography operation the information necessary has been described in several sources on the theoretical and design analysis of adsorption type chromatography [9, 10, 11]; it concerns characteristics of columns (size, geometry) and properties of gels and other adsorbants (binding capacity, dissociation constants, flow rate, half life, breakthrough curves).

Mathematical models and mathematical correlations of the operations will allow simulation of performance and also may be used to scale up individual operations. Computer simulations are a useful tool to optimize individual separations [12]. Examples of useful downstream process simulations, and investigation of process conditions are microbial cell breakage and selective product release using enzymes [13, 14] and investigation of the affinity and ion-exchange chromatography of proteins [15, 16].

Process Synthesis and Optimization

The problems that have to be solved in process synthesis and optimization of downstream protein separations, appear to be of two types: (i) choosing between alternative operations (e.g. homogenizer vs. bead mill or centrifugation vs. cross flow microfiltration) and (ii) the design of an optimal chromatographic sequence with maximum yield and minimum number of steps (1, 2 or 3), a problem that is combinatorial in nature. These problems can be solved by either finding a rigorous solution using numerical methods like mathematical programming techniques (eg. resolution of 'tree structure' in chemical engineering) or by using an Expert Systems (ES) approach [1]. The first solution by itself has limited use in biotechnology due to a lack of useful design equations and data bases. The second approach appears more attractive since it allows the use of empirical knowledge which is typical of that used by experts. Computer based expert systems are an important tool in the field of Artificial Intelligence (AI). Efforts have been made to develop AI systems for this purpose [17, 18, 19] or to adapt existing systems (called 'shells') [20] both for the manipulation of heuristics (rules of thumb), databases and simple algebraic design equations.

The optimization of a process design is carried out after one or more purification strategies have been chosen and performed and thus process conditions are known. The interest at this point is to find the optimal operating conditions for specific separation operations so that their performance can be compared with alternative operations. It would also be desirable to compare alternative chromatographic sequences giving a similar final product in terms of overall economics [21, 22].

ARTIFICIAL INTELLIGENCE -EXPERT SYSTEMS-

Developments in the field of artificial intelligence (AI) in the last few years, particularly in building and using expert systems (ES), have made this a potentially important tool in the area of computer based process design and process synthesis. In general, AI is concerned with the development of computer-based programmes that emulate the reasoning of humans. This requires the understanding of human problem-solving methods in areas such as those where the amount of knowledge to be manipulated is large and/or there are significant uncertainties. An expert will narrow down the search by recognizing patterns and using appropriate heuristic rules. Designing an interactive computer programme to do this is the study of knowledge based Expert Systems (ES), which has important potential in engineering design and operation [23]. An ES, shown diagramatically in Fig. 1, has two main parts: a 'knowledge base' and an 'inference engine'. To create an ES, the domain or body of knowledge specific to the class of problem must be organized into a 'knowledge base'. The mechanism that performs the inference procedure is the 'inference engine'.

Conventional scientific and engineering computer programmes consist of a set of statements where the order of execution is predetermined. These programmes are rather rigid and updating them needs considerable effort (the programmer has to locate the appropriate place to update in the predefined sequence). The programmer must at all times ensure complete specification of the problem and uniqueness of the solution. ES alleviate this strictness by making a clear distinction between the knowledge base and the control strategy. This partition allows for incremental addition of knowledge without manipulating the overall programme structure, and, by using confidence factors with the IF-THEN rules, the system can be made to provide a number of ranked alternative solutions.

Process synthesis is one area where expert systems can be of important help, particularly in the selection of plant equipment and in the design of separation trains or sequences, especially in the field of biotechnology.

Development of an Expert System for Protein Purification

The major steps involved in the development of an ES are shown in Figure 2 and consist of:

1. Identification of experts and resources as well as knowledge engineering concepts. It will only include a small part of the knowledge.

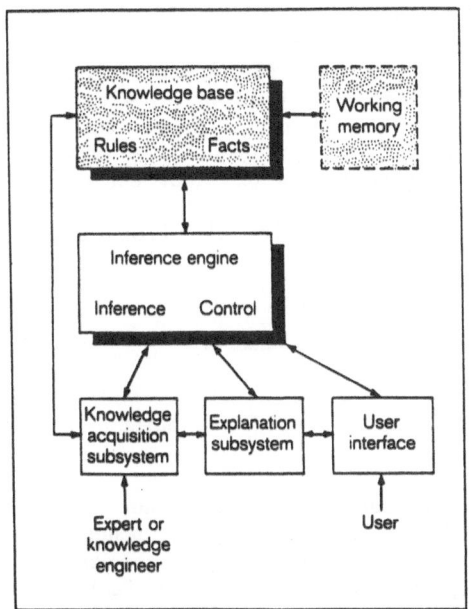

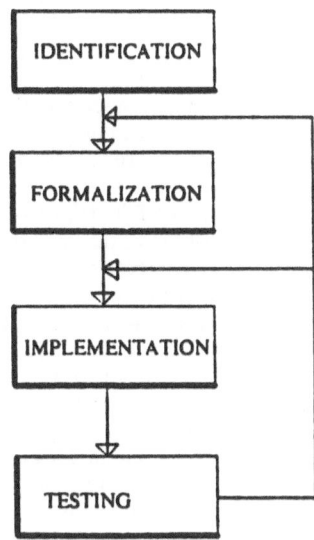

Fig.1. The architecture of a knowledge based expert system (The knowledge base is shaded for emphasis) [24].

Fig.2. Development process of a knowledge based expert system (ES) [23].

2. Formalization involves the selection of appropriate tools to build the ES. The knowledge engineer becomes familiar with the domain; he performs a few preliminary interviews with the expert.

3. Implementation consists of encoding the knowledge developed in the previous stage into a prototype system.

4. Testing and Refinement. The prototype system is taken to the expert and tested. The weaknesses in the knowledge base and the inference mechanism are identified.

Currently, the major bottleneck in the development of ES's is the knowledge acquisition process which is mainly carried out in stage 3: and to some extent in stage 4.

Today there are well developed expert systems software systems or "shells" (e.g., Personal Consultant Plus (PC Plus, Texas Instruments) or Expert Systems Environment (ESE, IBM)) that help develop an organised knowledge base from the domain knowledge and also provide the inference engine. Asenjo, et al [20] implemented a prototype ES with a second generation of protein purification rules in two shells ESE and PC Plus. Expert knowledge was obtained partially from the literature but mainly from industrial experts working on the large scale separation and purification of therapeutic, diagnostic and analytical proteins.

The knowledge was expressed in around 65 heuristic (expert) rules, some of which carry a degree of uncertainty [20]. The downstream process was divided into two distinct subprocesses in order to structure the knowledge. A first subprocess called Recovery after which the total protein concentration is 60 - 70 g/1, and a second subprocess called Purification.

The Recovery subprocess comprises harvesting, cell disruption, separation of solid debris and precipitation of nucleic acids (these last three steps only if disruption is required) and concentration.

The Purification subprocess takes the 60 – 70 g/l protein solution and purifies the individual protein product to a high purity with a high yield. It comprises preconditioning or cleaning, high resolution purification, which can be carried out in one, two or three steps and polishing, when necessary, usually to remove traces of minor contaminants such as for therapeutical applications.

The various parameters used to characterize the broth and the culturing system are:
- MICROBIAL SOURCE (Bacteria, Yeast, Fungus, Mammal)
  PRODUCT OF INTEREST: Name
- Cellular Location (Intracellular, Extracellular, Unknown)
- Molecular Weight
- Aqueous Two-Phase Separation data base
- Titration curve (charge as a function of pH), Isoelectric point
- Surface Hydrophobicity
- Biospecificity data base

The proposed process consists of a sequence of operations to obtain the stated design objective. There might be several different sequences of operations that will accomplish the same objective. In those cases, a quantitative degree of performance (given by the 'certainty factor') of each operation is assigned by the expert and carried by the system into the proposed design. In the development of the prototype ES it was found that the overall process was satisfactorily divided into two subprocesses with clear objectives: recovery and purification. Selection of operations in the recovery subprocess could be well structured. In the second subprocess (purification) the structuring of the knowledge was more difficult. The main deficiency of available information was found to be in that required for the selection of high resolution purification operations.

The work carried out on expert systems (ES) [1, 20] briefly reviewed in this paper clearly shows that properly developed ES can be a vital tool to assist in solving the knowledge intensive and heuristic based problem of process synthesis in biotechnology. Rigorous methods will not be appropriate to solve the overall synthesis problem as rigorous information and mathematical correlations are not as readily available as they are in chemical process engineering. The overall downstream process synthesis problem in biotechnology does not have a strict combinatorial nature whereas the high resolution purification stages within the purification subprocess (1, 2 or even 3 purification stages where several alternatives in different order combinations can be used) do. However rigorous models have a very important role in the simulation of individual operations [2] (e.g., for process evaluation and comparison of performance and cost of individual operations). It clearly appears that the limiting factor in the development of ES's for protein purification is not the implementation of new AI programmes but the acquisition, clarification, formalization and structuring of the domain of expert knowledge.

Hybrid Systems

It is clear that heuristic methods (such as those that have been implemented in expert systems) are one end of the spectrum of available process synthesis techniques, the other end consists of rigorous methodologies such as mathematical programming techniques. A second stage in the development of expert systems should consider the introduction of quantitative models (mathematical correlations, design equations and short-cut methods) for the design and evaluation of individual operations and their alternatives, and, for introducing basic cost calculations into the selection procedure of alternative processes. Such a hybrid system that will include heuristic rules, rigorous information and design correlations is particularly attractive for process biotechnology as rigorous correlations and detailed information are not readily available. At this point it will also be possible to establish evolutionary methods to improve the design of an existing or initial process sequence.

Evolutionary Methods

Evolutionary methods have been developed in chemical process engineering to overcome some of the uncertainty of the heuristic methods and to use, whenever available, simulation tools to improve the design. The basic idea of this approach is to modify substructures of an existing design to improve the objective function. The first successful attempts in this direction were made by Westerberg and Stephanopoulos [25]. The evolutionary synthesis of separation sequences includes the following subtastks: (i) generate an initial separation sequence, (ii) identify the rules which will govern the evolutionary steps, and (iii) determine the evolutionary strategy to use for comparing different candidate sequences.

CONCLUSIONS

The design of an optimal protein separation sequence is an important problem, which will gain in importance as biotechnology products become more widespread and competitive. There are several modern chemical engineering and computer tools that can be used for optimizing such design and selection processes. These include the use of rigorous methods as well as Expert Systems (ESs) that can use both numerical, quantitative, algorithmic and heuristic information. A hybrid approach appears particularly suitable for the rational design of large scale protein separation processes.

In order to advance the further development of this field there is an important need for generating databases for protein products, fermentation streams and contaminants as well as mathematical correlations and models and short-cut methods for simulating operations (interpreting their behaviour). It is also necessary to develop more specific recovery, separation and purification operations specifically designed for the large scale recovery of proteins in biotechnology.

In the development of a prototype ES the overall process was satisfactorily divided into two subprocesses with clear objectives: recovery and purification. Selection of operations in the recovery subprocess could be well structured. In the second subprocess (purification) the structuring of the knowledge was more difficult. The main deficiency of available information was found to be in that required for the selection of high resolution purification operations (mainly because the information does not exist). For selection of optimal protein separation sequences the ES will give the user a number of process alternatives which will be chosen based on extensive data back-up on protein sources and unit operations as well as algebraic correlations. Such a system will clearly constitute 'expert amplification' and not 'expert replacement'.

REFERENCES

1. Prokopakis, G.J. and Asenjo, J.A. Synthesis of Downstream Processes, in "Separation Processes in Biotechnology". Ed.: J.A. Asenjo, Marcel Dekker, N.Y., 1990, p. 571-601.
2. Asenjo, J.A., The Rational Design of Large Scale Protein Separation Sequences. 32 International IUPAC Congress, Stockholm, 2-7 August, 1989.
3. Asenjo, J.A. and Patrick, I., Large Scale Protein Purification, in: "Protein Purification Applications: a Practical Approach", Eds.: E.L.V. Harris and S. Angal, IRL press, U.K., 1990, p.1 - 29.
4. Eveleigh, J.W. Fluorocarbon liquid and solid affinity supports, 5th Int. Conf. on Partition in Aqueous Two-phase Systems, Oxford, U.K., August 1987.
5. Pungor, E., Afeyan, N.B., Gordon, N.F. and Cooney, C.L. Bio/Technology, 1987, 5, 604- 608.
6. Andrews, B.A. and Asenjo, J.A., Aqueous Two-Phase Partitioning. In Protein Purification Methods, a Practical Approach, eds. E.L.V. Harris and S. Angal, IRL Press, Oxford, 1989, pp. 161 - 174
7. King, C.J., "Separation Processes", Mc Graw Hill, N.Y., 1980.
8. Shutte, H., Kroner, K.H., Husted, H. and Kula, M.R. Enzyme Microb. Technol., 1983, 5, 143- 148.
9. Yamamoto, S., Nomura, M. and Sano, Y. AIChE Journal, 1987, 33, 1426.
10. Yamamoto, S., Nakanishi, K. and Matsuno, R. Ion-Exchange Chromatography of Proteins, Marcel Dekker, N.Y., 1988.
11. Wang, L. Ion-exchange in Purification, in "Separation Processes in Biotechnology". Ed.: J.A. Asenjo, Marcel Dekker, N.Y., 1990,p. 359-400.
12. Hedman, P., Janson, J.C., Arve, B., and Gustafsson, J.G., Large Scale Chromatography-Optimization of Preparative Chromatographic Separations. Proc. of 8th Int. Biotechnol. Symp., 1, eds. G. Durand, L. Bobichon, J. Florent, Societe Francaise de Microbiologie, 1989, pp. 612 - 622.
13. Hunter, J.B. and Asenjo, J.A. Biotechnol. Bioeng., 1988, 31, 929.
14. Liu, L.C., Prokopakis, G.J. and Asenjo, J.A. Biotechnol. Bioeng., 1988, 32, 1113-1127.
15. Chase, H.A. Affinity separations using immobilized antibodies. Symp. on Antibodies for Purification, SCI, London, March 1988.
16. Arve, B., Simulation and Modelling of Chromatographic Processes. 32 International IUPAC Congress, Stockholm, 2-7 August, 1989.
17. Siletti, C.A. and Stephanopoulos, G., Computer Aided Design of Protein Recovery Processes, 192nd ACS National Meeting, Anaheim, CA, Sept,1986.
18. Wacks, S. Design of Protein Separation Sequences and Downstream Processes in Biotechnology; Use of Artificial Intelligence, M.Sc. Thesis, Columbia University, New York, 1987.
19. Siletti, C.A., Computer Aided Design of Protein Recovery Processes, Ph.D. thesis, MIT, USA, 1989.
20. Asenjo, J.A., Herrera, L. and Byrne, B., J. Biotechnol., 11, 275-298, 1989.
21. Duffy, S.A., et al. Optimal Large Scale Purification Strategies for the Production of Highly Purified Monoclonal Antibodies for Clinical Application. 196th ACS National Meeting, MBTD division, Los Angeles, CA, 25-30 Sept. 1988.
22. Kosti, R., Economic Evaluation of Large Scale Protein Purification Operations, M.Sc. thesis, University of Reading, 1989.
23. Banares-Alcantara, R., et al., CEP, Sept. 1985,p. 25 - 30
24. Harmon, P. and King, D. Expert Systems: Artificial Intelligence in Bussiness, Wiley, 1985.
25. Westerberg, A.W. and Stephanopoulos, G., Chem Eng Sci, 31, 195, 1976.

DOWNSTREAM PROCESSING OF EXTRACELLULAR ENZYMES OF ASPERGILLUS NIGER

MICHAEL J BAILEY AND HEIKKI OJAMO
VTT, Biotechnical Laboratory,
Tietotie 2, SF-02150 Espoo, Finland

ABSTRACT

Different strategies were compared for the concentration of polygalacturonase, ß-glucosidase and ß-galactosidase produced by two strains of *Aspergillus niger*. Ultrafiltration was more successful than vacuum evaporation for all three enzymes but produced concentrates with high solids contents. The use of mineral adsorbents enabled almost complete separation and concentration of polygalacturonase and ß-glucosidase activities from one culture filtrate. Concentration of all three enzymes using adsorbents led to preparations with higher specific activities and lower solids contents than the corresponding ultrafiltration concentrates.

INTRODUCTION

Filamentous fungi are well known as efficient producers of extracellular hydrolytic enzymes. The enzymes are secreted into the medium as dilute solutions, which must be concentrated and stabilized before marketing. Although much work has been published concerning production of enzymes in fermenter cultivations, rather less information is generally available about the downstream processing of these enzymes. Primary separation of fungal mycelium and medium solids from the liquid phase is usually achieved by vacuum drum filtration and the filtrate is then concentrated. Ultrafiltration is the standard technique for concntration. Little attention is paid to quality properties of the product, such as concentrations of total dissolved solids and impurities, turbidity, colour and side activities. Advantages of ultrafiltration are that there is no need for phase change to remove water and that at least small molecular weight impurities are removed with the permeate. Single-stage vacuum evaporation at about 40°C can be used for concentration of some enzymes. Capacities per unit surface area may be quite high, but concentration of all the components of the culture filtrate with the product is often a disadvantage. Batchwise adsorption/desorption is a simple technique providing high capacities per unit area and often substantial product purification. Other interesting techniques for concentration of extracellular enzymes include precipitation with solvents, salts or polymers [1,2], aqueous two-phase extraction [3] and continuous affinity recycle extraction [4]. The aim of this work was to compare different methods for concentration of polygalacturonase, ß-glucosidase and ß-galactosidase produced by two strains of *Aspergillus niger*.

MATERIALS AND METHODS

Microbial strains and cultivation methods

The *Aspergillus niger* mutant strains VTT-D-86267 [5] and VTT-D-80144 [6] were used in this work. *A. niger* VTT-D-86267 was cultivated on 30 g l^{-1} beet pulping waste medium for production of polygalacturonase and ß-glucosidase as described previously [5]. ß-Galactosidase was produced by *A. niger* VTT-D-80144 on a medium containing 30 g l^{-1} wheat bran and 10 ml l^{-1} corn steep liquor concentrate (d.w. 45 %) with 5 g l^{-1} each of potassium dihydrogen phosphate and ammonium sulphate. The strains were cultivated in the 2 m^3 pilot fermenter at this laboratory with a working volume of 1200 l. Cultivation time was approximately 4 d for *A. niger* VTT-D-86267 and 3 d for VTT-D-80144. Cultivation conditions were: temperature 30°C, agitation 100 rpm, aeration 15 m^3h^{-1} for VTT-D-86267 and 30°C, 170 rpm and 18 m^3 h^{-1} for VTT-D-80144.

Downstream processing

Broth from the pilot fermentations was separated using a pilot scale vacuum drum filtration unit (Larox, Lappeenranta, Finland, model VF 8/1.0, surface area 1.0 m^2) with a 12 kg precoat of Celite 535 (Johns-Manville, USA). Clarified broth was divided into portions with varying volumes for concentration by different methods. Ultrafiltration experiments with 3.0 l batches of *A. niger* VTT-D-86267 culture filtrate were carried out using laboratory scale hollow fibre membranes (Romicon PM5, PM10, PM30 and PM100, A = 0.1 m^2). Larger volumes were concentrated using Paterson Candy (UK) tubular membranes (PCI ES 625, A = 2.4 m^2). Vacuum evaporation was carried out using a Hackman-MKT (Finland) falling film evaporator, model PE-1-1.6 (inner heating surface area 5.1 m^2) with a maximum temperature of 42°C.

The adsorbents used for concentration of individual enzymes were titanium oxyhydrate (Kemira Oy, Finland) for polygalacturonase [7,8] and bentonite (Na-Al-silicate, Berkbond No. 2, Steeley Minerals Division, Milton Keynes, UK) for ß-glucosidase [7] and ß-galactosidase [9].

Polygalacturonase of *A. niger* VTT-D-86267 was adsorbed on 0.8 g l^{-1} titanium oxyhydrate at pH 4.4. Under these conditions neither the ß-glucosidase nor the cellulase activity in the culture filtrate was adsorbed. After separation of the enzyme-adsorbent mass in an Alfa Laval BTPX-205 SGD separator, ß-glucosidase in the clarified liquid phase was adsorbed by contacting with 8 g l^{-1} bentonite at pH 3.0 and separating from the liquid phase. Desorption of polygalacturonase from titanium oxyhydrate at pH 7.5 and of ß-glucosidase from bentonite at pH 6.0 to greatly reduced liquid volumes was carried out as described previously [7], with final separation of the desorbed enzyme concentrate from the adsorbent in a Cryofuge 8000 laboratory centrifuge (Hereaus Christ, FRG) at 8000 x g for the titanium oxyhydrate slurry and 3000 x g for the bentonite.

ß-Galactosidase of *A. niger* VTT-D-80144 culture filtrate was adsorbed on 3 g l^{-1} bentonite at pH 3.3. The enzyme-adsorbent mass was separated on the drum filter and resuspended in sodium phosphate buffer (< 10 % of the original volume), pH 7.2, to give a final pH of 6.8-7.0 or in water with adjustment of pH to 6.8-7.0 with sodium hydroxide solution. Separation of desorbed enzyme from the adsorbent slurry was by drum filtration or centrifugation as indicated.

Analyses

Assays of polygalacturonase at 30°C and of ß-glucosidase and cellulase (against hydroxyethyl cellulose, HEC) at 50°C were carried out as described previously [5]. ß-Galactosidase was assayed using 1.8 ml 0.17 mM o-nitrophenyl-ß-D-galactopyranoside (Merck 6791) in 0.05 M sodium citrate buffer, pH 4.5 as substrate. Reaction conditions were: enzyme volume 200 μl (dilutions in citrate buffer); temperature 45°C; reaction time 600 s; stopping reagent 1.0 ml 1.0 M sodium carbonate. Activities were calculated as katals on the basis of liberation of o-nitrophenol (measurement at 420 nm). Soluble protein was measured by the method of Lowry et al. [10] after precipitation with three volumes of acetone. Dry weights of culture filtrates and their concentrates were measured after drying for 24 h at 103°C.

RESULTS

Culture filtrates and primary separation

Enzyme activities and soluble protein in the fermentation broths used for downstream processing experiments are presented in Table 1. Results of drum filtration for clarification of the broths of *A. niger* VTT-D-86167 cultivated on beet pulping waste and of *A. niger* VTT-D-80144 on wheat bran are displayed in Table 2. The overall rate of clarification of 1200 l of broth was 1.0 $m^3m^{-2}h^{-1}$ for *A. niger* VTT-D-86267 and 0.75 $m^3m^{-2}h^{-1}$ for *A. niger* VTT-D-80144 (calculated from Table 2). No problems were encountered in maintaining a firm cake on the drum and only a small part of the filter aid (12 kg precoat) was used in both cases. The dry weights of the filtrates were low (< 20 g l^{-1}).

TABLE 1

Final protein concentrations and enzyme activities in cultivations of *A. niger* VTT-D-86267 on beet pulping waste medium and *A. niger* VTT-D-80144 (2 broths, a and b) on wheat bran medium. Activities assayed were polygalacturonase (PG), ß-glucosidase (ßGLU), cellulase against hydroxyethyl cellulose (HEC) and β-galactosidase (β-GAL). nd = not determined.

Culture filtrate		Protein	PG	ßGLU	HEC	β-GAL
		g l^{-1}		nkat ml^{-1}		
VTT-D-86267		0.8	2800	35	45	nd
VTT-D-80144	(a)	1.0	nd	22	nd	120
	(b)	1.0	nd	24	nd	100

TABLE 2.

Drum filtration for clarification of pilot scale fermentation broths (1200 l) of *A. niger* VTT-D-86267 on beet pulping waste medium and of *A. niger* VTT-D-80144 on wheat bran medium. The surface area of the drum was 1.0 m^2.

Broth	Run time	Filter aid consumed	Filtrate volume	Filtrate dry weight
	min	kg	litres	g l^{-1}
VTT-D-86267	55	2.0	950	15
VTT-D-80144	80	1.5	1000	20

Concentration: *A. niger* VTT-D-86267

Ultrafiltration. The molecular weight of the polygalacturonase of *A. niger* VTT-D-86267 was known to be about 35 kDa, whereas that of ß-glucosidase was above 200 kDa [5]. Preliminary investigation of ultrafiltration of culture filtrate containing both enzymes was carried out using the laboratory scale ultrafiltration unit and hollow fibre modules with different nominal cut-off rates. Considerable leakage of polygalacturonase activity occurred through all the membranes with the exception of the PM5 membrane (cut-off 5 kDa), whereas the membranes rated PM5-PM30 retained the larger ß-glucosidase protein with approximately equal efficiency (Table 3). In contrast with these results, permeate from the pilot scale ultrafiltration of 468 l of *A. niger* VTT-D-86267 culture filtrate using PCI membranes (ES-625, nominal cut-off 25 kDa) contained less than 5 % of the polygalacturonase activity (results not shown). The concentrate (15.7 l) from the pilot scale ultrafiltration contained 75 % of the original polygalacturonase activity (64000 nkat ml^{-1}) and 90 % (925 nkat ml^{-1}) of the ß-glucosidase (Table 4). Permeate flux during the ultrafiltration was maintained at a high level for the first 7.5 h but decreased rapidly thereafter (Fig. 1).

TABLE 3

Concentration of *A. niger* VTT-D-86267 culture filtrate in the hollow fibre laboratory ultrafiltration unit using membranes with nominal cut-off values between 5 (PM 5) and 100 kDa (PM 100). Abbreviations: CF, culture filtrate; P, permeate; C, concentrate; PG, polygalacturonase; ßGLU, ß-glucosidase.

Membrane	Run time	Fraction	Fraction volume	PG	ßGLU
	min		ml	%	%
-	-	CF	3000	100	100
PM 100	36	P	2800	70	43
		C	190	46	68
PM 30	45	P	2800	18	<1
		C	180	79	95
PM 10	43	P	2820	19	<1
		C	165	77	100
PM 5	63	P	2760	5	<1
		C	220	99	98

Evaporation. Concentration by evaporation caused significant loss in the yields of all three enzymes assayed (Table 4). In the case of polygalacturonase this result was anticipated due to the poor heat stability of this enzyme [5,11]. Taking into account the different final volume concentrations in the evaporation and ultrafiltration concentrates, the former contained an even higher relative amount of solids on the basis of dry weight measurements (Table 4), as would be expected due to the nature of the process.

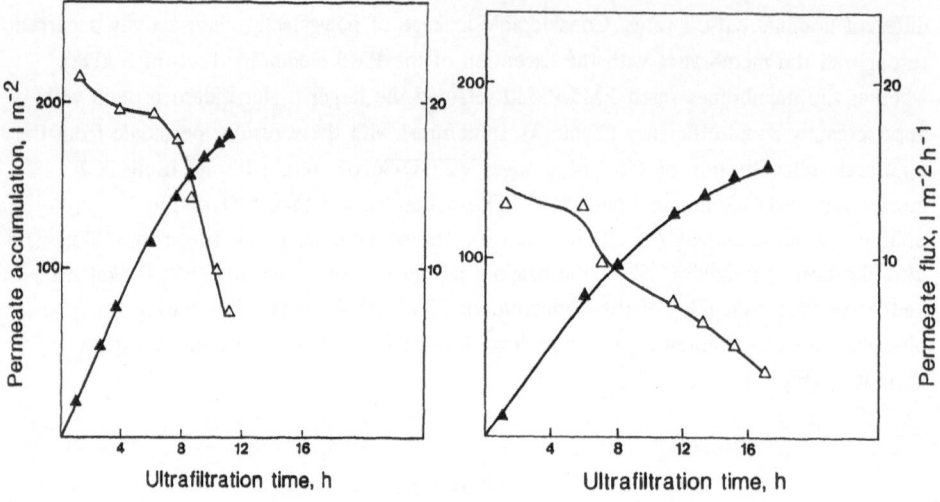

Figure 1. Permeate accumulation and permeate flux during pilot scale ultrafiltration of clarified fermentation broth of *A. niger* VTT-D-86267 (left) and A. niger VTT-D-80144 (right). Values for permeate flux were calculated for each sampling interval separately from the accumulated permeate volume. Symbols:▲ , permeate accumulation;△ , permeate flux.

TABLE 4

Data from the concentration of enzymes in clarified fermentation broth from pilot scale cultivation of *A. niger* VTT-D-86267 on beet pulping waste medium. Abbreviations as for Table 1.

Sample	Initial volume	Volume concen- tration	Dry weight	Activity yields %		
	litres	fold	g l^{-1}	PG	ßGLU	HEC
Broth		1	15	100	100	100
Concentrates:						
-ultrafiltration	468	30	60	75	90	85
-evaporation	190	12	130	55	70	65
-Ti-oxyhydrate	140	25	34	80	1	1
-bentonite	140	25	27	<1	75	20

Concentration using mineral adsorbents. Polygalacturonase of *A. niger* VTT-D-86267 culture filtrate was adsorbed on titanium oxyhydrate (0.8 g l^{-1}) at pH 4.4 as described previously [7,8]. The enzyme-adsorbent suspension (140 l) was separated in the Alfa-Laval separator and resuspended in 0.2 M sodium phosphate buffer, pH 8.0 for desorption of polygalacturonase (final pH of the stirred suspension was 7.4). The desorbed enzyme was separated from the adsorbent by centrifugation and the pH of the concentrated enzyme solution was adjusted to 6.0. The final volume of this concentrate was 5.7 l and the level of polygalacturonase activity was 55000 nkat ml^{-1} (80 % yield). Dry weight in this concentrate was lower than in the ultrafiltration and evaporation concentrates.

The concentrate contained only a very low proportion of the original cellulase (HEC) and β-glucosidase activities of the culture filtrate (Table 4). In keeping with earlier results [7,8], the specific activity of polygalacturonase in the concentrate prepared using titanium oxyhydrate was approximately tenfold compared with the specific polygalacturonase activity of the culture filtrate (results not shown).

ß-Glucosidase activity in the supernatant from separation of polygalacturonase by adsorption on titanium oxyhydrate at pH 4.4 was identical with that in the original culture filtrate, 35 nkat ml^{-1}. This activity was adsorbed on bentonite (8 g l^{-1}) at pH 3.0, separated and desorbed at pH 6.0 using 0.2 M sodium phosphate buffer as described previously [7]. ß-Glucosidase activity in the centrifuged concentrate was 660 nkat ml^{-1} (75 % yield) and the dry weight of the concentrate was low (Table 4).

Concentration: *A. niger* VTT-D-80144

Ultrafiltration. The concentrate (27.2 l) obtained from pilot scale ultrafiltration of 418 litres of *A. niger* VTT-D-80144 culture filtrate contained 1710 nkat ml^{-1} ß-galactosidase activity (93 % yield) and the activity in the permeate was below 0.1 nkat ml^{-1}. Ultrafiltration was also effective in the concentration of ß-glucosidase in this culture filtrate (Table 5). Permeate flux during the ultrafiltration was lower than that during the corresponding concentration of *A.niger* VTT-D-86267 culture filtrate (Fig. 1). This result is in keeping with the higher dry weight of the *A. niger* VTT-D-80144 culture filtrate after drum filtration (Table 2).

Evaporation. As in the case of the concentration of *A. niger* VTT-D-86267 culture filtrate described above, evaporation of the clarified broth of *A. niger* VTT-D-80144 led to considerable loss in the yields of both the activities assayed (Table 5). The dry weight of the evaporation concentrate was higher than that of the ultrafiltration concentrate, although the degree of concentration was lower.

TABLE 5

Data from the concentration of enzymes in clarified fermentation broth from pilot scale cultivation of *A. niger* VTT-D-80144 on wheat bran medium. Abbreviations as for Table 1.

Sample	Initial volume	Volume concen- tration	Dry weight	Activity yields %	
	litres	fold	g l^{-1}	β-GAL	β-GLU
Broth		1	20	100	100
Concentrates:					
-ultrafiltration	418	15	39	93	100
-evaporation	200	12	150	52	57
-bentonite[a]	650	13	48	44	3
-bentonite[b]	650	13	24	71	4

[a] Desorbed concentrate clarified by drum filtration
[b] Desorbed concentrate clarified by centrifugation

Concentration with bentonite. Adsorption of at least 90 % of the ß-galactosidase activity in culture filtrates of *A. niger* VTT-D-80144 required 2 to 3 g l^{-1} of bentonite at pH values between 3.2 and 3.5 and desorption was optimal at pH 6.5-7.5 (results not shown). In a first attempt to concentrate β-galactosidase using bentonite, the pH of the enzyme-adsorbent slurry was increased from 3.3 to 7.0 using sodium phosphate buffer as for the β-glucosidase of *A. niger* VTT-D-86267 described above. This resulted in a poor yield in the concentrate (20 %).

A second pilot cultivation of *A. niger* VTT-D-80144 was carried out to provide more culture filtrate for experimentation. After drum filtration, 650 l of clarified broth was contacted with 3 g l^{-1} bentonite at pH 3.3 and stirred for 2 h. The enzyme-adsorbent suspension was then separated on the drum filter coated with Celite 535 filter aid. The solids fraction (15.6 kg, dry weight 35 %), containing part of the filter aid in addition to the bentonite, was now resuspended in water instead of phosphate buffer and the pH of the mixed suspension (52 l) was increased to 7.0 with sodium hydroxide solution for desorption of ß-galactosidase activity. The major part of this suspension was again separated on the drum filter, using a fresh precoat of filter aid prepared from a suspension in

0.1 M ammonium sulphate, pH 6.5 in an attempt to avoid re-adsorption of β-galactosidase on the precoat material. A one litre sample of the suspension at pH 7.0 was separated in the laboratory centrifuge (3000 x g, 20 min) for comparison. The yield of β-galactosidase in this sample of concentrate was 71 % (890 nkat ml^{-1}), compared with only 44 % (550 nkat ml^{-1}) in the bulk of the concentrate separated on the drum filter (Table 5).

DISCUSSION

Clarification of both fermentation broths by drum filtration was successful and rapidly effected. Due to the high total solids contents of both broths, this was in fact the only practical method for primary separation with the pilot equipment available.

The experiments with the laboratory ultrafiltration unit (Table 3) demonstrated that nominal cut-off values can be used only as an approximate guideline. Although some differences were observed between leakage of ß-glucosidase and polygalacturonase through the PM10 and PM30 membranes (Table 3), ultrafiltration could not be used for efficient separation of these enzymes despite their molecular weight difference in excess of 100 kDa. Leakage of both proteins through the PCI membrane (nominal cut-off 25 kDa) was insignificant. Evaporation was clearly unsatisfactory for concentration of the fungal extracellular enzymes studied in this work.

Several advantages were gained by using adsorbents for the concentration of specific enzyme activities. In the case of the enzyme mixture produced by *A. niger* VTT-D-86267 it was possible, using different adsorbents and pH conditions, to produce two separate enzyme concentrates from a single culture filtrate. The polygalacturonase concentrate produced using titanium oxyhydrate was almost completely devoid of cellulase activity, which could confer advantages for applications aiming at enzymatic treatment of plant biomass without damage to cellulose fibres. The specific activities of the main enzymes in all three concentrates prepared using adsorbents were 5 to 10-fold compared with the corresponding culture filtrates, whereas the specific activities in the evaporation and ultrafiltration concentrates were not significantly increased (results not shown).

Final separation of the concentrated, desorbed β-galactosidase of *A. niger* VTT-D-80144 was unsuccessful on the drum filter (Table 5), due apparently to adsorption on the filter aid despite preliminary suspension of the material in buffer at pH 6.5. Separation of the viscous slurry in the separator could not be contemplated due to the possibility of uneven balancing during solids rejection. Pressure filtration of the slurry without the use of filtering aid was not attempted in this work but should be investigated. A Sharples centrifuge was used earlier for separation of bentonite in the concentration of β-glucosidase [7], but removal of the packed adsorbent was rather laborious.

ACKNOWLEDGEMENTS

This work was supported by TEKES, the Finnish Technology Development Centre. The authors express their gratitude to Tuomo Kuusela for expert assistance in the pilot hall and to Tarja Hakkarainen and Leila Pietikäinen for analyses in the laboratory.

REFERENCES

1. Przybycien, T.M. and Bailey, J.E. Solubility-activity relationships in the inorganic salt-induced precipitation of α-chymotrypsin. Enzyme Microbial Technol., 1989, 11, 264-276.

2. Fisher, R.R. and Glatz, C.E. Polyelectrolyte precipitation of proteins: I. The effects of reactor conditions. Biotechnol. Bioeng., 1988, 32, 777-785.

3. Hustedt, H., Kroner, K.H., Menge, U. and Kula, M-R. Protein recovery using two-phase systems. Trends in Biotechnology, 1985, 3, 139-144.

4. Pungor, E., Afeyan, N.B., Gordon, N.F. and Cooney, C.L. Continuous affinity-recycle extraction: A novel protein separation technique. Bio/Technology, 1987, 5, 604-608.

5. Bailey, M.J. and Pessa, E. Strain and process for production of polygalacturonase. Enzyme Microbial Technol., 1990, 12, 266-271.

6. Nevalainen, K.M.H. Induction, isolation and characterization of *Aspergillus niger* mutant strains producing elevated levels of β-galactosidase. Appl. Environ. Microbiol., 1981, 41, 593-596.

7. Bailey, M.J. and Ojamo, H. Selective concentration of polygalacturonase and ß-D-glucosidase of *Aspergillus niger* culture filtrate using mineral adsorbents. Bioseparation, 1990, 1, in press.

8. Bailey M.J., Ojamo, H. and Evilampi, T. Finnish Patent Application no. 894235, 1989.

9. Harvey, M.S. US Patent 3,620,924, 1971.

10. Lowry, O.H., Roseborough, N.J., Farr, A.L. and Randall, R.J. Protein measurement with the Folin phenol reagent. J. Biol. Chem., 1951, 193, 265-275.

11. Rexova-Benkova, L. On the character of the interaction of polygalacturonase with cross-linked pectic acid. Biochim. Biophys. Acta., 1972, 276, 215-220.

ETHANOL PRODUCTION BY EXTRACTIVE FERMENTATION:

A NOVEL MEMBRANE EXTRACTION TECHNIQUE

S. BANDINI, C. GOSTOLI
Dipartimento di Ingegneria Chimica e di Processo
Università di Bologna
Viale Risorgimento 2, I-40136 Bologna, Italy

ABSTRACT

A new type of membrane extraction for *in situ* removal of ethanol from fermentation broth is presented. Aqueous solutions of Ethylene or Propylene Glycol are used as extractants. The extractant and the broth are separated by a microporous hydrophobic membrane which is not penetrated by the broth or by the extractant. As a consequence a thin gas layer, essentially air, is immobilized within the membrane pores and separates the two liquid phases (gas membrane). Glycols lower the relative volatility of ethanol with respect to water and as a consequence ethanol vapours preferentially diffuse through the stagnant gas layer.

INTRODUCTION

The production of ethanol to be used as alcohol fuel does not differ, at present, from the process already developed for producing alcoholic beverages. The conventional process is affected by low productivity and high energy consumptions.

The volumetric productivity is limited by the low cell density achieved in the fermenter and by product inhibition. A low ethanol level should be maintained in order to have high conversion rates; on the other hand the recovery cost by distillation rapidly increases as the ethanol concentration decreases.

Substantial benefits could be achieved by developing new strains of yeasts or bacteria less sensitive to ethanol inhibition, which can produce more concentrated broths. However strain development probably has its limits; another means of overcoming inhibitory effects is by continuously removing ethanol from the broth by on-line extraction procedures.

Vacuferm [1] and the gas stripping ethanol purifier [2] are two proposed techniques, which offer many potential advantages. However a number of physiological as well as engineering problems arise, preventing practical application.

Extractive fermentation employing solvents immiscible with water to extract the ethanol directly from the fermenter is another promising tecnique [3,4]. The potential improvement of extractive fermentation has been theoretically demonstrated: for a continuous stirred tank fermenter the ethanol productivity can be raised from 8 g/(lh) to 48 g/(lh) by using a solvent with a distribution coefficient of 0.5 [3].

Unfortunately the extractants tested up to now exhibit much lower distribution coefficients, or suffer from other problems. Generally speaking a good extractant should be i) selective for ethanol, ii) non-toxic to microorganisms, iii) immiscible with water, iv) less emulsible in the aqueous phase, v) commercially available at low cost. Several solvents, including dodecanol, dibutylphtalate, tributylphosphate and many others have been tested, but to date, no extractant having the required properties has been found.

Membrane technology offers techniques that could circumvent the difficulties of both vacuum and extractive fermentation.

Pervaporation [5,6] coupled with the fermenter is, in a sense, analogous to fermentation under vacuum, exept that the fermenter operates at normal pressure, while vacuum is confined to the vapour space of the membrane modulus.

A dispersion-free solvent extraction using porous hydrophobic membranes has been recently proposed [7]: the membrane is used to immobilize a liquid-liquid interface between the broth and an organic solvent, based on the fact that the solvent wets the membrane but water does not. The technique is very interesting and promising in view of the fact that any direct contact between living cells and solvent droplets is prevented.

In the present work a new type of membrane extraction is presented in which use has been made of aqueous mixtures of ethylene or propylene glycol to extract the ethanol from the fermentation broth.

The principle of the technique is presented in the following paragraph after a short review on other separation processes based on porous hydrophobic membranes.

GAS MEMBRANE SEPARATIONS

A variety of novel separation processes based on porous hydrophobic membranes have been recently proposed.

The crucial membrane property which the processes rely on is represented by the non-wettability of the membrane material, due to which a non-wetting liquid is prevented from entering the membrane pores so long as the liquid pressure does not exceed the minimum entry value.

There are several commercial membranes with adequate liquid repulsion properties when aqueous solutions are involved. Typical examples are PTFE and Polypropylene membranes, widely used for microfiltration purposes. With the pore diameter of the order of 0.1 μm a penetration pressure of several bars is typically obtained.

A porous hydrophobic membrane can be used for gas-liquid [8] or liquid-liquid [7,9] mass transfer operations. The interface is immobilized at the pore mouth provided that the pressure on the aqueous phase is higher than that of the gas (or organic) phase, but lower than the pressure needed for the aqueous phase to displace the gas (or organic) phase in the pores of the membrane.

In "gas membrane" separation processes [10,11] the hydrophobic membrane is used as a support for a gaseous phase trapped inside the membrane pores, while at both pore entrances a gas-liquid interface is located, separating the outer non-wetting liquids from the internal gaseous phase.

Volatile components are exchanged between the two liquid phases through the stagnant gas layer immobilized within the membrane pores. The concept is similar to the more familiar immobilized liquid membrane, but two important differences should be pointed out. First, gas membranes are intrinsically more stable than liquid ones. Second, a greater mass transfer rate can be expected, due to faster diffusion in gases with respect to diffusion in liquids.

When a temperature difference across the membrane is maintained as the driving force for mass transfer, the process is called Membrane Distillation [12]. The process has attracted wide attention for desalination purposes. Membrane distillation of dilute ethanol-water mixtures has also been investigated [13].

The membrane extraction considered in the present investigation is based on the "gas membrane" idea; it is a very peculiar "extraction", indeed use has been made of water soluble "extractants". The principle of the process is shown in Fig. 1: a porous membrane is in contact with ethanol-water mixtures on both sides; in addition to ethanol and water, a third component G of low volatility is present on the right hand side. The third component alters the vapour-liquid equilibrium properties of the ethanol-water mixture, as a consequence a partial pressure gradient of both water and ethanol vapours holds through the gas membrane entrapped within the pores.

In order to obtain more ethanol in the extract with respect to the feed, the third component should lower the relative volatility of ethanol with respect to water. Non-volatile electrolytes such as salts generally have the opposite effect, i.e. ethanol is salted out.

Ethylene and Propylene Glycols were identified as components having the desired properties; indeed the vapour-liquid equilibrium diagram for the ternary system, Ethanol-Water-Ethylene Glycol [14], shows that the relative volatility of ethanol is lowered with respect to the binary system Ethanol-Water by up to nearly 25 mol % of ethanol (on a glycol free basis). For higher alcohol contents, on the other hand, ethylene glycol enhances the relative volatility of ethanol.

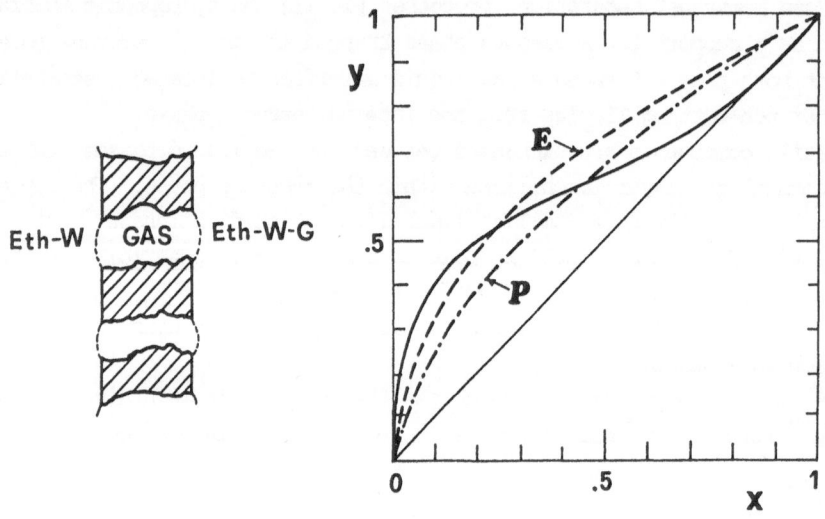

Figure 1. a) The principle of membrane extraction through a gas membrane.
b) The vapour-liquid equilibrium diagram of ethanol-water mixtures with ethylene glycol (----) and propylene glycol (-·-·--) 75% wt. The continuous line represents the vapour-liquid equilibrium without glycols.

Propylene Glycol exhibits a similar behaviour [15]: the relative volatility of ethanol is lowered with respect to the binary system up to nearly 45 mol % of ethanol (on glycol free basis). Furthermore, in the low ethanol concentration range of interest to us, propylene glycol is more effective than ethylene.

In addition to the effect on vapour-liquid equilibrium discussed above, ethylene and propylene glycols have all the required properties for the application, that is: i) high surface tension and contact angle with hydrophobic materials, ii)low volatility, iii)low toxicity, iv) moderate cost. As a consequence, glycols do not enter the membrane pores, the solvent leakage is negligible and the separation of ethanol and water from the extract is quite easy. The same solvent can eventually be used to produce absolute alcohol by extractive distillation .

In passing we observe that membrane extraction can be also performed by using membranes which are not porous and hydrophobic. Crucial points are: i) the membrane should be impermeable to the third component used, ii) the third component should lower the activity of ethanol much more than the activity of water. Under these conditions ethanol is extracted from the feed by the osmotic effect. Of course the use of dense membranes may lead to significantly lower fluxes, while the separation will depend also on membrane selectivity.

EXPERIMENTAL.

The membrane used in our experiments is a porous flat PTFE membrane manifactured by Gelman Instruments Inc. as TF200 with nominal pore size 0.2 μm. It is a composite membrane formed by a porous PTFE layer (the thickness of which was measured as 60 μm), supported by a polypropylene mesh, the nominal overall thickness is 175 μm. By using mercury porosimetry the void fraction $\epsilon = 0.60$ was measured. The membranes are water repulsing, i.e. water does not enter the pores unless the approximate threshold pressure of 2.8 bar gauge is exceeded [12].

The experimental apparatus is schematically shown in Fig. 2. Mass transfer took place across a membrane area of 34.2 cm². A continuous flow of water-ethanol mixtures was maintained through the bottom semicell; the upper semicell was initially filled by ethylene or propylene glycol. Both semicells were stirred. A constant rotation speed of 500 rpm was maintained in the lower compartment, while the upper compartment was agitated by an external impeller at speeds of 450 and 1200 rpm.

Feed and extract initial volumes were approximatively 5,000 and 100 cm³ respectively, the feed composition did not change appreciably during the experiments. Small samples were taken from the extract at various times and analysed by a gaschromatograph with a thermal conductivity detector.

The istantaneous fluxes were measured by the level decrease in the burette; to this purpose the feed reservoir was temporarily excluded by closing the valves V3 and V4 and opening the valve V1.

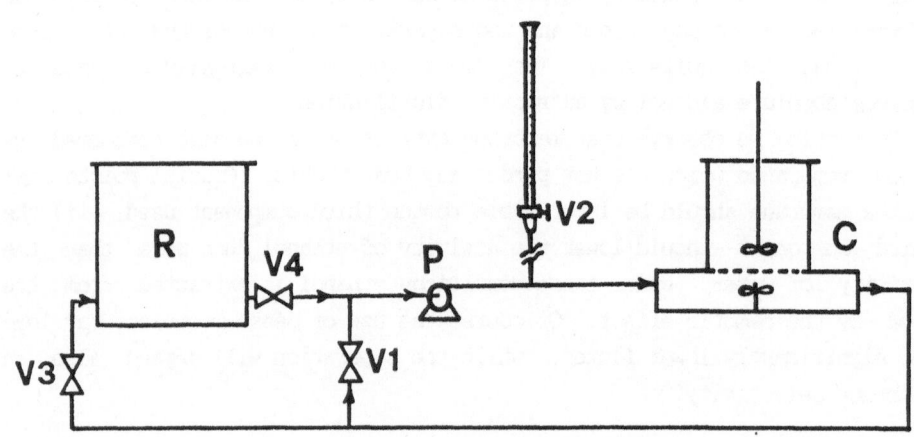

Figure 2. Schematic representation of the membrane extraction apparatus.

Preliminary runs were made using pure water as the feed; Fig. 3 shows the flux versus the glycol concentration in membrane extraction with propylene glycol initially 100% at different agitation speeds.

Apparently the flux is quite dependent on the agitation speed, indeed, due to the high viscosity of the extractant, large concentration polarization can be expected in the extract side, i.e. the glycol at the membrane face may be more dilute than in the bulk phase.

In extraction with ethylene glycol larger fluxes were observed. Of course this different behaviour is due to the greater hygroscopic character of ethylene glycol with respect to propylene.

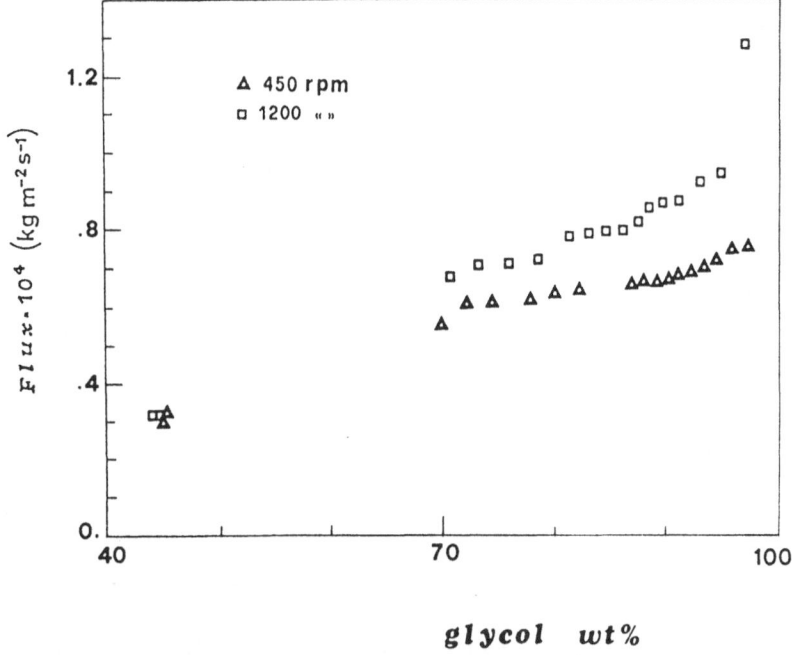

Figure 3. Transmembrane flux vs. glycol composition in the run with propylene glycol as extractant and pure water as feed.

The ethanol content of the extract as a function of glycol composition (both varying in time) is reported in Fig. 4. The extractant was in that case a 90% propylene glycol-water mixture. At the end of the experiment an alcohol content of nearly 7% (on a glycol free basis) was obtained starting from a feed with 4.7 % wt ethanol. This result shows that as expected ethanol is preferentially extracted.

The results obtained with ethylene glycol as extractant are reported in Fig. 5. In this case the extract ethanol composition (on a glycol free basis) is slightly larger than the feed composition; correspondingly the instantaneous separation factor is slightly larger than unity.

The ethylene glycol can thus be used as extractant, but it is much less effective than propylene glycol, which was anticipated considering their respective effects on the vapour-liquid equilibrium.

546

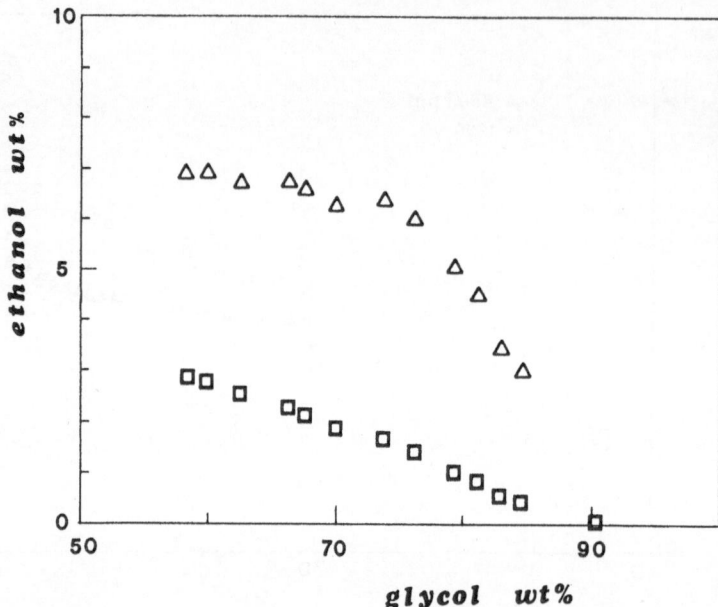

Figure 4. Membrane extraction with propylene glycol initially 90% wt from a 4.7% wt ethanol-water mixture: ethanol content (□) and ethanol content on glycol free basis (△) vs. the extractant composition.

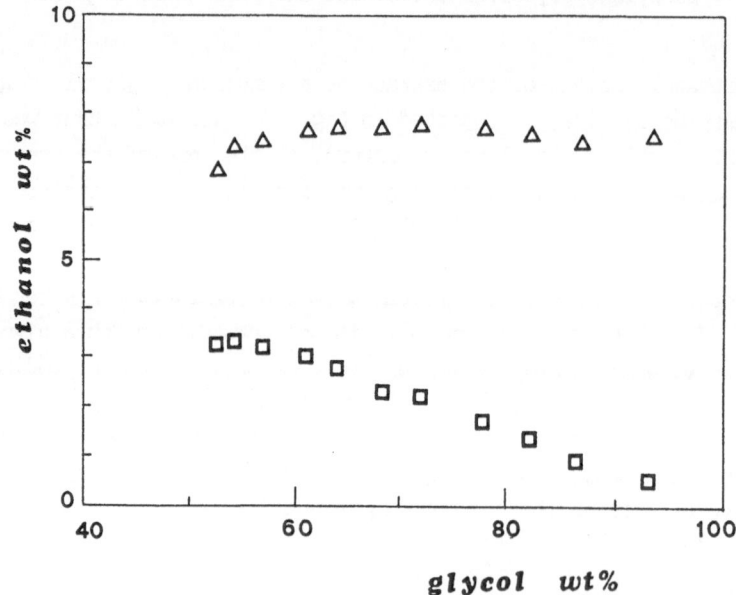

Figure 5. Membrane extraction with ethylene glycol initially 100% wt from a 6.9% wt ethanol-water mixture: ethanol content (□) and ethanol content on glycol free basis (△) vs the extractant composition.

CONCLUSIONS

The technical feasibility of membrane extraction with ethylene and propylene glycols has been established. Further studies are needed in order to establish the actual advantages of the process and to find the optimal conditions for the extraction. However some promising conclusions can be drawn from the few data obtained up to now.

Our membrane extraction results can be compared directly with published data on extractive fermentation. Kollerup and Daugulis [3] identified several biocompatible extractants which can be potentially used in conventional extractive fermentation. Typical values of the ethanol distribution coefficient of these solvents are 0.1-0.3.

The ethanol distribution coefficient is defined as the ratio between the ethanol content of the solvent phase and of the aqueous phase:

$$K = \frac{ethanol\ wt\ \%\ in\ the\ solvent\ phase}{ethanol\ wt\ \%\ in\ the\ aqueous\ phase}$$

Based on data reported in Fig. 4, the "distribution coefficient" for the extraction with propylene glycol is nearly 0.75. This K value is larger than the values observed for all the solvents considered so far for extractive fermentation purpose. Furthermore the technique here presented overcomes the various drawbacks of conventional extractive fermentation pointed out in the introduction.

Acknowledgements

This work was supported by the Commission of European Communities, D.G. XII/F, contract n. EN3B-0078-I. We wish to thank Dr. Gabriele Mei for his help in performing the experiments.

REFERENCES

1. Cysewski G.R., Wilke C.R., Process Design and Economic Studies of Alternative Fermentation Methods for the Production of Ethanol, Biotec. & Bioeng. 20(1978)1421-1444.

2. Mulholland D.L., Development of a low-cost process for ethanol recovery, NRC Final Report , July 1984.

3. Kollerup F., Daugulis A.J., A mathematical Model for Ethanol Production by Extractive Fermentation in a Continuous Stirred Tank Fermenter, Biotechnol. Bioeng.,27(1985)1335-1346.

4. Kollerup F., Daugulis A.P., Ethanol Production by Extractive Fermentation-Solvent Identification and Prototype Development, Can.J.Chem.Eng., 64 (1986) 598-606

5. Gudernatsch W., Kimmerle K., Stroh N., Chmiel H., Recovery and Concentration of High Vapor Pressure Bioproductsby means of controlled Membrane Separation, J.Membrane Sci.,36(1988)331-342.

6. Bandini S., Sarti G.C., Gostoli C., Vacuum Membrane Distillation: Pervaporation Through Porous Hydrophobic Membranes, Proc. 3rd Inter. Conference on Pervaporation in the Chemical Industry, Nancy (F), 19-22 Sept. 1988, Bakish R. Ed., 117-126.

7. Frank G.T.,Sirkar K.K., An Integrated Bioreactor-Separator: In Situ Recovery of Fermentation Products by a Novel Membrane-Based Dispersion Free Solent Extraction Tecnique, Biotechn. Bioeng. Symp. No.17(1986)303-316.

8. Zhang Qi, Cussler E.L., Microporous Hollow Fiber for gas Absorption, J.Membrane Sci., 23(1985)321-345.

9. Kiani A., Bhane R.R., Sirkar K.K., Solvent Extraction With Immobilized Interfaces in a Microporous Hydrophobic Membrane, J.Membrane Sci., 20 (1984) 125-145

10. Imai M., Furusaki S., Miyauchi T., Separation of Volatile Materials by Gas Membrane, I.E.C.Proc.Des.Dev., 21(1982)441.

11. Zhang Qi, Cussler E.L., Hollow Fiber Gas Membranes, AIChE Journal 31-9(1985)1548-1553.

12. Sarti G.C., Gostoli C., Matulli S., Low Energy Cost Desalination Processes Using Hydrophobic Membranes, Desalination, 56 (1985) 277-286

13. Sarti G.C., Gostoli C., Separation of Liquid Mixtures by Membrane Distillation, J.Membrane Sci.41(1989)211-224.

14. Ramanujam M., Laddha G.S., Vapor Liquid Equilibria for Ethanol-Wter-Ethylene Glycol System, Chem.Eng.Sci. 12(1960)65-68.

15. Ramanujam M., Laddha G.S., Relative volatility of Ethyl alcohol in aqueous solution in presence of propylene glycol and glycerol, Transaction I.I.Ch.Eng., XIII(1960-61)23-28.

CONTINUOUS CHROMATOGRAPHIC BIOREACTION-SEPARATION

BARKER P.E., GANETSOS G, AJONGWEN J and AKINTOYE A.
Department of Chemical Engineering and Applied Chemistry
Aston University, Birmingham B4 7ET

ABSTRACT

A Semi-Continuous Counter-Current Chromatographic Reactor-Separator (SCCR-S) has been used for the first time to carry out a biochemical reaction and product separation simultaneously. The continuous inversion of sucrose to glucose and fructose and the biosynthesis of dextran in the presence of the enzyme dextransucrase have been investigated. Sucrose conversions of up to 100% were achieved and product purities of up to 95% were shown to be possible. Substrate inhibition problems were overcome during the sucrose inversion due to the employment of the combined bioreaction-separation principle. Feed concentrations of up to 55% ^{w}/v sucrose were used.

INTRODUCTION

Since the early 1960's small scale chromatographic systems have been used as combined reactor-separators but their applications have been limited mainly in obtaining kinetic data for certain chemical reactions. The combined reaction and product separation principle has obvious advantages in reversible reactions where the removal of one or more of the products increases the overall conversion rates. Similarly the removal of acceptor products in polymerisation reactions gives improved yields of long chain polymers and the removal of poisons or inhibitors from the reaction mixture result in better conversions.

The application of chromatographic reactor-separators in the biochemical field was initiated 5 years ago by Barker and co-workers [1-3]. The biosynthesis of dextran, a polyglucose, from sucrose in the presence of the enzyme dextransucrase was studied initially in batch chromatographic columns with diameters and heights ranging from 0.9 to 5.4 cm and 30 to 200 cm respectively. The columns were packed with calcium charged polystyrene resin and the byproduct fructose was removed from the reaction mixture by complexing with the calcium ions of the resin. It was found that the instantaneous removal of the byproduct fructose, which is known to act as an acceptor from the reaction mixture led to the formation of up to twice the amount of high molecular weight dextran (over 160000) than a conventional fermentation process would produce [1-3]. In addition to the improved yields the combined operation is also expected to reduce the enzyme usage and lead to considerable reductions in the plant operating and capital costs.

The encouraging results obtained using batch chromatographic systems and the research group's long-standing experience in the development and scaling up of chromatographic systems, led us to investigate the possibility of carrying out continuously the combined bioreaction-separation using a Semi-continuous Counter-current Chromatographic Reactor-Separator (SCCR-S). Based on current literature this work is pioneering and the only other application which has some similarities is the work by Hashimoto *et al* [4].

Hashimoto *et al* studied the use of a continuous moving column chromatographic separator for the production of high fructose syrups (over 70% fructose) by combining the adsorption of fructose and isomerisation of the separated glucose fraction on alternatively arranged adsorption and bioreaction columns. Their system however does not fall exactly into the combined chromatographic bioreaction-separation mode of operation since separate columns were used as bioreactors while the separation was carried out inside the rotating bed of columns.

In a truely combined system the actual separation will be almost instantaneous, thereby maximising its benefits and capital cost should be lower. Mechanical limitations are apparent in a moving column system like that of Hashimoto *et al*. It is thus less flexible and more difficult to extend its separation length if required. The SCCR-S chromatographic system used in this work consisted of twelve 5.4 cm diameter by 70 cm long columns, interconnected to form a closed loop. The counter-current movement between the stationary (resin) and the mobile (deionised water) phases was achieved by employing the moving port principle. Each column has six pneumatic valves which control the flow of the various inlet and outlet streams. There are no moving parts as the function of these valves is governed by a pneumatic controller. The principle of operation has been explained in detail in reference 5.

SUCROSE INVERSION

In the sucrose inversion studies the SCCR-S system was packed with a calcium charged KORELA VO7C (FINN-SUGAR) resin having an average particle size of 270 μm. The enzyme invertase was added in the eluent stream (deionised water) at the required strengths. The bioreaction can be represented by:

$$C_{12}H_{22}O_{11} + H_2O \xrightarrow{\text{Invertase}} C_6H_{12}O_6 + C_6H_{12}O_6$$
$$\text{Sucrose} \qquad\qquad\qquad \text{Fructose} \quad\ \text{Glucose}$$

TABLE 1

Sucrose Inversion: - Experimental Conditions

EXPERIMENTAL RUN	FLOW RATES cm³/min			FEED SUCROSE CONC. % w/v	ENZYME ACTIVITY IN THE ELUENT STREAM U/cm³	ENZYME TO SUCROSE RATIO U/g	ENZYME USAGE ACTUAL THEORETICAL	SWITCH TIME mins
	FEED	ELUENT	PURGE					
1	9	31.5	76	20.77	60	1011	0.346	30.5
2	9	31.5	76	33.96	60	618	0.211	30.5
3	9	31.5	76	43.32	122	985	0.337	29.5
4	9	31.5	76	54.88	155	1007	0.338	29.5

KEY:

One unit of invertase activity (U) is the amount of invertase that will invert 1 μ mole of sucrose in 1 minute at 55°C and pH 5.2.

TABLE 2

Sucrose Inversion : - Experimental Results

| EXPERIMENTAL RUN | GLUCOSE RICH PRODUCT | | | | | FRUCTOSE RICH PRODUCT | | | | |
	Glucose Purity %	Product Conc. % w/v	Impurities % Sucrose	Impurities % Fructose	Fructose Purity %	Product Conc. % w/v	Impurities % Sucrose	Impurities % Glucose
1	95.8	2.31	-	4.2	91.4	0.93	-	8.6
2	89	4.2	-	10.1	94.8	1.18	-	5.2
3	95.6	3.54	-	4.4	82.5	2.85	-	17.5
4	\96	4.34	-	4.0	76.4	3.69	-	24.6

The operating temperature was kept at 55°C and the pH at 5.5. The operating conditions and results of four key experiments are shown in Tables 1 and 2. Complete sucrose conversions were obtained even at feed concentrations as high as 55% ^{w}/v sucrose. Fructose purities of up to 94.8% were achieved, however, the separation performance and hence the product purities reduced with concentration. This is expected due to the effects of background concentration [6], however some improvement is still possible and it can be achieved by appropriate adjustments to the switch time [5]. It is worth mentioning that during the inversion of sucrose substrate inhibition is known to occur at sucrose concentrations above 10% w/v. The combined chromatographic bioreaction-separation operation however has shown that such problems can be overcome and this is supported by the fact that complete conversions were obtained even at sucrose feed concentrations as high as 55% ^{w}/v.

DEXTRAN BIOSYNTHESIS

In the dextran biosynthesis studies viscosity problems associated with the formation of high molecular weight dextran and the relatively low operating temperature (25°C), led us to repack the SCCR-S system with calcium charged Purolite PCR563 resin having an average particle size of 450 μm. The column characterisation results carried out at infinite dilution conditions are shown in Table 3.

TABLE 3
Column Characterisation Results - Dextran Biosynthesis
(average bed height per column = 67 cm)

Column	Number of Theoretical Plates		
	w.r.t sucrose	w.r.t glucose	w.r.t fructose
1	27	18	11
2	28	30	11
3	18	21	8
4	29	30	11
5	26	14	9
6	20	31	6
7	16	15	8
8	24	20	8
9	18	17	8
10	29	24	8
11	19	14	9
12	15	13	8
AVERAGE	22.4	20.6	8.8

The number of theoretical plates (NTPs) represent the relative separating power of the packed column. As a result of the larger packing size the corresponding NTP values reduced by over 50% compared to those of the 270 μm resin. This reduction also leads to a reduction in the separation power of the system. However, this was acceptable since chromatographic separation of dextran and fructose is much easier than that of glucose and fructose. Also, the larger resin reduces the high pressure drop problems that were encountered with the smaller particle size resin. A dextran viscosity determination was carried out using a Haake Rotovisco RV12 viscometer fitted with a NV500 head/sensor to quantify how the viscosity increases with concentration and thus to identify the actual operating limitations. The results are summarised in Table 4, and the dextran biosynthesis is represented by;

$$n(C_{12}H_{22}O_{11}) \xrightarrow{\text{dextransucrase}} (C_6H_{10}O_5)n + n\ (C_6H_{12}O_6)$$
$$\text{sucrose} \qquad\qquad\qquad \text{dextran} \qquad \text{fructose}$$

TABLE 4
Viscosities of native dextran produced from
dextransucrase using different sucrose compositions

Sucrose concentration in reaction mixture (% w/v)	Viscosity at 25°C mPa.s
6	1205
12	2579
18	5117

In the biosynthesis reaction the enzyme dextransucrase was also added in the eluent stream at the required strengths. The optimum operating temperature was 25°C and the pH around 5.2. The experimental conditions and results of two runs that were carried out using purified dextransucrase enzyme (i.e. runs 5 and 6) are shown in Tables 5 and 6.

TABLE 5
Experimental conditions for dextran biosynthesis
in the SCCR-S System

Run No	5	6
Feed concentration (%w/v)	3	3
Feed flow rate (cm^3/min)	9	9
Eluent flow rate (cm^3/min)	30	32
Switch time (min)	30	25
Enzyme activity (DSU/cm^3)	14	50
Reaction temp (°C)	25	25
Purge rate (cm^3/min)	76	76

KEY : DSU - Dextransucrase Unit : The amount of enzyme required to convert 1mg of sucrose to dextran and fructose in 1 hr at 25°C and a pH of 5.2.

In Run 5, the activity of the enzyme entering the system in the eluent stream was only 14 DSU/cm^3. This low activity, coupled with dilution in the system, as well as losses due to enzyme denaturation gave only 53% conversion. Sucrose was present and was reacting throughout the system, leading to significant contamination of both the fructose rich product (FRP) and dextran rich product (DRP) streams (Table 6). The enzyme activity in Run 6 was much higher and led to higher sucrose conversion (87.9%) and purer FRP and DRP streams. However, the reduction in the switch time from 30 to 25 minutes led to a shorter time for reaction to take place in each one of the columns.

TABLE 6
Results of dextran biosynthesis in the SCCR-S

Run No.	5	6
Feed throughput (g/hr)	16.2	16.2
Overall sucrose conversion (%)	53.4	87.9
DRP : Dextran purity (%)	6.6	44.8
: % sucrose in feed recovered	43.8	12.1
: Total conc. (% w/v)	0.4	0.3
: Impurities (%) - S	69.5	27.5
G	16.2	17.2
F	7.6	10.3
FRP : Fructose purity (%)	43.8	62.3
: % sucrose in feed recovered	2.8	-
: Total Con (%w/v)	0.08	0.13
: Impurities (%) - D	12.5	37.7
S	31.3	-
G	12.5	-

KEY: D - Dextran, DRP - Dextran rich product, F - Fructose, FRP - Fructose rich product, G - Glucose and S - Sucrose.

Glucose was also found to be present almost throughout the system. Theoretically, glucose is not expected to be present because the reaction products do not include glucose. The presence of the glucose could be caused by the following: firstly, the bacteria which produce dextransucrase are also known to produce an enzyme of invertase activity. The presence of this might be responsible for some of the glucose formed. Also, the acidic conditions in the system (pH ~5.2) could be causing hydrolysis of the sucrose feed. Finally, the calcium ions present on the resin are known to cause sucrose inversion to glucose and fructose. These effects will be investigated further in the near future. These phenomena however were not apparent in the invertase reactions because the inversion reaction products are glucose and fructose anyway. It is hoped that an increase in the dextransucrase activity in the system would reduce these effects.

The above preliminary dextran biosynthesis results are encouraging and the future work will be concentrated in improving the sucrose conversion, minimising any side reactions, reducing enzyme usage and improving product purities and concentrations.

CONCLUSIONS

The above findings show a new very promising application for the SCCR-S type chromatographic systems. The employment of a continuous counter-current chromatographic system as a combined bioreactor-separator has shown for the first-time the possibility and advantages in carrying out simultaneously either the inversion of sucrose and the separation of the two products fructose and glucose or the biosynthesis of dextran with continuous fructose removal.

In the inversion studies complete sucrose inversions at feed concentrations of up to 55% were possible, while product purities of over 90% were readily obtainable with product concentrations and throughputs of up to 55 % w/v and 16 kg sugar solids/m^{3}resin/h respectively.

The advantages in carrying out the reaction and separation simultaneously and the inherent advantages of the continuous operation such as constant product quality, better utilization of the mass transfer area available and relative ease in operation, should be of particular interest to industry where it can be put into a variety of uses.

ACKNOWLEDGEMENT

The authors would like to thank the SERC (Biotechnology directorate) and FISONS Pharmaceuticals for the provision of a cooperative research grant.

REFERENCES

1.   Barker P.E., Zafar I and Alsop R M, "A novel method for the production of Dextran and Fructose", Bioreactors and Biotransformations, G.W. Moody and P.B.Baker (eds), Elsevier (publ), 1987, pp 141-157.

2.   Barker P.E., Zafar I and Alsop R.M., "Production of dextran and fructose in a chromatographic reactor-separator", Separation for Biotechnology, Verral MS and Hudson M J (eds), Ellis Horwood (publ), 1987, pp 127-151.

3.   Ganetsos G and Barker P.E., "The Biosynthesis of macromolecules using Chromatographic Biochemical Reactor-Separators", Bioreactors Downstream Processing, Dechema Biotechnology Conference Vol. 2, ACHEMA, June 1988, pp 91-95.

4.   Hashimoto K, Adachi S, Hirowitsu N and Veda Y, "A new process combining adsorption and enzyme reaction for producing higher-fructose syrup", Biotechnology and Bioengineering, Vol. XXV, 1983, pp. 2371.

5.   Barker P.E. and Ganetsos G, "Production of High Purity Fructose from Barley syrups using semi-continuous chromatography", J. Chem. Techn. Biotechnol., 35B, 1985, pp 217-228.

6.   Ganetsos G., "Prediction of the distribution coefficient (Kd) variation with operating conditions in chromatographic systems", J. of Chromatography, 411, Dec. 18, 1987, pp 81-94.

THERMODYNAMIC ANALYSIS OF SEPARATION PROCESS ALTERNATIVES

D. E. Essien[1] and D. L. Pyle[2]
1.Dept of Chemical Engineering, Imperial College, London SW7 2BY
and BP Research Centre, Sunbury on Thames, UK.
2.Biotechnology Group, Dept. of Food Science & Technology,
University of Reading, Reading RG6 2AP, UK.

ABSTRACT

Analysis of minimum work requirements, based on the concept of "exergy",
provides a formal and convenient thermodynamic basis for comparing process
schemes. The underlying principles of the analysis are summarised. The
method is applied to a comparison of four schemes for the
recovery of fuel-grade ethanol from fermentation broths. The schemes
analysed are conventional distillation, freeze crystallisation, solvent
extraction and reduced pressure fermentation coupled with distillation.
The latter scheme is shown to be superior for the conditions investigated.
The methodology is applicable to a wide range of bioseparations.

INTRODUCTION

In bioseparations dilute products have to be separated from aqueous
solution: thus it is not surprising that energy costs are significant in
the economics of recovery of many fermentation products. Recently, the
search for less energy-intensive production schemes for some large scale
products, such as ethanol, has been an key aim. However, it is difficult
to compare many existing analyses because of their different bases and
assumptions. Moreover, whilst the comparison of existing schemes is useful
and necessary, it is preferable to take a frame of reference which allows
comparison of alternatives with the thermodynamic ideal. This study had
that objective, with fermentation ethanol production in mind.

Currently, fuel ethanol production largely uses atmospheric fermentation
followed by atmospheric distillation to produce the azeotrope. However,
alternative schemes have been proposed and, in some cases, commercialised,
including vacuum fermentation, solvent extraction, coupled fermentation
and membrane separation etc. In this paper we report a thermodynamic
analysis of four downstream processing schemes: conventional distillation,
which serves as a convenient "engineering" reference,coupled reduced
pressure fermentation and distillation, solvent extraction, and freeze
crystallisation. Whilst the work provides some important results and
pointers, we are well aware that several promising separation schemes have
not been included here. In all the cases studied we assume typical broth
compositions and that the desired product is fuel-grade sub-azeotropic

ethanol. The method is widely applicable in analysing bioseparations.

THEORY : (A) - EXERGY

The exergy of a substance may be defined as the shaft work or electrical energy which is necessary to produce in a reversible way the material in a specified state (temperature, pressure, phase etc) from materials in the environment, heat being reversibly exchanged with the ambient during the process [1]. The exergy of a substance includes contributions from nuclear, magnetic, electrical and surface effects, in addition to potential and kinetic energy, and other physical and chemical energies.

In analysing conventional separation processes it is usual to assume that potential, kinetic and other effects are negligible in comparison to the physical and chemical energy contributions to total energy. Thus the molar exergy of a fluid in steady flow may then be expressed as:

$$Ex(T,P) = Ep(T,P) + Ec(T_o,P_o) \qquad (1)$$

where $Ex(T,P)$ is the molar exergy at an absolute temperature T and pressure P; the molar "physical" or 'sensible' (including phase changes) exergy $Ep(T,P)$ is given by:

$$Ep(T,P) = (H - H_o) - T_o (S - S_o) \qquad (2)$$

where H is molar enthalpy, S is molar entropy and subscript "o" signifies ambient conditions; the molar "chemical" exergy $Ec(T_o,P_o)$ is the contribution relating to the reversible, isothermal synthesis of the substance from the elements at ambient conditions.

When the fluid is a chemically inert mixture, it is easily shown that its molar exergy may be expressed as the sum of the contributions from the pure components plus a term due to mixing:

$$Ex(T,P) = \Sigma x_i Ex_i(T,P) + \Delta_m Ex(T,P) \qquad (3)$$

where x_i is the mole fraction of component i, $Ex_i(T,P)$ is the molar exergy of pure species i at (T,P) and $\Delta_m Ex(T,P)$ is the exergy change of mixing when one mole of mixture is formed from its components, each of which is at (T,P). This exergy of mixing is given by:

$$\Delta_m Ex(T,P) = (\Delta_m H - \Delta_m H_o) - T_o(\Delta_m S - \Delta_m S_o) + \Delta_m Ec(T_o,P_o) \qquad (4)$$

where $\Delta_m H$ and $\Delta_m S$ are respectively the molar enthalpy and entropy of mixing at (T,P); the molar chemical exergy change, $\Delta_m Ec(T_o,P_o)$, of mixing is equivalent to the Gibbs function change of mixing at (T_o,P_o):

$$\Delta_m Ec(T_o,P_o) = RT_o \Sigma \ln(f_i/f_i^o) = \Sigma RT_o \ln(x_i \gamma_i) \qquad (5)$$

where f_i is the fugacity of component i in the mixture; f_i^o is the fugacity of pure component i; γ_i is the activity coefficient of component in the mixture.

B : REVERSIBLE WORK

The concept of reversible work underlies much of the present study.
Briefly, it is the minimum possible gross work consumed (or the maximum
produced) by a system when it is changed from one state to another. This
situation is achieved only if the system is changed from the initial state
to the final state by a fully reversible process whilst reversibly
exchanging heat with an infinite thermal reservoir. For example, this
condition might be approached in a distillation column by maintaining
infinitely slow exchange processes with vanishingly small concentration
and thermal driving forces whilst having vanishingly small differences in
the compositions of streams mixing with each other. This might be
approximated in a binary separation with an infinite number of stages by
interchanging heat above and below the feed via a system of heat pumps and
exchangers with vanishingly small temperature differences.

The concepts of reversible work and exergy may be linked: for any fully
reversible process, the minimum work input (or maximum output) W^r is given
by the exergy difference between the specified final and initial states of
the system:

$$W^r = Ex_2 - Ex_1 \qquad (6)$$

where Ex_2 and Ex_1 are the final and initial exergies.

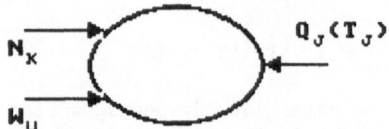

Figure 1.Control Volume

We can apply the concepts defined above to the general, non-reacting,
steady flow process depicted in figure 1 in which the control volume
receives and/or rejects j heat flows Q_j (each at temperature T_j), u power
flows W_u, and k material flows at molar flowrate N_k, molar enthalpy H_k and
molar entropy S_k. An exergy balance (ie eqn 4) applied to the process
yields:

$$\sum_u W_u + \sum_j Q_j (1 - T_o/T_j) - -\sum_k N_k Ex_k \qquad (7)$$

where, by convention, flows into the control volume are positive and flows
out are negative; $Q_j(1 - T_o/T_j)$ is the work potential (ie exergy) of Q_j.

C: THE IDEAL SEPARATION DEVICE

Since we are interested in determining the minimum work required for a
separation, we use the concepts of exergy and reversible work to define
the ideal separation device as:
* the device which receives material streams under ambient conditions,
  performs the required separation reversibly and yields a steady stream
  of products which are also at ambient temperature and pressure.
For such a device, the minimum work required to separate a chemically
inert mixture into its pure components may be obtained from eqns (3 - 7)
as:

$$W^{is} - - RT_o \sum f_i/f_i^o - - RT_o \sum x_i \gamma_i \qquad (8)$$

MINIMUM WORK PREDICTIONS

SEPARATION PROCESS ALTERNATIVES

There are many possible separation schemes (flowsheets) and sets of operating conditions which might form the basis of technically feasible processes for producing ethanol from a fermentation broth. The types of question one would like to answer include: what is the true thermodynamic minimum energy for the separation and how close can the best conceivable separation process approach this ideal? How do these different idealised processes compare with each other: eg, are ideal solvent extractions inherently better or worse than, say, idealised distillation-based processes? How far can real flowsheets approach these ideals? What types of process therefore might be preferred on energetic grounds? How sensitive are these answers to parameters such as the feed and product concentrations? What is the trade-off between thermodynamic and economic efficiency? We do not consider in any detail here the latter question, although we will refer in passing to the type of trade-off encountered. This study summarises some typical results from analysis of a very limited number of examples.

DISTILLATION

Idealised distillation

When the separation of a fluid is carried out by distillation we can postulate an idealised distillation scheme whose thermodynamic performance defines the limit for all practical distillation schemes designed to carry out the specified separation. The ideal distillation scheme will be reversible, and the work required for the separation will be given simply by the difference in steady-flow exergy between the products and feed, as for all reversible operations. The requirements for reversible distillation and the means by which these may theoretically be achieved have been discussed (e.g.[2],[3]). The ideal system does not correspond to the ideal isobaric-isothermal separation device at ambient which is defined above (ie eqn 8), because the driving force for distillation is generated by temperature and/or pressure differences in the system. Instead the minimum work requirements for the idealised distillation device are calculated from eqn (6), using eqns 3-5, noting that the entropy change of mixing for an isothermal process is:

$$\Delta_m S = \Delta_m H/T - R\sum x_i \ln(x_i \gamma_i) \qquad (9)$$

In the calculations reported here thermodynamic properties (eg enthalpies and activity coefficients) were calculated using UNIQUAC.

Distillation Process Alternatives

Apart from the ideal, the flowsheets studied here are:
* Adiabatic distillation: the feed enters the column at conditions as close to reversibility as possible; the reflux and reboil ratios are at their minima, but reversibilities still exist because of mixing under non-equilibrium conditions. This is the best possible simple distillation. Heat is supplied to the reboiler and removed at the condenser so that the work required is the net difference between the work equivalents (exergies) of the heat supplied and removed;

* A reversible flash fermentation (figure 2a) system in which the vapour from the flash vessel is separated in an ideal distillation column;
* Reduced pressure fermentation and separation: from the systems studied we discuss here a reversible vacuum fermentation (Fig 2b) in which an idealised membrane separates the carbon dioxide, which must be compressed up to atmospheric pressure, and an idealised low pressure distillation;
* A sparged CO_2 recycle system followed by ideal distillation: in this scheme ([4]) carbon dioxide is recycled around the process and the ensuing vapour is then fractionated; the process may be operated thermophilically and under vacuum.

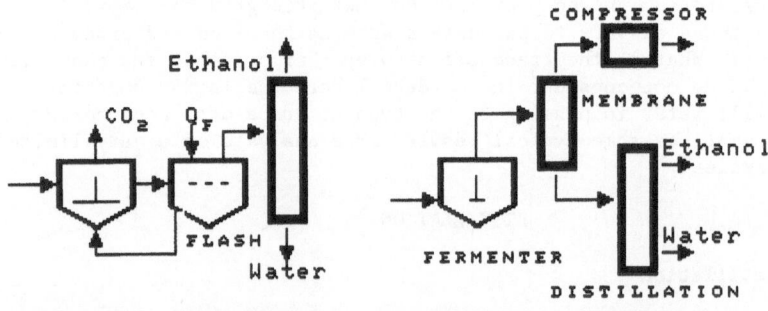

Figure 2a. Reversible Flash

Figure 2b. Reversible Vacuum

Comparison of distillation alternatives

Figures 3a,b show how the minimum ideal work of separation, W^{is}, the minimum work of distillation (which may not be achievable), W^{id}, and the work for an optimal adiabatic distillation, W^{ad}, vary with product concentration for two values of the feed (ie broth) concentration.

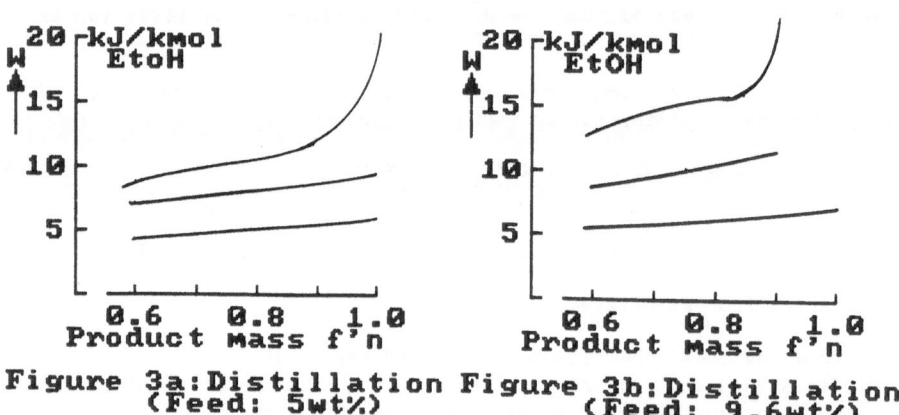

Figure 3a:Distillation (Feed: 5wt%)

Figure 3b:Distillation (Feed: 9.6wt%)

These graphs and other data not given here show that the best distillation is always less efficient than the thermodynamic ideal; both distillation efficiency measures are reduced with dilute feeds (the best region of operation is in the range 3-8 mol% ethanol) and near the azeotrope. The result shows the need to approach reversibility, especially with near-azeotropic products; the ideal distillation is fairly insensitive to feed concentration. If adiabatic distillation is dictated, it will be most appropriate for feed compositions between 2 and 6 mol % and distillates between 80 and 87 mol%. Non-idealities in the vapour liquid equilibrium of this system have a profound effect on performance. Figs 4a,b compare the minimum work requirements for ethanol recovery in sparged, flash, vacuum and conventional (ideal) distillation.

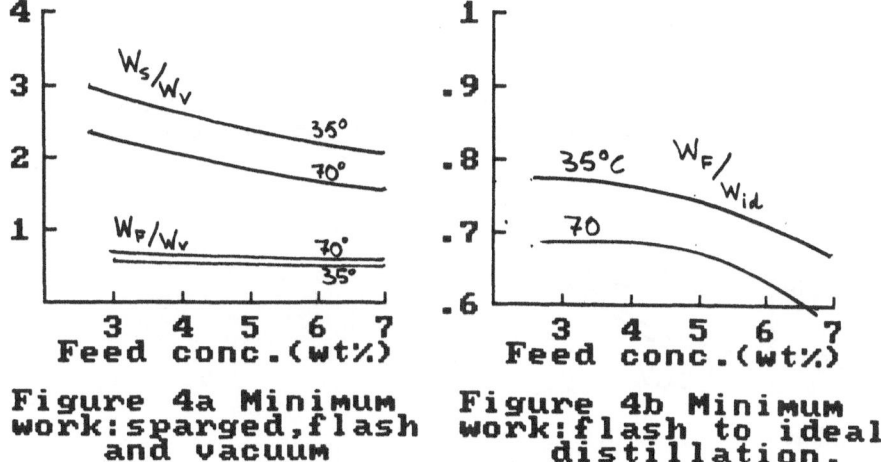

Figure 4a Minimum work:sparged,flash and vacuum

Figure 4b Minimum work:flash to ideal distillation.

Over the range of conditions explored the flash option is more efficient than the conventional option; relative performance improves with higher feed temperatures (eg following thermophilic fermentation) to the flash unit, increased feed composition and reduced product quality. The vacuum option requires approximately twice as much minimum work as the flash device. Sparging is the least attractive option, reflecting the carbon dioxide compression requirements. In absolute terms it is interesting to note that the minimum ideal work of separation is around 0.1 MJ/litre ethanol; the ideal distillation needs around 0.15 MJ/L; current best practice technology is achieved with a work requirement of around 3 MJ/L. The potential energy savings between flash and conventional distillation are equivalent to around 10% of total production costs.

SOLVENT EXTRACTION

There are many possible solvent/co-solvent systems for ethanol separation [5]. An ideal scheme is illustrated in fig 5.

This comprises an extraction column, for which the work requirements are assumed zero, an idealised membrane separator, a decanter and solvent recycle. The solvent flow assumed is the minimum requirement in order to approach reversibility. For these calculations heptanal was solvent; liquid/liquid equilibria and thermodynamic parameters were predicted from UNIFAC [6].

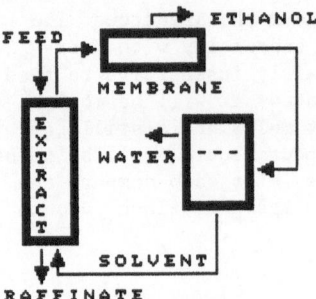

Figure 5: Ideal solvent extraction.

The ratio of the minimum work of separation of ethanol from the solvent to the minimum work of direct recovery of ethanol from the broth is plotted in figure 6.

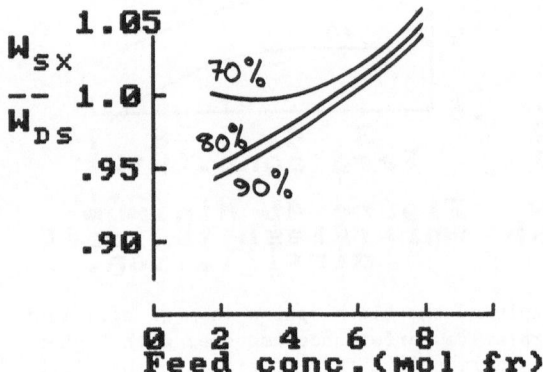

Figure 6: Work of solvent Extraction

The ratio is below 1 for dilute feeds, and depends on product concentration. The thermodynamic advantage is never large. The results are sensitive to solvent choice, and other systems may show greater benefits. A complete design and economic analysis of this system, in which heptanal extraction is followed by solvent recovery by distillation, shows that for dilute feeds (<10 wt%) conventional distillation is to be preferred, mainly due to the costs of solvent make-up. For more concentrated feeds (> 25 mol%) solvent extraction is preferred. The operating costs are most sensitive to solvent cost.

FREEZE CONCENTRATION

The potential advantages of freeze separation processes stem from the low enthalpies of freezing. For reversible operation a freezer would need to operate with equilibrium freezing temperatures throughout; heat removed along the freezers would be by a reversible infinite train of heat pumps. In this analysis the following assumptions were made: all sensible heat heat requirements are met by reversible exchange between products and feed; there are no heat leaks into the system; all mechanical devices operate reversibly; freeze concentration is carried out stagewise and the enthalpy of freezing is extracted at the equilibrium freezing point; ice and concentrated product are completely separated at no energy cost; extracted heat is reversibly pumped to 273K and used to melt recovered ice. Figure 7a shows how minimum work requirements vary with ideal stage number (1,2,3 and infinite stages); Fig 7b shows that there is a region of operation (ie of feed and product concentrations) in which three-stage freezing is more efficient than adiabatic distillation; the process is most attractive for concentrated products (because of the penalty paid by distillation close to the azeotrope). Whilst these results are very encouraging it must be noted that freeze cycles are very complicated; our estimate is that the installed capital costs of an efficient freeze separation system would be around 10 times that for distillation.

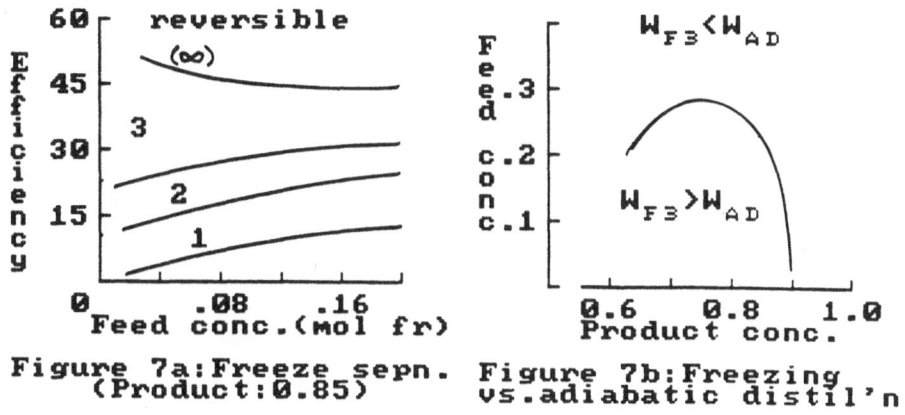

Figure 7a: Freeze sepn. (Product: 0.85)

Figure 7b: Freezing vs. adiabatic distil'n

CONCLUSION

Based on the concept of exergy, we have developed a method of thermodynamic analysis to calculate the absolute/ideal minimum work (or work equivalent) requirements for a given separation. Application of the method to analyse selected flowsheets for ethanol recovery illustrates the potential benefits from certain options. Flash distillation offers advantages; solvent extractions may be favoured on thermodynamic and economic grounds if make-up costs can be reduced; whilst freeze separations are energetically favoured for products close to the azeotrope their costs appear prohibitive.

REFERENCES

[1] Riekert, L. (1980), Energy, 5, 235.
[2] Flower, J.R. and Jackson, R. Trans. Inst. Chem. Engrs. (London), (1964), T249.
[3] King, C.J. Separation Processes, (1971), McGraw Hill, N.Y.

[4] Pyle, D.L., Hartley B.S., Mistri, P.M., Payton, M.A. and Shama, G.,
    Proc Biotech '83 (1983), On-Line Publications, London.
[5] Essien, D.E. and Pyle, D.L. Separations for Biotechnology (M.S.
    Verrall and M.J. Hudson, Eds), (1987), 320 - 332.
[6] Fredenslund, A., Mehling, J. and Rasmussen, P. Vapour Liquid
    Equilibria using UNIFAC. (1977), Elsevier Scientific
    Publns., Amsterdam.

SECONDARY METABOLITES BY EXTRACTIVE FERMENTATION

DIRK HOLLMANN, UTE MERRETTIG-BRUNS, ULRICH MÜLLER
AND ULFERT ONKEN
Universität Dortmund, Lehrstuhl Technische Chemie B
Emil-Figge-Str. 66
D-4600 Dortmund 50

ABSTRACT

Product yield in fermentation processes is often limited by product inhibition or degradation. This may be avoided by extracting the product from the broth during fermentation. Solvents to be used in these extractive fermentations must be selective for the product and non-toxic for the microorganism. For the formation of cycloheximide by Streptomyces griseus and gibberellic acid by Gibberella fujikuroi the procedure of solvent selection is exemplified. It is shown that substituted polyglycols are well suited for the requirements of extractive fermentations of secondary metabolites. Furthermore, process variants for performing an extractive fermentation are presented.

INTRODUCTION

In many fermentation processes product yield is decreased by product inhibition or degradation of the product in the broth. In ethanol fermentation e.g. the concentration of ethanol in the broth is limited, because metabolism of the producing microorganism is inhibited by ethanol. Product inhibition also plays an important role in the fermentation of secondary metabolites [1]. Furthermore, the degradation of high-value products like antibiotics, proteins or hormones during fermen-

tation can influence the yield of the micriobial process negatively. Examples are the hydrolyses of ß-lactam-antibiotics
like penicillin or cephalosporin [2]. The productivity in
these cases can be improved, if the product concentration in
the broth is maintained on a low level. For this a continuous
removal of the product is necessary, which can be realized by
liquid-liquid-extraction. Such a process is called "extractive
fermentation".

Up to now, several extractive fermentation systems have been
described in literature, but they mainly deal with the production of primary metabolites like ethanol or butanol. However,
secondary metabolites are also worth to be extracted from
broth during fermentation, if their biosynthesis or their stability decrease productivity of the process. The fermentation
of the antibiotic cycloheximide by Streptomyces griseus was
chosen as a model system (Figure 1). Even for production
strains the formation of this antibiotic is known to be pro-
duct-inhibited [3].

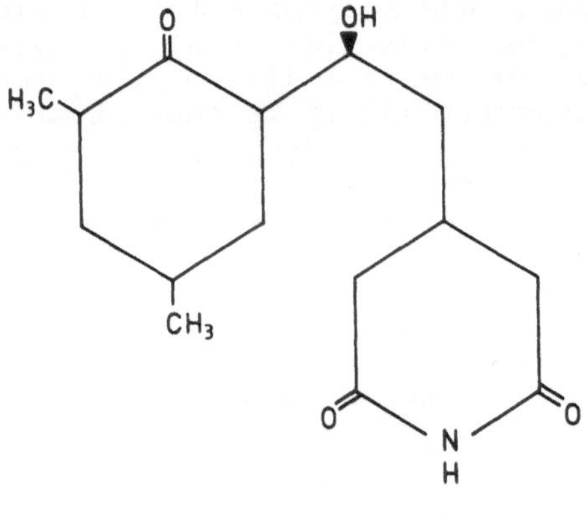

Figure 1. Structural formula of cycloheximide

Additionally, the production of the plant hormone gibberellic
acid by the fungus Gibberella fujikuroi has been investigated
(Figure 2). In this case a considerable amount of the product
is degraded during fermentation.

Gibberellic   acid

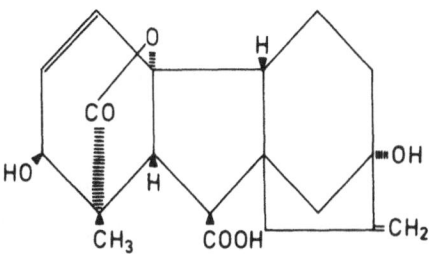

Figure 2. Structural formula of gibberellic acid

Both fermentation systems promise to be good tools for the ap
plication of the extractive fermentation process.

SELECTION OF EXTRACTION SOLVENTS

For extractive fermentation the selection of the appropriate
extraction solvent is of great importance (see Table 1).

TABLE 1
Requirements of extraction solvents for extractive
fermentations

- Product biosynthesis of the microorganism must not be
  influenced negatively;
- high selectivity for the product;
- Equilibrium concentration of the product in the extract
  phase must be considerably higher than in the raffinate
  phase, i.e. the fermentation broth;
  (Distribution coefficient K should be high, with:

$$K = \frac{c_P{}'}{c_P{}''}$$

$c_P{}'$: product concentration in extract phase
$c_P{}''$: product concentration in raffinate phase)

- The extracted product has to be recovered from the extract
  phase without difficulty;

First, organic solvents, which have been proposed for extraction of ethanol [4,5], have been tested for toxicity. This was done by adding the solvent to the medium before it was inoculated. But these solvents (dodecane, octanol, dodecanol, oleylalcohol, dibutylphthalate and tetra-chloromethane) showed toxicity to the bacterium Streptomyces griseus. There was no growth and production, respectively, measurable, when the culture was contacted with these organic solvents.

Polyethylene glycols and derivatives which can form two liquid phases with electrolytes and water-soluble polysaccharides and are applied for the enrichment of proteins from bioprocesses [6] also inhibited production of cycloheximide by Str. griseus completely.

Polyethylene glycols, etherified with an aliphatic alcohol or an alkylphenol, however, show limited miscibility with pure water. Above a certain temperature, which is dependent on the chain length of the hydrophobic aliphatic part and of the hydrophilic part of the molecule, two liquid phases with high water contents occur [7].

From alkanol and alkylphenol polyethylene glycols several compounds with different chain lengths have been tested for extractive fermentation of cycloheximide. Only alkylphenol-substituted polyethylene glycols did not affect growth of Str. griseus. Especially nonylphenol-polyglycolether proved to be well suited as extraction solvent, if the average number of ethyleneoxide-groups was 8.1 (see Figure 3).

NP-8,1:

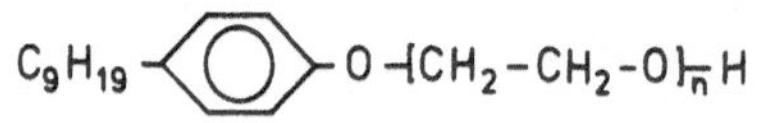

$$C_9H_{19} - \bigcirc - O-[CH_2-CH_2-O]_n H$$

$n \approx 8,1$

Genapol 2822:

$$C_nH_{2n+1}-O-[CH_2-CH_2-O]_4-[CH-CH_2-O]_4-H$$
$$\qquad\qquad\qquad\qquad\qquad\quad CH_3$$

$n \approx 10...12$

Figure 3. Structural formulae of nonylphenol-8.1-polyethylene glycol and Genapol 2822

For the production of gibberellic acid by the fungus Gibberella fujikuroi it was found, that a alkanol-substituted poly-
glycol with four ethyleneoxide and four propyleneoxide groups
(Genapol 2822, Figure 3) did not decrease growth and productivity of the organism.

PROPERTIES OF THE CHOSEN SOLVENTS

A solubility diagram for nonylphenol-8.1-polyethylene glycol
with water is given in Figure 4. It is obvious that at fermentation conditions (29 °C) the "organic" phase contains more
than 70 % (w/w) of water, whereas the second phase exhibits a
solvent concentration of about 0.005 % (w/w).

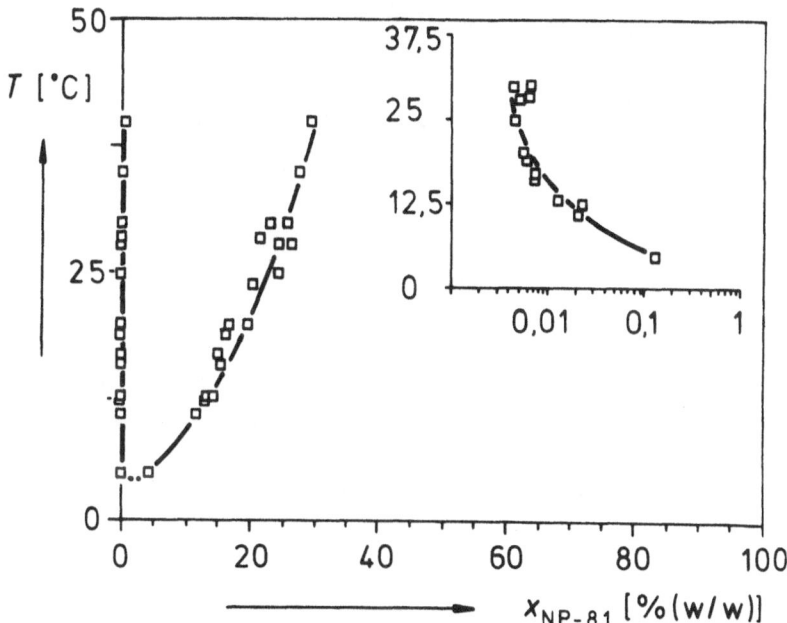

Figure 4. Solubility diagram for nonylphenol-8.1-polyethylene
glycol in water

The distribution coefficient K for cycloheximide in this system was found to be a function of temperature (Figure 5). As
can be seen, K increases with temperature. At fermentation
conditions it amounts to 10.1, which means that the equilibrium concentration of the antibiotic in the extract phase is

then ten-fold of the raffinate phase. The distribution coefficient for glucose is one, therefore only little glucose is extracted from the broth. The reextraction of cycloheximide from
the extract phase with water is possible by reducing the temperature.

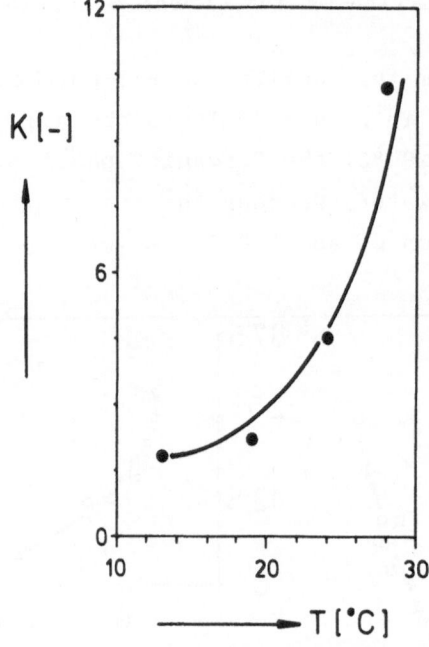

Figure 5. Distribution coefficient K for cycloheximide in the
system nonylphenol-8.1-polyethylene glycol/water as
a function of temperature

Nonylphenol-8.1-polyethylene glycol has also been tested for
several other bacterial and fungal strains. It proved not to
be toxic for the producers of many secondary metabolites like
oxytetracyclin, erythromycin or penicillin.
The solubility diagram for Genapol 2822 in water is comparable
to that of nonylphenol-8.1-polyethylene glycol. The distribution coefficient K for gibberellic acid, however, increases
with decreasing pH-value (see Figure 6).

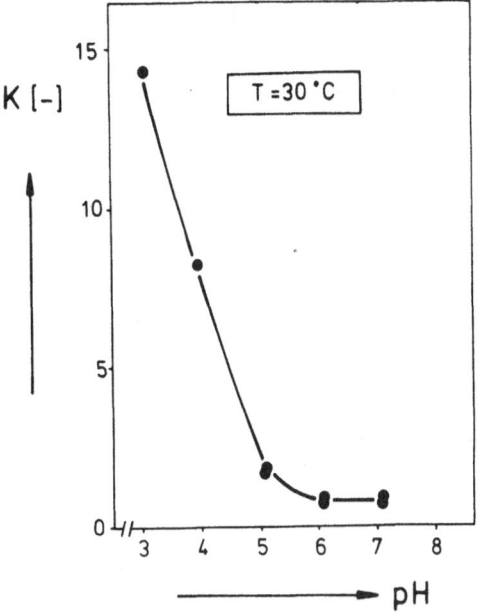

Figure 6. Distribution coefficient K for gibberellic acid in
the system Genapol 2822/water as function of pH
(T = 30 °C)

During fermentation of gibberellic acid pH falls to 3.5. At
fermentation conditions, therefore, a high distribution
coefficient for the hormone exists, whereas the reextraction
for recovery of the product can be done under less acidic
conditions.

PROCESS DESIGN FOR EXTRACTIVE FERMENTATION

For realization of an extractive fermentation several design
variants are possible.
1. Addition of the extraction solvent directly into the fermenter with continuous three-phase-separation: biomass/ extract phase/ raffinate phase (Figure 7.1)

2. Phase separation biomass/ broth and addition of the solvent
to the cell-free broth with two-phase-separation extract
phase/ raffinate phase (Figure 7.2)

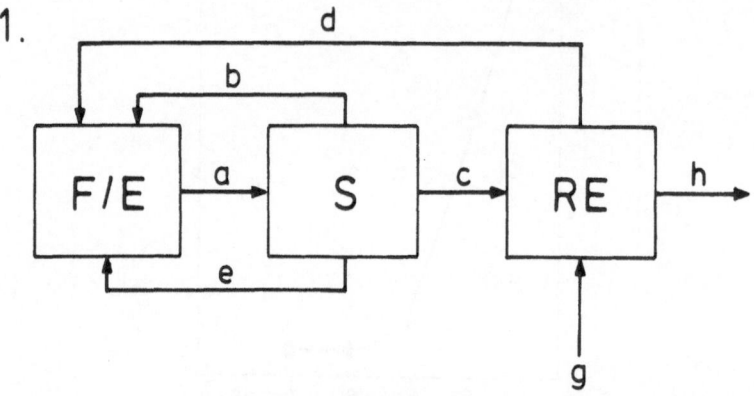

F :Fermentation
S :Separation
E :Extraction
RE:Reextraction

a:broth with/without organic phase
b:cell-free broth
c:extract phase
d:recycle of extraction solvent
e:biomass recycle
f:recycle of raffinate phase
g:reextraction solvent
h:reextraction solvent
   containing the product

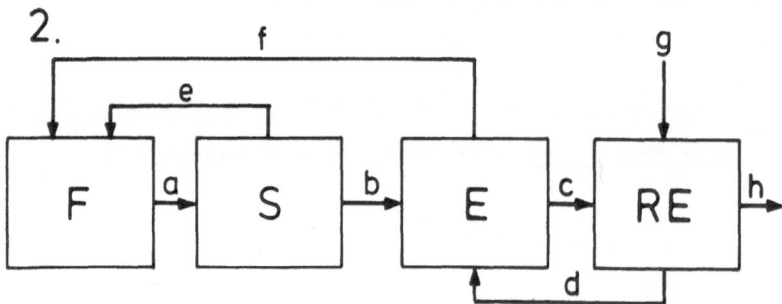

Figure 7. Design variants for extractive fermentation

In both cases biomass and raffinate phase have to be recycled
to the reactor, whereas the extract phase can undergo the
reextraction. For the production of secondary metabolites it

is important that the separation of biomass does not damage the culture, otherwise a reduction of productivity would be the consequence.

The problem of the first variant obviously is the separation of three phases. As there are only small differences in densities of the phases, quite a lot of time will be required for sedimentation. For the aerobic culture, however, this will lead to irreversible damages of the cells by the absence of oxygen in the settler. Centrifugal methods could be the solution, but a three-phase-centrifuge, which must be sterile and does not destroy the cells, has not been developed so far. Therefore the second variant seems to be more efficient, because separation steps are not as complicated. For biomass separation, for example, cross-flow-microfiltration can be applied. The separation of extract and raffinate phase can be done by sedimentation, as the long residence time in the settler does not affect the culture at all.

In model calculations, which were based on the properties of a production strain, it was found that the yield of a cycloheximide process can be doubled by extractive fermentation. However, a production strain of Streptomyces griseus was not available. At the moment the extractive fermentation of gibberellic acid with Genapol 2822 is performed on laboratory scale. The results of these investigations will be published in a different paper.

REFERENCES

1. Demain, A.L., Mutation and the production of Secondary Metabolites, Adv. Appl. Microbiol., 1973, 16, 177-202.

2. Usher, J.J., Lewis, M.A., Hughes, D.W. and Compton, B.J., Development of the Cephalosporin C Fermentation taking into account the Instability of Cephalosporin C, Biotechnol. Lett., 1988, 10, 543-548.

3. Kominek, L.A., Cycloheximide Production by Streptomyces griseus, Antimicrob. Agents Chemother., 1975, 7, 856-863

4. Diaz, M., Three phase extractive fermentation, <u>Trends Biotechnol.</u>, 1988, **6**, 126-130.

5. Roffler, S.R., Blanch, H.W. and Wilke, C.R., In situ recovery of fermentation products- comparison of flash vaccuum extractive and dialysis fermentation and adsorption and ion-exchange resin methods, <u>Trends Biotechnol.</u>, 1984, **2**, 129-136.

6. Albertsson, P.-A., Partition of Cell Particles and Macromolecules, 2nd Ed., John Wiley, New York, 1971.

7. Schönfeldt, N., Grenzflächenaktive Äthylenoxid-Addukte, Wissenschaftliche Verlagsgesellschaft, Stuttgart, 1976.

CLONING AND EXPRESSION AS A TOOL TO EASILY PURIFY A NEW HIGHLY THERMOSTABLE ARCHAEBACTERIAL β-GALACTOSIDASE

MARCO MORACCI, *MARIO DE ROSA, MOSE' ROSSI

Istituto di Biochimica delle Proteine ed Enzimologia, CNR

Dipartimento di Chimica Organica e Biologica e *Istituto di Biochimica

delle Macromolecole, Università di Napoli

ABSTRACT

Since enzymes from extremophiles are thermostable they could, in theory, be purified very easily after cloning and expression of their genes in mesophiles. This paper demonstrates that the gene of a new β-galactosidase isolated from the extreme thermophile Sulfolobus solfataricus can be cloned and expressed in E. coli under the control of the archaebacterial promoter. The recombinant enzyme is similar to the native and, as expected, can be easily purified by heat treatment and isoelectric point precipitation. The partial purification and the properties of the recombinant enzyme are described.

INTRODUCTION

Biotechnological applications of enzymes are often hampered by their low stability to heat, pH, organic solvents and proteolysis. For these reasons attempts are being made to improve the operational stability of current commercial enzymes by protein engineering, protein modification and immobilization, and in general to search for general rules to improve the enzyme's thermostability, or to discover enzymes with new properties (1, 2). In this context attention has been focused on enzymes extracted from some extreme thermophilic bacteria, archaebacteria, growing between 70 and 110°C (3, 4). Archaebacteria belong to a new taxonomic unit recently recognized as a kingdom in addition to eubacteria and eukariotes (3). Presumably thermophilic eubacteria evolved from mesophilic lines by devising different ways to turn mesophilic macromolecules into more thermophilic ones. In addition to this, the extreme thermophilic archaebacteria, according to Kandler "may have invented and further transmitted a unique, most successfull principle to construct thermostable proteins allowing them to grow even in boiling water" (5).

Enzymes isolated from extreme thermophiles are thermophilic and thermostable and exhibit an enhanced stability in the presence of the common protein denaturants such as sodium dodecyl sulfate, urea, guanidine hydrochloride, organic solvents and proteolitic enzymes.

Not only are these enzymes of high biotechnological interest but they are also natural examples of special proteins which can be used, because of their peculiar properties, as model for designing new enzymes.

To overcome the difficulties in obtaining adequate quantities of extremophiles biomass and enzymes it was considered that the expression of archaebacterial genes in mesophilic bacteria could enable the clarification of the archaebacterial gene regulatory mechanisms, and subsequent attainment of good quantities of enzymes with high biotechnological value.

A new β-galactosidase isolated from the extreme thermophile Sulfolobus solfataricus was chosen as a model because of the importance of the enzyme in food industry, where it is used for the production of lactose-fre milk and lactose hydrolysis in whey, and also where the possibility of operating at high temperature prevents bacterial contamination.

The enzyme was purified to homogeneity, the NH_2-terminal aminoacid sequence was determined, the corresponding oligonucleotides were synthesized and the gene from S. solfataricus λ libraries was isolated and sequenced (6,7). The DNA fragment containing the β-galactosidase gene was inserted into a plasmidic vector used for the transformation of a β-gal⁻ E. coli strain. In homogenates of transformed E. coli a thermophilic and thermostable β-galactosidase was found having clear similarities to the native enzyme.

RESULTS

Purification, structure and properties of β-galactosidase: From the crude extract of the extreme thermoacidophilic bacteria S. solfataricus a constitutive β-galactosidase activity was obtained in homogeneous form after a multistep purification with about 30% recovery (6). The molecular mass of the native β-galactosidase estimated either by gel filtration or sucrose gradient centrifugation was found to be 240 ± 8 KDa. SDS-PAGE electrophoresis with the cross-linked enzyme, and aminoacid analysis, gave clear evidence that the enzyme is a tetramer composed of four similar or identical subunits of 60 KDa. The enzyme had an optimal pH at 6.5 and its activity was independent of the presence of monovalent or divalent metal cations. The substrate specificity was quite broad with a tolerance regarding structural variation of the aglycon moiety.

The activity variations with temperature showed a continuous increase up to 95°C, while the thermal stability was greater than that reported for β-galactosidases from mesophilic and thermophilic bacterial sources. In addition the enzyme was very stable in the presence of certain water miscible solvents and detergents. Studies are in progress on their effect on both the activity and stability of the β-galactosidase. This aspect is very important for enzyme regeneration in bioreactors operating on milk or derivatives.

Cloning and expression of the gene in E. coli: Two genomic libraries from S. solfataricus in λgt11 and λEMBL3 vectors were screened using antibodies and a synthetic oligonucleotide mixture as probes which were constructed taking into account the NH_2 terminal sequence of the enzyme. A DNA fragment containing the gene coding for β-galactosidase and its 5' and 3' flanking regions were isolated and completely sequenced. The isolated gene encodes a protein of 489 aminoacids with a predicted molecular mass of 56,650 Da, which is in agreement with the molecular weight and aminoacids composition directly determined from purified protein. In addition the aminoacids preceding the termination codon correspond to the carboxyterminal residues of the purified protein determined by carboxypeptidase digestion.

The expression of the β-galactosidase gene in E. coli was achieved by inserting a DNA fragment of 3.0 Kb containing the ORF and its flanking regions in the plasmid pEMBL18 and transforming a β-gal⁻ E. coli, strain JM109. The expression of the gene was independent from the orientation of the insert in the plasmid suggesting that its transcription was not initiated from a vector promoter but from its own regulatory sequence.

Purification and properties of the recombinant enzyme: 5 grams of transformed E. coli were broken by sonication in 10 mM Tris-HCl pH 7.0, 1 mM EDTA and centrifuged at 8,000 rpm (SS34 Sorvall Rotor) for 20 min. The supernatant represented the crude extract and was assayed for β-galactosidase activity at 40°C. The crude extract was heated in a water bath for 30 min at 75°C cooled and centrifuged at 10,000 rpm for 30 min. The active supernatant was adjusted to pH 3.0 using 0.5 M HCl in an ice bath and stirred continuously for 10 min. The precipitate, collected by centrifugation at 10,000 rpm for 20 min, was dissolved in 10 mM Tris-HCl pH 8.0 and adjusted to this pH by addition of 0.5 M NaOH. Table 1 shows the results obtained with this procedure.

TABLE 1

	ml	Protein mg	U* tot	U*/ mg	U° tot	U°/ mg	Yield %	Purific. Fold
Crude extract	45	305	1,8	0,006	-	-	-	-
75°C supernatant	33	23	2,0	0,09	32	1,4	100	15
pH 3.0 pellet	8	16	1,9	0,12	31	1,95	95	20

U* = units at 40°C; U° = units at 75°C.

The expressed enzyme was purified 20 fold with a recovery of the initial activity of 95%. The molecular weight of the expressed enzyme determined by gel filtration on a Superose 12 column was 240 KDa in accordance with the molecular weight of the native enzyme.

The partial characterization of the β-galactosidase reported in Table 2 indicates that the expressed enzyme has properties similar to those of the native enzyme.

TABLE 2

Comparison of the native and expressed β-galactosidase

	Native enzyme	Expressed enzyme
K_m (β-ONPG)	0,23 mM	0,9 mM
Optimal pH	6,5	6,5
Optimal activity temperature	> 100°C	> 105°C
Isoelectric point	4.5	4.5
Residual activity after 2 h at 75°C	100%	100%
Molecular weight	240 KDa	240 KDa

DISCUSSION

The cloning and expression in mesophilic hosts of genes isolated from mesophilic microorganisms is an useful approach in the production of biocatalists for industrial use. With this strategy it is possible to increase the cell's content of the enzyme and to obtain its secretion into the growth medium. The technique also permits the selection of appropriate hosts, well characterized from microbiological and genetic point of view, in line with safety rules and requirements for industrial fermentation processes.

The cloning and expression in mesophilic hosts of genes coding for enzymes of thermophilic organisms offers additional advantages, particularly in relation to the down stream processing of enzyme purification. Generally the thermophilic proteins are inactive at the host optimal growth temperature, do not interfere with its metabolism and, consequently, the cells may permit the accumulation of higher quantities of the expression product. Moreover, the intrinsic thermostability of the expressed proteins make possible their selective purification from host mesophilic proteins by using a thermal denaturation step, which can be easily scaled up in an industrial down stream process.

The degree of enzyme purity needed for industrial applications of thermophilic enzymes expressed in mesophilic hosts is lower than that required for the native enzymes of mesophilic or thermophilic organisms, since mesophilic host enzyme contaminants are generally inactive at high temperature and do not interfere with the process catalyzed by the expressed thermophilic enzyme. In contrast, contaminants of a native enzyme isolated either from mesophilic or thermophilic sources are likely to be active in the same temperature range as the desired biocatalist and thus could produce indesired activities.

The enzyme that we have used as a model in our studies is an archaebacterial β-galactosidase which has a thermostability and a thermophilicity greater than that reported for this type of enzyme from other sources. The resistance to solvents detergents, proteolitic enzymes and the absence of product inhibition are other properties of this β-galactosidase that make it of great interest for milk and whey lactose hydrolysis.

The cloning and expression of the gene coding for this enzyme in E. coli, provides information on some aspects of the structure and regulatory mechanisms of archaebacterial genes and cloning strategies for such genes in eubacterial hosts. The system also offers specific advantages in the down stream processing of enzyme purification for potential industrial utilization.

The comparison of the purification scheme of the β-galactosidase from the

S. solfataricus homogenate with that of recombinant E. coli strain, confirms the advantages of the cloning strategy. A 20 fold purification of the enzyme with about 100% yeld was obtained from the recombinant E. coli strain by an heating step performed on a crude cell extract followed by an isoelectric precipitation. In contrast only a 10 fold purification with about 45% of recovery of the initial enzyme activity was obtained from S. solfataricus homogenate using a DEAE column and a pH precipitation of protein contaminants.

Work is in progress to purify the recombinant enzyme to homogeneity, to optimize its expression in E. coli and to clone it into yeast, with vectors having an high efficiency promoter and a secretion signal sequence.

It is obvious that the success of this methodology depends on the quantity of enzyme produced by the mesophilic host. Provided E. coli and/or yeast are capable of expressing foreign genes under the control of highly efficient promoters it will be possible to obtain highly purified thermophilic enzymes using these relatively simple procedures.

REFERENCES

1. Mozhaev, V.V., Berezin, I.V. and Martinek, K., Structure-stability relationship in proteins: fundamental tasks and strategy for the development of stabilized enzymes for biotechnology. CRC Critic Rev. Biochem., 1988, **23**, 235-281.

2. Fontana, A., Structure and stability of thermophilic enzymes. Biophys. Chem., 1988, **29**, 181-185.

3. Brock, T.D. and Zeikus, J.G., Thermophiles: general molecular and applied microbiology. Wiley & Sons, New York, 1986.

4. Rossi, M., Enzymes from extreme thermophilic bacteria as models of special catalists. European Conference on Biotechnology, Scientific Technical and Industrial Challanges, EIT and ZAI Verona Agricultural Italy, Nov. 7-8, 1988, 46-50.

5. Kandler, O., Archaebacteria: biotechnological implications. Proc. 3rd Eur. Congr. on Biotechnology, München 1984, Vol. IV, 551-560.

6. Pisani, F.M., Rella, R., Raia, C.A., Rozzo, C., Nucci, R., Gambacorta, A., De Rosa, M. and Rossi, M., Thermostable β-galactosidase from the archaebacterium S. solfataricus. Eur. J. Biochem., 1990, **187**, 321-328.

7. Cubellis, M.V., Rozzo, C., Montecucchi, P. and Rossi, M., Isolation sequencing and cloning in E. coli of a new β-galactosidase archaebacterial gene. Gene, 1990 (in press).

Part 6

Downstream Systems Design and Control

DOWNSTREAM PROCESS DESIGN AND BIOPRODUCT QUALITY CONTROL.

Henry Y. Wang.
Department of Chemical Engineering,
The University of Michigan,
Ann Arbor, Michigan 48109-2136.

Abstract.

Design and quality control of the bioseparation processes become ever more important as new and more complex bioproducts start to appear in the commercial market. New technologies and thinking must be developed concurrently so that issues such as quality, safety, and economics of these new bioproducts and bioprocesses can be addressed adequately from the very beginning. Some thoughts relevant to these topics are discussed here.

Introduction.

Recent advances in molecular biology and chemistry allow the scientist to use various cells created artificially through cell fusion and genetic engineering techniques to produce large amounts of biomolecules, particularly the protein based therapeutic drugs. A list of current and potential therapeutic protein drugs is shown in Table 1. Downstream processing becomes a much more important part of the process development because of the unique characteristics of these macromolecules and the stringent regulatory requirements for their therapeutic use. Translation of the traditional unit operations of isolating and purifying small molecular weight bioproducts such as antibiotics used in the traditional fermentation industries to these new bioprocesses is quite limited. Large throughput and efficiency are less of a concern than the quality and validation of the process and the product.

Table 1: A list of current and potential therapeutic protein
drugs in the market.

Name	Host	Status	Company	Comment
r-Insulin	E.coli	on market	Eli Lilly (1983)	
Human growth hormone (HGH)	E.coli	on market	Genentech (1985)	-100 million dollars (1988)
alpha-interferon	E.coli	on market	Schering-Plough Hoffman-LaRoche (1986)	-100 million dollars (1988)
OKT3Mab	Hybridoma	on market	Ortho Pharmaceutical (1986)	- 60 million dollars (1988)
HBsAg	Yeast	on market	Merck (1986)	
TPA	Mammalian Cells (CHO)	on market	Genentech (1987)	-200 million dollars (1988)
Erythropoetin	Mammalian Cells (CHO)	on market	Amgen (1989)	>150 million dollars (1989)
Interleukin-2	E.coli	NDA	Cetus (-1990)	
beta-interferon	Mammalian Cells	Clinical III	Cetus	
gamma-interferon	E.coli	Clinical III	Biogen (-1991)	
Factor VIII	Mammalian Cells (CHO)	Clinical III	Genetic Institute (-1991)	
Epidermal growth factor	E.coli	Clinical III	(-1991)	
G-CSF	E.coli	Clinical III	Amgen (-1991)	

Figure 1: Comparison of academic and industrial research.

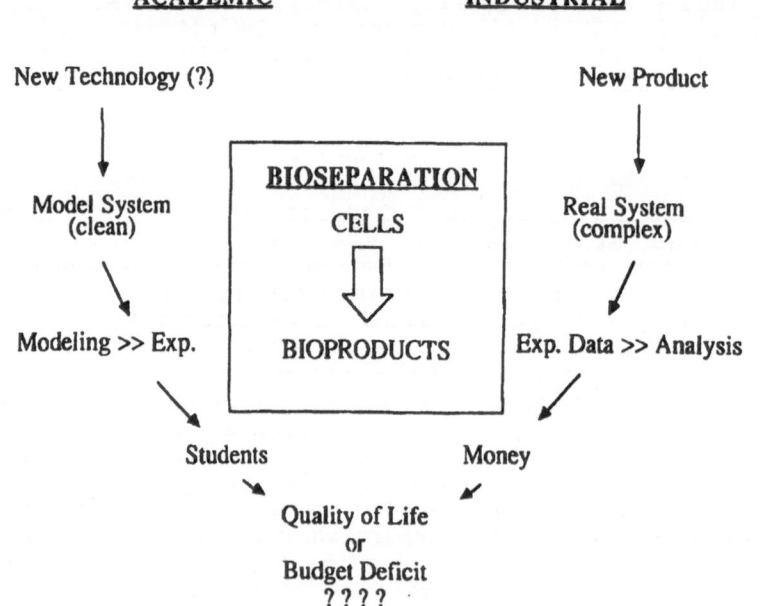

Since bioseparation processes are always product specific and to a lesser extent origin specific, we also need to know the physico-chemical characteristics of the materials we are dealing with. Most academic scientists do not wish to deal with multi-component systems because of the difficulties in modelling and analysis, not to mention those with unknown quantities. On the other hand, industrial scientists seldom have the time or means to understand the fundamentals of the technologies used in their daily routine. Changes are usually resisted because of lack of fundamental understanding of the used processes. We need to reconcile these different approaches undertaken by academic scientists and industrial practitioners and learn from each other (Figure 1).

On top of the problems among academic and industrial scientists there are different views on bioseparation research by various basic scientists versus applied technologists. There is no major disagreement that any applied problems can be reduced down to basic and fundamental principles. On the other hand, any basic research must have some long term impact in terms of its benefit to society and in this case, to the processing industries. The question is how to formulate the basic research agenda so that new fundamental information can be generated and be relevant to the application at the time. Given the diversity of biological systems, translation of information from one system to another is often difficult. Any generalisation of a biological phenomenon can always have some exceptions. Computers may eventually help us in this respect to expand beyond our meagre mental capacity. Various computer databases must be generated and used to digest the enormous amount of information in biological systems. The human genome project is such an area. New knowledge in protein structure and function will eventually generate another massive sum of information that is beyond the mental ability of the common man.

Downstream Processing System Design.

There are some fundamental differences between these new biological products and materials that need to be reckoned with when we try to develop a suitable bioseparation scheme. The physico-chemical properties of many protein based bioproducts are intrinsically different from the traditional small molecular weight products (less than 1,000 m.w.). What is good for one may not be good for the other during bioprocessing. New bioseparation techniques such as chemical solubilisation of cells and proteins, membrane processing, partitioning of biomolecules in aqueous two-phase systems and biospecific chromatographies are being developed and evaluated in addition to the traditional unit operations. We also need to learn how to integrate these new and powerful technologies in addition to the old ones for the design of new bioseparation schemes to meet the regulatory and technical needs for those anticipated new bioproducts (Table 1).

Most bioseparation schemes can roughly be divided into two general areas: isolation and purification. The initial isolation scheme should emphasise volume reduction and partial purification. The second part of the bioseparation scheme emphasises more on purification of the bioproduct and trace removal of the contaminants. Is the product extracellular or intracellular? This also influences the amount and nature of the background contaminants that the entire bioseparation scheme needs to focus on. Multi-component mixtures reduce our ability to quantify these procedures. Technologists prefer to emphasise the physical functionality of the unit operations which include solid-liquid separation, precipitation and membrane processing. Actually, the biochemistry and molecular biology of the cells and proteins should also be examined before emphasising only the physical aspects of the processes. This is particularly true if different options are available in using molecular genetics to produce a particular bioproduct.

The choice of an ideal host will have an impact on the downstream processing steps. The issue of where to express intracellular protein products also dictates the sequence and approach of the downstream processing. For example, recombinant human growth hormone (r-HGH) and other animal growth hormones such as recombinant bovine somatotrophin (r-BST) require very different approaches in their isolation and purification procedures. r-HGH is expressed in an active form within the periplasmic space of *E.coli* cells. The active proteins are released through osmotic shock and processed accordingly afterwards. r-BST is over-expressed as inclusion bodies in the same host. Over expression of eukaryotic proteins inside prokaryotic hosts frequently results in accumulation of these protein products in the form of insoluble aggregates or inclusion bodies. These aggregates can be be quantitatively recovered in the pellet fraction after cell lysis. Then, various means have been used to refold or renature the proteins into their native and active forms. Typically refolding processes include chemical solubilisation, dilution and filtration. Many of these approaches are very much product specific and fundamental understanding of these protein molecules is limited. Many of the cytokines and growth factors are produced in such a manner. Part of the problem has been the inability to evaluate their biological activities using established laboratory assays and means to separate the active forms from the inactive ones.

The purification steps are more focused. Usually, the system can be reduced down to two component studies if a key micro-contaminant is known. Significant reduction in volume allows the use of smaller scale chromatographic and crystallisation techniques. Less engineering and more chemistry seem to play a significant role here.

Affinity chromatography is a very powerful technique and has been mostly used in the purification steps. The concern of ligand leakage and lower production cost of the affinity ligands have induced most bioseparation scientists to suggest the use of bio-affinity separation during earlier isolation steps. Other chromatographic steps are usually labelled as size exclusion, hydrophobic interaction, or ion-exchange based on the chromatographic medium used. Even though this may be the dominant mechanism of the chromatographic medium, most of these media tend to have some degree of the other types of molecular interactions that may contribute to the success and failure of the chromatographic operations.

Membrane processes are increasingly being used in various bioseparation processes. Yet, misunderstanding of the fundamental knowledge of membrane materials and chemistry, including the ever present fouling phenomenon, in combination with a lack of simple and constructive guidelines for operating these devices, hinders their acceptance in many bioprocessing applications.

There are many different alternatives to achieve the isolation and purification objectives of a particular bioproduct. Since the benefit of one route over another tends to be quite small, whoever develops the original route tends to have a certain advantage. Introduction of any new technology is only favoured during the early stage of product development. Although there are many references in the literature on one step isolation and purification methodology, it is clear in practice that one step methods of isolation and purification usually cannot be used to achieve the necessary objectives. Rational approaches are needed to identify and to integrate each isolation and purification step into the overall downstream processing scheme for a particular bioproduct.

Bioseparation Process Validation and Control.

Most protein products produced using animal tissues or cell culture sources need to address a variety of issues to assure the safety of such highly complex protein molecules. The stability of the recombinant hosts, control, and reproducibility of the cell culture conditions are the most important of all. The isolation and purification processes must be validated to guarantee the purity, safety and stability of the final products. An important requirement of process validation is to demonstrate, to a high degree of assurance, that the process is not only capable of producing a pure, consistent and safe product, but also the ability to remove various trace contaminants such as endotoxins, cellular DNA and retroviruses. Inclusion of immuno-affinity chromatography in the processing train must also demonstrate minimal leakage of the affinity ligands into the product stream after repeated use.

Endotoxins have potential adverse biological effects in humans and in many animal species. The presence of endotoxins in any protein product poses a significant risk in therapeutic use. Many chromatographic media have a tendency to retain endotoxins and should be evaluated accordingly. Depyrogenation should be integrated into the separation process if this poses a significant problem. Various new micro-separation devices have been advocated for use in depyrogenation.

There will be a small amount of cellular DNA produced from normal cells death and lysis. During the purification process, steps must be included to eliminate such trace contaminants if they are present at detectable levels. To ensure the complete removal of cellular DNA, both process validation and final product testing are applied to give a high degree of assurance of batch reproducibility and final product purity. Apart from relying on the purification procedures already in use, some investigators advocate the use of additional enzymatic and physical treatments to eliminate any contaminating cellular DNA. One commercial device (Threshold Bioanalytical System) has been developed by Molecular Devices (Menlo Park, California) that can quantify total DNA with a detection limit of 2 pg within several hours. Devices with similar sensitivity and for other contaminants should be developed.

Viruses also represent serious potential contaminants in parenteral drugs derived from tissue or cell culture sources. Currently several chemical and physical methods exist to inactivate viruses. None of these methods is generic to all viruses and some reduce labile protein products. Again, bioprocess scientists and engineers would like to have additional arsenals to physically remove and inactivate viral contaminants during the final steps of the bioprocess. Membrane processing and affinity adsorption should be investigated for this purpose.

Most pharmaceutical bioproducts are highly priced and of low volume. For example, the total amount of Factor VIII used in the world is less than 0.5 Kg. The quality of the end product will make or break a company over its competitors. Quality control and analysis will dominate the bioseparation part of the process. Therefore, in the long run, process monitoring during various bioseparation steps will increasingly be an important component of process validation and control. We ought to develop the necessary monitoring and analytical tools to fulfil these needs.

Conclusions.

If there is any danger looming in our field, it is the number of bioproducts that can actually be economically produced. On the other hand, we need the necessary technologies to generate these new bioproducts and to produce them economically. Without a dynamic and profitable biotechnology industry, the field will relegate itself into the existing pharmaceutical industries. Our need to function as a single and separate group will be severely tested.

BIOSAFETY IN DOWNSTREAM PROCESSING

ALLAN M. BENNETT, JOHN E. BENBOUGH & PETER HAMBLETON.
Division Of Biologics, PHLS CAMR,
Porton Down, Salisbury, Wiltshire.

ABSTRACT

Separation processes in biotechnology such as centrifugation and homogenisation may pose a serious health risk to personnel due to their potential to produce aerosols of potentially allergenic biological material. Bioprocessing equipment should be monitored for released biological aerosols to ensure its safe operation. Manufacturers of downstream processing equipment should consider the need to design equipment so as to minimise biological aerosol release during operation. Methods for testing the biosafety performance of downstream processing equipment are described.

INTRODUCTION

Generally, industrial biotechnology has an excellent health and safety record with no recorded fatalities resulting from the use of microorganisms or their products in exposed workers. However in recent years an increasing number of incidents involving respiratory problems in workers exposed to microorganisms and their products have been reported (1). The majority of these incidents have occurred during downstream processes such as centrifugation and homogenisation or during finishing operations (1). Both centrifuges and homogenisers exert forces on fluids which can result in the formation of small droplet aerosols of diameters that allow their access to all parts of the human respiratory tract. This paper describes a range of incidents involving health problems related to downstream processing equipment, reviews the data available concerning aerosol production by downstream processing equipment, reports on an ongoing project monitoring release from in-place homogenisers and centrifuges and discusses integrity testing of centrifuges and cell separators.

HEALTH PROBLEMS ASSOCIATED WITH BIOSEPARATION EQUIPMENT

Bioseparation processes can expose personnel to a wide range of potentially hazardous chemicals and biochemicals. These include solvents used in extraction and chromatography, and potentially allergenic packing materials used for filtration and ion exchange. However, this review is concerned with the exposure of personnel to biological materials since this can constitute a significant health hazard. For example, allergic respiratory symptoms, such as asthma, have been caused by exposure to viable microorganisms (2), non-viable microorganisms (3) or microbial products (4). Also, non-allergic symptoms caused by Gram negative lipopolysaccharide have occurred (5-8).

It has been found from experience with both laboratory and industrial accidents involving microorganism or their products that the major route of transmission into the human body of biological material released into the air is by inhalation (1,9). Both skin contact and ingestion may occur but these can normally be prevented by good laboratory practice and any incident of ill health caused by these routes is normally the result of gross carelessness. Inhalation, however, is difficult to prevent as the presence of microorganism or product in the atmosphere may not be noticeable to the person exposed.

When dealing with large quantities of "harmless" microorganisms or "biologically inactive" products it is important to realise that biological material of molecular weight greater than a few hundred can produce severe allergic asthma if inhaled regularly at high enough doses. For example, a survey of allergic symptoms in farmers has suggested that exposure to airborne actinomycetes or fungi at levels of between 10^6 and 10^8 organisms per cubic meter was the cause of their respiratory symptoms (10).

Illness caused by centrifugation of biological material

Many incidents of infection or allergic asthma in both laboratories and biotechnology plants have been shown to have resulted from centrifugation of biological materials. In 1939 at the Michigan State University, the improper use of an enclosed tubular bowl centrifuge generated an aerosol containing Brucella abortus causing an outbreak of brucellosis in the college in which 27% of the occupants were infected and one person died (11). In a French pharmaceutical factory severe allergic effects in four workers were caused by the generation of a tuberculin aerosol by a centrifuge (3). In University College, London an incident in which 5 laboratory workers suffered stomach and kidney pains was linked to the tubular bowl centrifugation of Pseudomonas aeroginosa cell debris (7).

Illness caused by sonication of biological material
Sonication caused a laboratory-acquired infection in a pharmaceutical worker who was testing Pseudomonas strains for enzyme activities. Unfortunately one of the strains was an improperly identified Pseudomonas pseudomallei and caused a form of melidiosis in the exposed worker (12).

Illness caused by drying and packing of biological material
Spray drying and packaging of single cell protein (SCP) has caused influenza-like symptoms, dermatitis and conjunctivitis in exposed workers in Britain, Sweden and the U.S.S.R. (5,6,13). In the Swedish incident it was found that symptoms could be prevented by granulating the protein product to particle sizes of greater than 20 microns, which were unable to enter the respiratory tract and give rise to symptoms (14).

BIOLOGICAL AEROSOL PRODUCTION BY BIOPROCESSING EQUIPMENT

The formation of microbial aerosols by downstream processing equipment has been studied in a variety of different scales of operation from laboratory scale to production scale in university, governmental research and industrial facilities. At PHLS CAMR a study is being carried out to assess the production of aerosols containing microorganisms, endotoxin and products during all stages of a variety of bioprocesses (15). The objective of this study is to assess the degree of hazard resulting from different bioprocessing steps and to identify means of reducing the hazard. The use of air sampling equipment allows the amount of product or organism inhaled by operators exposed to them to be calculated . Details of air sampling data obtained by this study or published in the literature are given below for various separation devices.

Laboratory bucket or rotor type centrifuges
There are two basic designs of laboratory centrifuge; the angle head rotor and swing out bucket types. Because of health and safety concerns rotors and buckets incorporate seals designed to prevent the release of material to the chamber and thence from the centrifuge to the environment. In addition a sealed lid to the rotor chamber may be incorporated.
   Swing out bucket-type centrifuges are, by their intrinsic design, generally unlikely to generate and release aerosols to the environment. Centrifugal force is applied at right angles to the liquid surface which is perpendicular to the bucket bottom such that any aerosol droplet formed would tend to be forced back into the body of the liquid which is in turn forced against the vessel bottom. However, with angle head rotors, the liquid surface is not perpendicular to the chamber bottom and so with an

overfilled bucket liquid may be forced directly by centrifugal force against the seal. Leakage past this seal could readily provide an opportunity for aerosol production.

Aerosolisation of microorganisms may also be caused by spills onto the centrifuge rotor. It has been shown that when culture fluid was spilt onto a spinning rotor the resulting aerosol had a mean particle size of between 3-5 microns (16),well within the range which could be inhaled and retained by the human lung. When a tube containing a suspension of <u>Salmonella indica</u> was intentionally broken in a bucket type centrifuge an aerosol was generated containing 4,000 <u>S. indica</u> particles per cubic metre (17). A review of a large range of sealed buckets and sealed rotor laboratory centrifuges showed that 21% of the buckets and 43% of the rotors were unable to contain microorganisms (18). The author of this particular study recommended that safe sealed buckets should be circular, both cap and bucket should preferably be made from stainless steel, should seal by all round pressure on a recessed moulded O-ring seal, the sealing cap should screw into the bucket (not close over the outside of the bucket) and should withstand repeated autoclaving (18).

Tubular Bowl centrifuges

Tubular bowl centrifuges are continuous feed centrifuges which accumulate the microbial paste on the walls of the bowl and continuously discharge the supernatant fluid. At the end of the process the centrifuge bowl is dismantled and the paste is removed. Centrifuges of this type are widely used in the biotechnology industry despite being renowned for their ability to generate aerosols during operation.

When tubular bowl centrifuges are used to harvest pathogenic microorganisms they should be contained inside specially designed class III cabinets to prevent operator exposure. When this type of device was operated with a broth suspension of <u>Pseudomonas aeroginosa</u> inside a tented enclosure (19) the aerosol concentration of the process organism was found to be as high as 500 colony forming units per cubic meter(c.f.u./m^3).

Tubular bowl centrifugation has been studied at PHLS CAMR (15). The particular device studied was sited in an enclosed unoccupied ventilated (22 air changes an hour) room. It was found that, when operated, the tubular bowl centrifuge generated an aerosol containing between 50,000 and 90,000 c.f.u/m^3 and that 90% of these particles were of sizes below 3 microns. This type of centrifuge produced particles within the size range that are deposited in the alveoli of the lungs, which could explain why it has a history of causing ailments of the lower respiratory tract. The manual removal of cell paste from the tubular bowl created an aerosol containing approximately 1000 c.f.u/m^3 with a wide particle size distribution.

Disc-Stack centrifuges

Disc-stack centrifuges are the type most commonly used in biotechnology processes (20). There are two main types of these centrifuge, the disc bowl type and the solids ejecting type.

The disc bowl centrifuge consists of a solid bowl having a number of cone shape inserts. The suspension to be harvested is run continuously through the device and solids are deposited on the bowl and the inserts by centrifugal force. Like the tubular bowl centrifuge the device has to be stopped and dismantled for manual removal of the recovered paste. The potential for aerosol generation by this type of centrifuge has been studied during its use for harvesting microbial biomass (14). The particular device studied is sited in a contained negative pressure room ventilated by 22 air changes/hour. Personnel entering the room wear gowns, rubber boots and gloves. Air sampling showed that the disc bowl centrifuge did not generate any aerosol during operation apart from one incident in which there was a leak in the inlet tube to the device and an aerosol of up to 4,000 c.f.u/m^3 was produced. When this fault was repaired no aerosol production was detected during the subsequent operation.

The highest levels of airborne production organisms were found during the removal of microbial cell paste from the bowl and its subsequent bagging and weighing. This part of the process was monitored during two different production runs involving different microorganisms. In the first process the aerosol levels of production strain organism reached a maximum value of approximately 4,000 c.f.u/m^3 while for the second process the maximum aerosol concentration exceeded 20,000 c.f.u/m^3. The probable reason for this difference was that the cell paste produced in the second process was less viscous and so was more prone to aerosolisation. It was calculated that during the second process the operators may have inhaled in excess of 10,000 organisms over the 1.5 hour duration of the process.

Solids-ejecting centrifuges are of similar design to disc bowl centrifuges but differ in that they allow controlled intermittent discharge of solids. In one study (8), airborne levels of 3,000 Escherichia coli c.f.u./m^3 and a ten fold increase in the concentration of airborne product has been detected in the vicinity of this device. In another study it was found that air levels of 1000 process organisms/m^3 were produced by this type of device but that none were released into the work environment when the device was placed within a plexiglass containment unit (21). Disc stack centrifuges have been designed so as to limit the possibility of aerosol generation (22,23).

Centritech cell separator

The Centritech cell separator (Alpha Laval) is a unique design of centrifuge used to remove shear-sensitive animal cells from their culture media. This device is designed to a high level of containment, with the only possible areas of environmental contamination being the connections to the feed and collection vessels. This novel device has been shown by studies at PHLS CAMR to be effective in containing materials that might be released within the device during its operation.

Filtration equipment

Filtration is the most commonly used method of biomass/mother liquor separation in fungal and actinomycete bioprocessing. Two methods of filtration that have been investigated for their potential to generate aerosols are rotary vacuum filters and filter presses. An air sampling programme carried out on a commercially used rotary vacuum filter used in the United States found that it did not generate a significant aerosol of the production organism. This was explained by stress caused by the processing and lack of aeration reducing the aerostability of the organism. Also the effective use of a local exhaust hood and retention by vacuum of the biomass mat onto the filter served to limit the released material (24). However, air sampling undertaken during operation of a filter press showed continuous generation of aerosols of 3,910 c.f.u./m^3. In addition discrete releases of up to 10,600 c.f.u./m^3 were caused by the operator having to "knock"the cake off the filter cloth when the press was opened (25). The main danger of aerosol production during filter operation is due to cake removal and cleaning and maintenance.

Cell Disruption Equipment

Such devices use physical or mechanical methods to allow disruption of microbial cell walls with release of intracellular contents. Since a large degree of microbial inactivation may occur during the operation of these devices anticipation of high levels of aerosolised viable cells may be unfounded. Despite this as many as 4,000 c.f.u./m^3 have been detected around a homogeniser (15). In addition it should be remembered that such devices can create aerosols of non-living materials, such as endotoxin, and biologically active cell-free products. At PHLS CAMR procedures to monitor airborne levels of endotoxin and microbial protein products at all stages of downstream processing including cell disruption are being developed.

INTEGRITY TESTING OF BIOPROCESSING EQUIPMENT

Over the last 11 years CAMR has tested a wide range of biotechnological and medical equipment for potential to generate microbial aerosols. The equipment tested has included a wide range of centrifuges, cell separation devices and fermenter sampling valves (18,26). Integrity testing of a device uses a microbiological tracer to assess whether it is suitable to be used with pathogenic or genetically manipulated microorganisms in the worst case situation.

Integrity tests are performed using a concentrated suspension of <u>Bacillus subtilis</u> var <u>niger</u> in place of the process fluid i.e in centrifuge buckets. The apparatus to be tested is placed in a special clean room facility with filtered inlet and outlet air. This room also contains remotely controlled air sampling equipment, video link and radio communication. The airborne and surface concentration of microorganisms is reduced to zero by fumigation with formaldehyde vapour. Also any personnel entering the room after fumigation wear full clean room clothing to prevent shedding of contaminated particles from the body.

The device to be tested is operated according to the manufacturers instructions and air samples are taken before, during and after operation. The presence of airborne <u>B.subtilis</u> var <u>niger</u> spores observed during or after operation may indicate that containment has not been achieved. The absence of airborne tracer may be taken to indicate that the device is safe for the handling of hazardous biological material. This type of test can be carried out on a wide variety of bioseparation equipment operating at laboratory or production scales.

CONCLUSIONS

Professor Dunnil of University College, London has stated that"the greatest demands in terms of biosafety occur from the time the broth leaves the fermenter through to the post-precipitation step" due to the presence of large quantities of cell wall materials and other potentially bioactive products (7). The incidents of health problems in biotechnology detailed above show that this is the case. Therefore, great care should be taken to ensure that the release of aerosolised microorganisms or other biological material must be minimised by good design of equipment, or by regularly tested secondary containment or as a last resort by ensuring that any exposed operator wears well maintained personal protective equipment.

Acknowledgement
Studies described in this paper were carried out within the Industrial Biosafety Project sponsored by PHLS CAMR and the Department of Trade and Industry, Warren Spring Laboratory.

REFERENCES

1. Bennett, A.M. and Norris, K.P., <u>Evaluation of Hazards From Exposure to Microorganisms and Their Products.</u>, State of The Art Report No. 2, Industrial Biosafety

Project, Warren Spring Laboratory, Stevenage, 1988.

2. Topping, M.D., Scarisbrick, D.A., Luczynska, C.M., Clarke, E.C. and Seaton A., Clinical and immunological reaction to Aspergillus niger among workers at a biotechnological plant. British J. Industrial Medicine., 1985, 42, 312-318.

3. Harris-Smith, R. and Evans, C.G.T., Bioengineering and protection during hazardous microbiological processes. Biotechnology and Bioengineering Symposium., 1974, 4, 837-855.

4. Flindt, M.H.L., Pulmonary disease due to inhalation of derivatives of Bacillus subtilis containing proteolytic enzymes. Lancet., 1969(i), 1177-1180.

5. Ekenvall, L., Dolling, B., Gothe, C.J., Ebbinghaus, L., Von Stedingk,L-V and Wasserman,J.M., Single cell protein as an industrial hazard. British J. Industrial Medicine., 1983, 40, 212-215.

6. Mayes, R.W, Lack of allergic reaction in workers exposed to Pruteen (bacterial single-cell protein). British J. Industrial Medicine., 1982, 39, 183-186.

7. Dunnil, P., Biosafety in the large-scale isolation of intracellular microbial enzymes. Chemistry & Industry., 1982, 22, 877-879.

8. Gibson, D.E., Hygiene and Safety in Biotechnology - A Case History at University College., M.Sc. Thesis, London School of Hygiene and Tropical Medicine, 1972.

9. Collins,C.H., Laboratory-acquired infections., Butterworths, 1983, pp 31-35.

10. Lacey, J. and Crook, B., Fungal and actinomycete spores as pollutants of the workplace and occupational allergens. Annals of Occupational Hygiene., 1988, 32, 515-533.

11. Isreali,E. Biosafety in biotechnological processes. Advances In Biotechnology., 1986, 6, 1-20.

12. Schlech, W.F., Turchik, J.B., Westlake, R.E., Klein, G.C., Band, J. and Weaver, R.E., Laboratory-acquired infection with Pseudomonas pseudomallei (melidiosis). New England J. Medicine., 1981, 305, 1133-1135.

13. Rimmington, A., The Release of Microorganisms and Other Pollutants from Soviet Microbiological Facilities - The Political and Environmental Fallout., University of Birmingham Report, 1988.

14. Muir, D.F.C., Deposition and clearance of inhaled particles. In Clinical Aspects of Inhaled Particles. ed. D.F.C Muir, William Heineman Medical Books, London, 1972, pp 1-21.

15. Bennett, A.M., Hill, S.E., Benbough, J.E. and Hambleton, P., Monitoring process safety in biotechnology. J. Applied Bacteriology Technical Series. (in press).

16. Kenny, M.T and Sabel, F., Particle size distribution of Serratia marcescens aerosols created during common laboratory processes and simulated laboratory accidents. Applied Microbiology., 1968, 16, 1146-1150.

17. Reitman,M. and Wedum,A.G.,Microbiological safety. Public Health Reports., 1956, 71, 659-665.

18. Harper, G.J., Evaluation of sealed containers for use in centrifuges by a dynamic microbiological test method. J. Clinical Pathology., 1984, 37, 1134-1139.

19. Veale, D.E., Personal communication., 1988.

20. Salusbury, T.T., Containment of Centrifuges and Cell Disrupters in Biotechnology. State of the Art Report No. 5, Industrial Biosafety Project, Warren Spring Laboratory, Stevenage, 1989.

21. Elliot, L.J., Carson, G., Parker, S., Wallingford, K. and Greife, A., Industrial Hygiene Characterisation of Commercial Applications of Genetic Engineering and Biotechnology., U.S Department of Commerce, 1983.

22. Aronsson, G., Plant equipment: Scaling down for downstream scale-up. Biotechnology., 1987, 5, 394-395.

23. Brunner, K.-H., van Hemert, P., Kohlstette, W and Tiesjma, R.H., Continuous Centrifugation in Large-Scale Vaccine Production., Westphalia Separator AG, Odele, FRG, 1984.

24. Martinez, K., In-depth Survey Report - Control Technology Assessment of Enzyme Fermentation Processes at Novo Biochemical Industries Incorporated Franklintown, North Carolina., National Institute for Safety and Health, Cincinnati, Ohio, U.S.A., 1986.

25. Wickramanayake, G.B., Assessment of Decontamination Technologies for Release from Large-Scale rDNA Processing Facilities., Battelle report commissioned by the U.S Environmental Protection Agency, 1987.

26. Cameron, R., Hambleton, P. and Melling J., Assessing the microbiological integrity of biotechnology equipment. In Separations for Biotechnology., eds. M.S. Verrall and M.J. Hudson, Society of Chemical Industry, London, 1987.

AUTOMATED MULTI-STAGE CHROMATOGRAPHIC SYSTEMS
FOR PRODUCTION OF BIOTHERAPEUTICS

GEORGE E. CHAPMAN[1,2], PAUL MATEJTSCHUK[1], JOHN E. MORE[1]
and PATRICIA PILLING[2]
[1]Bio Products Laboratory,
Dagger Lane, Elstree, Herts WD63BX, UK
and
[2]Garvan Institute of Medical Research,
St Vincent's Hospital, Darlinghurst, NSW 2010, Australia

ABSTRACT

Systems for multi-stage chromatographic purification of biotechnology-sourced therapeutic proteins have been designed and built using commercially available components, modified where necessary. The systems operate as sterile self-contained units totally under computer control. The chemistries of the individual stages are integrated so that the product peak from one stage is loaded directly on to the next stage. Automatic CIP can be added and programmed as part of the process cycle. Two such systems are in operation: the first is for the purification of human growth hormone from a recombinant mammalian cell line; the second is for the purification of therapeutic monoclonal antibodies. Both systems have been used for the production of material for clinical trials.

INTRODUCTION

Regulatory requirements for biotechnology-sourced therapeutic proteins demand minimal levels of potentially harmful contaminants in the product. Contaminants such as host cell proteins and DNA, present in the cell culture supernatants must be removed during purification of the active protein. The overall purification process must also be shown to be capable of removing and/or inactivating adventitious agents such as viruses which may be present but undetected in the supernatants. These stringent requirements have led to the virtually universal use of multi-stage liquid chromatography in the downstream processing of bio-technology-source therapeutics.

Automation of single-stage chromatographic processes is now widely used. This automation usually consists of: (i) automated buffer change-over for column regeneration and equilibration, sample loading and step-wise elution; (ii) two-pump control for gradient elution; (iii) absorbance sensing of the column effluent and automated collection of the product peak. Partial automation of a multi-stage chromatographic process has been reported [1], where a peak is collected from one stage and the pooled fraction is loaded on to the next stage.

We have developed a design concept whereby multi-stage chromatographic purification processes are integrated into sealed sterile systems total-ly under computer control. The chemistries of the chromatographic stages are where possible matched so that the peak from one column is directly loaded on to the next column as it is eluted, instead of pool-ing it and then loading. The direct load approach results in systems which are simpler, easier to sterilise (and to maintain sterile) and have shorter process times. Also, chromatographic performance may be enhanced in some applications.

We report here the design, construction and operation of two quite different multi-stage chromatographic systems based on these principles. The first is a pilot production system for final purification of recom-binant human growth hormone expressed by an engineered CHO cell line [2]. The second is a system for purification of a therapeutic monoclon-al antibody from a cell culture supernatant. Both systems have been validated and used for the production of material for clinical trials.

MATERIALS AND METHODS

Components
The systems are based on Fast Protein Liquid Chromatography (FPLC) components, manufactured by Pharmacia-LKB, Uppsala, Sweden. The princi-pal modifications are to the motor-driven valves. Systems of the type that we wished to build would have required a large number of production standard solenoid valves and motor-driven valves to control liquid flows within the system, more than a single LCC-500 FPLC computer can drive. This can in principle be overcome by using a two computer master-slave arrangement, but it is an expensive and inelegant solution. We have found that by customising the motor-driven valves, i.e. modifying the ports and flow paths within the valves, they can be made to perform the functions of two or more production standard valves in specific situa-tions. An example of this is the "horseshoe" valve in fig.1. This has general application in multi-stage systems, allowing the effluent from a

selected column to be directed to a monitor while simultaneously allowing the effluent from other columns in the system to flow to waste. It
enables one column to be eluted while others are loading, regenerating
or equilibrating, with consequent savings in total process time. The
"horseshoe" valve takes the place of n solenoid valves in an n column
system.

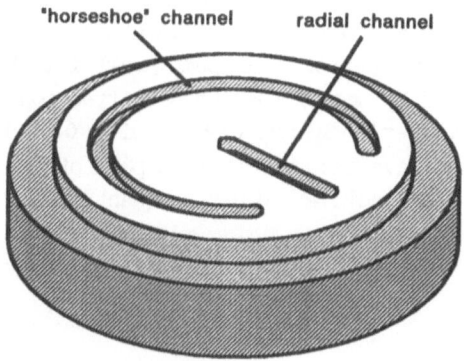

Figure 1. "Horseshoe" valve channel plate: an example of a modification
to a Pharmacia MV-8 valve.

Sterility Considerations

All solutions enter the system through 0.22μm filters. The flow paths in
the system are arranged so that these filters can be changed and sterilised in situ without compromising system sterility. Effluent lines are
automatically isolated from the system when there is no effluent flow.
Great care has been taken to eliminate dead-legs in the system. The
concept of a sterile sealed system was originally devised to obviate the
need for CIP after every process cycle, thereby increasing chromatographic media lifetime and increasing product throughput. However, it
is very difficult to demonstrate in such systems that contaminants do
not build up on the media from one cycle and slowly leach off in subsequent cycles. Hence we have opted for CIP as part of the process cycle
in our latest systems, on regulatory grounds. A short (3 hour) CIP is
more than adequate to achieve sterility as well as inactivate viruses
and endotoxins and chemically decontaminate the system. The summed
time of exposure of the media to CIP solutions is not significantly
greater with a short CIP after every process cycle than for a lengthy
CIP after five to ten process cycles.

Human Growth Hormone Purification System

Human growth hormone is concentrated from the conditioned cell culture
medium by a chromatographic capture stage prior to final purification in
the system. The prototype system is shown in fig.2 and the flow diagram
in fig.3. The system consists of four non-affinity chromatography
stages. The third stage is a high pressure/performance ion exchange
chromatography column. Loading of this stage with the product peak from
the second stage results in a high static pressure in the second stage
column, due to the flow resistance of the third stage bed. The elution
rate of the second stage is therefore reduced when loading the third
stage, to avoid exceeding the pressure limit of the second stage column.
On/off control of the peristaltic pump used for loading the system
feedstock is provided by one of the (unused) fraction collecter outports
on the LCC-500 computer. System CIP is with 0.5M NaOH. The system
illustrated does not have automatic CIP as part of the process cycle but
this facility has been included in a subsequent production system.

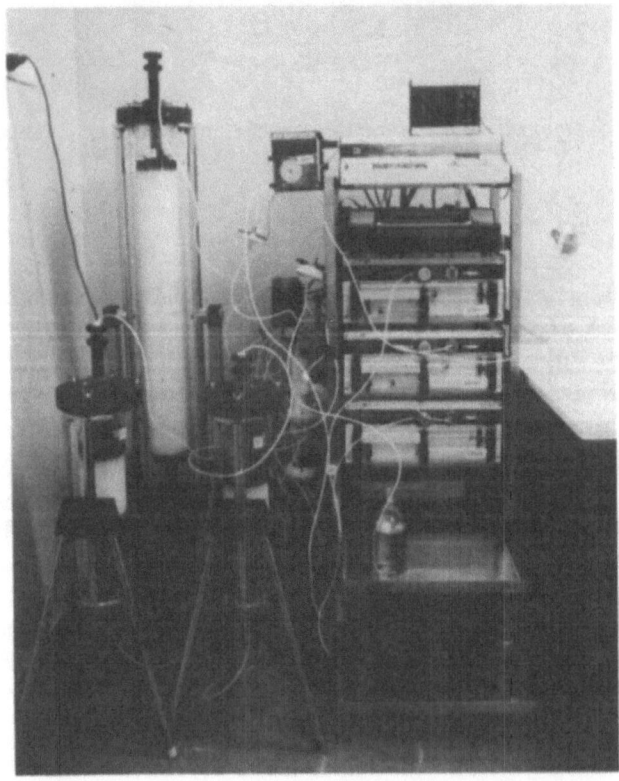

Figure 2. The prototype human growth hormone purification system.

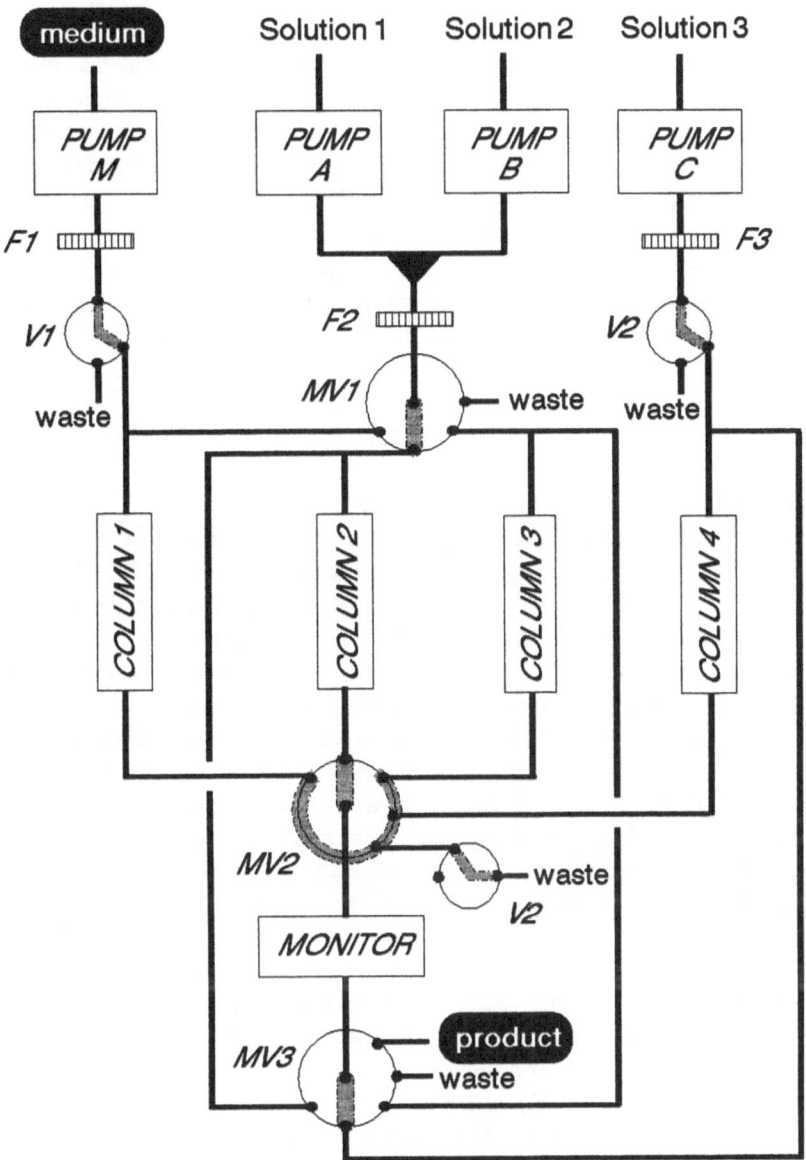

Figure 3. Flow diagram of the prototype human growth hormone purification system illustrated in figure 2

Monoclonal Antibody Purification System

This is a three stage chromatographic system for the purification of therapeutic monoclonal antibodies from supernatants produced by hollow fibre bioreactor cell culture. The system flow diagram is shown in fig.4. The first stage is protein G affinity chromatography with reverse elution, followed by high performance cation exchange and gel filtration. The eluting peak from the first stage is continuously mixed with the second stage equilibration buffer. The system has automated CIP: 0.5M NaOH is used for all parts except the affinity column, which uses 70% ethanol. There is also provision for automated filling of the entire system with 20% ethanol for extended shutdowns. Four of the six motor-driven valves (MV2, MV3, MV5 and MV 6 in fig.4) are customised.

RESULTS

Human Growth Hormone Purification System

The system was originally designed to run as a completely sterile system, with multiple production cycles in between system CIP. Samples of column effluent were taken at the start of each run and cultured on agar plates. The presence of any colonies was a criterion for rejection of the product from that run. In practice, Pseudomonas colonies were observed after about five runs, indicating the presence of a system deadleg which was not completely sterilised during CIP and from which contamination slowly leaked into the system. We suspect that the unused blocked-off ports of the "horseshoe" valve are the source of the contamination, and are experimenting with modified valves in which only the used ports are bored out in the distributing plate. The system has been validated for the production of 1 gram of purified recombinant human growth hormone from the concentrate resulting from chromatographic capture of the cell culture supernatant in a 15 hour process cycle, though higher throughputs have been achieved in practice. Overall system yield is in excess of 50%. System clearance/inactivation factors for host protein, host DNA, a model retrovirus and a model envelope virus have been determined, and meet regulatory requirements. A production system with automated CIP is now in operation.

Monoclonal Antibody Purification System

The system has been used for the purification of a number of human monoclonal antibodies produced by transformed human and human/mouse hybridoma cell lines in both suspension culture and hollow fibre bioreactors. The total process cycle time, including CIP, is about 24 hours, depending on the time taken to load the cell culture medium. The system is thus conveniently geared to the routine processing of daily harvests

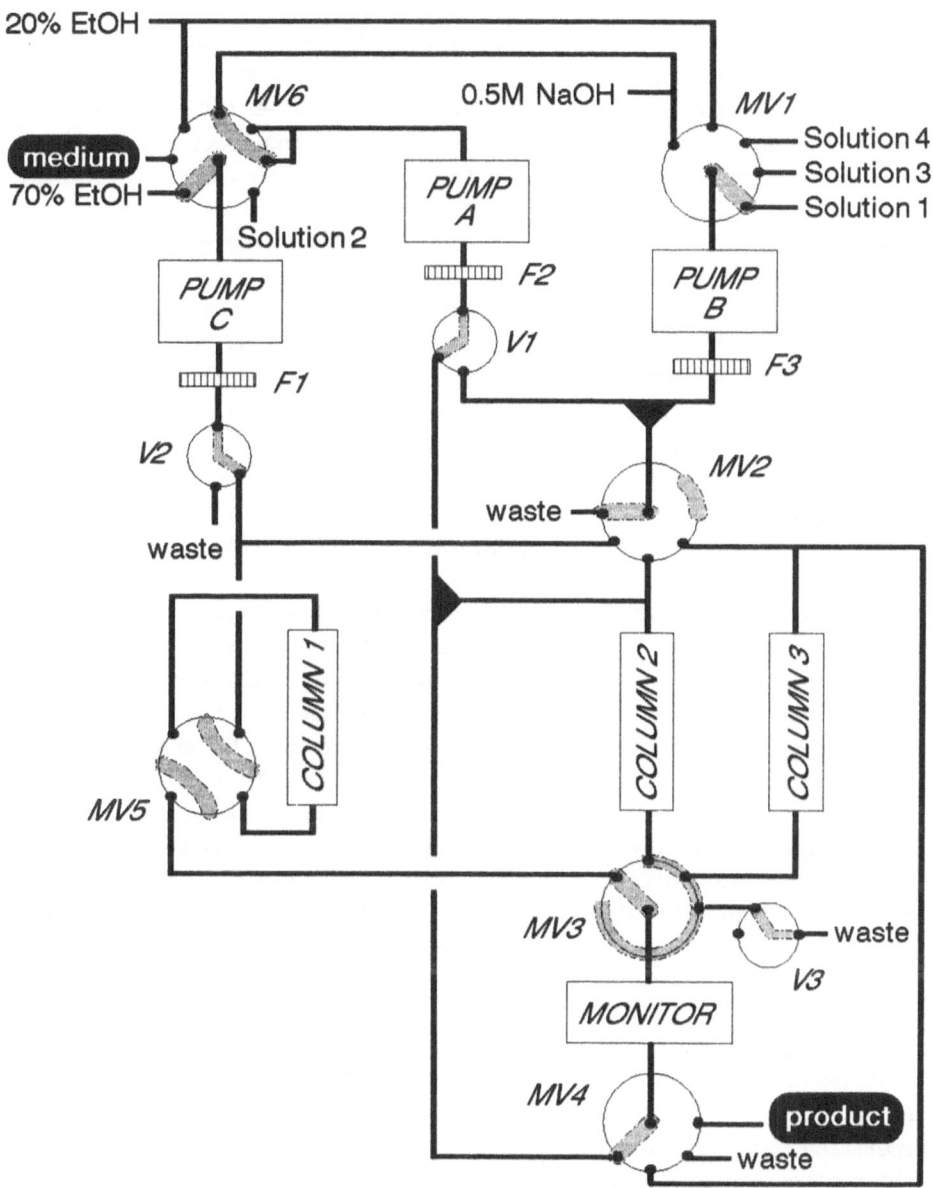

Figure 4. Flow diagram of the monoclonal antibody purification system

from a continuous cell culture system. In its present scale the system has achieved production of 50mg purified antibody per cycle at yields in excess of 70%. The system has capacity for at least 10x scale-up, merely by increasing column cross-sectional areas, before the system component design becomes limiting. The system chemistry is, with minor modifications, suitable for the purification of a wide spectrum of monoclonal antibodies of differing isotype and isoelectric point from the most commonly used species. Clearance factors for host protein, host DNA and model viruses meet current regulatory requirements for biotechnology-source therapeutics. The use of reverse elution for the affinity chromatography stage, the control of elution rate and in-line pH adjustment of the eluted peak has enabled us to control precisely the length of time of exposure of the antibody to the low pH required for elution. We are thus able to balance the loss of yield due to acid-induced denaturation against the viral inactivation factors resulting from the low pH, for a particular antibody.

DISCUSSION

Integrated operation of multi-stage chromatographic processes has obvious advantages with respect to process hygiene: in some cases chromatographic performance may also be enhanced. Direct loading of a chromatographic peak on to a subsequent non-absorption chromatographic stage (such as gel filtration) gives sharper eluted peaks off the second stage than loading the corresponding pooled fraction. The computer simulations in fig.5 demonstrate the point. This phenomenon can be used to obtain greater selectivity for the second stage. Alternatively, resolution can be traded off against the volume of the gel filtration column: a larger percentage volume can be loaded as a peak than as a pooled fraction before the resolution degrades, hence a smaller column may be used when directly loading a peak.

Process validation in self-contained multi-stage systems such as we have described must contain an extra component over and above more traditional approaches to downstream processing; namely validation of the system operation. Programming of the system should include event markers on the monitor trace to demonstrate that the program flow has operated as intended. Chromatographic peaks should elute in quite closely defined time windows: if they are outside such windows, the system should give an indication of this. It is of course possible to program the system to take samples of intermediate peaks for testing, though we have not felt the need for such measures in production situations we have yet encountered. The final assurance of correct system operation for a

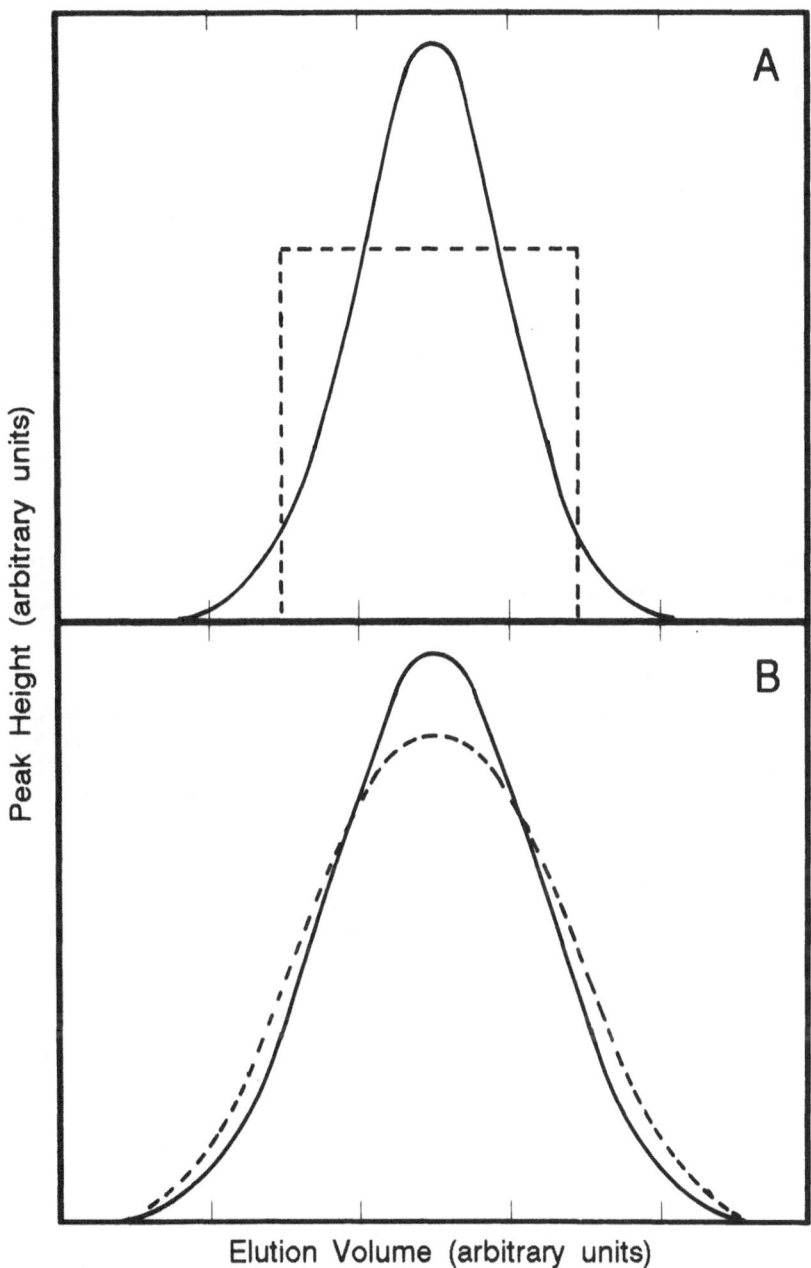

Figure 5. Computer simulation of a two-stage chromatographic process. A) Gaussian chromatographic peak cut for 95% yield and the corresponding pooled fraction. B) Simulated gel filtration, loading the peak and pooled fraction from A, obtained by Gaussian convolution, using a line broadening equal to that of peak A.

process cycle is examination of the monitor trace and comparison with previous process cycles. This is a pattern recognition exercise which the human brain is better equipped to analyse than any machine.

The complete automation of multi-stage chromatographic downstream processing of biotherapeutics vis a vis manual and partially automated processing has significant implications for both product quality and process economics. Process hygiene and consistency are improved and the likelihood of operator error is decreased. The role of the operator is changed to that of a machine overseer, enabling him/her to concentrate on the critical issues of Good Manufacturing Practice and product safety.

ACKNOWLEDGMENTS

We gratefully acknowledge the technical assistance of Andrzej Surus, Erefili Valiontis, Jane Fellows and Steven Price and the technical cooperation of Pharmacia-LKB.

REFERENCES

1. Daniels, A.I., Pettersson, N.T. and Scandella, C.J., An automated purification system for the production of recombinant human superoxide dismutase. First Conference on Advances in Purification of Recombinant Proteins, Interlaken, Switzerland, March 1989.

2. Friedman, J.S., Cofer, C.L., Anderson, C.L., Kushner, J.A., Gray, P.P., Chapman, G.E., Stuart, M.C., Lazarus, L., Shine, J. and Kushner, P.J., High expression in mammalian cells without amplification. Biotechnology, 1989, 7, 359-362.

APPLICATIONS OF MOLECULAR SIZE DETECTION IN COLUMN CHROMATOGRAPHY

Paul Claes, Sue Fowell, Andrew Kenney, Penny Vardy and Caroline Woollin.
Oros Instruments Ltd., 715, Banbury Avenue., Slough.
Berkshire. England. SL1 4LJ.

ABSTRACT

A compact, economical detector has been developed for rapid, on-line analysis of the molecular size of macromolecules in liquid streams. This instrument has many applications in column chromatography and is compatible with both low and high pressure systems. The detector uses photon correlation spectroscopy to measure the hydrodynamic size of macromolecules in solution and provides an estimation of molecular weight, polydispersity of the major component and purity. The basis of operation of the detector is described and the particular application of the instrument in protein chromatography is illustrated with examples of peak identification and aggregate detection.

INTRODUCTION

Column chromatography plays a major role in protein purification [1]. The identification of components eluted from columns can be complex and time-consuming. At present the available techniques fall into two main categories; specific methods such as enzyme assays and ELISA's (enzyme linked immunosorbent assays) or the determination of molecular size. Techniques based on molecular size include analytical ultracentrifugation and the estimation of molecular weight by gel filtration or electrophoresis. These methods have the advantage that they are applicable to all proteins regardless of class or activity, but require off-line and time-consuming manipulations. An on-line technique for determining molecular size would obviously be extremely useful in analysing protein chromatography. One potential method for the on-line analysis of molecular size relies on the use of light scattering to determine the hydrodynamic size of particles in suspension. Measurement of particle size by dynamic light scattering or photon correlation spectroscopy is a well established technique [2], however, until recently, this has not been of general use to biochemists because of the complexity and cost of the necessary instrumentation [3].

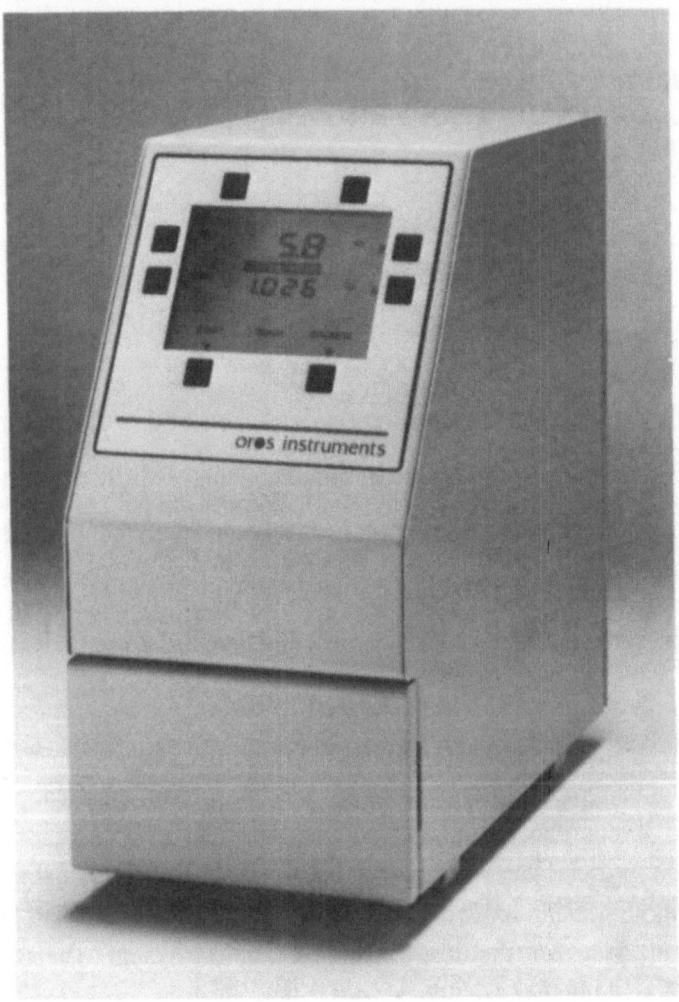

Figure 1. Model 801 Molecular Size Detector.

We have developed an economical molecular size detector which uses photon correlation spectroscopy to measure the hydrodynamic size of macromolecules in a flowing liquid stream. The Model 801 Molecular Size Detector is straightforward to operate and can be used for preparative protein chromatography with both low and high pressure systems. The instrument can be used for peak identification, peak cutting decisions and detection of aggregation or fragmentation in quality assurance procedures. It is small and compact and typically produces a read out of molecular size within 6 to 30 seconds.

Use of Photon Correlation Spectroscopy to Determine Molecular Size.
The Model 801 Molecular Size Detector (Figure 1) uses photon correlation spectroscopy to determine the translation diffusion coefficient (D_T) of particles in solution. This technique is based on the measurement of fluctuations in scattered light intensity caused by the relative movements of macromolecules in solution. The organisation of the instrument and its basis for operation are shown in Figure 2. Light from a near infra-red semiconductor laser diode passes through a flow cell and is scattered by molecules moving under Brownian motion. Photons which are scattered at 90° to the incident beam are collected by a lens and directed to an Avalanche Photo Diode (APD) via an optical fibre. The APD produces a single pulse for each photon detected and these pulses are stored by an integral computer. The basis of photon correlation spectroscopy is described below.

Light scattered by macromolecules interferes either constructively or destructively and the manner of interference changes as molecules move, causing a fluctuation in scattered light intensity. The time scale of these fluctuations depends on the speed of movement of molecules, hence D_T can be obtained from measurements of these time scales. Comparatively small, fast moving particles such as macromolecules scatter very little light and cause rapid intensity fluctuations, and it is necessary to count individual scattered photons to obtain measurements of D_T. Photons are counted into time windows or channels and the time constant of the intensity fluctuation is obtained by autocorrelation of this data.

The autocorrelation function (R) for an ideal monodisperse suspension of particles can be represented by:

$$R(\tau) = 1 + A\ e^{(-2\Gamma\tau)} \qquad (1)$$

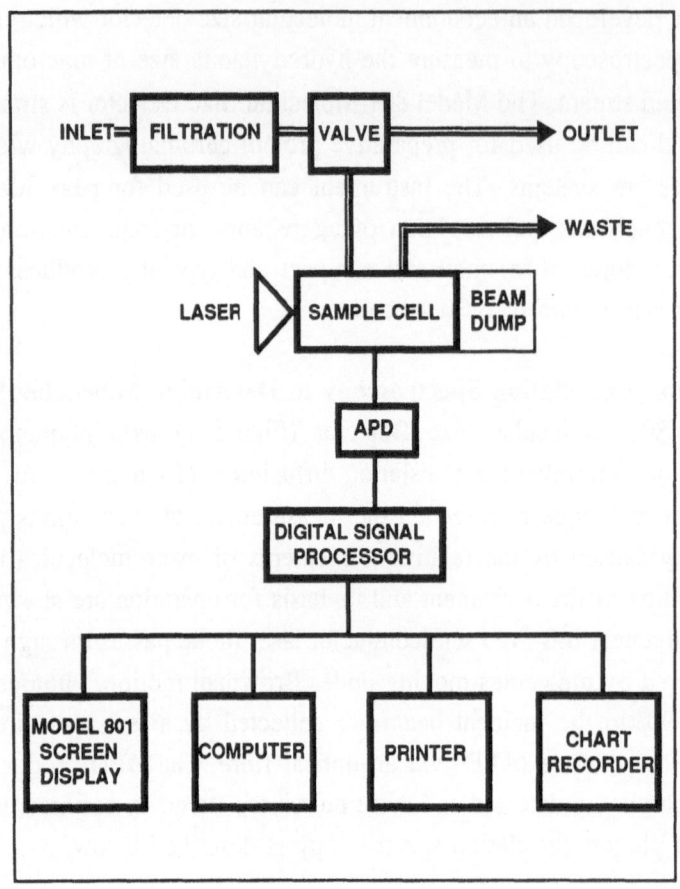

Figure 2. Schematic diagram of organisation and the basis of operation of the Model 801 Molecular Size Detector.

Samples enter the inlet from a chromatography stream via an in line filter or can be injected manually with or without filtration. The solenoid operated 3-way diverter valve opens at an interval determined by the set flow rate to allow a sample (40 μl) to enter the sample cell (7 μl volume). This interval is set such that the sample cell is filled and sufficient liquid passes through the cell between readings to ensure that the cell is swept out. Laser light passes through the sample, non-scattered light is absorbed by the beam dump, whilst light scattered at 90° to the incident beam is collected by a lens and transmitted via a fibre optic cable to an actively quenched solid state avalanche photo detector (APD) where the photons are converted to electrical pulses. These pulses are then passed to the digital signal processor where they are counted and auotocorrelated. When autocorrelation is complete, the results are displayed on the screen and can be passed to a printer, a chart recorder and a personal computer as required.

where A is the amplitude of the scattered light intensity and the time scale of the intensity fluctuations is given by Γ, the exponential decay time constant. Physically, Γ represents the time taken for a particle to move the distance which gives rise to an optical phase change of π radians in its scattered light field at the detector. The exact value of this distance depends on the scattering angle θ, the free space wavelength of the incident light, λ_0, and the refractive index of the solution, n. The amplitude of the scattering vector, q, is defined as:

$$q = \frac{4\pi n \sin(\theta/2)}{\lambda_0} \qquad (2)$$

D_T is then given by:

$$D_T = \frac{\Gamma}{q^2} \qquad (3)$$

and the hydrodynamic radius, R_H, can be calculated from D_T using the Stokes-Einstein equation:

$$D_T = \frac{kT}{6\pi\eta R_H} \qquad (4)$$

where k is Boltzmann's constant, T is the absolute temperature and η is viscosity. The molecular weights of molecules detected are derived using the relationship between molecular weight and measured values of R_H determined for a series of standard proteins by the Model 801. Values of R_H and D_T are listed in Table 1. The estimation of molecular weight is based on two assumptions. First, to determine the molecular volume from R_H it is assumed that proteins are approximately spherical in shape and, secondly, it is assumed that all proteins have a constant density relation to their size in order to calculate mass from the molecular volume. These assumptions breakdown in certain cases, for example non-globular proteins have a R_H which is greater than that predicted from the molecular weight, and literature values for molecular weights are often based on DNA or amino acid sequence data and do not take into account the contribution to molecular weight from glycosylation. Heavily glycosylated proteins will thus

Table 1.

Translational diffusion coefficients (D_T) and hydrodynamic radii (R_H) of standard proteins of known molecular weight.

Protein	Molecular weight	D_T (x $10^{-13}m^2s^{-1}$)	R_H (nm)
Ribonuclease A	13700	1171	2.2
Lysozyme	14300	1278	1.9
Myoglobin	16900	1164	2.2
Chymotrypsinogen	25000	1075	2.4
Carbonic anhydrase	29000	954	2.6
Ovalbumin	43000	751	3.0
Hexokinase sub-unit	51000	746	3.3
Haemoglobin	65000	700	3.5
Bovine serum albumin	67000	628	3.0
Transferrin	76000	583	4.2
Horse alcohol dehydrogenase	80000	651	3.7
Amyloglucosidase	99000	638	3.9
Hexokinase	102000	564	4.3
Yeast alcohol dehydrogenase	150000	512	4.9
Immunoglobulin G	150000	344	7.1
Apoferritin	443000	287	8.2
Thyroglobulin	669000	244	10.1

also display a R_H which is larger than expected. Figure 3 shows a plot of $\log_{10} R_H$ against $\log_{10}$ molecular weight superimposed on the line generated by a plot of the relationship used by the Model 801 to estimate molecular weight from R_H. The data obtained for most of the proteins examined lies on or close to the line, indicating that the assumptions used provide a good estimate of molecular weight in most cases.

Operation of the Model 801 Molecular Size Detector.

The Model 801 Molecular Size Detector takes a 40 μl sample from a flowing liquid stream, which can either be the outlet of a chromatography system or a sample injected directly into the instrument. If there are enough molecules in the sample to produce a result, photon correlation continues until sufficient data have been gathered to provide 99% accuracy in R_H.

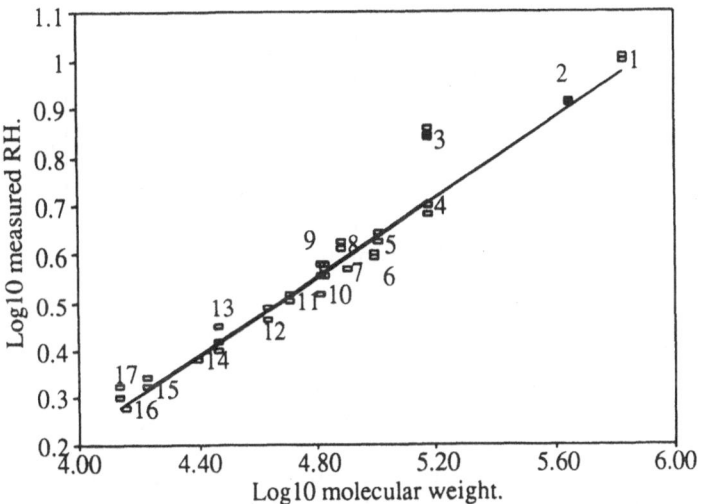

Figure 3. Plot of $\log_{10} R_H$ against $\log_{10}$ molecular weight.

Proteins measured were: 1/. thyroglobulin; 2/. apoferritin; 3/. IgG; 4/. yeast alcohol dehydrogenase; 5/. hexokinase; 6/. amyloglucosidase; 7/. horse alcohol dehydrogenase; 8/. transferrin; 9/. bovine serum albumin; 10/. haemoglobin; 11/. hexokinase sub-unit; 12/. ovalbumin; 13/. carbonic anhydrase; 14/.chymo-trypsinogen; 15/. myoglobin; 16/. lysozyme; 17./ ribonuclease A. The relationship between $\log_{10} R_H$ and $\log_{10}$ molecular weight used in the estimation of molecular weight from measured R_H is also shown (_____).

Autocorrelation of data proceeds as the scattered photons are counted. After autocorrelation is complete a new sample is taken and the autocorrelation function obtained analysed to provide D_T, R_H, molecular weight, polydispersity or purity and statistical parameters indicating the reliability of the measurement. This data can be printed out, transferred to a personal computer via a serial link and/or output to a chart recorder.

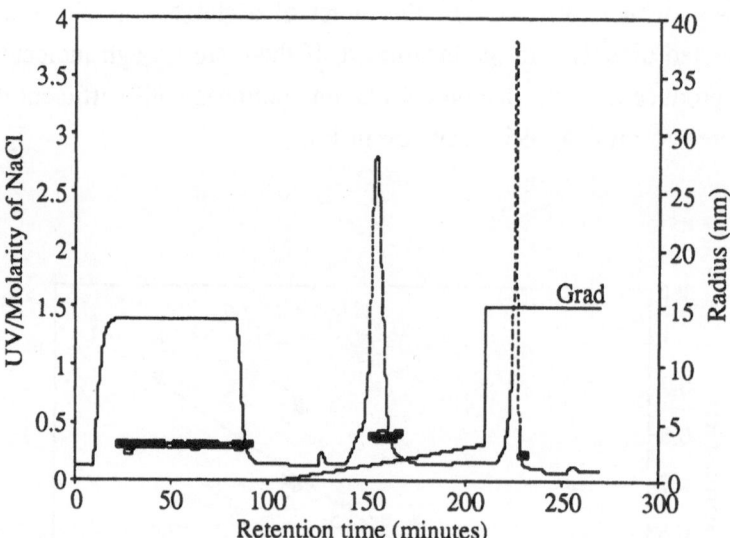

Figure 4. Model 801 molecular size analysis during cation exchange of egg white proteins.

Chromatography was carried out on a Oros Instruments Model 100 Touch Screen Laboratory Scale Chromatograph. Prepared egg white sample (diluted 6-fold in Buffer A, 5 mM NaH_2PO_4; pH 6.0 and gel filtered to remove aggregated material; 60 ml volume) was loaded at a flow rate of 2 ml/min on to a pre-equilibrated 1.6 cm x 32 cm column of S-Sepharose Fast Flow. After sample loading the column was washed with 70 ml 5 mM NaH_2PO_4; pH 6.0. Bound proteins were eluted with a gradient of 100% Buffer A to 68% Buffer A, 32% Buffer B (1 M NaCl, 5 mM NaH_2PO_4; pH 6.0) over 96 min followed by a step to 1.5 M NaCl, 5 mM NaH_2PO_4; pH 6.0. The measured R_H in nm (□), molarity of NaCl in the elution buffer (____) and the measured count rate per second (·····) are shown plotted against the retention time in minutes.

Chromatographic Applications of the Model 801 Molecular Size Detector.
The benefit of using the Model 801 Molecular Size Detector for peak identification
is illustrated by the separation of egg white proteins by cation exchange, shown
in Figure 4. It was known that the sample contained three major components,
ovalbumin (molecular weight 43,000), lysozyme (molecular weight 14,000) and
ovotransferrin (molecular weight 76,000), but their elution positions in the gradient
could not be predicted in the absence of knowledge of their pI's. Different
components were identified by the Model 801 Molecular Size Detector in each of
the 3 peaks. The first peak (unbound material) had an average R_H of 3.1 nm,
estimated molecular weight 45,000, peak 2 had an average R_H of 3.8 nm, estimated
molecular weight 74,000 and peak 3 had an average R_H of 2.3 nm, estimated
molecular weight 22,000. This suggested that ovalbumin passed through without
binding to the column and that the first bound and eluted peak contained
ovotransferrin, and the second contained mainly lysozyme. Subsequent gel
electrophoresis confirmed this (the lysozyme peak was contaminated by larger
molecular weight material, leading to higher values for R_H and molecular weight
than expected).

Another advantage of molecular size analysis during chromatographic
separations is illustrated by the affinity purification of mouse IgG_1. The instrument
detects particles, and is very sensitive to small concentrations of large particles
including non-UV absorbing material which would not be detected by conventional
UV detectors. Figure 5 shows the gradient elution stage of affinity adsorption of
IgG_1 on immobilised Protein A. The main eluted peak has an average R_H of 5.7
nm. At the end of the gradient, there is a count rate burst (increase in scattering),
caused by a small concentration of large particles with R_H = 21-32 nm. Detection
of this peak by UV adsorption requires a high sensitivity setting compared to the
main peak. Subsequent analysis of this material by gel electrophoresis has shown
that this material is composed of aggregated proteins.

DISCUSSION AND CONCLUSIONS.

Molecular size analysis of macromolecules is an extremely useful technique for use
with chromatographic separations and has many potential applications. The role of
molecular size analysis in rapid peak identification and detection of aggregates has
been illustrated. Light scattering systems available for molecular analysis have,
until recently, been cumbersome, expensive and complicated to operate. The Model

801 Molecular Size Detector represents a significant advance in this area, because the instrument is small, compact and relatively inexpensive. Operation of the instrument is straightforward and does not require a thorough understanding of the physics involved in photon correlation spectroscopy or the data analysis.

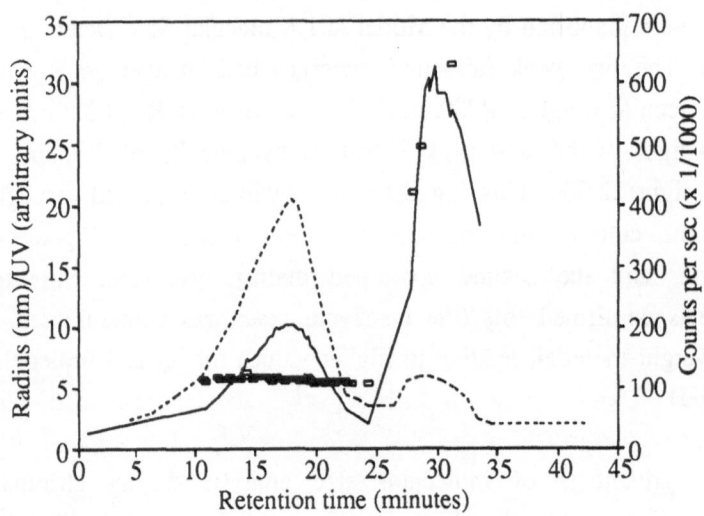

Figure 5. Model 801 molecular size analysis of the elution of IgG$_1$ from immobilised Protein A.

Concentrated hybridoma cell culture (30 ml) was diluted with an equal volume of running buffer (0.5 M NaCl, 50 mM Tris; pH 8.8) and loaded onto a 1 x 6 cm column of immobilised Protein A at 2 ml/min. After sample loading, non-specifically bound proteins were removed by washing with run buffer. Bound proteins were eluted with a pH gradient from pH 8.0 to pH 2.0, decreasing at 0.2 pH units per minute, formed using the Model 100 system's facility to mix 0.2 M Na$_2$HPO$_4$ and 0.1 M citric acid in predefined ratios. UV absorbance (⋯), measured counts per second (—) and measured R$_H$ in nm (□) are shown plotted against the retention time in minutes for the elution stage.

The instrument can be used over a flow rate range of to 0.1 to 9.9 ml/min and is compatible with high or low pressure and preparative chromatography systems. The Model 801 Molecular Size Detector can be used to analyse separations based on most chromatographic methods in addition to those illustrated here.

ACKNOWLEDGEMENTS

The Model 801 Molecular Size Detector includes technology developed by the UK Ministry of Defence Royal Signals and Radar Establishment and licensed to Oros Instruments Limited by Defence Technology Enterprises Limited (DTE).

REFERENCES

1. Bonnerjea, J., Oh., S., Hoare, M. and Dunnill, P., Protein purification: The right step at the right time. Bio/Technology, 1986, **4**, 954-958.

2. Berne, B.J. and Pecora, R. Dynamic Light Scattering. Wiley, New York, 1976.

3. Brown, R.G.W., On-line analysis of proteins using optical techniques. In Separations For Biotechnology, ed. Verrall, M.S. and Hudson, M.J., Ellis Horwood Ltd., Chichester. 1987, pp.430-435.

THE MONITORING AND CONTROL OF PROTEIN PURIFICATION AND RECOVERY PROCESSES

MARIA NIKTARI, STEPHEN CHARD, PHILLIP RICHARDSON and MICHAEL HOARE
SERC Centre for Biochemical Engineering
Department of Chemical and Biochemical Engineering,
University College London, Torrington Place,
London WC1E 7JE, U.K.

ABSTRACT

A novel method for the on-line monitoring of fractional precipitation of proteins has been developed using a combination of an automated microcentrifuge and flow injection analysis for enzyme and total protein analysis. The optimisation of a fractional precipitation process is discussed as is also the use of the on-line monitoring equipment for control to predetermined set-points of the fraction of the enzyme remaining soluble. The on-line measurement of precipitate concentration and size distribution using a combination of turbidity and light diffraction measurements and its application to improving centrifugal recovery of precipitates is discussed. One such improvement is the use of these measurement techniques to control the in-line low-frequency conditioning of the precipitate to improve particle characteristics such as size distribution, density and resistance to shear-related or mechanical break-up.

INTRODUCTION

In a typical sequence of downstream processing for the recovery of intracellular products, shown in Figure 1, cells are harvested in a high speed disk stack centrifuge and subsequently disrupted in a high pressure homogeniser. The cell debris is removed to yield a process stream suitable for subsequent purification by fractional precipitation and, if necessary, chromatography-based techniques.

The process streams in the early stages of downstream processing of a fermentation or tissue culture broth contain a wide range of biological particle types and soluble constituents. The particles in suspension may include whole cells, cell debris, inclusion bodies or protein precipitates. It is necessary to be able to monitor on-line the biological particles and soluble components of interest if operation under optimal conditions and on-line control is to be applied.

One example is the need to optimize yield versus purification during fractional protein precipitation in such a way that maximum yield for a set purity of the product is achieved. However, the performance of the fractional precipitation process will be determined by the purification achieved in the earlier process stages, e.g. cell debris removal, and by the ability to recover the precipitate from the mother liquor and effectively dewater the sediment in the subsequent centrifugation stage.

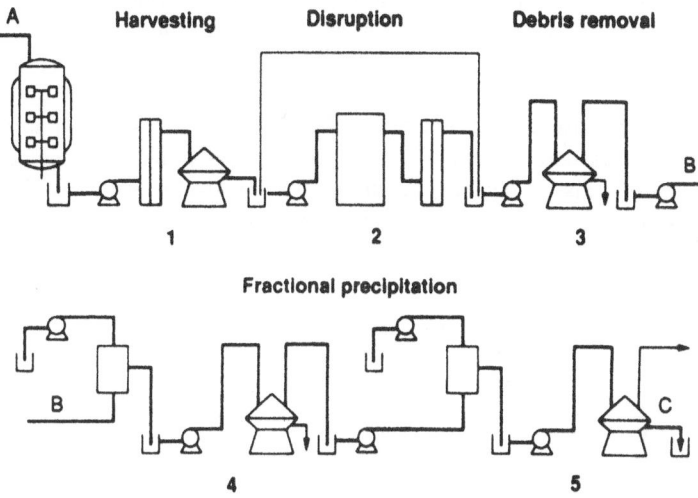

Figure 1.   Flowsheet of unit operations sequence for the purification of intracellular products.

ON LINE MONITORING AND CONTROL OF SOLUBLE PROTEIN PRECIPITATION

The mechanisms which determine the solubility characteristics of proteins are not well understood even for solutions of a single protein species. Typically for salting out of proteins the semi-empirical Cohn equation [1] is used.

$$\log S = kI + b \tag{1}$$

where S is the solubility of the protein at ionic strength I, k is the salting-out constant and b is a constant expressing solubility at $I=0$. While it is possible to relate these constants to a combination of hydrophobic and electrostatic interaction effects [2] a great deal needs to be known of the protein structure and surface effects to allow their prediction. The effects of simple operating variables such as pH and temperature on k and b are well known although again only general trends are available rather than specific relationships. For more complex mixtures of proteins in the presence of many soluble and insoluble components it is generally not possible to predict the protein solubility characteristics. For example, Foster et al., [3] have shown the complex changes which occur in precipitation profiles of intracellular enzymes from complex mixtures such as a clarified yeast homogenate as a result of variations in the means and scale of the precipitation process.

An example of the effect of variations in process feed stream composition on the fractional precipitation of proteins is given in Figure 2. The solubility profiles of salicylate hydroxylase from a *Pseudomonas* species, using polyethylene glycol as the precipitating agent, when cell debris and/or broth are present, are compared with a control where neither of these contaminants are present. In this graph the fraction of the enzyme remaining in solution is plotted against the concentration of the precipitant. The presence of cell debris alone serves to reduce the amount of polyethylene glycol required to achieve complete product recovery while the effect of soluble broth constituents arising from poor washing of the cell paste prior to disruption results in an increase in polyethylene glycol required. Such changes are difficult to predict and would vary in their extents depending on the fermentation operation and the performance of earlier purification steps. Maintaining the same cut positions as those used in the control for 100% enzyme recovery leads to a 20% enzyme loss in the presence of soluble broth constituents or in the presence of cell debris. In both cases there is also a decrease in the purification factor achieved [5]. Since in a purification chain there are a sequence of steps where such losses may occur it is necessary to optimize individual steps such as fractional precipitation to operate at yields approaching 100%.

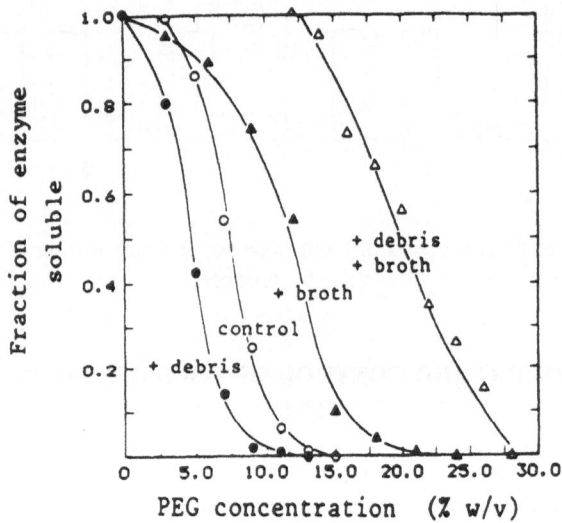

Figure 2.   Effect of cell debris and broth constituents on the solubility profiles of salicylate hydroxylase from a *Pseudomonas* species.

A system for the on-line analysis of soluble enzyme and total protein concentration requires firstly the rapid and reproducible removal of the precipitate phase without any resolubilization and then the use of on-line analysis of the soluble component.

To this effect a high speed microcentrifuge, operating at 40,000g, capable of separating the precipitate reproducibly and rapidly has been developed [5]. The microcentrifuge in conjunction with the Flow Injection Analyser (FIA) and a computer controlled precipitation rig can generate on-line solubility profiles. The system developed is illustrated in Figure 3. The flow of the process stream containing the enzyme and contaminant proteins and that of the precipitant are controlled using a set-point and flowrate ratio controller and variable speed gear pumps linked to magnetic flowmeters. A small sample (~0.5 mL) of the resulting suspension is delivered to the microcentrifuge where it is centrifuged. Typically, 60 s are sufficient to clarify fine precipitates such as those resulting from ammonium sulphate precipitation. The supernatant is subsequently delivered to the FIA for measurement of enzyme activity and total protein concentration. The microcentrifuge bowl is then washed automatically to be prepared for the initiation of the next cycle. A cycle of the above operations, from the mixing of the two streams to the logging in the computer of the concentrations measured, requires ~200 s. The microcentrifuge timings and operating conditions as well as the operation of the FIA are controlled by the supervisory computer.

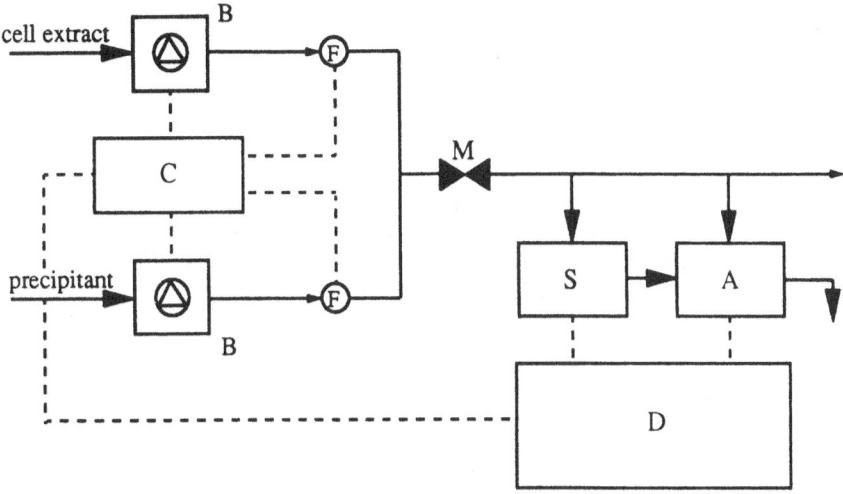

Figure 3. Experimental system for the on-line monitoring of enzymes and protein. A:FIA system; B:variable speed gear pumps; C:dual loop controller; D:supervisory microcomputer; F:flowmeters; M:mixing element; S:microcentrifuge.

Figure 4 shows typical solubility profiles of alcohol dehydrogenase and total protein from baker's yeast which have been generated on-line by the system described above. The precipitant in this case was ammonium sulphate. The solubility curves are fitted by applying linear regression to the experimental data using the linearised form of equation (2):

$$E = \frac{1}{1 + \left(\frac{C}{\alpha}\right)^{\eta}} \tag{2}$$

where E is the fraction of enzyme remaining soluble, C is the concentration of ammonium sulphate expressed as (%) saturation, and α and η are constants. Such a description applies for the whole curve whereas the Cohn equation (1) describes only the salting-out part of the curve. Equation (2) can therefore be used as a model of the solubility profiles of enzymes for control purposes. The above equation is less successful in describing the fraction of contaminant protein remaining in solution. In the case presented in Figure 4, the fraction of protein in solution has been described with two different curves of the form of equation

(2), one below and one above 50% ammonium sulphate saturation. Also solubility profiles such as those in Figure 4 can be used for the on-line optimisation of the precipitation cuts in order to achieve maximum yield for a given purity or vice versa [4].

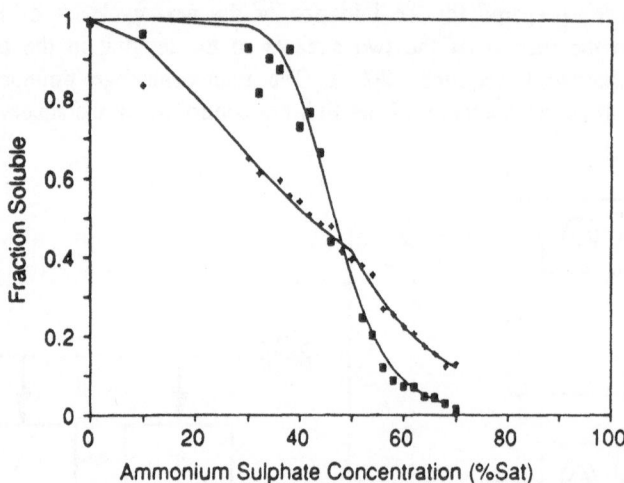

Figure 4.   Solubility profiles diagram generated on-line. Alcohol dehydrogenase and total protein fractions remaining soluble versus ammonium sulphate concentration.

After the precipitation cuts have been selected the process may then be controlled in an on-line rather than retrospective fashion using the same instrumentation. Such a system has been used to control the ADH recovery from total protein by adjustment of the precipitant concentration to give the desired set-point of enzyme remaining soluble [5]. The state variables of the process, i.e. enzyme fraction remaining soluble and precipitant concentration were estimated with an adaptive algorithm relying initially on data from previous runs. The parameters of the model are calculated after each sample on the basis of the last and previous measurements using the simplex method of Nelder and Mead [5]. A modified action on the control variable, namely precipitant concentration is then taken. One major problem arises from the high measurement noise (assay error for ADH remaining soluble ~10.3%) and constraints have been incorporated in the control algorithm to account for uncertainty in the measurements.

ON-LINE MONITORING AND CONTROL OF CENTRIFUGAL RECOVERY OF PROTEIN PRECIPITATES

The recovery of protein precipitates by centrifugation presents a number of problems representative of the challenges involved in dealing with biological particles. Centrifuge types commonly used in biotechnology-based industries include tubular bowl, scroll discharge, multichamber and disc centrifuges. The design and operation of these centrifuges for the industrial recovery of biological particles has been reviewed elsewhere [6]. For the contained recovery of particles from relatively dilute feed streams the intermittent disc stack centrifuge represents one major option.

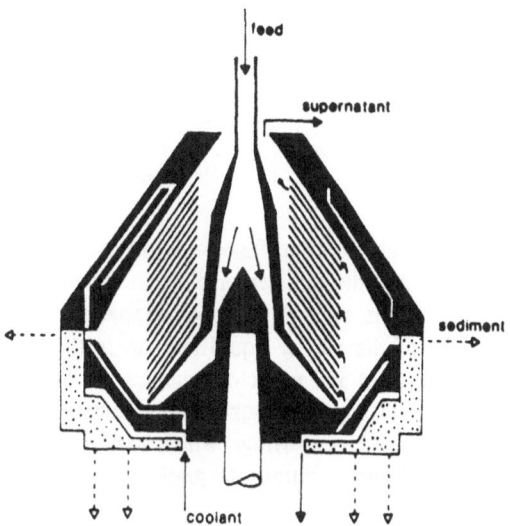

Figure 5.   Industrial high speed, continuous flow disc stack centrifuge. Cross section of the cooled rotor. The bottom section of the rotor is lowered hydraulically to release sediment at intervals.

Figure 5 shows an intermittent discharge disc-stack centrifuge with direct cooling of the centrifuge bowl. Such cooling is specified to allow rigourous study of the operation of disc stack centrifuges at low throughputs necessary for the recovery of biological particles. For example, even small temperature rises on passage through the centrifuge can lead to redissolution of the protein precipitate. The suspension is fed via a rotating pipe into the settling space of the centrifuge. A set of truncated, conical discs ensures that the distance which the solids must traverse before settling on a surface, i.e., the underside of a conical disc, is minimised. The solids are collected at the periphery of the bowl and are intermittently released by the operation of a hydraulically driven piston. The supernatant leaves via the top of the centrifuge.

Biological particle aggregates resulting from protein precipitation are subject to break-up when they are exposed to a high level of shear rate or shear stress such as those imposed on them in the feed zone of industrial centrifuges and especially at the entry point to the centrifuge settling zone. The strength, size and density of the flocs are the most important factors determining the extent of break-up and resulting particle size distribution to be separated in the conical discs [7]. However it is not possible to measure directly the break-up which occurs in the feed zone and indirect estimation of this parameter is important to evaluate the importance of its effect on centrifuge performance.

The performance of a centrifuge is usually described by the grade efficiency curve. The grade efficiency, $T(d)$, is the fraction of particles of diameter d recovered in the sediment, and when plotted against the dimensionless size d/d_c yields the grade efficiency curve. The critical diameter, d_c, is the diameter of the smallest particle that may be recovered by the centrifuge. This critical diameter is predicted by Stokes Law. For a disc stack centrifuge the critical diameter is given by:

$$d_c = \left\{ \frac{18\mu Q \tan\theta}{K\,\Delta\rho\,\omega^2\,z\,\frac{2}{3}\pi\left[R_a^3 - R_i^3 - \frac{3z_L b_L}{4\pi}(R_a^2 - R_i^2)\right]} \right\}^{0.5} \tag{3}$$

where Q is the flowrate through the centrifuge, ω is the angular velocity of the bowl, θ is the half disc opening angle, z is the number of discs in the disc stack, R_a is the outer disc radius, R_i is the inner disc radius, z_L is the number of spacer caulks, b_L is the caulk width, μ is the dynamic viscosity of the carrier liquid and $\Delta\rho$ is the density difference between solids and liquid phase. The fluid and particle dynamics are extremely complex within the centrifuge and especially within the disc stack section. Equation 3 is based on a simplified analysis of particle recovery in the centrifuge which assumes fluid plug flow through the conical discs. For these ideal flow conditions the grade efficiency curve follows Stokes Law:

$$T(d) = (d/d_c)^2 \qquad \text{for} \quad d/d_c < 1$$
$$T(d) = 1 \qquad \text{for} \quad d/d_c > 1$$

In practice, such a curve is not followed due for example to flow disturbances causing resuspension of particles and the parabolic flow profiles in the disc stack. The actual grade efficiency curve of an industrial disc-stack centrifuge for the recovery of robust latex particles, i.e., no shear break up effects, is shown in Figure 6 and is compared with the theoretical curve based on Stokes Law and ideal flow conditions. In practice it can be seen that $T(d)$ approaches 1, i.e., complete particle recovery, when the particle diameter is approximately twice the critical particle diameter, i.e. $d/d_c{\sim}2$ [8].

Biological particles subject to break-up during centrifugation, exhibit a distinctly differing solids recovery efficiency as the comparison shows between the grade efficiency curves for robust latex particles, i.e., not subject to shear break-up, and protein precipitate recovery efficiency (Figure 6). While there is an apparent improvement in the grade efficiency curve at high values of d/d_c, this is in fact an artifact due to the break up of the large aggregates rather than their improved recovery. The tendency towards negative values of $T(d)$ at lower d/d_c is a result of this break up leading to the generation of large quantities of fine particles. Hence, the new form of grade efficiency curve is unsuitable for the specification of the centrifuge and it is necessary to break the process into two parts:
a) the feed of the material into the centrifuge settling zone,
b) the recovery of the material in the settling zone.
While the performance of the centrifuge for part (b) is known, i.e., the measured grade efficiency curve for latex particles, the effect of part (a) is not characterised and an indirect estimation technique is necessary to predict the extent of break up in the feed zone. Once known, then the effect of preconditioning of the precipitate to minimize such break up effects or the redesign of the centrifuge feed zone can be considered.

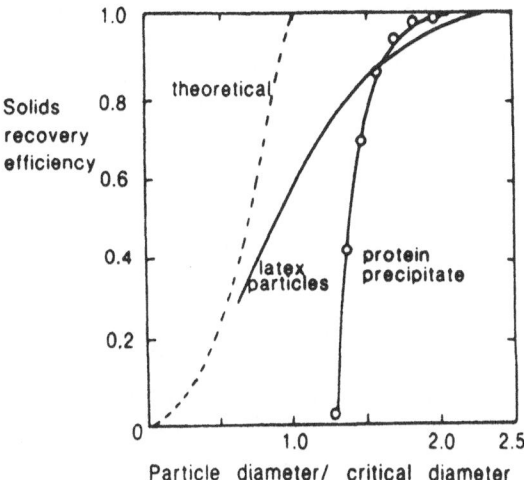

Figure 6.   The effect of particle break-up in the feed zone of a centrifuge on solids recovery efficiency.

An example of such a preconditioning process which has been recently developed at UCL is that of low-frequency conditioning of a flowing precipitate suspension with the precipitate in the nascent state [9]. A suspension of protein precipitate is subjected to low frequency vibration (5 Hz). As the acceleration of the fluid increases, collisions between particles are more frequent and aggregation is enhanced resulting in a decrease in the number of small particles by preferential aggregation with large particles. This leads to a difference in the cumulative size at the fine end of the size distribution, e.g. D_{95} value. D_{95} denotes the particle diameter above which 95% of the particles are larger by volume. S:microcentrifuge;There is an optimum fluid acceleration as, at high levels, resultant shear forces prevent aggregation. The observed improvement in particle size is maintained, albeit to a lesser extent, even after the aggregate suspension was subjected to capillary shear typical of that to which precipitate particles are exposed in a centrifuge feed zone (Figure 7). Low-frequency conditioning also has a positive effect on the density of the particles by enhancing the packing of the aggregate structure and therefore increasing the density difference between solids and fluid, thus further facilitating separation.

Figure 8 shows a system, currently under development, in which particle size distribution and concentration are monitored on-line to estimate particle break-up in a disc-stack centrifuge. This information is then used to determine the appropriate values of the operating variables of low frequency conditioning, namely frequency and amplitude of vibration.

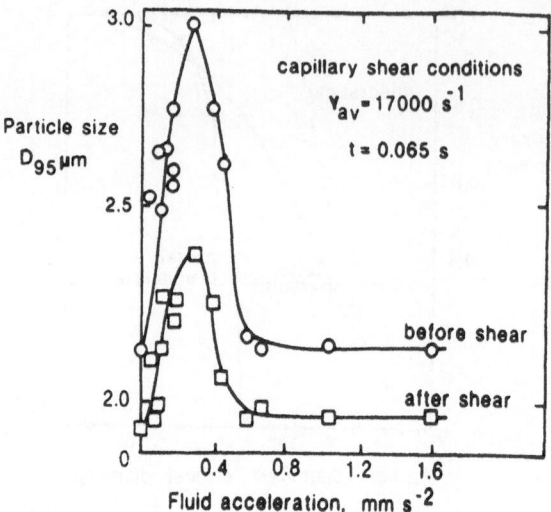

Figure 7. The effect of low frequency conditioning on precipitate size and resistance to shear. D_{95} denotes the particle diameter from which 95% of particles are larger by volume. (Titchener-Hooker N.J. et al, to be published).

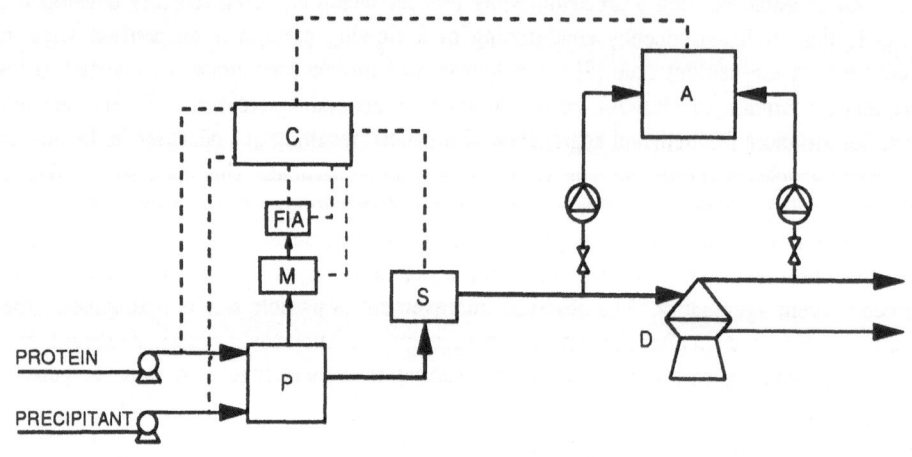

Figure 8. On-line monitoring of particle size and concentration for the control of shear break-up in a disc stack centrifuge. A:Size distribution and concentration analysis; D:disc-stack centrifuge; C:supervisory microcomputer; P:precipitation reactor; S:low frequency conditioning; M: microcentrifuge; FIA:flow injection analysis.

CONCLUSIONS

On-line monitoring of biological particle concentration and size distribution has proved a valuable tool for the estimation of particle break-up in the feed zone of industrial centrifuges. The method can be used for the on-line feedback control of a low-frequency conditioning aiming at producing precipitate particles of improved size, density and resistance to shear break- up.

The on-line monitoring of enzyme and total protein remaining soluble has been achieved using a microcentrifuge for automated sample preparation and flow injection analysis. The resultant solubility profiles generated on-line are used for process optimisation and as the basis for on-line control of fractional precipitation.

Both the FIA method used here and the novel microcentrifuge have a wider applicability on the monitoring and control of other processes in which soluble enzymes and/or proteins are of importance.

REFERENCES

1. E.J. Cohn, Strong L.E., Hughes W.L., Mulford D.J., Asworth J.N., Melin M. and Taylor J., "Preparation and Properties of Serum and Plasma Proteins. IV. A System for the Separation into Fractions of the Protein and Lipoprotein Components of Biological Tissues and Fluids", J. Amer. Chem. Soc., 1946, 68, 459-475.

2. Melander, W. and Horvath C., "Salt Effects on Hydrophobic Interactions in Precipitation and Chromatography of Proteins: An Interpretation of the Lyotropic Series ", 1977, Arch. Biochem. Biophys., 183, 200-215.

3. P.R. Foster, Dunnill P., Lilly M.D., "The Kinetics of Protein Salting-out: Precipitation of Yeast Enzymes by Ammonium Sulphate", Biotech. Bioeng., 28, 1976, 545-580.

4. P. Richardson, Hoare M. and Dunnill P., Optimisation of Fractional Precipitation for Protein Purification, Chem. Eng. Res. and Design, 1989, 67, 273-277.

5. Niktari M., Richardson P., Ravenhall R., Flanagan M.T., Molloy J., and Hoare M., The Modelling and Control of Fractionation Processes for Enzyme and Protein Purification, Proc. American Control Conf., 1989, 3, Omnipress Inc., USA, 2436-2440.

6. Bell D. J., Hoare M. and Dunnill P., The Formation of Protein Precipitates and their Centrifugal Recovery, Advances in Biochem. Eng./Biotech., 1983, 26, 1-72.

7. Bell D. J. and Dunnill P., The Influence of Precipitation Reactor Configuration on the Centrifugal Recovery of Isoelectric Soya Protein Precipitate, Biotech. Bioeng., 1982, 24, 2319-2336.

8. Mannweiler K., Titchener-Hooker N.J. and Hoare M., Biochemical Engineering Improvements in the Centrifugal Recovery, presented at IChemE Advances in Biochemical Engineering, 11-13 April 1989, University of Newcastle.

9. Hoare M., Titchener N.J. and Foster P.R., Improvement in Separation Characteristics of Protein Precipitates by Acoustic Conditioning, Biotech. Bioeng., 1987, 29, 24-32.

A STUDY OF PERMEABILITY AND HYDRODYNAMIC DISPERSION UNDER CONDITIONS OF CHROMATOGRAPHIC FLOW

KARIN ÖSTERGREN AND CHRISTIAN TRÄGÅRDH

Division of Food Engineering, Lund University
P.O.Box 124, S–221 00 Lund, Sweden

ABSTRACT

The influence of particle size and particle size distribution on permeability and hydro–dynamic dispersion was studied as the first stage in the development of a numerical model of chromatographic processes. Permeability and hydrodynamic dispersion were measured and models were tested against the data obtained. The influence of the chromatographic column wall was minimized by using a wide column ($d = 0.113$ m). Nonporous glass beads of different sizes (d_p(mean) = 125–250 μm) and size distributions (SD/d_p(mean) = 0.06–0.21) were used as packing material. The longitudinal dispersion was measured between Pe_p numbers 3 and 430 by recording the response of a sodium chloride step input at four radial positions at each of three levels of electrodes inside the column. Permeability was measured at the same time; the Re_p numbers lay between 0.004 and 0.6.

NOTATION

C	= concentration	Re_p	= particle Reynolds number, $u_o d_p \rho / \mu$
D	= dispersion coefficient		
D_L	= longitudinal dispersion coefficient	S	= source term
		t	= time
$\mathscr{D}$	= molecular diffusion coefficient	u	= interstitial velocity
		u_o	= superficial velocity
$\mathscr{D}_e$	= effective molecular diffusion coefficient	x	= axial position
d	= column diameter	Greek letters	
d_p	= particle diameter	ε	= porosity (void fraction)
d_{pore}	= pore diameter	μ	= viscosity
F	= force	ν	= kinematic viscosity
k	= permeability	ρ	= density
ℓ	= characteristic length, d_p for particles	σ	= standard deviation, zone spreading
p	= pressure	σ_p	= standard deviation, particle size distribution
Pe_p	= particle Peclet number, $d_p u / \mathscr{D}$	σ_r	= standard deviation, pore size distribution
r	= pore radius		

INTRODUCTION

Models are needed for efficiently scaling up and optimizing chromatographic processes. Numerous models have been developed to predict the performances of chromatographic processes [1,2,3,4,5,6,7,8,9,10,11,12,13,14] and in addition Verhoff et al. [15] have derived a model for calculating the permeability in a compressible bed.

Neither the chromatographic zone spreading in a compressible bed nor the influence of radial variations arising from proximity to the column wall are described in any of these models. We are in the course of developing a two–dimensional numerical mass transport model that is able to predict effects of compression and radial variations. The model is based on the following equations:

$$\frac{\partial(\rho\varepsilon)}{\partial t} + \nabla(\rho u_o) = 0 \tag{1}$$

$$\rho\frac{D(u)}{Dt} = -\Delta p + \mu\nabla^2 u + \Sigma\rho F \tag{2}$$

$$\frac{\partial C}{\partial t} + \nabla(uC) = \nabla(D\Delta C) + S \tag{3}$$

Equation (1) is the continuity equation for flow through a porous medium and Equation (2) is the momentum equation where $\Sigma\rho F$ represents the body forces. Equation (3) is the mass conservation equation for dissolved species where D is the dispersivity tensor and S the source term which links these equations to the set of equations that describe the mass exchange between interparticle space and the chromatographic medium. Under isotropic conditions and at low Re_p numbers the continuity equation and the momentum equation can be reduced to Darcy's law [16]

$$u_o = k\frac{\rho}{\mu}\frac{dP}{dx} \tag{4}$$

On compression permeability (k) becomes a function of the position in the porous bed.

So that a mass transport model can be developed it is necessary to calculate permeability (k) and interparticle dispersion (D), both of which are related to the packing structure of the chromatographic bed, and in the case of the interparticle dispersion the fluid velocity and molecular diffusion have also to be taken into consideration. The packing structure is influenced by particle size and size distribution, porosity, particle shape, etc.

We have investigated the influence of particle size and particle size distribution on permeability and hydrodynamic longitudinal dispersion (interparticle dispersion) under chromatographic flow conditions to obtain information about how these parameters are related to the packing material and to find suitable models for predicting permeability and longitudinal hydrodynamic dispersion.

Permeability

A number of advanced models for porous media have been developed [16,17,18]. They describe the hydrodynamics well but are not suitable for practical calculations.

Various types of simplifications have given rise to different models for predicting permeability based on measurable properties of a porous medium. We have used models based on conduit flow in capillaries.

A common approach is to consider the porous medium as a bundle of straight capillaries and the pressure drop is calculated from Hagen–Poiseuille's law. Scheidegger [cited in ref. 18] suggested the following model, where $\alpha(d_{pore})$ is the normalized distribution of the pore diameters based on the pore volume distribution:

$$k = \frac{\varepsilon}{96} \int_0^\infty d^2_{\text{pore}}\alpha(d_{\text{pore}})dd_{\text{pore}} \tag{5}$$

Marshall et al. [19] developed another model also based on Hagen–Poiseuille's law and they proposed the following relation between the permeability and the pore radius distribution:

$$k = \frac{\varepsilon^2}{8n^2} (r_1^2 + 3r_2^2 + 5r_3^2 + ... + (2n - 1)r_n^2) \tag{6}$$

where r represents the mean radius of equal fractions of pore space in decreasing order from r_1 to r_n.

One of the most frequently used equations for predicting the permeability is the Kozeny–Carman equation [cited in ref. 16]. It was derived by solving the Navier–Stokes equations simultaneously for all channels passing through a cross–section normal to the flow [16]. Introduction of the hydraulic radius, defined as the ratio of volume of fluid to the wetted surface area, leads to [16]:

$$k = \frac{\overline{d_p}^2}{C_{kc}} \frac{\varepsilon^3}{(1-\varepsilon)^2} \tag{7}$$

where C_{kc} is equal to 180. In the hydraulic radius models the average particle diameter characterizes the porous medium. The value for k varies according to how the mean particle diameter is defined. The most consistent values are achieved when the average diameter is defined according to [18]:

$$\overline{d_p} = \left[\int_0^\infty d_p(d_p^2)n(d_p)dd_p \right] \Big/ \left[\int_0^\infty d_p^2 n(d_p)dd_p \right] \tag{8}$$

where $n(d_p)$ is the particle size distribution.

Rumpf and Gupte [20] pointed out that the "constant" in the Kozeny–Carman model varies with porosity and presented the following relation:

$$k = \frac{\overline{d_p}^2 \varepsilon^{5.5}}{5.6} \tag{9}$$

Interparticle dispersion

The dimensionless coefficient of hydrodynamic dispersion $(D/\mathscr{D})$ is usually correlated to the Pe_p number. The Pe_p number for granular materials is defined according to $Pe_p = d_p u/\mathscr{D}$. Mechanisms for the longitudinal hydrodynamic dispersion (D_L) vary with the Pe_p number [16,17]. The molecular diffusion predominates when the Pe_p number is low; $(D_L/\mathscr{D})$ is constant. Between Pe_p numbers c. 0.4–5 the molecular diffusion is of the same order of magnitude as the convective diffusion, and the mechanisms are additive. Between the Pe_p numbers c. 5–50 the longitudinal hydrodynamic dispersion is caused by convective diffusion combined with transverse molecular diffusion. The two mechanisms interfere with each other and are not additive. The influence of molecular diffusion can be disregarded when Pe_p–numbers are higher than 50, but at very high Pe_p numbers (c.10,000) the effects of turbulence and inertia interfere. In chromatographic applications the convective diffusion is usually the major mechanism; molecular diffusion can usually be disregarded.

Assumptions about the nature of the porous media and the flow through it have given rise to a number of models for predicting hydrodynamic dispersion [16,17,18]. Some are purely geometric continuum models and others are statistical. There are also several empirical relations based on correlation of data.

Saffman developed a statistical model for longitudinal hydrodynamic dispersion when molecular diffusion and convective diffusion interact [21]. The model consists of a random network of capillaries and is described by

$$DL = \frac{ul}{6}\left[\ln\frac{3}{2}\frac{ul}{\mathcal{D}} - \frac{17}{12} - \frac{1}{8}\frac{r^2}{l^2}\frac{ul}{\mathcal{D}}\right] + \mathcal{D}_e + \frac{4}{9}\mathcal{D} + 0\left[\frac{\mathcal{D}^2}{ul}\right] \tag{11}$$

where the pore radius is calculated from the permeability ($r^2 = 24k/\varepsilon$), l is the characteristic dimension and $\mathcal{D}_e$ the effective molecular diffusion which is typically $0.67\,\mathcal{D}$ in granular beds. The model applies as long as the following assumption is valid:

$$\frac{ul}{\mathcal{D}} \ll \frac{8l^2}{r^2} \tag{12}$$

Earlier Saffman used the random walk method making the assumption that the dispersion is primarily due to macroscopic mixing [22]. He derived the following relation:

$$DL = \frac{ul}{6}\left[\ln\frac{3}{2}\frac{ul}{\mathcal{D}} - \frac{1}{4}\right] \tag{13}$$

Harleman et al. [cited in ref. 18] has made a correlation between the permeability and the longitudinal hydrodynamic dispersion coefficient

$$DL/\nu = C\,Re_k^{1.2} \tag{14}$$

where Re_k varies with permeability according to $Re_k = u\sqrt{k}/\nu$. For spheres the constant C is equal to 54. This type of relation only exists for equal porosity, equal grain size distribution and equal particle shape. The model does not take into account the influence of molecular diffusion.

An empirical expression for longitudinal dispersion in beds of spheres is given by Hilby [cited in ref. 21]

$$DL = 0.67\,\mathcal{D} + \frac{0.65ul}{1 + 6.7(\mathcal{D}/ul)^{1/2}} \tag{15}$$

Niemann [cited in ref. 21] has shown that the hydrodynamic dispersion varies with particle size distribution.

MATERIAL AND METHODS

The column was a Pharmacia BP113 with an inner diameter of 0.113 m and the height was 0.3 m. Three levels of concentric conductivity electrodes were mounted inside the column at distances of 1.8 cm, 11.6 cm and 22.3 cm from the inlet. The electrodes were placed 0.3 cm, 1.3 cm, 2.4 cm, 3.5 cm and 4.6 cm from the column wall. Figure 1 shows the experimental set-up.

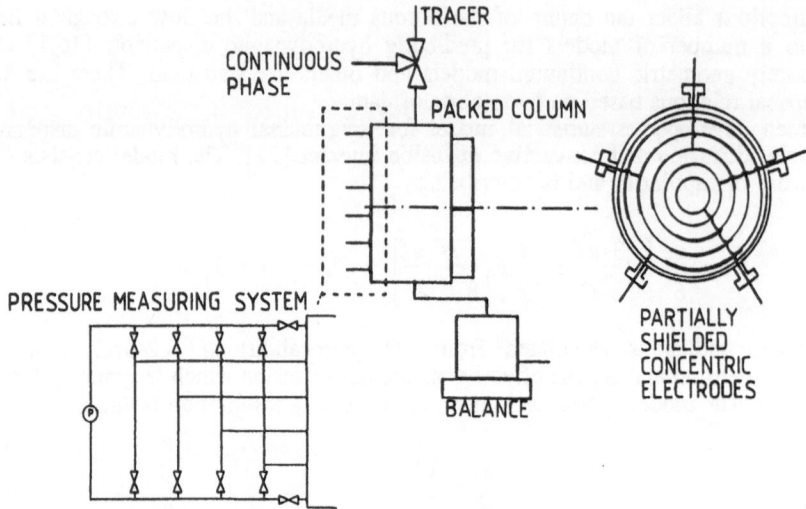

Figure 1. Experimental apparatus

The design assured fully developed dispersion in all experiments. According to Han et al. [23] this is achieved when:

$$\frac{x}{d_p}\left[\frac{1}{Pe_p}\right] \geq 0.3 \tag{16}$$

The electrodes were made of stainless steel (diam =1 mm). They were partially shielded so that a uniform electrode constant was achieved. The concentrations of NaCl were 1 mg/l and 9 mg/l respectively. It was confirmed that the electrodes had a linear response within this interval. The pressure was measured by a difference pressure transducer at levels of 2.7 cm, 6.0 cm, 11.5 cm, 16.3 cm, 21.0 cm and 26.1 cm. The flow was kept constant by applying a constant pressure along the column. The glass beads used as packing material had average diameters of 125–250 μm. The particle size distributions were determined by photomicrography and the pore size distribution by mercury porosimetry on a Micromeritics pore sizer 9310.

Preparation of packing material

Different size distributions were obtained by sieving. The size distributions lay within the range of most of the common size distributions used in low pressure process chromatography.

Preparation of column

To ensure that none of the beads stuck together we dried and sieved them before packing. Small portions of dry beads (about 50 g) were poured into the column and were closely packed by gently tapping the column wall. The column was then filled with degassed continuous phase, care being taken to leave no air in the column.

Experiments

The response of a sodium chloride step, pressure drop, flow and temperature were measured in all experiments. The step response was monitored at all positions along the column during each experiment, by using a scanner. The interstitial velocity range was 0.003–0.22 cm/s and Re_p was 0.004–0.6.

Calculations

Porosity was calculated from $\varepsilon = u_o/u$. The interstitial velocity was obtained from the dispersion measurements and the superficial velocity from the flow measurements.

The permeability was calculated from the slope of a curve where pressure was plotted against velocity (according to Darcy's law).

When the hydrodynamic dispersion is not extensive it can be calculated from

$$\frac{\sigma_2^2 - \sigma_1^2}{\bar{t}^2} = 2\left[\frac{D_L}{ux}\right] \tag{17}$$

where σ^2 is the variance measured at two positions, $\bar{t}$ the time interval between the measurements and x the distance between the probes [24]. The step response was analysed by calculating the internal residence time distribution by numerical integration. The first to the third moments were calculated. The first moment is equal to the mean time, the second is equal to the variance and the third is a measurement of the skewness of the distribution and was mainly used as a check on how well the bed was packed.

RESULTS

Porosity

The porosity lay within the range of 0.363–0.370. This is close to random close–packing, which is 0.36 for equal–sized spheres [17]. The particle size distribution affected the porosity only slightly (Figure 2).

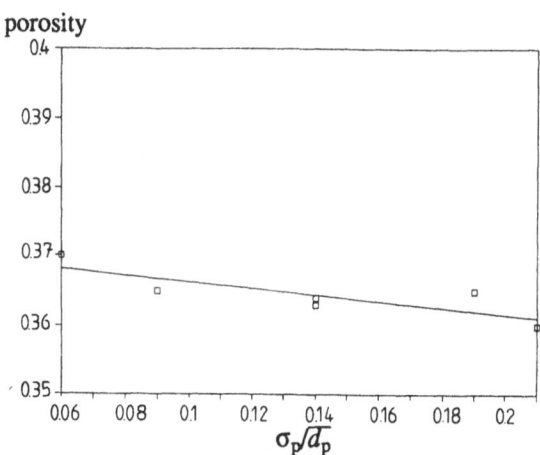

Figure 2. The porosity at different σ/d_p. σ_p = standard deviation of particle size distribution.

The result is comparable with the results reported by Sohn et al.[25], who have experimentally investigated the relation between particle size distribution and porosity.

Flow pattern

The flow pattern in the column was investigated. Figure 3 shows the average velocity profile for all experiments and the radial distribution of the measured dispersion coefficients.

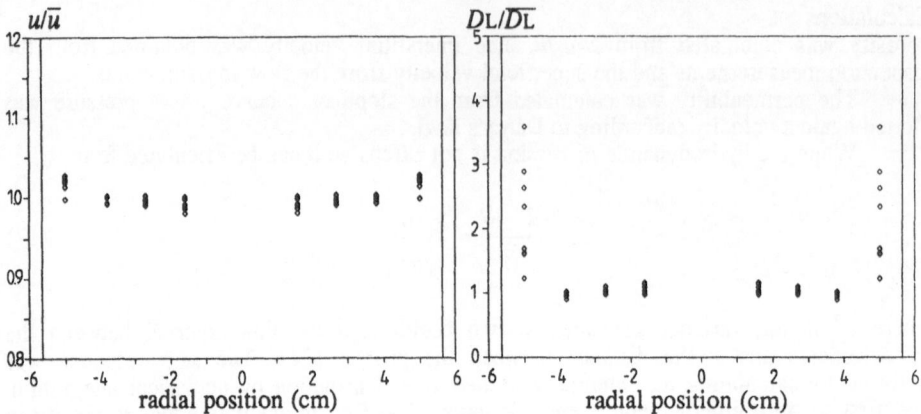

Figure 3. The velocity profile ($u = x/\bar{t}$) and the variation of D_L (calculated from Eq. 17) across the column. Each point represents an average of one experimental series.

Each point represents an average of one experimental series. Neither the velocity profile nor the dispersion profile showed any dependence on velocity or packing material. The velocity profile is fairly flat, as would be expected since porosity variations arising from proximity to the column wall will only affect a region of approximately six particle diameters [26]. However, Figure 4 shows that the flow pattern was disturbed close to the wall. In experiments involving a temporarily longer column and change in inlet flow direction, the measurements indicated that the disturbances were results of uneven packing, probably because the outer electrodes had been placed too close to the column wall.

Permeability
The measured permeabilities and the permeabilities calculated from Equations (5), (6),(7) and (9) are summarized in Table 1.

TABLE 1
Experimental and calculated permeability

Bed characteristics Particle range				Permeability, k, (cm^2*10^{-6})				
(μm)	ε	$\sigma_p/\bar{d}_p$	$\sigma_r/\bar{r}$	k_{exptl}	k(Eq. 5)	k(Eq. 6)	k(Eq. 7)	k(Eq. 9)
125–140	0.370	0.06	0.35	0.108	0.105	0.089	0.116	0.123
100–160	0.365	0.09	0.38	0.106	0.101	0.080	0.109	0.113
125–224	0.363	0.14	0.38	0.224	0.207	0.147	0.190	0.196
100–250	0.365	0.19	0.35	0.249	0.230	0.179	0.241	0.250
100–330	0.360	0.21	0.32	0.338	0.349	0.273	0.369	0.378
130–330	0.364	0.14	0.31	0.53	0.44	0.376	0.424	0.441

ε = porosity; d_p = average particle diameter; σ_p = standard deviation, particle size distribution; r = average pore radius; σ_r = standard deviation, pore radius

Table 1 shows that the permeability can be satisfactorily predicted both by using models based on pore size distribution and models based on hydraulic radius. The empirical correlation described by Equation 9 is also satisfactory. Scheidegger's model (Eq. 5) and the Kozeny–Carman equation (Eq. 7) give values closest to our measured values. It is pointed out in the literature [16] that the numerical coefficient in the equations has to be adjusted for different porous materials. According to our measurements the coefficient of equation 5 should be adjusted from 96 to 90 and the Kozeny–Carman constant from 180 to 174.

Dispersion

The measured values of the dimensionless longitudinal dispersion, $DL/\mathcal{D}$, at different Pe_p numbers are shown in Figure 4.

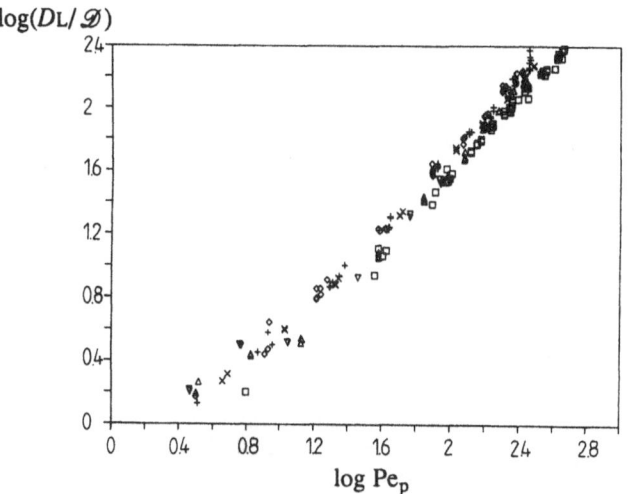

Figure 4. Average dimensionless dispersion coefficient ($\mathcal{D} = 1.545*10^{-6}cm^2/s$ at $25^{\circ}C$) measured at the two middle radial positions. □ $\bar{d}_p = 244$ μm, $\sigma_p/\bar{d}_p = 0.19$; ◊ $\bar{d}_p = 164$ μm, $\sigma_p/\bar{d}_p = 0.14$; △ $\bar{d}_p = 126$ μm, $\sigma_p/\bar{d}_p = 0.09$; × $\bar{d}_p = 224$ μm, $\sigma_p/\bar{d}_p = 0.21$; ▽ $\bar{d}_p = 127$ μm, $\sigma_p/\bar{d}_p = 0.06$

The models (Equations (11),(13)–(15)) were evaluated graphically. An approximate measurement of the accuracy of the models was obtained by calculating the average relative deviations from the measured values at two different Pe_p intervals (Table 2).

TABLE 2
Average relative deviation of calculated dispersion coefficients

Model		Relative deviation DL/DL_{exptl} ($Pe_p<50$)	DL/DL_{exptl} ($Pe_p>50$)
Saffman	(Eq. 11)	1.12	0.95
Saffman	(Eq. 13)	1.406	1.53
Harleman et al.	(Eq. 14)	1.10	1.06
Hilby	(Eq. 15)	0.92	0.74

The graphical evaluation showed that Saffman's statistical model (Eq. 11) was the most successful model for low Pe_p numbers and that the model developed by Harleman et al. (Eq. 14) was the better model for larger Pe_p numbers (Figure 5).

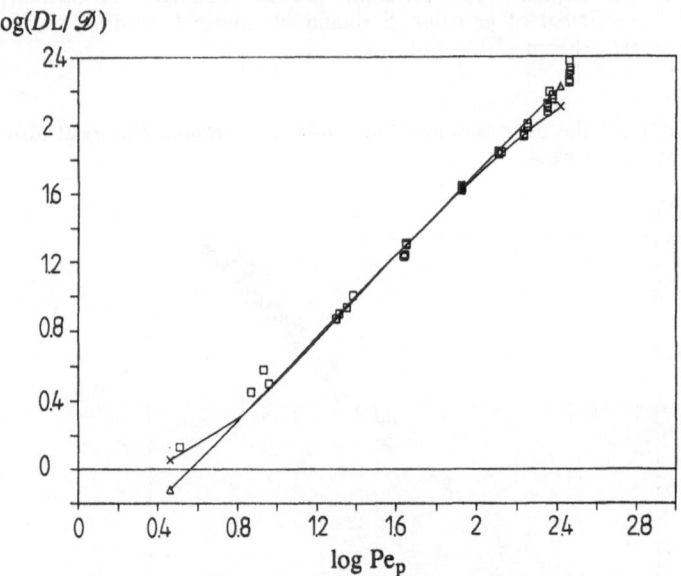

Figure 5. Measured dispersion coefficients and theoretically calculated values. □ measured values ($\bar{d}_p$ = 177 µm, $\sigma_p/\bar{d}_p$ = 0.19); Δ–Δ model according to Harleman et al. (Eq. 14); x–x model according to Saffman (Eq. 11)

CONCLUSIONS

We have constructed and tested experimental equipment for studying hydrodynamics in two dimensions in a chromatographic column.

The influence of particle size and particle size distribution was investigated and the results showed that size distribution has very little influence on the porosity and on interparticle dispersion under chromatographic flow conditions.

As a next step the influence of compression on porosity, characteristic length and pore size distribution will be investigated experimentally and theoretically so that the permeability and interparticle dispersion can be predicted by relating the changes in the chromatographic bed to the models described here.

ACKNOWLEDGEMENT

This work is being financially supported by the Swedish National Board for Technical Development.

REFERENCES

1.   Giddings, J.C., <u>Dynamics of Chromatography, Part 1: Principles and theory,</u> Marcel Dekker, New York, 1965
2.   De Ligny, C.L.,The contribution of eddy diffusion and of the macroscopic mobile

phase velocity profile to plate height in chromatography, A literature investigation, J. Chromatogr., 1970 49 393–401

3. Weiss, G.H., Contributions to the stochastic theory of chromatographic kinetics, Separ. Sci., 5 1970 51–62

4. Chen, J–C. and Weber, S.G., Theoretical and Experimental determination of band broadening in liquid chromatography, Anal. Chem., 1983 55.127–134

5. Gut, A., and Ahlberg, P., On the theory of chromatography based upon renewal theory and a central limit theorem for randomly indexed partial sums of random variables, Chemica Scripta, 1981 18 248–245

6. Rasmuson, A., Exact solution of a model for diffusion and transient adsorption in particles and longitudinal dispersion in packed beds, AICHE. J., 1981 27 1032–1035

7. Arnold, F.H., Blanch, H.W. and Wilke, C.R., Analysis of affinity separations I: Predicting the performance of affinity adsorbers, Chem Eng.J. 1985 30 B9–B25

8. Arnold, F.H., Blanch,H.W. and Wilke, C.R., Analytical affinity chromatography, 1:Local equilibrium theory and the measurement of association and inhibition constants, J.Chromatogr. 1986 355 1–12

9. Smit, J.C. and Smit, H.C., Computer implementations of simulation models for non–linear, non–ideal chromatography, Anal. Chim. Acta, 1980 122 1–26

10. Chilocote, D.D. and Scott, C.D., A theoretical derivation of the axial diffusion coefficient in chromatography, J. Chromatogr., 1973 87 315–323

11. Giddings, J.C., Schettler; P.D., General nonequilibrium theory of chromatography with complex flow transport, J. Phys. Chem., 1969 73 2577–2589

12. Kubin, M., Beitrag zur Theorie der Chromatographie Collection Czechoslov. Chem. Commun., 1965 30 1104–1118

13. Kucera, E., Contribution to the theory of chromatography linear non–equilibrium elution chromatography, J. Chromatogr., 1965 19 237–248

14. Cownan, G.H., Gosling, I.S., Laws, J.F. and Sweetenham, W.P.,Physical and mathematical modelling to aid scale–up of liquid chromatography, J. Chromatogr., 1986 363 37–55

15. Verhoff, F.H.and Furjanic Jr, J.J., Compressible packed bed fluid dynamics with applications to a glucose isomeras reactor, Ind. Eng. Process Des. Dev., 1983 22 192–1983

16. Bear, J., Dynamics of Fluids in Porous Media, American Elsevier, New York,1972

17. Greenkorn, R.A., Flow phenomena in porous media Fundamentals and applications in petroleum, water and food production, Marcel Dekker, New York, 1983

18. Dullien, F.A., Porous Media Fluid Transport and Pore Structure, Academic Press, New York, 1979

19. Marshall, T.J., A relation between permeability and size distribution of pores, J. Soil Sci., 1958 9 1–8

20. Rumpf. H. and Gupte A.R., Einflüsse der Porosität und Korngrössenverteilung im Widerstandgesetz der Porenstömung, Chemie. –Ing. –Techn. 1971 43 367–375

21. Saffman,P.G., Dispersion due to molecular diffusion and macroscopic mixing in flow through a network of capillaries, J. Fluid Mech., 1960 7 194–208

22. Saffman, P.G., A theory of dispersion in a porous medium, J. Fluid Mech. 1959 6 321–349

23. Han,N–W., Bhakta,J., Carbonell,R.G., Longitudinal and lateral dispersion in packed beds: Effect of column Length and particle size distribution, AIChE J, 1985 31 277–287

24. Levenspiel, O., Chemical Reaction Engineering, John Wiley & Sons Inc., New York, 1972, pp. 253–315

25. Sohn, H.Y., Moreland, C., The effect of particle size distribution on packing density. Can. J. Chem. Eng., 1968 46 162–167

26. Vortmeyer, D., Die mathematische Modellierung von Reaktions– und Austauschprozessen in durchströmten Festbetten unter Berüksichtigung von ungleichmäßigen Strömungsverteilungen, Wärme–Stoffübertag., 1987 21 247–257

SIMPLE DIAGNOSTIC GUIDELINES FOR ADSORPTION COLUMN PERFORMANCE

NEALE THOMAS
FRED Ltd, R & D Institute, Vincent Drive, Birmingham B15 2SQ.
Correspondence to:
FAST Team, School of Chemical Engineering,
Birmingham University, Edgbaston, Birmingham B15 2TT, UK.

ABSTRACT

An algebraic approximation for constant pattern concentration profiles is found in terms of two dimensionless groups representing the ratios of diffusive to convective flux and desorption to adsorption flux. The analysis is developed without appealing to any particular kinetic formulation although linear kinetics are shown to be incompatible with constant pattern behaviour. Explicit formulae are given for the breakthrough profiles in the case of Langmuir isotherms and a compact statement of the exact solution for non-diffusive conditions is recovered in that limit. The approximate solution is shown in another paper [1] to be asymptotically exact in the limits of small and large values of the diffusivity. The coupling of diffusion and desorption causes displacement and distortion of the breakthrough profile, an effect which suggests a simple method for in situ estimation of the effective diffusivity.

The present algorithm and its practical benefits are contrasted with BIOSEP's FACSIMILE code reported in Cowan et al [2]. Operational implications of the present scaling groups are compared with Chase's [3] conclusions based on his numerical survey of Thomas' [4] solution for non-diffusive transport. Recent work on the elucidation of optimal performance conditions, to be incorporated into a software package 'FACT' (Fast Analysis of Chromatography Tubes), are mentioned. The analysis and diagnostic implications will be reported at a later date.

1. INTRODUCTION

An impressive hierarchy of theoretical and computational tools is now available to investigate the factors governing performance of packed bed adsorption equipment. At the most fundamental level, Koch et al [5] described a first principles derivation for the effective diffusivity of disordered matrices in terms of the coupling between flow dispersion and molecular diffusion. Of more immediate practical interest is the simplified treatment by Clark, Bailey & Do [6] for restricted diffusion by macromolecules entering porous supports. At the level of applicable guidelines, Arnold et al [7] have provided a semi-empirical link between the microstructure transport and the macroscale parameters of operational significance. However, there are some drawbacks and uncertainties in implementing their formulae: firstly, prior knowledge of the performance limiting factor (ie dispersion, diffusion or kinetics); secondly, semi-empirical parameterisation (eg pore diffusion number); thirdly, enumeration of Hiester & Vermeulen's 'J' functions; lastly, coupling of transport coefficients is uncertain, as Koch et al have demonstrated.

Until these newer formulations are more securely established, there remains a useful practical role for traditional modelling in terms of macroscale transport parameters. However, although conceptually simple, these models have not been widely adopted for everyday exploitation in the laboratory because the existing algorithms are either analytically awkward or computationally intensive. For example, Chase's [3] evaluation of Thomas' [4] exact solution for the evolving profiles of nondispersive transport with Langmuir kinetics involves the enumeration of 'J' function asymptotics. The latter obscure the essentially simple scalings which govern the ultimate shapes of the concentration profiles: ie constant pattern behaviour which, for practical purposes, might reasonably be assumed ab initio because the initial zone of profile establishment is no more than about one column diameter. Moreover, the absence of dispersion in Thomas' model is a real shortcoming, as Chase conceded in commenting on the discrepancies between his measured and calculated breakthrough profiles.

Cowan et al [2] described a computational scheme which incorporated longitudinal dispersion in the one-dimensional evolution equations with Langmuir kinetics. However, their 'FACSIMILE' code, promoted by the commercial club 'BIOSEP' at Harwell Laboratories, must seem equally daunting to biological practitioners. The numerical procedure was essentially the method of lines: a fourth order library routine based on Gear's method to solve the ordinary differential equations resulting from first order spatial discretisation of the convective-diffusive transport terms. To minimise the number of ordinary differential equations, their mesh travelled at the constant pattern wave speed given by the asymptotic state of Thomas' nondiffusive solution. They then estimated the packed bed diffusivities in Chase's experiments by minimising mismatches between the computed and measured breakthrough profiles, an exercise which is notoriously prone to numerical errors as has long been recognised in the literature on gasdynamic calculations of shock-wave thickening in the presence of viscous diffusion.

The present paper seeks to demonstrate an elementary algebraic approximation under the constant pattern approximation. In section 2 we show how the governing equations can be rescaled onto two dimensionless groups. In section 3 we comment on the qualitative implications for column operation and demonstrate that linear kinetics are incompatible with constant pattern behaviour. In section 4 we identify explicit approximations for the diffusive profiles and give simple algebraic formulae for the particular case of Langmuir kinetics. The operational guidelines suggested by Chase [3] are reviewed in the light of an exact solution recovered in the limit zero diffusivity. In section 5 we comment on the practical implications for in situ estimation of the effective diffusivity.

2. CONSERVATION EQUATIONS AND SCALING ANALYSIS

The material transport equations (Chase [3]; Cowan et al [2]) for free and bound adsorbate are of the general form (see figure 1):

$$\partial c/\partial t + u\partial c/\partial x - D\partial^2 c/\partial x^2 = -e \quad ; \quad \partial q/\partial t = e \quad ; \quad e = e(c, q; q^*) \tag{1}$$

Here t is time, x is distance measured in the flow direction along the device and u is the interstitial convection speed within a dispersive packed bed, effective diffusivity D; c and q are the volume concentrations of free and bound adsorbate. Notice that e, the rate of exchange from c to q is taken to depend on c, q, and some limiting concentration q^* of bound adsorbate, the 'maximal capacity' in Chase's terminology. Although no specific functional form need be assigned to e for purposes of developing our general procedure, we shall eventually focus on Langmuir kinetics as advocated by Chase and adopted by Cowan et al, namely:

$$e = k^+c(1 - q/q^*) - k^-q \qquad (2)$$

where k^+ is the forward rate constant, here scaled on q^* for dimensional consistency (ie with $k^+ = Kq^*$, where K would be k_1 in Chase's notation) and k^- is the backwards rate constant (ie k_2 in Chase's notation). Clearly the units of k^+ and k^- are both inverse time, as befits rate constants for first order processes. The limitations of this 'lumped parameter' approach (eg no mass transfer limitations) are surveyed in the literature review by Cowan et al [3].

Constant pattern behaviour
A steady-state description is secured via the travelling coordinates:

$$\xi = x - Ut \ , \ \tau = t \qquad (3)$$

where U is a kinematic speed determined from the compatibility condition that far behind the frontal wave the concentration approaches inlet value $(c \to \hat{c}$, say), and $q \to \hat{q}$ (the 'attainable capacity'), such that:

$$U\hat{q} = (u-U)\hat{c} \ , \ \text{ so that } \ U = u\hat{c}/(\hat{c} + \hat{q}) \ . \qquad (4)$$

The attainable capacity $\hat{q}$ follows from the equilibrium condition, so that

$$e(\hat{c}, q; q^*) = 0 \quad \text{ defines } \quad q(\hat{c}, q^*) = \hat{q} \ . \qquad (5)$$

Far ahead of the front $c \to 0$ and $q \to 0$, such that the equilibrium condition

$$e(0, q; q^*) = 0 \quad \text{ implies } \quad q(0; q^*) = 0 \qquad (6)$$

These points are best illustrated with reference to a specific exchange function: of main interest here are Langmuir kinetics whose exchange function (equation 2) obviously satisfies the equilibrium condition (equation 6) and yields an attainable capacity given by

$$k^+\hat{c}(1 - \hat{q}/q^*) - k^-\hat{q} = 0 \ , \ \text{ ie } \quad \hat{q} = \hat{c}(k^-/k^+ + \hat{c}/q^*)^{-1} \qquad (7)$$

For this particular case, the frontal speed is given by:

$$U = u[1 + (k^-/k^+ + \hat{c}/q^*)^{-1}]^{-1} \qquad (8)$$

and so is now explicitly identified in terms of the empirically known kinetic parameters and prescribed $\hat{c}$.

Two-parameter description

The elementary scaling ideas reported here are most straightforwardly revealed if the exchange function is represented as an explicit combination of forward and backward rate processes, namely

$$e = k^+ c e^+ - k^- q e^- \tag{9}$$

in which $e^{\pm}$ are dimensionless functions of c, q, and q^*. For Langmuir kinetics, then, $e^+ = 1 - q/q^*$ and $e^- = 1$ and we see that for $q/q^* \ll 1$ – which may often be the case in practical applications, the exchange function is approximately linear in c and q. Although tempting on grounds of analytical simplification, in fact the linearised exchange function (constant $e^{\pm}$) is excluded on grounds of self-consistency.

We can minimise the number of parameters appearing in the transport equations by scaling according to the following dimensionless quantities:

$$q' = q/q^* \quad , \quad c' = (\hat{q}/q^*)c/\hat{c} \quad , \quad k' = (\hat{q}/\hat{c})k^-/k^+ \quad , \tag{10a}$$

$$\xi' = (1 + \hat{c}/\hat{q}) k^+ \xi/u \quad , \quad D' = (1 + \hat{c}/\hat{q})^2 k^+ D/u^2 \quad . \tag{10b}$$

With these substitutions, it can be shown (as described in [1]) how the equations and their far-field conditions become

$$q' = c' - D'dc'/d\xi' \quad , \quad dq'/d\xi' = -e'(c',q') \tag{11a}$$

$$e' = c'e^+ - k'q'e^- \quad , \quad e^{\pm} = e^{\pm}(c',q') \quad , \tag{11b}$$

$$q' \to 0 \quad \& \quad c' \to 0 \quad , \quad \text{so} \quad e' \to 0 \quad \text{as} \quad \xi \to \infty \tag{11c}$$

$$c' \to q' \quad \& \quad e' \to 0 \quad , \quad \text{ie} \quad e^+ = k'e^- \quad \text{as} \quad \xi' \to -\infty \quad . \tag{11d}$$

This last condition is more usefully expressed as

$$c' \to \hat{c}' \quad , \quad q' \to \hat{q}' \quad \& \quad \hat{c}' = \hat{q}' \quad \text{as} \quad \xi' \to -\infty \tag{11e}$$

ie $e^+(\hat{c}',q') = k'e^-(\hat{c}',q')$ defines $q'(\hat{c}',k') = \hat{q}' = \hat{c}'$.

Judicious selection of scaling units has reduced the original set of seven dimensional parameters (ie q^*, $\hat{q}$, $\hat{c}$, k^+, k^-, u and D) to just two independent groups D' and k'. Physically, D' measures the ratio of diffusive to advective fluxes of free material and k' the ratio of desorptive to adsorptive fluxes of bound material. Notice that the equilibrium constraint (equation 6), here $e' = 0$ when $c' = q' = 0$, is now explicitly accomodated in formulating equation 11b for the exchange function.

The dimensionless group D' can be regarded as an inverse Peclet number in which the length scale is taken as u/k^+: ie the convection distance in a time scale representative of the adsorption rate. The same length scale appears for dimensionless distance ξ', in both cases simply because there is no externally imposed length in the problem as posed here. D' is akin to a "Hatta" number: refer to the chemical engineering literature for details.

3. QUALITATIVE IMPLICATIONS

Favourable operating conditions

A useful indication of favourable operating conditions can be had simply in terms of this scaling analysis. Indeed, equation 8 for the frontal speed suffices to demonstrate that increasing $\hat{c}$ and decreasing q^* together favour efficient practical operations. Equation 8 demonstrates how both conditions act to increase the frontal speed and hence to reduce the delay time for 'break-through' of the frontal profile. More particularly, it is their ratio $\hat{c}/q^*$ which is the governing factor here, together with the ratio k^-/k^+ of the rate constants. Increasing the latter also reduces the delay time, but this is clearly not an attractive option with regard to the attainable capacity; see equation 7. In fact, Chase [3] advocated the condition $\hat{c}/q^* > k^-/k^+$ and we see immediately from equation 7 that his guideline will furnish an attainable capacity greater than one half the maximal capacity. Note that for the biological affinity systems studied by Chase, both k^-/k^+ and $\hat{c}/q^*$ are very much less than unity and typically comparable, say 1/100 in magnitude order. Under these circumstances, low frontal speeds $(U/u \sim 1/100)$ and hence long delay times are inevitable; only when $\hat{c}/q^*$ or k^-/k^+ is large compared with unity does the frontal speed approach the flow speed; see equation 8. These arguments all still hold even when we relax the assumption of constant D.

Existence of optimal conditions

Without recourse to detailed analysis, we can also identify an implication of these scaling arguments for the way in which practical performance may already be anticipated to depend on u, the interstitial flow rate per unit area. According to our definitions of equation 10, ξ' varies as u^{-1}, D' as u^{-2} and all other quantities are independent of u. Decreasing u thus encourages not only a sharpening of the q' profile (because ξ' increases) but also a broadening of the c' profile (because D' increases faster than ξ'); see equations 10b. Thus, for any prescribed diffusivity D (presumed to be independent of u, at least in part - viz the molecular component; see also immediately below) there must obviously be an 'optimal' value of u which represents the best practical compromise between these competing effects. Only when D is identically zero does profile sharpening continue indefinitely with diminishing u, as Chase concluded from his numerical survey of Thomas' solution for non-diffusive equations.

It is also interesting to link this implication with the theoretical findings reported by Koch et al [5] for the effective diffusivity of particulate media. Briefly, they reported several component contributions and emphasised some crucial distinctions between ordered and disordered media. Thus Taylor dispersion $D \sim u^2 l^2/\varkappa$ and boundary-layer dispersion $D \sim ul\{Ln(ul/\varkappa)\}$ occur in both, mechanical dispersion $D \sim ul$ arising from the random transport occurs only in disordered media and an enhanced molecular diffusion $D \sim \varkappa$ occurs in both. Here l is a microstructure length scale and $\varkappa$ is the molecular diffusivity. For sufficiently small values of ul, then, the implications for D' above still hold; indeed, they broadly hold unless the Taylor component dominates (ie $D' \sim$ constant).

Exclusion of linear kinetics

The far-field conditions of equations 11e impose some restriction on the class of physically admissable formulations of $e^{\pm}$ that are consistent with our travelling steady-state description: linear kinetics, for example, are definitely excluded. To see this, consider that with constant e^{+} and e^{-} in the exchange function of equation 11b, the far-field condition of equation 11d would then require $e^{+} = k'e^{-}$ for all c' and q'. Setting $e^{+} = k'e^{-} = e^{\bullet}$ say, we would have $e' = e^{\bullet}(c' - q')$ in equation 11b and transport equations 11a would then appear as

$$D'dc'/d\xi' = c' - q' \quad \& \quad dq'/d\xi' = -e^{\bullet}(c'- q') .$$

With constant $D' > 0$ and constant $e^{\bullet} > 0$, this linear formulation clearly cannot furnish both c' and q' positive and increasing from their zero values in the far-field $\xi \to \infty$. Reassuringly, there is no problem with Langmuir kinetics, for which

$$e^{+}= 1 - q' \quad \& \quad e^{-}= 1 \tag{12}$$

when the far-field conditions (equation 11d) require

$$\hat{q}' = \hat{c}' = 1 - k' \tag{13}$$

In this case, the transport equations 11a are now

$$D'dc'/d\xi' = c' - q' \quad \& \quad dq'/d\xi' = -c'(1 - q') + k'q' \tag{14}$$

and these can clearly deliver solutions with both c' and q' positive and monotonically increasing as ξ' goes from $+\infty$ to $-\infty$ (ie with both $dc'/d\xi' < 0$ and $dq'/d\xi' < 0$ everywhere) providing only that

$$c' - q' < 0 \quad \& \quad -c'(1 - q') + k'q' < 0 \quad , \text{ or } \quad c' < q' < c'/(k' + c')$$

which is certainly so for all values of c' in the range

$$0 < c' < \hat{c}' \quad \text{when} \quad 0 < q' < \hat{q}'$$

ie throughout the entire solution domain $-\infty < \xi' < +\infty$. As a check on these sums we note that equation 13 is equivalent to equation 7, as is readily shown by elementary algebraic manipulation with the aid of the definitions of equation 10a. Notice also that

$$k' = k^{-}/k^{+}(k^{-}/k^{+} + \hat{c}/q^{\bullet})^{-1} \quad ; \text{ ie } k' < 1 \text{ as required.} \tag{15}$$

4. BREAKTHROUGH PROFILE

As described in [1], an adequate approximation to the solution of equations 11 can be constructed which is robust enough to cope with most relevant situations, yet flexible enough for on-the-spot evaluations. The key to this approximation comes from recognising that the profile possesses the following limit behaviours at small and large D' :

$$c'/q' = 1 + O(D') \quad \text{as} \quad D' \to 0, \quad c'/q' = k'e^-/e^+ + O(1/D') \quad \text{as} \quad D' \to \infty. \quad (17)$$

These limit terms can be accommodated within a single algebraic expression

$$c'/q' = [1 + D'k'e^-]/[1 + D'e^+] \quad (18)$$

which represents the leading order composite expansion with residuals at most $O(D')$ as $D' \to 0$ and $O(1/D')$ as $D' \to \infty$. In fact, the residual as $D' \to 0$ is strictly $O(D'^2)$ so our formula offers a reliable guideline of diffusion effects in the limit of most practical interest.

Application with Langmuir kinetics

All we need to do here is substitute the Langmuir exchange functions $e^\pm$ from equation 12 into equation 18 to yield an algebraic expression for $c'(q')$ and hence obtain $e'(q')$, ie $e'(c'(q'),q')$, by equation 11b. This result can then be substituted in the second of equations 11a and integrated to yield the inverse profile $\xi'(q')$ and thence $\xi'(c')$, under the far field conditions of equation 13. Doing this, we obtain (see [1])

$$\xi'(q') = \{(1 + D'k') \, Ln(1 - k' - q') - (1 + D') \, Ln(q')\}/(1 - k') . \quad (19)$$

For $\xi'(c')$, we return to $c'(q')$ and obtain

$$q' = (1 + D')c'/[1 + D'(k' + c')] \quad (20)$$

from which $\xi'(c')$ follows immediately by substitution in equation 19. Unlike the non-diffusive solution given below, these relations do not admit any obvious rescaling into universal coordinates (ie single profile representation). Practically adequate approximations to such a universal representation for purposes of parameter estimation will be reported later.

Of fundamental interest for purposes of verifying our algorithm is the case $D' = 0$, for which we can immediately confirm that equations 19 & 20 recover the exact solution:

$$c' = q' \quad \& \quad \xi'(c') = [1/(1 - k')] \, Ln\{(1 - k' - c')/c'\} \quad \text{for} \quad D' = 0 .$$

In this non-diffusive limit, we can invert the formulae for ξ' to obtain

$$c'(\xi') = q'(\xi') = (1 - k')/\{1 + Exp[(1 - k')\xi']\} \quad \text{when} \quad D' = 0 . \quad (21)$$

This simple algebraic expression, rescaled as shown in equation 22, is graphed on figure 2. It represents an exact solution to the travelling steady-state equations with Langmuir kinetics in the absence of diffusion.

Comparison with previous guidelines

There are two facets to the practical implications: the attainable concentration of bound material and the sharpness of the breakthrough profile. Both are enhanced by decreasing k' and the latter is favoured also by increasing ξ' . From equation 15 we see that k' decreases with

increasing $\hat{c}$ and k^+ , also with decreasing q^* and k^- . From
equation 10a we see that ξ' increases with decreasing $\hat{q}$ and u , also with
increasing $\hat{c}$ and k^+ . In words then, favourable operation is promoted by
the following conditions:
- low flow rate (improves sharpness)
- high adsorbate concentration (improves sharpness and recovery)
- low density of adsorbent sites (improves both)
- fast adsorption, slow desorption (improves both).
All four conclusions accord with the practical guidelines suggested by
Chase following his numerical survey of Thomas' asymptotic approximation
for the nondiffusive equations. The last of Chase's deductions (that large
adsorbent volume improves recovery) presumably corresponds physically to
increasing length of packing when all other parameters are held constant.
Whilst our present considerations have been confined to the nature of the
travelling front, clearly the total quantity of bound material goes in
proportion with the length of the tube.

     This last conclusion only holds exactly if the diffusivity is zero,
when the centroids of the travelling profiles are located precisely in the
plane where $c' = q' = \frac{1}{2}(1-k')$ and the profiles are antisymmetrical about
this plane. The antisymmetry is readily demonstrated if we write equation
21 in the form:

$$2c''-1 = 2q''-1 = (1-Exp(\xi''))/(1+Exp(\xi'')) = -(1-Exp(-\xi''))/(1+Exp(-\xi'')) \qquad (22)$$

where $c'' = c'/(1-k')$, $q'' = q'/(1-k')$ and $\xi'' = \xi'(1-k')$. However, when
the diffusivity is non-zero, the centroids are shifted and the profiles are
no longer antisymmetric. This feature provides a first guideline diagnostic
of diffusional effects, as indicated below.

5. DIAGNOSTIC IMPLICATIONS

Global conservation principle
There is an undetermined integration constant associated with ξ' which
has been arbitrarily neglected above. As described in [1], we can appeal to
global conservation of adsorbate so as to ensure that the origin $\xi'=0$ is
assigned such that translation and spreading of the c' and q' profiles
are correctly apportioned. With $\xi'=0$ located in the centroid of the
travelling profile of free-plus-bound adsorbate at all later times, the
undetermined integration constant can be expressed symbolically as:

$$\xi_c^0 = -\frac{1}{2}D' Ln (1 - k') \quad , \quad \xi_q^0 = \frac{1}{2}D'[1 - Ln (1 - k')] \quad , \text{ so}$$

$$\xi^0 = \xi_c^0 + \xi_q^0 = \frac{1}{2}D'[1 - 2 Ln (1 - k')] .$$

Notice that $0 < k' < 1$, so $Ln (1 - k') < 0$, and hence $\xi^0 > 0$, because
$D' > 0$. The effect of diffusion is thus to shift the centroid of both the
c' and q' and profiles forward of where they lie in the absence of
diffusion. This result accords with prima facie expectations that diffusion
should extend the leading edge of the frontal profile, so making the free
adsorbate available for binding earlier than with purely convective

transport. Interestingly the shift in the centroid of q' is always larger than that of c' , an effect which becomes increasingly striking as $k' \to 0$: when $k' = 0$ we find $\xi_q^o = \frac{1}{2}D'$ but $\xi_c^o = 0$. Again it is intuitively sensible that the margin by which ξ_q^o exceeds ξ_c^o should increase with decreasing k' because the latter corresponds physically to diminishing desorption.

Profile tailing behaviour

The fore and aft tailings of our one-term composite expansion for the bound adsorbate profiles both scale in linear proportion to D'. We see from equation 19 that the leading zone $(q' \to 0)$ of ξ' is dominated by $Ln(q')$ with coefficient $(1+D')/(1-k')$, whereas the trailing zone $(q' \to 1-k')$ of ξ' is dominated by $Ln(1-k'-q')$, with coefficient $(1+D'k')/(1-k')$. That D' alone appears in the former whereas $D'k'$ appears in the latter is nicely consistent with a simple physical picture that whilst diffusion enhances the free concentration ahead of the travelling front and hence also the flux from free to bound adsorbate, it also lessens the free concentration behind the front and so encourages the desorption from bound to free adsorbate. Of greater significance than $\xi'(q')$ for the practicalities of parameter characterisation are the tailings of $\xi'(c')$, described here by equation 20.   Although this equation contains a more complicated mixture of linear and logarithmic scalings on D' , our interpretations still carry over because the dominant contributions to the leading edge and trailing edge behaviours (ie $c' \to 0$ & $c' \to 1-k'$) remain as for q'.

ACKNOWLEDGEMENT

My thanks to Dr Andy Lyddiatt for introducing me to the practical needs, jargon and recent developments in this area.

REFERENCES

1. Thomas, N H, Algebraic approximation for advective-diffusive adsorbtion chromatography. Biochemical Engineering J, 1990 (to appear).
2. Cowan, G H, Gosling, I S, Laws, J F & Sweetenham, W P, Physical and mathematical modelling to aid scale-up of liquid chromatography. J Chromatography 1986, 363, 37.
3. Chase, H A, Prediction of the performance of preparative affinity chromatography. J Chromatography, 1984, 297, 179.
4. Thomas, H C, Heterogeneous ion exchange in a flowing system. J American Chemical Soc, 1944, 66, 1664.
5. Koch, D L, Cox, R G, Brenner, H & Brady, J F, The effect of order on dispersion in porous media. J Fluid Mechanics, 1989, 200, 173.
6. Clark, D S, Bailey, J E & Do, D D, A mathematical model for restricted diffusion effects on macromolecule impragnation in porous supports. Biotechnology and Bioengineering, 1985, 27, 208.
7. Arnold, F H, Blanch, H W & Wilke, C R, Analysis of affinity separations - Part 1: Predicting the performance of affinity adsorbers. Chemical Engineering J, 1985, 30, B9.

NOTATION

c free adsorbate concentration

D effective diffusivity

e rate of exchange from c to q

k sorption rate constant

q bound adsorbate concentration

t elapsed time

u interstitial axial velocity

U constant pattern velocity

x axial distance

ξ travelling coordinate

τ elapsed time

$\hat{c}, \hat{q}$ values far behind profile

$e^{\pm}, k^{\pm}$ forward and backward values

q^{*} maximal capacity (Langmuir)

q', c', k' dimensionless groups

ξ', D', e' (equation 10 and below)

$$q' = q/q^{*} \quad , \quad c' = (\hat{q}/q^{*})c/\hat{c} \quad , \quad k' = (\hat{q}/\hat{c})k^{-}/k^{+} \quad ,$$
$$\xi' = (1 + \hat{c}/\hat{q}) k^{+}\xi/u \quad , \quad D' = (1 + \hat{c}/\hat{q})^{2}k^{+}D/u^{2} \quad , \quad e' = c'e^{+} - k'q'e^{-} .$$

FIGURES

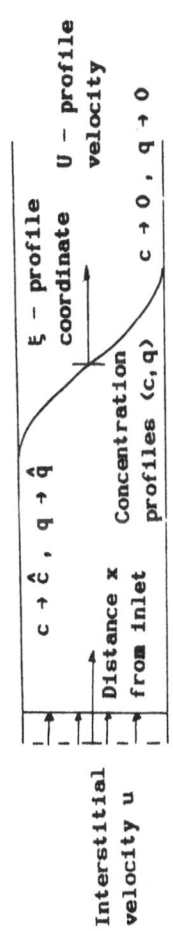

Figure 1. Definition schematic
illustrating the travelling
steady-state profiles.

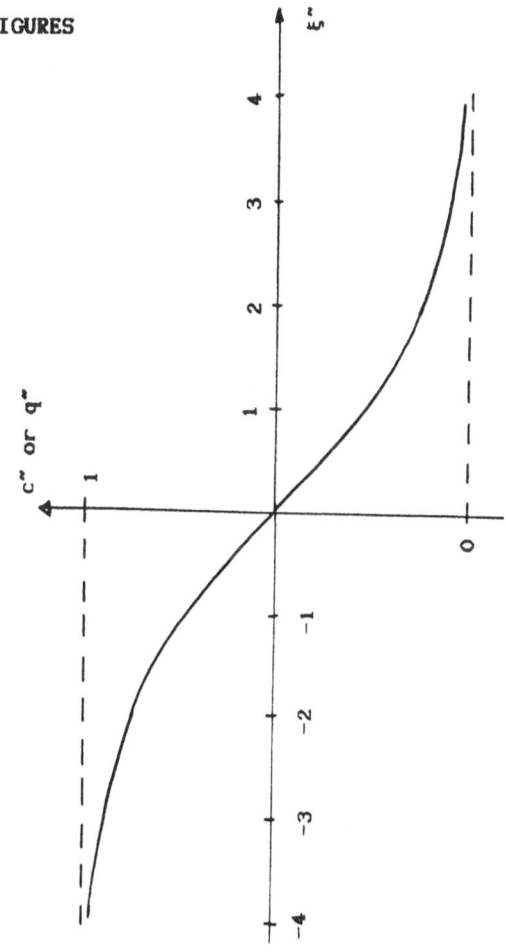

Figure 2. Exact solution for non-
diffusive profiles in terms of
universal coordinates (equation 22).

INDEX OF CONTRIBUTORS

SUBJECT INDEX.